Antennas
and
Wave Propagation

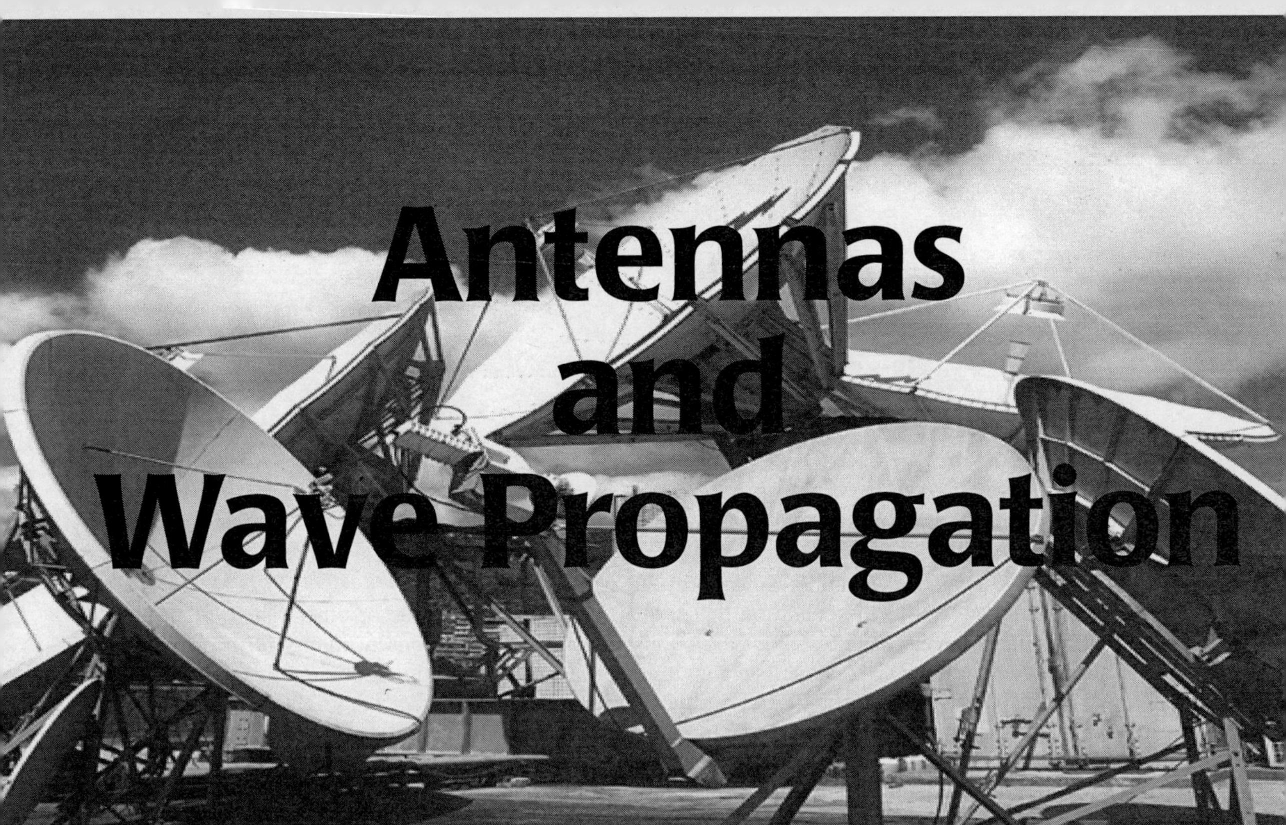

G.S.N. Raju

M.E., Ph.D (IIT-KGP), FIE, FIETE

Department of Electronics and Communications Engineering

College of Engineering, Andhra University

Visakhapatnam

India

ISB 978-81-317-0184-3

First Impression, 2006
Seventeenth Impression, *2015*
Eighteenth Impression, 2016

Published by Pearson India Education Services Pvt. Ltd, CIN: U72200TN2005PTC057128.

Head Office: 1st Floor, Berger Tower, Plot No. C-001A/2, Sector 16B, Noida - 201 301, Uttar Pradesh, India.

Registered Office: 6th Floor, Tower A, Unit A and B, International Tech Park, CapitaLand Chennai, 200 Feet Radial Road,Zamin Pallavaram, Old Pallavaram Chennai – 600117, Tamil Nadu, India

Website: in.pearson.com; Email: companysecretary.india@pearson.com

Digitally printed in India by Trinity Academy for Corporate Training Ltd, New Delhi in the year of 2026

To

My Brother

Prof KRISHNAM RAJU, M.Sc. (Math. Phy), Ph.D.

*a man of academic excellence with impeccable character and
a great administrator
who made me what I am today*

Antenna is an essential terminal device in all types of communication and radar systems. If the antenna is well designed, it is possible to send communication signals from one place to any other place and anywhere on the globe. If the antenna is not well designed, it is not possible to send a signal even beyond the premises of the compound of the transmitting antenna. All communication and radar systems require an antenna in one form or the other. Without an antenna, there would be no communication system and no radar. The antenna is a source as well as a sensor of electromagnetic waves. The behaviour of electromagnetic waves between a transmitter and a receiver is important as the received power depends on the propagation characteristics.

Antennas and Wave Propagation is a core subject for all students of B.E./B.Tech. in Electronics and Communications Engineering, Electronics Engineering, M.Sc. (Electronics), M.Sc. (Applied Physics), AMIETE and AMIEs, etc. throughout the world. While teaching Antennas and Wave Propagation, I have referred to books written by several experts like Kraus, Balanis, Jordan, Collin, Terman, Elliot, Skolnik, Silver and Kreyszig. However, I often felt that no single available book caters to the needs of the complete course required for undergraduate and post-graduate programmes. In view of this, an attempt has been made to bring out a comprehensive book on the subject.

The book is divided into nine chapters. It begins with an **Introduction,** listing frequency spectrum of electromagnetic waves, systems of units, notation of scalar parameters, notation of vector parameters, small value representation, large value representation and frequency ranges of TV channels.

Chapter 1 introduces students to the **Mathematical Preliminaries** required for a clear understanding of the mathematical concepts needed for a course on Antennas and Wave Propagation.

The core text begins with Chapter 2 on **Maxwell's Equations and Electromagnetic Waves.** Chapter 3 discusses **Antenna Fundamentals and Radiation,** Chapter 4 deals with **Analysis of Linear Arrays** and Chapter 5 explains **Array Synthesis** in detail.

Chapter 6 deals with **HF, VHF, UHF Antennas** and Chapter 7 talks about **Microwave Antennas** at length.

The various antenna-related measurement techniques are expounded in Chapter 8 on **Antenna Measurements**.

The book concludes with a detailed discussion in Chapter 9 on **Wave Propagation**.

An exclusive question bank of objective type questions is provided at the end of the book for practice.

It is hoped that this book will be extremely useful for students, teachers, professionals, engineers, technicians, designers and also for short-term course organizers.

Please inform me of any error that may have crept in inadvertently, as well as suggestions to improve the book.

G.S.N. RAJU

I am grateful to my gurus Prof. B.N. Das and Prof. Ajoy Chakraborty of IIT Kharagpur under whose guidance, I developed motivation and keen interest to write this book. I thank all my colleagues, Prof. Madhusudhan Rao, Venkata Rao, Mrs. Veena Kumari, Prof. Raja Rajeswari, Prof. Satyanarayana Reddy, Dr. Mallikarjuna Rao, Mrs. Santa Kumari and Dr. Gopala Rao for extending their cooperation while writing this book.

I am extremely grateful to Sri K. Raghu, Chairman, Raghu Engineering College, and Sri S.V.S.S. Rama Chandra Raju, Former PF commissoner and director, Raghu Engineering College for their motivation and great appreciation of my teaching and research efforts.

I extend my thanks to Mr. Narayana, Balasubramanyam, Prof. Srinivasa Baba and Mr. Chandra Bhushana Rao for assisting in proof reading.

I cherish the association of all my research scholars P.M. Rao, Sridevi, Y. Gopala Rao, P. Mishra, Sudhakar, Baba, Srinivasa Rao, Padma Raju, Chandra Bhushana Rao, Prasad, Habibullah Khan, Subrahmanyam, Narayana and Sadasiva Rao in putting their efforts for their Ph.D. degrees. I thank my friends, Prof. Appa Rao and Gopala Krishna Raju, for giving me moral support. I love all my students who are always fond of my teaching, guidance and discipline.

I extend my regards to all great authors and experts in the field of antennas for their contributions.

I am extremely grateful to the Andhra University for the encouragement given throughout my teaching and research career.

I am also thankful to Mr. K. Srinivas, Commissioning Editor, and Mr. Vishnu Kumar for their keen interest and confidence shown in me to bring out this book quickly.

I am extremely grateful to the reviewers of my book for their excellent and positive feeback.

I am also thankful to M. Sahu and Sankar for assisting me in preparing the manuscript. I am grateful for the support and respect extended by our technical and office staff, Koteswara Rao, Somayajulu, Prasada Reddy, Appla Raju, Srinivas, Babji, Ramesh, Adilakshmi, Nara Hari, Ramana, Tavudu and Kondamma.

Finally, I convey my respects to my parents Venkatrama Raju and Sitama and thanks to my wife Kanaka Durga, daughter Narmada Devi and son Venkata Krishna Varma for their love, affection, help and patience throughout the preparation of the book.

I am especially joyful with our loving German Shepherd Opy, playing with whom has always relieved stress after hectic academic and administration work.

G.S.N. RAJU

Brief Contents

Contents

Introduction

Antenna or aerial means the same. It is used in several domestic, civilian and military communication and radar applications. It is also used in microwave radiation therapy systems.

> An **antenna** is transducer, an impedance matching device, a radiator and a sensor of electromagnetic waves. It is an essential device/element in all types of communication and radar systems. It can be considered as a source of electromagnetic waves.

Antennas are used in houses, cars, trains, ships, spacecrafts, aircrafts, satellites and radars etc. There is no place where there is no antenna for communications. That is, telecommunication is not possible without one type of antenna or other.

FREQUENCY SPECTRUM OF ELECTROMAGNETIC WAVES

Electromagnetic waves produced by the antennas have a wide range of frequencies. The wavelength of the EM wave depends on its velocity and frequency. That is, the wavelength λ is

$$\lambda = \frac{v_0}{f}$$

Here v_0 = velocity of propagation of EM waves in free space

$$= 3 \times 10^8 \text{ m/s}$$

$$f = \text{frequency in Hz.}$$

Example 1 If the frequency of an EM wave in free space is 30 MHz, the wavelength is

$$\lambda = \frac{v_0}{f} = \frac{3 \times 10^8}{30 \times 10^6} = 10 \text{ m}$$

Example 2 If the frequency of an EM wave in free space is 300 MHz, the wavelength is

$$\lambda = \frac{v_0}{f} = \frac{3 \times 10^8}{300 \times 10^6} = 1 \text{ m}$$

Example 3 If the frequency of an EM wave in a medium whose relative permittivity is 4 is 300 MHz, its velocity is given by

$$v = \frac{1}{\sqrt{\mu \epsilon}} = \frac{1}{\sqrt{\mu_0 \epsilon_0 \epsilon_r}} = \frac{v_0}{\sqrt{\epsilon_r}}$$

$$= \frac{1}{2} v_0 = \frac{1}{2} \times 3 \times 10^8 = 1.5 \times 10^8 \text{ m/s}$$

The corresponding wavelength at 300 MHz is

$$\lambda = \frac{1.5 \times 10^8}{300 \times 10^6} = 0.5 \text{ m}$$

A typical frequency band representation is shown in the following table:

Table 1

Band Name	Abbreviated Band Name	Frequency	Wavelength
Extremely low frequency	ELF	30 - 300 Hz	10 mm - 100 km
Very low frequency	VLF	3 - 30 kHz	100 - 10 km
Low frequency	LF	30 - 300 kHz	10 - 1 km
Medium frequency	MF	300 - 3,000 kHz	1 km - 100 m
High frequency	HF	3 - 30 MHz	100 - 10 m
Very high frequency	VHF	30 - 300 MHz	10 - 1 m
Ultra high frequency	UHF	300 - 3,000 MHz	1 m - 10 cm
Super high frequency	SHF	3 - 30 GHz	10 - 1 cm
Extremely high frequency	EHF	30 - 300 GHz	1 cm - 1 mm

Radar bands specified by IEEE (USA) are shown in the following table:

Table 2

Name of the band	Frequency range (GHz)	Wavelength range (cm)
L	1 - 2	30 - 15
S	2 - 4	15 - 7.50
C	4 - 8	7.50 - 3.75
X	8 - 12	3.75 - 2.50
Ku	12 - 18	2.50 - 1.67
K	18 - 27	1.67 - 1.11
Ka	27 - 40	1.11 - 0.75
mm	40 - 300	0.75 - 0.01

Examples

- If an antenna is operated at 100 Hz, it is called an ELF antenna.
- If an antenna is operated at a wavelength of 20 km, it is called a VLF antenna.
- If the frequency of operation of an antenna is 200 kHz, it is in LF range.
- A frequency of 10 MHz is in HF range.
- If the operating wavelength of an antenna is 2 m, it is in VHF range.
- 1 GHz frequency is within UHF range.
- If an antenna is operated at 3.5 GHz, it is in SHF range or in S-band.
- If the operating wavelength of an antenna is 3 cm, it is in X-band.
- If an antenna is designed at a frequency of 15 GHz, it is a Ku band antenna.
- A wavelength of 0.5 cm is said to be a part of millimeter (mm) band.

Systems of Units

The common systems of units are CGS system, MKS system and SI system. These are given in the following table:

Table 3

System of Units	Quantities and Units						
	Length (l)	Mass (m)	Time (t)	Force (F)	Current	Temperature	Luminous Intensity
CGS	cm	gm	sec	dyne			
MKS	meter	kg	sec	newton			
SI	meter	kg	sec	newton	ampere (A)	°K	candela (cd)

In fact, the International System (SI) of units is the extended version of the metric system. In SI units, current, temperature and luminous intensity are also fundamental quantities.

Electric current is expressed in ampere (A).

Temperature is in degree kelvin ($^{\circ}$K).

Luminous intensity is in candela (cd).

The units for other quantities are only units derived from the fundamental units.

Notation of Scalar Parameters

Table 4

Parameter	Notation/Symbol	Definition	Unit name
Frequency	f	It is the reciprocal of one time period of a periodic waveform.	Hertz (Hz) 1 Hz = 1 cycle/sec
Energy	W	It is the work done when force is exerted through a distance of one meter.	joule (J) 1 joule = 10^7 ergs
Power	P	It is the time rate of energy.	watt (W) 1 W = 1 joule/sec or 1 W = 1 volt $\times$ 1 amp
Charge	Q	It is the product of current and time.	coulomb (C) 1 C = 1 A-sec
Resistance	R	It is the ratio of voltage and current.	ohm (Ω) 1 Ω = 1 volt/1 amp
Conductance	G	It is the reciprocal of R.	Mho $1 \text{ Mho} = \dfrac{1 \text{ amp}}{1 \text{ volt}}$
Resistivity	ρ	It is the resistance measured between two parallel faces of a unit cube.	Ohm-meter
Conductivity	σ	It is the reciprocal of resistivity.	Mho/meter
Electromotive force	Emf, V	It is the ratio of power to current.	volt 1 volt = 1 J/C or watt/amp
Electric flux	ψ	It is nothing but displaced charge.	coulomb (C)
Magnetic flux	ϕ	$\phi = -\int_0^t V dt$	weber (wb) 1 wb = 1 volt-sec
Magnetomotive force	V_m (mmf)	$V_m = \int_A^B \mathbf{H} \cdot d\mathbf{L}$	amp (A)
Capacitance	C	$C = \dfrac{Q}{V}$	farads (F) 1 F = 1 C/1 volt
Inductance	L	$L = \dfrac{N\phi}{I}$	henry (H)
Mutual inductance	M	$M = N_2 \phi_{12}/I_1$	henry (H) 1 H = 1 wb/amp

Parameter	Notation/Symbol	Definition	Unit name
Permittivity	ϵ	$\epsilon = \dfrac{D}{E}$	farad/meter (F/m)
Permeability	μ	$\mu = \dfrac{B}{H}$	henry/meter (H/m)
Permittivity of free space	ϵ_0	8.854×10^{-12} F/m	F/m
Permeability of free space	μ_0	$4\pi \times 10^{-7}$ H/m	H/m
Relative permittivity of a medium	ϵ_r	$\epsilon_r = \dfrac{\epsilon}{\epsilon_0}$ $\epsilon_r = 1$ for free space	No units
Relative permeability of a medium	μ_r	$\mu_r = \dfrac{\mu}{\mu_0}$ $\mu_r = 1$ for free space	No units
Electric susceptibility	χ_e	$\chi_e = \epsilon_r - 1$ χ is pronounced as Chi	No units
Magnetic susceptibility	χ_m	$\chi_m = \mu_r - 1$	No units
Steradian	Str	It is a measure of solid angle	Steradian
Intrinsic impedance of free space	η_0	$\eta_0 = \sqrt{\dfrac{\mu_0}{\epsilon_0}}$	$120\pi\,\Omega$
Differential length	dL	Small length	m
Area	S	Product of two lengths	m^2
Differential area	dS	Product of two differential lengths	m^2
Differential volume	dv	Product of differential length, width and breadth	m^3
Volume	v	Product of length, width and breadth	m^3
Angular frequency	ω	$2\pi f$	rad/sec
Wavelength	λ	v/f	m
Electric potential	V	$-\int \mathbf{E} \cdot d\mathbf{L}$	volt
Magnetic scalar potential	V_m	$-\int \mathbf{H} \cdot d\mathbf{L}$	ampere
Surface charge density	ρ_s	Q/S	c/m^2

Parameter	Notation/ Symbol	Definition	Unit name
Line charge density	ρ_L	Q/L	c/m
Volume charge density	ρ_v	Q/v	c/m^3
Propagation constant	γ	$\alpha + j\beta$	dB/m
Attenuation constant	α	It is a measure of reduction of EM wave as it progresses.	dB/m
Depth of penetration or skin depth	δ	It is the depth in which an EM wave is attenuated to 37% of original value.	$\delta = 1/\alpha$ (m)
Phase constant	β	It is a measure of phase shift of EM wave.	rad/m
Group velocity	v_g	It is the velocity with which the energy propagates in a guided structure.	m/sec
Phase velocity	v_p	It defines a point of constant phase.	m/sec
VSWR	S	V_{max}/V_{min}	No units
Reflection coefficient	ρ	$\dfrac{\text{reflected wave}}{\text{incident wave}} = \dfrac{V_r}{V_i}$	No units

Notations of Vector Parameters

Table 5

Parameter	Notation/ Symbol	Definition	Unit name
Force	**F**	It is the product of mass and acceleration.	newton (N) $1 \text{ newton} = \dfrac{kg - m}{sec^2}$
Electric field strength	**E**	It is the force per one coulomb.	volt/m or newton/C
Conduction current density	$\mathbf{J}_c$	It is defined as the ratio of current to area.	A/m^2
Displacement electric flux density	**D**	$\mathbf{D} = \epsilon \mathbf{E}$	C/m^2
Displacement current density	$\mathbf{J}_d$	$\mathbf{J}_d = \dfrac{\partial \mathbf{D}}{\partial t}$	A/m^2
Magnetic flux density	**B**	$\mathbf{B} = \mu \mathbf{H}$	wb/m^2 or tesla

Parameter	Notation/ Symbol	Definition	Unit name
Magnetic field strength	**H**	It is the current per meter width.	A/m
Velocity of EM wave	V_0	λ/f	3×10^8 m/s
Velocity	**V**	Rate of displacement	m/sec
Vector	Bold faced letter	—	—
Unit vector	Bold faced small letter	—	—
Electric dipole moment	**p**	$Q\mathbf{d}$	columb-m
Magnetic dipole moment	**m**	$I\mathbf{A}$	$A - m^2$
Polarisation	**P**	$\chi_e \, \varepsilon_0 \, \mathbf{E}$	columbs/m^2
Magnetisation	**M**	$\chi_m \, \mathbf{H}$	amperes/m
Torque	**T**	$\mathbf{R} \times \mathbf{F}$	N-m
Surface current density	$\mathbf{J}_s$	Current per meter	ampere/m
Poynting vector	**P**	$\mathbf{E} \times \mathbf{H}$	watts/m^2
Tangential component of **E**	$\mathbf{E}_t$	Tangential component of **E**	volt/m
Tangential component of **H**	$\mathbf{H}_t$	Tangential component of **H**	amperes/m
Normal component of **D**	$\mathbf{D}_n$	Normal component of **D**	c/m^2
Normal component of **B**	$\mathbf{B}_n$	Normal component of **B**	Wb/m^2

Small Value Representation

Table 6

Value	Prefix	Symbol	Length
10^{-1}	deci	d	dm
10^{-2}	centi	c	cm
10^{-3}	milli	m	mm
10^{-6}	micro	μ	μm
10^{-9}	nano	n	nm
10^{-12}	pico	p	pm
10^{-15}	femto	f	fm
10^{-18}	atto	a	am

Large Value Representation

Table 7

Value	Prefix	Symbol	Length	Frequency
10	deka	*da*	dam	daHz
10^2	hecto	*h*	hm	hHz
10^3	Kilo	*K*	Km	kHz
10^6	Mega	*M*	Mm	MHz
10^9	Giga	*G*	Gm	GHz
10^{12}	Tera	*T*	Tm	THz

Frequency Ranges of TV Channels

Table 8

Channel Number	Frequency Band (MHz)
2	54 - 60
3	60 - 66
4	66 - 72
5	76 - 82
6	82 - 88
7	174 - 180
8	180 - 186
9	186 - 192
10	192 - 198
11	198 - 204
12	204 - 210
13	210 - 216
UHF Band	470 - 806

Mathematical Preliminaries

"Mathematics is the backbone of all branches of Sciences and Engineering."

CHAPTER OBJECTIVES

This chapter discusses

- ✦ Coordinate systems
- ✦ Vector algebra and calculus
- ✦ Series and identities
- ✦ Determinants and matrices
- ✦ Trigonometry
- ✦ Differential and Integral formulae
- ✦ Antenna related scalar and vector quantities
- ✦ Objective questions and solved problems useful for class tests, final examinations and also for competitive examinations
- ✦ Exercise problems to develop self problem solving skills

1.1 FUNDAMENTALS OF SCALARS AND VECTORS

A **scalar** has magnitude and an algebraic sign.
For example, temperature, mass, charge, work and so on.
A **vector** has both magnitude and direction.
For example, velocity, force, electric field, magnetic field and so on.

In this book, scalar is represented by simple letters like A or B and a vector is represented by bold letters like **A** or **B**. Unit vectors are represented by small bold letters like **a**, **b**.

The vector **A** is expressed in two forms:

(*i*) $\mathbf{A} = (A_x, A_y, A_z)$. A_x, A_y, A_z are known as the components of vector **A**.

(*ii*) $\mathbf{A} = A_x \mathbf{a}_x + A_y \mathbf{a}_y + A_z \mathbf{a}_z$. $\mathbf{a}_x$, $\mathbf{a}_y$, $\mathbf{a}_z$ are unit vectors along the coordinate axes.

The magnitude of **A** is written as A
or $A = |\mathbf{A}|$

The unit vector of **A** is **a** and it is given by

$$\mathbf{a} = \frac{\mathbf{A}}{A}$$

The sum and difference of two vectors are given by

$$\mathbf{A} + \mathbf{B} = (A_x + B_x)\,\mathbf{a}_x + (A_y + B_y)\,\mathbf{a}_y + (A_z + B_z)\,\mathbf{a}_z$$
$$\mathbf{A} - \mathbf{B} = (A_x - B_x)\,\mathbf{a}_x + (A_y - B_y)\,\mathbf{a}_y + (A_z - B_z)\,\mathbf{a}_z$$

The dot product is denoted by

$$\mathbf{A} \cdot \mathbf{B} \quad \text{or} \quad \mathbf{B} \cdot \mathbf{A}$$

$$\mathbf{A} \cdot \mathbf{B} = \mathbf{B} \cdot \mathbf{A} = AB \cos \theta$$
$$= A_x B_x + A_y B_y + A_z B_z$$

where θ is the angle between the vectors **A** and **B**.

Dot product of two vectors is a scalar.

The cross product is denoted by $\mathbf{A} \times \mathbf{B}$

$$\mathbf{A} \times \mathbf{B} = AB \sin \theta \, \mathbf{a}_n$$

where $\mathbf{a}_n$ is the unit vector perpendicular to **A** and **B**.

$$\mathbf{A} \times \mathbf{B} = \begin{vmatrix} \mathbf{a}_x & \mathbf{a}_y & \mathbf{a}_z \\ A_x & A_y & A_z \\ B_x & B_y & B_z \end{vmatrix}$$

$$= \mathbf{a}_x \left[A_y B_z - A_z B_y \right] + \mathbf{a}_y \left[A_z B_x - A_x B_z \right] + \mathbf{a}_z \left[A_x B_y - A_y B_x \right]$$

where $\qquad \mathbf{A} = A_x \mathbf{a}_x + A_y \mathbf{a}_y + A_z \mathbf{a}_z$

and
$$\mathbf{B} = B_x\,\mathbf{a}_x + B_y\,\mathbf{a}_y + B_z\,\mathbf{a}_z$$

The cross product of two vectors is a vector.

1.2 COORDINATE SYSTEMS

Coordinate system is defined as a system used to represent a point in space. Basically coordinate systems are of three types:
1. Cartesian coordinate system
2. Cylindrical coordinate system
3. Spherical coordinate system

1.2.1 Cartesian Coordinate System

In this system, a point P is represented by $P\,(x,\ y,\ z)$. The variables are x, y, z.

A point is obtained by the intersection of three planes given by
$$x = k_1 \ (\text{constant})$$
$$y = k_2 \ (\text{constant})$$
$$z = k_3 \ (\text{constant})$$

The unit of $x,\ y$ and z is meter.

The Cartesian Coordinates are represented as in Fig. 1.1.

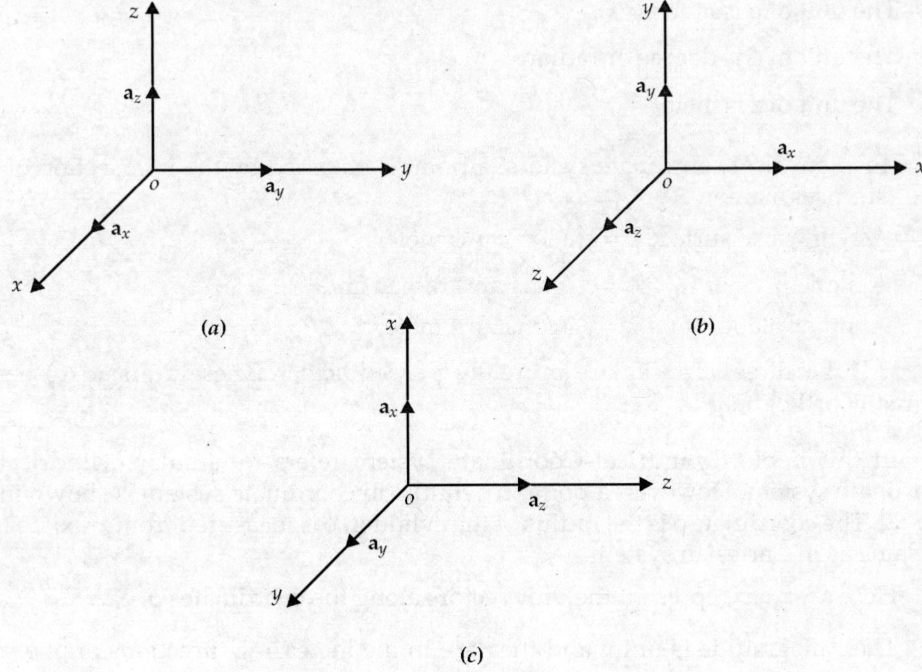

Fig. 1.1 *Cartesian coordinate system*

The three axes x, y, z are mutually perpendicular. These are said to be orthogonal to each other.

The unit vectors along the coordinate axes are represented by $\mathbf{a}_x$, $\mathbf{a}_y$ and $\mathbf{a}_z$. Their magnitude is unity and they are in the increasing directions of x, y and z axes respectively.

Properties of unit vectors

$$\mathbf{a}_x \cdot \mathbf{a}_x = 1 \qquad\qquad \mathbf{a}_y \cdot \mathbf{a}_z = 0$$
$$\mathbf{a}_y \cdot \mathbf{a}_y = 1 \qquad\qquad \mathbf{a}_x \times \mathbf{a}_y = \mathbf{a}_z$$
$$\mathbf{a}_z \cdot \mathbf{a}_z = 1 \qquad\qquad \mathbf{a}_y \times \mathbf{a}_z = \mathbf{a}_x$$
$$\mathbf{a}_x \times \mathbf{a}_x = 0 \qquad\qquad \mathbf{a}_z \times \mathbf{a}_x = \mathbf{a}_y$$
$$\mathbf{a}_y \times \mathbf{a}_y = 0 \qquad\qquad \mathbf{a}_y \times \mathbf{a}_x = -\mathbf{a}_z$$
$$\mathbf{a}_z \times \mathbf{a}_z = 0 \qquad\qquad \mathbf{a}_z \times \mathbf{a}_y = -\mathbf{a}_x$$
$$\mathbf{a}_x \cdot \mathbf{a}_y = 0 \qquad\qquad \mathbf{a}_x \times \mathbf{a}_z = -\mathbf{a}_y$$
$$\mathbf{a}_x \cdot \mathbf{a}_z = 0$$

1.2.2 Cylindrical Coordinate System

In this system, a point P is represented by $P\,(\rho, \phi, z)$. ρ represents radius of cylinder, ϕ is called azimuthal angle and z is the same as in Cartesian coordinate system.

The unit of ρ is meter.

The unit of ϕ is degree or radian.

The unit of z is meter.

In cylindrical coordinate system, a point is obtained by the intersection of three surfaces, namely,

A cylindrical surface $\rho = k_1$ (constant) meter

A plane $\qquad\qquad \phi = \alpha$ (constant) radian and

Another plane $\qquad z = k_2$ (constant) meter

All the three surfaces are mutually perpendicular. These are said to be mutually orthogonal.

In this book, Cylindrical Coordinate system refers to circular cylindrical coordinate system. However, a point in cylindrical coordinate system is shown in Fig. 1.2. The coordinate ρ is the radius of the cylinder. ϕ is measured from x-axis. z is the same as in Cartesian system.

Here $\mathbf{a}_\rho$, $\mathbf{a}_\phi$, $\mathbf{a}_z$ represent the unit vectors along the coordinates ρ, ϕ, z.

Their magnitude is unity and they are in the increasing directions of ρ, ϕ, z respectively.

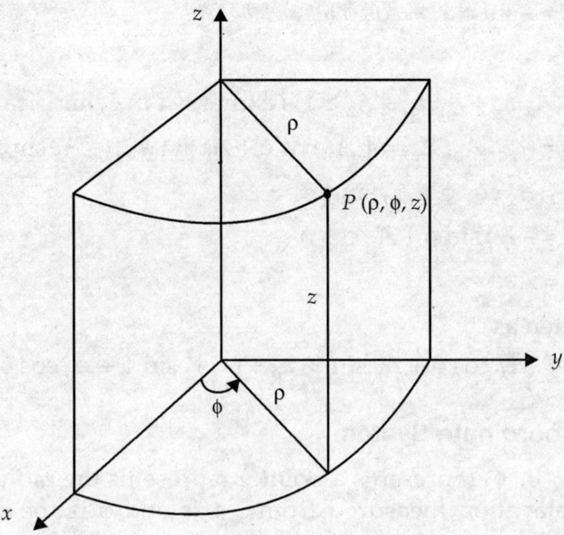

Fig. 1.2   *A point in cylindrical coordinates*

It is obvious that increase in ρ results in cylinders of greater radius. φ increases in anti-clockwise direction. z is the same as in Cartesian system.

The relations between x, y, z and ρ, φ, z:

$$x = \rho \cos \phi$$
$$y = \rho \sin \phi$$
$$z = z$$

and

$$\rho = \sqrt{x^2 + y^2}, \quad 0 \leq \rho < \infty$$

$$\phi = \tan^{-1} \frac{y}{x}, \quad 0 \leq \phi < 2\pi$$

$$z = z, \quad\quad 0 \leq z < \infty$$

Dot products of $\mathbf{a}_x$, $\mathbf{a}_y$ and $\mathbf{a}_z$ with $\mathbf{a}_\rho$, $\mathbf{a}_\phi$ and $\mathbf{a}_z$ are given by

$$\mathbf{a}_x \cdot \mathbf{a}_\rho = \cos \phi$$
$$\mathbf{a}_x \cdot \mathbf{a}_\phi = -\sin \phi$$
$$\mathbf{a}_y \cdot \mathbf{a}_\rho = \sin \phi$$
$$\mathbf{a}_y \cdot \mathbf{a}_\phi = \cos \phi$$
$$\mathbf{a}_z \cdot \mathbf{a}_\rho = 0$$
$$\mathbf{a}_z \cdot \mathbf{a}_\phi = 0$$

The unit vectors of cylindrical coordinates in terms of Cartesian coordinates are given by

$$\mathbf{a}_\rho = \cos \phi \, \mathbf{a}_x + \sin \phi \, \mathbf{a}_y$$

$$\mathbf{a}_\phi = -\sin\phi\,\mathbf{a}_x + \cos\phi\,\mathbf{a}_y$$

$$\mathbf{a}_z = \mathbf{a}_z$$

A vector $\mathbf{A} = (A_x\,\mathbf{a}_x + A_y\,\mathbf{a}_y + A_z\,\mathbf{a}_z)$ is expressed in cylindrical coordinates as:

$$\mathbf{A} = (A_\rho, A_\phi, A_z) = [(A_x\cos\phi + A_y\sin\phi),\ (-A_x\sin\phi + A_y\cos\phi),\ A_z]$$

That is
$$A_\rho = A_x\cos\phi + A_y\sin\phi$$

$$A_\phi = -A_x\sin\phi + A_y\cos\phi$$

$$A_z = A_z$$

$\mathbf{A}$ is also written as

$$\mathbf{A} = (A_x\cos\phi + A_y\sin\phi)\,\mathbf{a}_\rho + (-A_x\sin\phi + A_y\cos\phi)\,\mathbf{a}_\phi + A_z\,\mathbf{a}_z.$$

1.2.3 Spherical Coordinate System

In this system, $P(r,\ \theta,\ \phi)$ represents a point. r represents the radius of the sphere, θ is the angle of elevation measured from z-axis and ϕ is the azimuthal angle measured from x-axis.

A point is obtained by the intersection of three surfaces, namely,

a spherical surface $r = k$ (constant) meter

a cone $\qquad\qquad \theta = \alpha$ (constant) radian and

a plane $\qquad\qquad \phi = \beta$ (constant) radian

The three surfaces are mutually perpendicular. These are said to be orthogonal.

$\mathbf{a}_r,\ \mathbf{a}_\theta,\ \mathbf{a}_\phi$ represent unit vectors along the coordinate axes. Their magnitude is unity and they are in the increasing directions of r, θ and ϕ axes.

A point in spherical coordinate system is shown in Fig. 1.3.

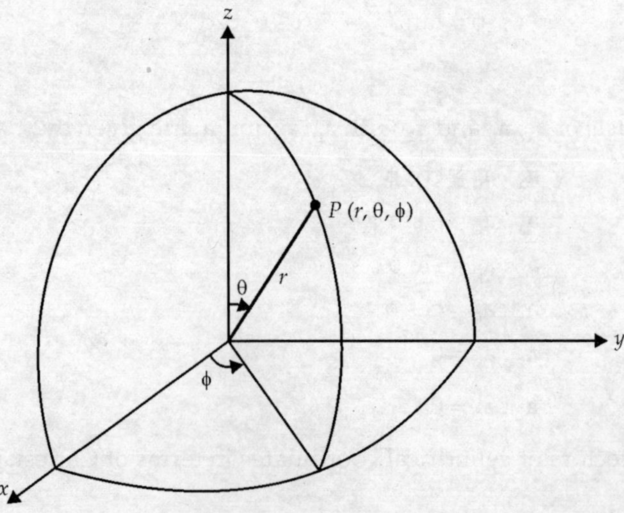

Fig. 1.3 *A point in spherical coordinate system*

Increase of r results in spheres of larger radius. θ increases in clockwise direction and ϕ increases in anti-clockwise direction.

The variables of Cartesian and spherical coordinates are related by

$$x = r \sin \theta \cos \phi, \qquad -\infty < x < \infty$$

$$y = r \sin \theta \sin \phi, \qquad -\infty < y < \infty$$

$$z = r \cos \theta, \qquad -\infty < z < \infty$$

and

$$r = \sqrt{x^2 + y^2 + z^2} \qquad 0 \le r \le \infty$$

$$\theta = \cos^{-1} \frac{z}{\sqrt{x^2 + y^2 + z^2}} \qquad 0 \le \theta \le \pi$$

$$\phi = \tan^{-1} \frac{y}{x} \qquad 0 \le \phi \le 2\pi$$

The relations between the variables of cylindrical and spherical coordinates are given by

$$\rho = r \sin \theta$$

$$\phi = \phi$$

$$z = r \cos \theta$$

$$r = \sqrt{\rho^2 + z^2}$$

$$\theta = \tan^{-1} \frac{r}{z}$$

$$\phi = \phi$$

The unit vectors of spherical coordinates in terms of Cartesian coordinates are given by

$$\mathbf{a}_r = \sin \theta \ \cos \phi \ \mathbf{a}_x + \sin \theta \ \sin \phi \ \mathbf{a}_y + \cos \theta \ \mathbf{a}_z$$

$$\mathbf{a}_\theta = \cos \theta \ \cos \phi \ \mathbf{a}_x + \cos \theta \ \sin \phi \ \mathbf{a}_y - \sin \theta \ \mathbf{a}_z$$

$$\mathbf{a}_\phi = - \sin \phi \ \mathbf{a}_x + \cos \phi \ \mathbf{a}_y$$

A vector $\mathbf{A} = (A_x, A_y, A_z)$ is expressed in spherical coordinates as

$$\mathbf{A} = (A_r, A_\theta, A_\phi)$$

$$= [(A_x \sin \theta \ \cos \phi + A_y \sin \theta \ \sin \phi + A_z \cos \theta),$$

$$(A_x \cos \theta \ \cos \phi + A_y \cos \theta \ \sin \phi - A_z \sin \theta),$$

$$(- A_x \sin \phi + A_y \cos \phi)]$$

The dot products of $\mathbf{a}_x$, $\mathbf{a}_y$ and $\mathbf{a}_z$ with $\mathbf{a}_r$, $\mathbf{a}_\theta$ and $\mathbf{a}_\phi$ are given by

$$\mathbf{a}_x \cdot \mathbf{a}_r = \sin \theta \ \cos \phi$$

$$\mathbf{a}_x \cdot \mathbf{a}_\theta = \cos \theta \ \cos \phi$$

$$\mathbf{a}_x \cdot \mathbf{a}_\phi = - \sin \phi$$

$$\mathbf{a}_y \cdot \mathbf{a}_r = \sin \theta \ \sin \phi$$

$$\mathbf{a}_y \cdot \mathbf{a}_\theta = \cos \theta \ \sin \phi$$

$$\mathbf{a}_y \cdot \mathbf{a}_\phi = \cos \phi$$

$$\mathbf{a}_z \cdot \mathbf{a}_r = \cos \theta$$

$$\mathbf{a}_z \cdot \mathbf{a}_\theta = -\sin \theta$$

$$\mathbf{a}_z \cdot \mathbf{a}_\phi = 0$$

Here,

$$A_r = A_x \sin \theta \, \cos \phi + A_y \sin \theta \, \sin \phi + A_z \cos \theta$$

$$A_\theta = A_x \cos \theta \, \cos \phi + A_y \cos \theta \, \sin \phi - A_z \sin \theta$$

$$A_\phi = -A_x \sin \phi + A_y \cos \phi$$

The point $A\,(x,\,y,\,z) = A\,(\rho,\,\phi,\,z) = A\,(r,\,\theta,\,\phi)$ in Cartesian, cylindrical and spherical coordinate systems is shown in a single Fig. 1.4 to explain the concept at a glance.

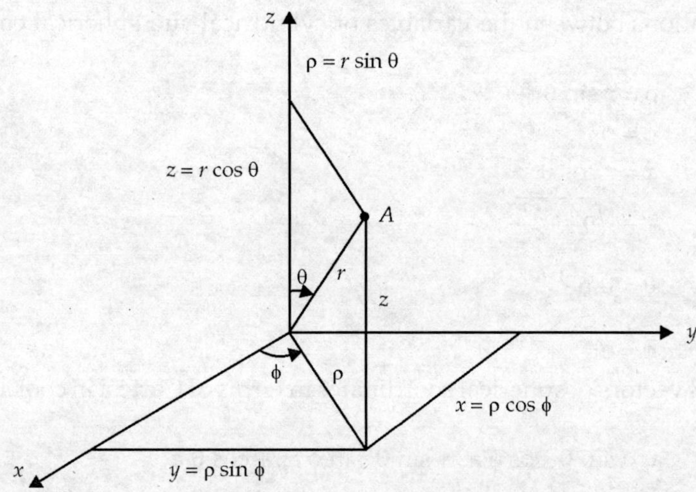

Fig. 1.4 *Coordinates in all the three systems*

The relations for differential length, area and volume are given in Table 1.1.

Table 1.1   Differential quantities in different coordinates

Coordinate System	dL	dS	dυ
Cartesian	$dx\mathbf{a}_x + dy\mathbf{a}_y + dz\mathbf{a}_z$	$dxdy\mathbf{a}_z$ or $dydz\mathbf{a}_x$ or $dzdx\mathbf{a}_y$	$dx\,dy\,dz$
Cylindrical	$d\rho\,\mathbf{a}_\rho + \rho\,d\phi\,\mathbf{a}_\phi + dz\,\mathbf{a}_z$	$\rho\,d\rho\,d\phi\,\mathbf{a}_z$ or $\rho\,d\phi\,dz\mathbf{a}_\rho$ or $d\rho\,dz\mathbf{a}_\phi$	$\rho\,d\rho\,d\phi\,dz$
Spherical	$dr\mathbf{a}_r + rd\theta\mathbf{a}_\theta + r\sin\theta\,d\phi\mathbf{a}_\phi$	$rdr\,d\theta\mathbf{a}_\phi$ or $r\sin\theta\,drd\phi\mathbf{a}_\theta$ or $r^2\sin\theta\,d\theta\,d\phi\mathbf{a}_r$	$r^2\sin\theta\,dr\,d\theta\,d\phi$

The relations between the polar coordinates $(\rho,\,\phi)$ and $(x,\,y)$ of Cartesian Coordinates are

$$x = \rho \cos \phi$$

$$y = \rho \sin \phi$$

$$\rho = \sqrt{x^2 + y^2}$$

$$\phi = \tan^{-1}\left(\frac{y}{x}\right)$$

$$dx\,dy = \rho\,d\rho\,d\phi.$$

1.3 DEL (∇) OPERATOR

The **del** or **nabla** is known as differential vector operator and is defined as

$$\nabla \equiv \mathbf{a}_x \frac{\partial}{\partial x} + \mathbf{a}_y \frac{\partial}{\partial y} + \mathbf{a}_z \frac{\partial}{\partial z}$$

Del has units of 1/meter (1/m)

Del operates in three ways.

1.4 GRADIENT OF A SCALAR V ($= \nabla V$)

Gradient of a scalar is a vector and is defined as

$$\nabla V \equiv \frac{\partial V}{\partial x}\,\mathbf{a}_x + \frac{\partial V}{\partial y}\,\mathbf{a}_y + \frac{\partial V}{\partial \mathbf{a}_z}\,\mathbf{a}_z$$

Examples are gradient of temperature, gradient of electric potential and so on.

It gives the maximum space rate of change of the scalar. The scalar can be a temperature, potential and so on.

1.5 DIVERGENCE OF A VECTOR A ($= \nabla \cdot A$)

Divergence of a vector is a scalar and is defined as

$$\nabla \cdot \mathbf{A} \equiv \operatorname{div} \mathbf{A} \equiv \frac{\partial A_x}{\partial x} + \frac{\partial A_y}{\partial y} + \frac{\partial A_z}{\partial z}$$

The divergence indicates the spreading or diverging of a quantity from a point. It is applicable to vectors only. The divergence of a vector indicates the net flow of quantities like gas, fluid, vapour, electric and magnetic flux lines and so on. In other words, it is a measure of the difference between outflow and inflow.

The divergence of a vector is positive if the net flow is outward.

It is negative if the net flow is inward.

The fluid is said to be incompressible if the divergence is zero.

That is, $\nabla \cdot \mathbf{A} = 0$ is the condition of incompressibility.

Examples and features of divergence

1. Leaking air from a balloon yields positive divergence.

2. The rushing of air into the drum under the carriage of a train yields negative divergence.

3. The divergence of water or oil is almost zero and hence they are incompressible.

4. The divergence of electric flux density is equal to volume charge density.

   That is, $\qquad \nabla \cdot \mathbf{D} = \rho_v.$

5. The divergence of magnetic flux density is equal to zero.

   That is, $\qquad \nabla \cdot \mathbf{B} = 0.$

6. The divergence of the gradient of scalar electric potential is equal to Laplacian of the scalar.

   That is, $\qquad \nabla \cdot \nabla V = \nabla^2 V.$

1.6 CURL OF A VECTOR ($\equiv \nabla \times \mathbf{A}$)

The **Curl of a vector** is a vector and is defined as

$$\text{Curl } \mathbf{A} = \nabla \times \mathbf{A} = \begin{vmatrix} \mathbf{a}_x & \mathbf{a}_y & \mathbf{a}_z \\ \dfrac{\partial}{\partial x} & \dfrac{\partial}{\partial y} & \dfrac{\partial}{\partial z} \\ A_x & A_y & A_z \end{vmatrix}$$

$$= \mathbf{a}_x \left[\frac{\partial}{\partial y} A_z - \frac{\partial}{\partial z} A_y \right] + \mathbf{a}_y \left[\frac{\partial}{\partial z} A_x - \frac{\partial}{\partial x} A_z \right] + \mathbf{a}_z \left[\frac{\partial}{\partial x} A_y - \frac{\partial}{\partial y} A_x \right]$$

It indicates a measure of the tendency of a vector quantity to rotate or twist or curl.

As the curl of a vector represents rotation, it is also written as

$$\text{curl } \mathbf{A} = \text{rot } \mathbf{A} = \nabla \times \mathbf{A}$$

It may be noted that curl (gradient of a scalar)

$$= \nabla \times (\nabla V) \text{ is zero.}$$

This means that the gradient of fields is irrotational. Also div (curl) = 0.

Examples

- When a leaf floats in sea water and its rotation is about z-axis, the Curl of velocity $\mathbf{V}$ is in z-direction. When $(\nabla \times \mathbf{V})_z$ is positive, it represents rotation from x to y.

- For a rotating rigid body, the Curl of velocity is in the direction of the axis of rotation. Its magnitude is equal to two times the angular speed of rotation.

   In Cartesian coordinates system,

$$\nabla V = \frac{\partial V}{\partial x} \mathbf{a}_x + \frac{\partial V}{\partial y} \mathbf{a}_y + \frac{\partial V}{\partial z} \mathbf{a}_z \text{ is a vector}$$

$$\nabla \cdot \mathbf{A} = \frac{\partial A_x}{\partial x} + \frac{\partial A_y}{\partial y} + \frac{\partial A_z}{\partial z} \text{ is a scalar}$$

$$\nabla \times \mathbf{A} = \left(\frac{\partial A_z}{\partial y} - \frac{\partial A_y}{\partial z} \right) \mathbf{a}_x + \left(\frac{\partial A_x}{\partial z} - \frac{\partial A_z}{\partial x} \right) \mathbf{a}_y + \left(\frac{\partial A_y}{\partial x} - \frac{\partial A_x}{\partial y} \right) \mathbf{a}_z \text{ is a vector.}$$

In cylindrical coordinates

$$\nabla V = \frac{\partial V}{\partial \rho} \mathbf{a}_\rho + \frac{1}{\rho} \frac{\partial V}{\partial \phi} \mathbf{a}_\phi + \frac{\partial V}{\partial z} \mathbf{a}_z \text{ is a vector}$$

$$\nabla \cdot \mathbf{A} = \frac{1}{\rho} \frac{\partial A_\rho}{\partial \rho} + \frac{1}{\rho} \frac{\partial A_\phi}{\partial \phi} + \frac{\partial A_z}{\partial z} \text{ is a scalar}$$

$$\nabla \times \mathbf{A} = \left[\frac{1}{\rho} \frac{\partial A_z}{\partial \phi} - \frac{\partial A_\phi}{\partial z} \right] \mathbf{a}_\rho + \left[\frac{\partial A_\rho}{\partial z} - \frac{\partial A_z}{\partial \rho} \right] \mathbf{a}_\phi + \frac{1}{\rho} \left[\frac{\partial (\rho A_\phi)}{\partial \rho} - \frac{\partial A_\rho}{\partial \phi} \right] \mathbf{a}_z$$

$$\text{is a vector}$$

In spherical coordinates

$$\nabla V = \frac{\partial V}{\partial r} \mathbf{a}_r + \frac{1}{r} \frac{\partial V}{\partial \theta} \mathbf{a}_\theta + \frac{1}{r \sin \theta} \frac{\partial V}{\partial \phi} \mathbf{a}_\phi$$

$$\nabla \cdot \mathbf{A} = \frac{1}{r^2} \frac{\partial}{\partial r} (r^2 A_r) + \frac{1}{r \sin \theta} \frac{\partial}{\partial \theta} (\sin \theta \, A_\theta) + \frac{1}{r \sin \theta} \frac{\partial A_\phi}{\partial \phi}$$

$$\nabla \times \mathbf{A} = \frac{1}{r \sin \theta} \left[\frac{\partial}{\partial \theta} (A_\phi \sin \theta) - \frac{\partial A_\theta}{\partial \phi} \right] \mathbf{a}_r + \frac{1}{r} \left[\frac{1}{r \sin \theta} \frac{\partial A_r}{\partial \phi} - \frac{\partial (r A_\phi)}{\partial r} \right] \mathbf{a}_\theta$$

$$+ \frac{1}{r} \left[\frac{\partial}{\partial r} \left(r A_\theta - \frac{\partial A_r}{\partial \theta} \right) \right] \mathbf{a}_\phi.$$

1.6.1 Vector Identities

$$\mathbf{B} \cdot \mathbf{B} = B^2$$

$$\mathbf{B} \cdot \mathbf{B}^* = B^2$$

$$\mathbf{B} + \mathbf{C} = \mathbf{C} + \mathbf{B}$$

$$\mathbf{B} \cdot \mathbf{C} = \mathbf{C} \cdot \mathbf{B}$$

$$\mathbf{C} \times \mathbf{B} = - \mathbf{B} \times \mathbf{C}$$

$$(\mathbf{D} + \mathbf{B}) \cdot \mathbf{C} = \mathbf{D} \cdot \mathbf{C} + \mathbf{B} \cdot \mathbf{C}$$

$$(\mathbf{D} + \mathbf{B}) \times \mathbf{C} = \mathbf{D} \times \mathbf{C} + \mathbf{B} \times \mathbf{C}$$

$$\mathbf{D} \cdot \mathbf{B} \times \mathbf{C} = \mathbf{B} \cdot \mathbf{C} \times \mathbf{D} = \mathbf{C} \cdot \mathbf{D} \times \mathbf{B}$$

$$\mathbf{D} \times (\mathbf{B} \times \mathbf{C}) = (\mathbf{D} \cdot \mathbf{C}) \mathbf{B} - (\mathbf{D} \cdot \mathbf{B}) \mathbf{C}$$

$$(\mathbf{E} \times \mathbf{B}) \cdot (\mathbf{C} \times \mathbf{D}) = \mathbf{E} \cdot \mathbf{B} \times (\mathbf{C} \times \mathbf{D})$$

$$= (\mathbf{E} \cdot \mathbf{C}) \, (\mathbf{B} \cdot \mathbf{D}) - (\mathbf{E} \cdot \mathbf{D}) \, (\mathbf{B} \cdot \mathbf{C})$$

$$(\mathbf{E} \times \mathbf{B}) \times (\mathbf{C} \times \mathbf{D}) = (\mathbf{E} \times \mathbf{B} \cdot \mathbf{D}) \mathbf{C} - (\mathbf{E} \times \mathbf{B} \cdot \mathbf{C}) \mathbf{D}$$

$$\nabla (V_m\, V) = V_m\, \nabla V + V\nabla V_m$$

$$\nabla . (\nabla \times C) = 0$$

$$\nabla \times \nabla V = 0$$

$$\nabla (V_m + V) = \nabla V_m + \nabla V$$

$$\nabla . (C + B) = \nabla . C + \nabla . B$$

$$\nabla \times (C + B) = \nabla \times C + \nabla \times B$$

$$\nabla . (VC) = C . \nabla V + V\nabla \times C$$

$$\nabla \times (VC) = \nabla V \times C + V\nabla \times C$$

$$\nabla (C . B) = (C . \nabla)\, B + (B . \nabla)\, C + C \times (\nabla \times B) + B \times (\nabla \times C)$$

$$\nabla . (C \times B) = B . \nabla \times C - C . \nabla \times B$$

$$\nabla \times (C \times B) = C\, (\nabla . B) - B\, (\nabla . C) + (B . \nabla)\, C - (C . \nabla)\, B$$

$$\nabla \times \nabla \times C = \nabla\, (\nabla . C) - \nabla^2 C$$

$$\oint_L C . d\mathbf{L} = \int_s (\nabla \times C) . d\mathbf{s}$$

$$\oint_s C . d\mathbf{s} = \int_v (\nabla . C)\, dv$$

$$\oint_s (\mathbf{a}_n \times C)\, ds = \int_v (\nabla \times C)\, dv$$

$$\oint_s V\, d\mathbf{s} = \int_v \nabla V\, dv$$

$$\oint_L V\, d\mathbf{L} = \int_s \mathbf{a}_n \times \nabla V\, ds.$$

1.7 LAPLACIAN OPERATOR (∇^2)

Laplacian Operator is defined as $\nabla . \nabla$. Its unit is $\dfrac{1}{m^2}$. It is a scalar differential operator. It operates on a scalar as well as a vector.

$$\nabla^2 \equiv \frac{\partial^2}{\partial x^2} + \frac{\partial^2}{\partial y^2} + \frac{\partial^2}{\partial z^2}$$

Laplacian of a scalar electric potential, ($\nabla^2 V$), is a scalar.

Laplacian of an electric field, ($\nabla^2 E$), is a vector.

Laplacian of a scalar and a vector in different coordinate systems.

$$\nabla^2 V = \frac{\partial^2 V}{\partial x^2} + \frac{\partial^2 V}{\partial y^2} + \frac{\partial^2 V}{\partial z^2} \quad \text{(Cartesian)}$$

$$\nabla^2 V = \frac{1}{\rho}\frac{\partial}{\partial\rho}\left(\rho\frac{\partial V}{\partial\rho}\right)+\frac{1}{\rho^2}\frac{\partial^2 V}{\partial\phi^2}+\frac{\partial^2 V}{\partial z^2}\quad\text{(Cylindrical)}$$

$$\nabla^2 V = \frac{1}{r^2}\frac{\partial}{\partial r}\left(r^2\frac{\partial V}{\partial r}\right)+\frac{1}{r^2\sin\theta}\frac{\partial}{\partial\theta}\left(\sin\theta\frac{\partial V}{\partial\theta}\right)+\frac{1}{r^2\sin^2\theta}\frac{\partial^2 V}{\partial\phi^2}\quad\text{(Spherical)}$$

$$\nabla^2 \mathbf{A} = \frac{\partial^2 \mathbf{A}}{\partial x^2}+\frac{\partial^2 \mathbf{A}}{\partial y^2}+\frac{\partial^2 \mathbf{A}}{\partial z^2}\quad\text{(Cartesian)}$$

$$=\frac{\partial}{\partial x^2}(A_x\,\mathbf{a}_x+A_y\,\mathbf{a}_y+A_z\,\mathbf{a}_z)+\frac{\partial}{\partial y^2}(A_x\,\mathbf{a}_x+A_y\,\mathbf{a}_y+A_z\,\mathbf{a}_z)$$

$$+\frac{\partial}{\partial z^2}(A_x\,\mathbf{a}_x+A_y\,\mathbf{a}_y+A_z\,\mathbf{a}_z).$$

1.8 DIRAC DELTA

The Dirac Delta is used to represent very short pulse of high amplitude. It is also sometimes called unit pulse function.

The properties of Dirac Delta are:
$$\delta(t-t_0)=\infty,\quad\text{if } t=t_0$$
and
$$\delta(t-t_0)=0\quad\text{otherwise.}$$

The Dirac Delta is shown in Fig. 1.5.

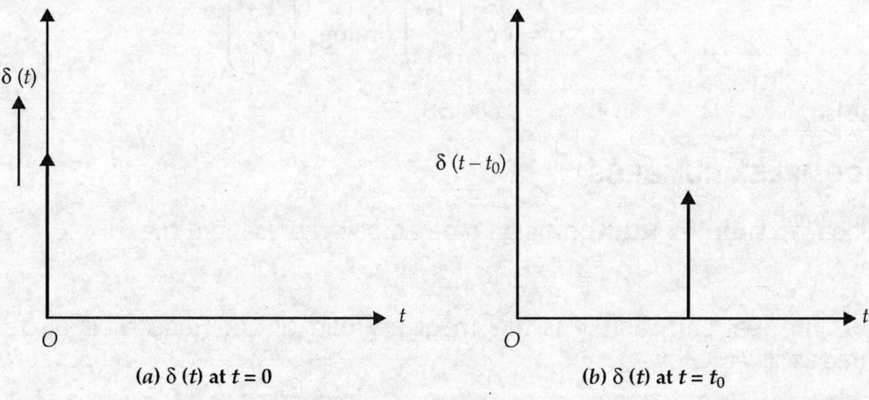

Fig. 1.5 *The Dirac Delta*

1.9 DECIBEL AND NEPER CONCEPTS

Decibel (dB) is defined as ten times the common logarithm of the power ratio. That is, $1\,\mathrm{dB} \equiv 10\log_{10}\left(\dfrac{P_1}{P_2}\right)$

If P_1 and P_2 are input and output powers respectively of an electric circuit, then dB is negative if $P_2 > P_1$. This indicates power loss. If $P_1 > P_2$, dB is positive. This indicates power gain.

Decibel has no dimensions and it is used to express the ratio of two powers, voltages, currents or sound intensities.

One Bel (B) is equal to 10 decibels.

When input and output currents and voltages are known, we have

$$1 \text{ dB} = 20 \log_{10} \left(\frac{V_2}{V_1} \right)$$

$$= 20 \log_{10} \left(\frac{I_2}{I_1} \right)$$

Neper (NP) It has no dimensions and it is used to express the ratio of powers in communications.

A **Neper** is defined as the natural logarithm of the square root of the power ratio.

$$\text{That is,} \quad 1 \text{ Np} \equiv \log_e \sqrt{\frac{P_2}{P_1}} = \frac{1}{2} \log_e \left(\frac{P_2}{P_1} \right)$$

In terms of input and output currents or voltages, it is given by

$$1 \text{ Np} = \log_e \left(\frac{I_2}{I_1} \right) \text{ or } \log_e \left(\frac{V_2}{V_1} \right)$$

Also, $\qquad\qquad$ 1 Np = 8.686 dB.

1.10 COMPLEX NUMBERS

By definition, a **complex number** is an ordered pair represented by

$$A = (x, y)$$

where x is the real part, and y is the imaginary part of A. Hence it is also represented as $A = x + jy$

Here $\qquad\qquad$ $j \equiv \sqrt{-1} \equiv (0, 1)$ is called as imaginary unit.

1.10.1 Properties of Complex Numbers

1. Two complex numbers are equal if their real parts are equal and their imaginary parts are also equal.
2. If $A_1 = x_1 + jy_1$, $A_2 = x_2 + jy_2$, the sum of the two complex numbers A_1 and A_2 is given by $A = A_1 + A_2 = (x_1 + x_2) + j\,(y_1 + y_2)$.

3. The subtraction of A_2 from A_1 is given by

$$B = A_1 - A_2 = (x_1 - x_2) + j\,(y_1 - y_2).$$

4. The product of A_1 and A_2 is

$$C = A_1 A_2 = (x_1 + jy_1)\,(x_2 + jy_2)$$

$$= x_1 x_2 + j x_1 y_2 + j x_2 y_1 - y_1 y_2$$

$$= (x_1 x_2 - y_1 y_2) + j\,(x_1 y_2 + x_2 y_1).$$

5. The division of A_1 by A_2 is given by

$$D = \frac{A_1}{A_2} = \frac{x_1 + jy_1}{x_2 + jy_2}$$

Multiplying numerator and denominator by $(x_1 - jy_2)$, we get

$$D = \frac{A_1}{A_2} = \frac{(x_1 + jy_1)}{(x_2 + jy_2)}\,\frac{(x_2 - jy_2)}{(x_2 - jy_2)}$$

$$= \frac{x_1 x_2 - j x_1 y_2 + j x_2 y_1 + y_1 y_2}{(x_2^2 + y_2^2)}$$

Hence,
$$D = \frac{(x_1 x_2 + y_1 y_2) + j\,(x_2 y_1 - x_1 y_2)}{(x_2^2 + y_2^2)}.$$

1.11 LOGARITHMIC SERIES AND IDENTITIES

$$\log (1 + x) = x - \frac{x^2}{2} + \frac{x^3}{3} - \frac{x^4}{4} + \cdots$$

This series converges for

$$-1 < x < 1$$

It diverges for
$$x = -1$$

$$\log (1 + x) = -\left(x - \frac{x^2}{2} + \frac{x^3}{3} - \frac{x^4}{4} + \cdots \right)$$

$$\log_a (xy) = \log_a x + \log_a y$$

$$\log_a \left(\frac{x}{y} \right) = \log_a x - \log_a y$$

$$\log_a \left(\frac{1}{y} \right) = -\log_a y$$

$$\log_a (x^n) = n \log_a x$$

$$\log_a (x^{1/n}) = \frac{1}{n} \log_a x$$

$$\log_a x = \log_b x \cdot \log_a b$$

$$= \log_b x / \log_b a$$

$$\log_e x = \log_{10} x \cdot \log_e 10 = 2.30 \log_{10} x$$

$$\log_{10} x = \log_e x \cdot \log_{10} e = 0.434 \log_e x.$$

1.12 QUADRATIC EQUATIONS

A quadratic equation is represented by $ax^2 + bx + c = 0$, where a, b and c are constants. Its roots are given by

$$x_1 = \frac{-b + \sqrt{b^2 - 4ac}}{2a}$$

and
$$x_2 = \frac{-b - \sqrt{b^2 - 4ac}}{2a}$$

The roots may be real or imaginary and may be equal or unequal depending on whether the quantity $(b^2 - 4ac)$ is positive, zero, or negative.

1.13 CUBIC EQUATIONS

It is given by
$$f(x) = x^3 + ax^2 + bx + c = 0$$

By substituting $\left(y - \dfrac{a}{3} \right)$ for x, $f(x)$ becomes

$$y^3 + dy + e = 0$$

Here,
$$d = \frac{1}{3}(3b - a^2)$$

and
$$e = \frac{1}{27}(2a^2 - 9ab + 27c)$$

If
$$A = \left(-\frac{e}{2} + \sqrt{\frac{e^2}{4} + \frac{d^3}{27}} \right)^{1/3}$$

and
$$B = \left(-\frac{e}{2} - \sqrt{\frac{e^2}{4} + \frac{d^3}{27}} \right)^{1/3},$$

then the three values of y are

$$y = A + B$$

$$y = -\frac{A + B}{2} + \frac{(A - B)\,(-3)^{1/2}}{2}$$

and
$$y = -\frac{A + B}{2} - \frac{(A - B)\,(-3)^{1/2}}{2}$$

If $\left(\dfrac{e^2}{4}+\dfrac{d^2}{27}\right)<0$, the trigonometric solution is found out by evaluating the cube roots of complex quantities. Then the angle ϕ is evaluated from

$$\cos\phi=-\frac{\dfrac{e}{2}}{\left(-\dfrac{d^3}{27}\right)^{1/2}}$$

The values of y will be

$$y=2\left(-\frac{d}{3}\right)^{1/2}\cos\frac{\phi}{3}$$

$$y=2\left(-\frac{d}{3}\right)^{1/2}\cos\left(\frac{\phi}{3}+120°\right)$$

and

$$y=2\left(-\frac{d}{3}\right)^{1/2}\cos\left(\frac{\phi}{3}+240°\right).$$

1.14 DETERMINANTS

A quantity $\begin{vmatrix} a_1 & a_2 \\ b_1 & b_2 \end{vmatrix}$ is defined as a **determinant**.

It is a second order determinant. Its value is given by

$$\begin{vmatrix} a_1 & a_2 \\ b_1 & b_2 \end{vmatrix}=a_1 b_2-a_2 b_1.$$

- **Example**   The value of a determinant

$$\begin{vmatrix} 1 & 2 \\ 3 & 5 \end{vmatrix}\text{ is }1\times5-2\times3=5-6=-1$$

Similarly, $\begin{vmatrix} a_1 & a_2 & a_3 \\ b_1 & b_2 & b_3 \\ c_1 & c_2 & c_3 \end{vmatrix}$ is called a third order determinant.

- **Example**   The value of a third order determinant $\begin{vmatrix} 1 & -2 & -3 \\ -4 & 5 & -6 \\ -7 & 8 & -9 \end{vmatrix}$

$$= 1 \times \begin{vmatrix} 5 & -6 \\ 8 & -9 \end{vmatrix} - (-2) \times \begin{vmatrix} -4 & -6 \\ -7 & -9 \end{vmatrix} + (-3) \times \begin{vmatrix} -4 & 5 \\ -7 & 8 \end{vmatrix}$$

$$= 3 - 12 - 9$$

$$= -18.$$

Application of determinants They are widely used in solving simultaneous equations and are also useful in developing the theory of matrices.

1.14.1 The Minor of Determinant

In a determinant $\begin{vmatrix} a_1 & a_2 & a_3 \\ b_1 & b_2 & b_3 \\ c_1 & c_2 & c_3 \end{vmatrix}$, the minor of b_3 is given by $\begin{vmatrix} a_1 & a_2 \\ c_1 & c_2 \end{vmatrix}$ and the

minor of c_2 is given by $\begin{vmatrix} a_1 & a_3 \\ b_1 & b_3 \end{vmatrix}$.

That is, the minor of an element in a determinant is a determinant obtained by deleting the row and the column which intersect at that element.

The cofactor of an element in a determinant is its minor with a suitable sign. The sign in the ith row and j^{th} column is given by $(-1)^{i+j}$. Then the cofactor B_3 of b_3 is $(-1)^{2+3} \times$ minor of b_3.

That is, $B_3 = -\begin{vmatrix} a_1 & a_2 \\ c_1 & c_2 \end{vmatrix}$

1.14.2 Properties of Determinants

(*i*) The value of the determinant remains the same after changing its rows into columns and columns into rows.

(*ii*) The value of the determinant remains the same if any two rows or columns are interchanged but the sign changes.

(*iii*) If two rows or columns are identical, the value of the determinant is zero.

(*iv*) If each element of a row or column is multiplied by a factor, the determinant is multiplied by the same factor.

(*v*) If each element of a row consists of n terms, the determinant can be expressed as the sum of n determinants.

(*vi*) When equi-multiples of the corresponding elements of one or more parallel lines are added to each element of a line, the determinant value remains the same.

1.15 MATRICES

A **Matrix** is defined as a rectangular array of numbers of functions enclosed in brackets. These numbers or functions are known as the elements of the matrix.

Example $\begin{bmatrix} 1 \\ 3 \end{bmatrix}$, $\begin{bmatrix} 2 & 1 & 4 \\ 3 & 0 & 1 \end{bmatrix}$, $[a \quad b \quad c]$, $\begin{bmatrix} a & b \\ c & d \end{bmatrix}$, $\begin{bmatrix} e^{-x} & x \\ x^2 & e^x \end{bmatrix}$ and so on are called matrices.

$\begin{bmatrix} 1 \\ 3 \end{bmatrix}$ is a matrix containing two rows and one column.

Similarly, $\begin{bmatrix} 2 & 1 & 4 \\ 3 & 0 & 1 \end{bmatrix}$ is a matrix containing two rows and three columns.

Here, row means a horizontal line and column means a vertical line.

1.15.1 Applications of Matrices

Matrices are widely used to solve linear systems of equations. They often appear as models of different problems. For example, in electrical circuits and numerical methods they are used for solving differential equations.

1.15.2 Types of Matrices

(*i*) Row Matrix like $[1 \quad 2 \quad 3]$.

(*ii*) Column Matrix like $\begin{bmatrix} 1 \\ 2 \\ 3 \end{bmatrix}$.

(*iii*) Square Matrix like $\begin{bmatrix} 1 & 2 \\ 3 & 4 \end{bmatrix}$.

(*iv*) Diagonal matrix like $\begin{bmatrix} 1 & 0 & 0 \\ 0 & -2 & 0 \\ 0 & 0 & 3 \end{bmatrix}$ and $\begin{bmatrix} 2 & 0 & 0 \\ 0 & 1 & 0 \\ 0 & 0 & 3 \end{bmatrix}$.

(*v*) Unit Matrix like $\begin{bmatrix} 1 & 0 & 0 \\ 0 & 1 & 0 \\ 0 & 0 & 1 \end{bmatrix}$.

(vi) Null Matrix like $\begin{bmatrix} 0 & 0 & 0 \\ 0 & 0 & 0 \end{bmatrix}$.

(vii) Symmetric Matrix like $\begin{bmatrix} 1 & 2 & 3 \\ 2 & 1 & 6 \\ 3 & 6 & 7 \end{bmatrix}$.

$(viii)$ Skew Symmetric Matrix like $\begin{bmatrix} 0 & 1 & -2 \\ -1 & 0 & 3 \\ 2 & -3 & 0 \end{bmatrix}$.

(ix) Upper Triangular Matrix like $\begin{bmatrix} 1 & 3 & 4 \\ 0 & 2 & 5 \\ 0 & 0 & 6 \end{bmatrix}$.

(x) Lower Triangular Matrix like $\begin{bmatrix} 1 & 0 & 0 \\ 3 & 5 & 0 \\ 4 & 6 & 7 \end{bmatrix}$.

1.15.3 Properties of Matrices

(i) Two matrices A and B are equal if and only if they are of the same order and each element of A is equal to the corresponding element of B.

(ii) The sum of A and B exists only when they are of the same order.

(iii) The difference of A and B exists only when they are of the same order.

$$\text{If} \qquad A = \begin{bmatrix} a_1 & b_1 \\ a_2 & b_2 \\ a_3 & b_3 \end{bmatrix}, \; B = \begin{bmatrix} c_1 & d_1 \\ c_2 & d_2 \\ c_3 & d_3 \end{bmatrix},$$

$$A + B = \begin{bmatrix} a_1 + c_1 & b_1 + d_1 \\ a_2 + c_2 & b_2 + d_2 \\ a_3 + c_3 & b_3 + d_3 \end{bmatrix}$$

$$A - B = \begin{bmatrix} a_1 - c_1 & b_1 - d_1 \\ a_2 - c_2 & b_2 - d_2 \\ a_3 - c_3 & b_3 - d_3 \end{bmatrix}.$$

(iv) Addition of matrices is commutative. That is,
$$A + B = B + A.$$

(v) Addition and subtraction of matrices are associative.

That is, $(A + B) - C = A + (B - C)$
$$= B + (A - C).$$

(vi) The multiplication of a matrix A by a scalar K gives a matrix each element of which is K times the corresponding element of A. The distributive law is valid for such a product.

That is, $K (A + B) = KA + KB.$

(vii) All the laws of ordinary algebra hold good for the addition or subtraction of matrices and their multiplication by scalars.

($viii$) Two matrices can be multiplied only when the number of columns in the first is equal to the number of rows in the second. Such matrices are said to be conformable.

(ix) $AB \neq BA.$

Sometimes if AB exists, BA may not exist at all.

(x) $IA = AI = A$ if A is a square matrix which has the same order as that of I.

(xi) $OA = AO = O$, where O is a null matrix.

(xii) If $AB = O$, it does not mean that A or B is a null matrix.

($xiii$) Multiplication of matrices is associative. That is, $(AB) C = A (BC)$ provided A, B are conformable for the product AB and B, C are conformable for the product BC.

(xiv) Multiplication of matrices is distributive. That is, $A (B + C) = AB + AC$ provi-ded A, B are conformable for AB and A, C are conformable for AC.

(xv) If A is a square matrix, then $A . A = A^2$.

(xvi) The matrix obtained from a given matrix A by interchanging rows and columns is called a Transpose of A denoted by A'.

($xvii$) For a symmetric matrix, $A' = - A$.

($xviii$) $(AB)' = B' . A'$.

(xix) Adjoint of a matrix A is the transposed matrix of cofactors of A.

(xx) If A is any matrix, then a matrix B, if it exists such that $AB = BA = I$, is called the Inverse of A. Both the matrix and its inverse must be non-singular. Inverse of a matrix A is denoted by A^{-1} so that

$$A . A^{-1} = A^{-1} . A = I.$$

Also, $$A^{-1} = \frac{\text{Adj } A}{|A|}.$$

(xxi) Inverse of a matrix is unique.

($xxii$) $(AB)^{-1} = B^{-1} A^{-1}$.

($xxiii$) A matrix is said to be of rank r when it has at least one non-zero minor of order r and every minor of order higher than r vanishes. This means that the rank of a matrix is the largest order of any non-vanishing minor of the matrix.

1.16 FACTORIAL

Factorial of n is defined as the product of the first n natural numbers.

Factorial of n is represented by $n!$ or $\angle n$.

$$\text{Factorial of } n = n! = 1 \times 2 \times 3 \times \ldots \times n$$

Using Stirling's approximation, $n!$ is evaluated from

$$n! = e^{-n}\, n^n\, (2\pi n)^{1/2}$$

$$0! = 1.$$

1.17 PERMUTATIONS

Permutations are defined as different possible arrangements which can be formed with a given number of items taking some or all of them at a time.

The number of permutations of n different things taken r at a time is given by np_r and

$$np_r = \frac{n!}{(n-r)!}$$

If p, q, r types of elements exist among n elements, then the number of permutations are given by

$$n!/(p! + q! + r! + \ldots), \quad n = p + q + r$$

The number of permutations of n different things taken all at a time is np_n.

1.18 COMBINATIONS

Combinations are defined as the number of possible groups which can be formed by taking some or all the things at a time irrespective of the order.

The number of combinations or groups of n different elements taken r at a time is given by nC_r where

$$nC_r = \frac{n!}{r!\,(n-r)!}$$

$$(n+1)\,C_r = nC_r + nC_{r-1}$$

$$nC_1 + nC_2 + nC_3 + \ldots + nC_n = 2^n - 1.$$

1.19 BASIC SERIES

Binomial Series   These are given by

$$(a \pm b)^n = a^n \pm na^{n-1} b + \frac{n\,(n-1)}{2!}\, a^{n-2}\, b^2 \pm \frac{n\,(n-1)\,(n-2)}{3!}\, a^{n-3}\, b^3$$

$$+ (\pm 1)^r\, \frac{n\,(n-1)\,\ldots\,(n-r+1)}{r!}\, a^{n-r}\, b^r$$

Here the last term shown is $(r+1)^{\text{th}}$ term.

If n is a positive integer, the series is finite and the last term is b^n. If n is less than unity, the series is infinite and converges only when $b < a$. For any value of n, and when x^2 is less than unity, the following series are convergent.

$$(1 \pm x)^n = 1 \pm nx + \frac{n\,(n-1)}{2!}\, x^2 \pm \frac{n\,(n-1)\,(n-2)}{3!}\, x^3 +$$

$$\frac{n\,(n-1)\,(n-2)\,(n-3)}{4!}\, x^4 \pm \ldots$$

$$(1 \pm x)^{-n} = 1 \mp nx + \frac{n\,(n-1)}{2!}\, x^2 \mp \frac{n\,(n-1)\,(n+2)}{3!}\, x^3 +$$

$$\frac{n\,(n-1)\,(n+2)\,(n-3)}{4!}\, x^4 + \ldots$$

$$(1 \pm x)^{-1} = 1 \mp x + x^2 \mp x^3 + x^4 \mp x^5 \ldots$$

$$(1 \pm x)^{1/2} = 1 \pm \frac{1}{2}\, x - \frac{1}{2.4}\, x^2 \pm \frac{1.3}{2.4.6}\, x^3 - \ldots$$

$$(1 \pm x)^{-1/2} = 1 \mp \frac{1}{2}\, x + \frac{1.3}{2.4}\, x^2 \pm \frac{1.3.5}{2.4.6}\, x^3 + \ldots$$

$$\left.\begin{aligned} (1 + x)^{-1} &\approx 1 - x \\ (1 - x)^{-1} &\approx 1 + x \end{aligned}\right\rvert \quad \text{if } |x| \ll 1.$$

1.20 EXPONENTIAL SERIES

$$\sum (n^2) = 1^2 + 2^2 + \ldots + n^2 = \frac{n\,(n+1)\,(2n+1)}{6}$$

$$\sum (n^3) = 1^3 + 2^3 + \ldots + n^3 = \frac{n^2\,(n+1)^2}{4}$$

$$e^x = 1 + \frac{x}{1!} + \frac{x^2}{2!} + \frac{x^3}{3!} + \ldots$$

This series converges for all values of x.

For $x = 1$,

$$e^x = e^1 = 1 + \frac{1}{1!} + \frac{1}{2!} + \frac{1}{3!} + \ldots = 2.718$$

$$e^x e^y = e^{x+y}, \quad e^x/e^y = e^{x-y}, \quad (e^x)^y = e^{xy}.$$

1.21 SINE AND COSINE SERIES

If x is a variable,

$$\sin x = x - \frac{x^3}{3!} + \frac{x^5}{5!} - \ldots \infty$$

$$\cos x = 1 - \frac{x^2}{2!} + \frac{x^4}{4!} - \ldots \infty.$$

1.22 SINH AND COSH SERIES

$$\sinh x = x + \frac{x^3}{3!} + \frac{x^5}{5!} + \ldots \infty$$

$$\cosh x = 1 + \frac{x^2}{2!} + \frac{x^4}{4!} + \ldots \infty$$

1.23 HYPERBOLIC FUNCTIONS

Hyperbolic sine of $x = \sinh x = \dfrac{(e^x - e^{-x})}{2}$

Hyperbolic cosine of $x = \cosh x = \dfrac{(e^x + e^{-x})}{2}$

Hyperbolic tan of $x = \tanh x = \dfrac{(e^x - e^{-x})}{(e^x + e^{-x})}$

Hyperbolic cot of $x = \coth x = \dfrac{(e^x + e^{-x})}{(e^x - e^{-x})} = \dfrac{1}{\tanh x}$

Similarly, $\operatorname{sech} x = \dfrac{1}{\cosh x} = \dfrac{2}{(e^x + e^{-x})}$

$$\operatorname{cosech} x = \dfrac{1}{\sinh x} = \dfrac{2}{(e^x - e^{-x})}$$

It may be noted that

$$\sinh 0 = 0, \quad \cosh 0 = 0 \quad \text{and} \quad \tanh 0 = 0$$

$$\cosh x + \sinh x = e^x$$

$$\cosh x - \sinh x = e^{-x}$$

$$\cosh^2 x - \sinh^2 x = 1$$

$$\sinh^2 x = \frac{1}{2}(\cosh 2x - 1)$$

$$\cosh^2 x = \frac{1}{2}(\cosh 2x + 1)$$

$$\sinh(x \pm y) = \sinh x \ \cosh y \pm \cosh x \ \sinh y$$

$$\cosh(x \pm y) = \cosh x \ \cosh y \pm \sinh x \ \sinh y$$

$$\tanh(x \pm y) = \frac{\tanh x \pm \tanh y}{1 \pm \tanh x \ \tanh y}.$$

1.24 SINE, COSINE, TAN AND COT FUNCTIONS

For all values of θ

$$\sin \theta = \frac{e^{j\theta} - e^{-j\theta}}{2j}$$

$$\cos \theta = \frac{e^{j\theta} + e^{-j\theta}}{2}$$

If $\theta = jx$

$$\sin jx = \frac{(e^{-x} - e^{x})}{2j} = -\frac{(e^{x} - e^{-x})}{2j}$$

$$= j^2 \frac{(e^{x} - e^{-x})}{2j} = j \frac{(e^{x} - e^{-x})}{2}$$

$$= j \sinh x$$

and
$$\cos jx = \frac{e^{-x} + e^{x}}{2} = \cosh x$$

Similarly,
$$\tan jx = j \tanh x$$
$$\sinh jx = j \sin x$$
$$\cosh jx = \cos x$$
$$\tanh jx = j \tan x$$
$$\sin(-x) = -\sin x$$
$$\cos(-x) = \cos x$$
$$1 \text{ degree} = 0.01745 \text{ radian}$$
$$1 \text{ radian} = 57.29577 \text{ degrees}$$
$$\sin^2 x + \cos^2 x = 1$$
$$\sin(x + y) = \sin x \ \cos y + \cos x \ \sin y$$
$$\sin(x - y) = \sin x \ \cos y - \cos x \ \sin y$$
$$\cos(x + y) = \cos x \ \cos y - \sin x \ \sin y$$

$$\cos (x - y) = \cos x \ \cos y + \sin x \ \sin y$$

$$\sin 2x = 2 \sin x \ \cos x$$

$$\cos 2x = \cos^2 x - \sin^2 x$$

$$\sin x = \cos \left(x - \frac{\pi}{2} \right) = \cos \left(\frac{\pi}{2} - x \right)$$

$$\cos x = \sin \left(x + \frac{\pi}{2} \right) = \sin \left(\frac{\pi}{2} - x \right)$$

$$\sin (\pi - x) = \sin x$$

$$\cos (\pi - x) = - \cos x$$

$$\cos^2 x = \frac{1}{2} (1 + \cos 2x)$$

$$\sin^2 x = \frac{1}{2} (1 - \cos 2x)$$

$$\sin x \ \sin y = \frac{1}{2} [- \cos (x + y) + \cos (x - y)]$$

$$\cos x \ \cos y = \frac{1}{2} [\cos (x + y) + \cos (x - y)]$$

$$\sin x \ \cos y = \frac{1}{2} [\sin (x + y) + \sin (x - y)]$$

$$\sin a + \sin b = 2 \sin \frac{a + b}{2} \cos \frac{a - b}{2}$$

$$\cos a + \cos b = 2 \cos \frac{a + b}{2} \cos \frac{a - b}{2}$$

$$\cos a - \cos b = - 2 \sin \frac{a + b}{2} \sin \frac{a - b}{2}$$

$$A \cos x + B \sin x = \sqrt{A^2 + B^2} \ \cos (x \pm \delta)$$

$$\tan \delta = \frac{\sin \delta}{\cos \delta} = \frac{B}{A}$$

$$A \cos x + B \sin x = \sqrt{A^2 + B^2} \ \sin (x \pm \delta)$$

$$\tan \delta = \frac{\sin \delta}{\cos \delta} = \pm \frac{A}{B}$$

$$\tan x = \frac{\sin x}{\cos x}$$

$$\cot x = \frac{\cos x}{\sin x}$$

$$\sec x = \frac{1}{\cos x}$$

$$\csc x = \frac{1}{\sin x}$$

$$\tan (x + y) = \frac{\tan x + \tan y}{1 - \tan x \ \tan y}$$

$$\tan (x - y) = \frac{\tan x - \tan y}{1 + \tan x \ \tan y}.$$

1.25 SOME SPECIAL FUNCTIONS

1.25.1 Gamma Function

$\Gamma (\alpha)$ is defined as

$$\Gamma (\alpha) \equiv \int_0^\infty e^{-t} \, t^{\alpha - 1} \, dt$$

This is true if $\qquad \alpha > 0$

$$\Gamma \left(\frac{1}{2} \right) = \sqrt{\pi}.$$

1.25.2 Beta Function

$$B (x, \ y) \equiv \int_0^1 t^{x-1} \, (1 - t)^{y-1} \, dt, \quad (x > 0, \ y > 0)$$

$$= \frac{\Gamma (x) \ \Gamma (y)}{\Gamma (x + y)}.$$

1.25.3 Error Function

$$\operatorname{erf} x = \frac{2}{\sqrt{\pi}} \int_0^x e^{-t^2} \, dt$$

$$\operatorname{erf} (\infty) = 1.$$

1.25.4 Bessel Function

Bessel function is a set of solutions for differential equations of certain type. These are useful to obtain propagation characteristics of electromagnetic waves in circular waveguides.

Considering circular waveguides, the solutions of the axial component is written conveniently in terms of Bessel function. For transverse magnetic wave, the solution is

$$E_{z,\,nm} = J_n\,(K_c\,r)\ (A\cos n\phi + B\sin m\phi)$$

and for transverse electric wave, the solution is

$$H_{z,\,nm} = J_n\,(K_c\,r)\ (A'\cos n\phi + B'\sin m\phi)$$

where $\qquad\qquad J_n\,(K_c\,r) =$ Bessel function of the first kind

$$r = \text{radius of the guide}$$

$$K_c = \text{cut-off wave number}$$

$A,\ B,\ A',\ B'$ are constants.

The solutions for the Bessel function are obtained for certain values of K_c. These values of K_c are known as eigen values. If K_c is to produce solution of the Bessel function, $(K_c\,r)$ must be the roots of the Bessel function. Then

$$J_n\,(K_c\,r) = 0$$

The propagation parameters for nm^{th} mode of TM waves are

Phase constant,

$$\beta_{nm} = (K^2 - K_{c,\,nm}^2)^{1/2}$$

where $\qquad\qquad K_{c,\,nm} = \dfrac{p_{nm}}{r}$

and $\qquad\qquad \beta_{nm} = \left[\,K^2 - \left(\dfrac{p_{nm}}{r}\right)^2\,\right]^{1/2}$

where $\qquad\qquad p_{nm} = $ the roots of the Bessel function

$$K = \text{free space wave number} = \frac{2\pi}{\lambda}$$

The cut-off wavelength for TM wave,

$$\lambda_{c,\,nm} = \frac{2\pi}{K_{c,\,nm}} = \frac{2\pi r}{p_{nm}}$$

The roots of the Bessel function for TM mode are shown in the following Table 1.2.

Table 1.2 Roots of Bessel function for TM mode

Order n	First order p_{n_1}	Second order p_{n_2}	Third order p_{n_3}
0	2.405	5.520	8.654
1	3.832	7.016	10.174
2	5.135	8.417	11.620

The roots of the Bessel function for TE modes are shown in the following Table 1.3.

Table 1.3   Roots of Bessel function for TE modes

Order n	First order p'_{n_1}	Second order p'_{n_2}	Third order p'_{n_3}
0	3.832	7.016	10.174
1	1.841	5.331	8.536
2	3.054	6.706	9.970

The propagation parameters for TE_{nm} mode are

$$\beta_{nm} = [K^2 - K'^2_{c,\,nm}]^{1/2}$$

where

$$K'^2_{c,\,nm} = \frac{p'_{nm}}{r}$$

Hence,

$$\beta_{nm} = \left[K^2 - \left(\frac{p'_{nm}}{r} \right)^2 \right]$$

$$\lambda_{c,\,nm} = \frac{2\pi r}{p'_{nm}}$$

Guide wavelength,

$$\lambda_g = \frac{\lambda}{\left[1 - \left(\frac{\lambda}{\lambda_{c,\,nm}} \right)^2 \right]^{1/2}}$$

The variation of the first three orders of the Bessel function and their roots are shown in Fig. 1.6.

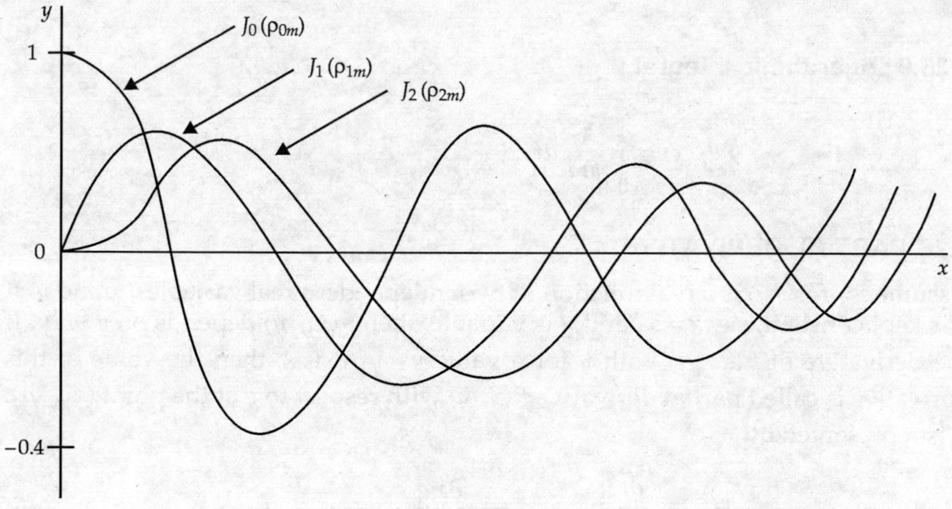

Fig. 1.6   *Bessel function*

1.25.5 Fresnel Integral

$$C(x) = \int_0^x \cos(t^2)\, dt$$

$$S(x) = \int_0^x \sin(t^2)\, dt$$

$$C(\infty) = \sqrt{\pi/8}$$
$$S(\infty) = \sqrt{\pi/8}.$$

1.25.6 Sine Integral

$$s_i(x) = \int_0^x \frac{\sin t}{t}\, dt$$

$$s_i(\infty) = (\pi/2).$$

1.25.7 Cosine Integral

$$C_i(x) = \int_x^\infty \frac{\cos t}{t}\, dt.$$

1.25.8 Exponential Integral

$$E_i(x) = \int_x^\infty \frac{e^{-t}}{t}\, dt.$$

1.25.9 Logarithmic Integral

$$l_i(x) = \int_0^\infty \frac{1}{\ln t}\, dt.$$

1.26 PARTIAL DERIVATIVE

Assume $z = f(x, y)$ is a real function of two independent real variables, x and y. If x is kept constant, say $x = x_1$ and y is variable, then $f(x_1, y)$ depends on y only. If the derivative of $f(x, y_1)$ with y for a value $y = y_1$ exists, then the value of this derivative is called partial derivative $f(x, y)$ with respect to y at the point (x_1, y_1). It is repreasented by

$$\left. \frac{\partial f}{\partial x} \right|_{(x_1,\, y_1)} \quad \text{or} \quad \left. \frac{\partial z}{\partial x} \right|_{(x_1,\, y_1)}.$$

1.27 SOME DIFFERENTIATION FORMULAE

Assume u, v to be function of x and let a, n be constants. Then,

$$\frac{d\,(a)}{dx} = 0$$

$$\frac{d\,(u+v)}{dx} = \frac{du}{dx} + \frac{dv}{dx}$$

$$\frac{d\,(au)}{dx} = a\,\frac{du}{dx}$$

$$\frac{d\,(u \cdot v)}{dx} = u\,\frac{dv}{dx} + v\,\frac{du}{dx}$$

$$\frac{d}{dx}\left(\frac{u}{v}\right) = \frac{v\,\dfrac{du}{dx} - u\,\dfrac{dv}{dx}}{v^2}$$

$$\frac{du^n}{dx} = nu^{n-1}\,\frac{du}{dx}$$

$$\frac{d}{dx}\log_e(u) = \frac{1}{u}\,\frac{du}{dx}$$

$$\frac{d}{dx}(x^n) = n \cdot x^{n-1}$$

$$\frac{d}{dx}(e^x) = e^x$$

$$\frac{d}{dx}(a^x) = a^x \ln(a)$$

$$\frac{d}{dx}(\sin x) = \cos x$$

$$\frac{d}{dx}(\cos x) = -\sin x$$

$$\frac{d}{dx}(\tan x) = \sec^2 x$$

$$\frac{d}{dx}(\cot x) = -\operatorname{cosec}^2 x$$

$$\frac{d}{dx}(\sinh x) = \cosh x$$

$$\frac{d}{dx}(\cosh x) = \sinh x$$

$$\frac{d}{dx}(\ln(x)) = \frac{1}{x}$$

$$\frac{d}{dx}(\log_a x) = \frac{\log_a e}{x}$$

$$\frac{d}{dx}(\sin^{-1} x) = \frac{1}{\sqrt{1-x^2}}$$

$$\frac{d}{dx}(\cos^{-1} x) = -\frac{1}{\sqrt{1-x^2}}$$

$$\frac{d}{dx}(\tan^{-1} x) = \frac{1}{1+x^2}$$

$$\frac{d}{dx}(\cos^{-1} x) = -\frac{1}{1+x^2}$$

$$\frac{d}{dx}(\log_e(u)) = \frac{1}{u}\frac{du}{dx}$$

$$\frac{d}{dx}(\sin u) = \cos u\,\frac{du}{dx}$$

$$\frac{d}{dx}(\cos u) = -\sin u\,\frac{du}{dx}$$

$$\frac{d}{dx}(\tan u) = \sec^2 u\,\frac{du}{dx}$$

$$\frac{d}{dx}(\cot u) = -\csc^2 u\,\frac{du}{dx}$$

$$\frac{d}{dx}(\sec u) = \sec u\,\tan u\,\frac{du}{dx}$$

$$\frac{d}{dx}(\csc u) = -\csc u\,\cot u\,\frac{du}{dx}$$

$$\frac{d}{dx}(\sinh^{-1} u) = \frac{1}{\sqrt{u^2+1}}\,\frac{du}{dx}$$

$$\frac{d}{dx}(\cosh^{-1} u) = \frac{1}{\sqrt{u^2-1}}\,\frac{du}{dx}$$

$$\frac{d}{dx}(\tanh^{-1} u) = \frac{1}{1-u^2}\,\frac{du}{dx}$$

$$\frac{d}{dx}(\coth^{-1} u) = \frac{1}{1-u^2}\,\frac{du}{dx}$$

$$\frac{d}{dx}(\operatorname{csch}^{-1} u) = -\frac{1}{u\sqrt{1+u^2}}\,\frac{du}{dx}$$

$$\frac{d}{dx}(\operatorname{sech}^{-1} u) = -\frac{1}{u\sqrt{1-u^2}}\,\frac{du}{dx}\,.$$

1.28 SOME USEFUL INTEGRATION FORMULAE

$$\int a\,dx = ax, \text{ 'a' being a constant}$$

$$\int (u+v)\,dx = \int u\,dx + \int v\,dx, u \text{ and } v \text{ are variables}$$

$$\int u\,dv = uv - \int v\,du$$

$$\int \log_e x\,dx = x \log_e x - x$$

$$\int x^n\,dx = \frac{x^{n+1}}{n+1} \qquad (n \neq -1)$$

$$\int \frac{dx}{x} = \log_e x + c \qquad (c = a \text{ constant})$$

$$\int e^{ax}\,dx = \frac{1}{a} e^{ax}$$

$$\int a^x\,dx = \frac{a^x}{\log_e a}$$

$$\int a^x \log_e a\,dx = a^x$$

$$\int \sin ax\,dx = -\frac{1}{a} \cos ax$$

$$\int \cos ax\,dx = \frac{1}{a} \sin ax$$

$$\int \tan ax\,dx = -\frac{1}{a} \log_e \cos ax$$

$$\int \cot ax\,dx = \frac{1}{a} \log_e \sin ax$$

$$\int \sec ax\,dx = \frac{1}{a} \log_e (\sec ax + \tan ax)$$

$$\int \csc ax\,dx = \frac{1}{a} \log_e (\csc ax - \cot ax)$$

$$\int \frac{dx}{x^2 + a^2} = \frac{1}{a} \tan^{-1} \frac{x}{a}$$

$$\int \frac{dx}{\sqrt{a^2 - x^2}} = \sin^{-1} \frac{x}{a}$$

$$\int \frac{dx}{\sqrt{x^2 + a^2}} = \sinh^{-1} \frac{x}{a}$$

$$\int \frac{dx}{\sqrt{x^2 - a^2}} = \cosh^{-1}\frac{x}{a}$$

$$\int \sin^2 x \, dx = \frac{1}{2}x - \frac{1}{4}\sin 2x$$

$$\int \cos^2 x \, dx = \frac{1}{2}x + \frac{1}{4}\sin 2x$$

$$\int \tan^2 x \, dx = \tan x - x$$

$$\int \cot^2 x \, dx = -\cot x - x$$

$$\int \ln x \, dx = x \ln x - x$$

$$\int e^{ax} \sin bx \, dx = \frac{e^{ax}}{a^2 + b^2}(a \sin bx - b \cos bx)$$

$$\int e^{ax} \cos bx \, dx = \frac{e^{ax}}{a^2 + b^2}(a \cos bx + b \sin bx)$$

$$\int \cos^3 ax \, dx = \frac{1}{a}\sin ax - \frac{1}{3a}\sin^3 ax$$

$$\int \sin^3 ax \, dx = -\frac{1}{a}\cos ax + \frac{1}{3a}\cos^3 ax.$$

1.29 RADIAN AND STERADIAN

Radian is a measure of a plane angle. One radian is defined as the plane angle with its vertex at the centre of a circle of radius r which is subtended by an arc whose length is equal to r.

This is shown in Fig. 1.7. Since circumference of a circle is $2\pi r$, there are 2π radians in a full circle.

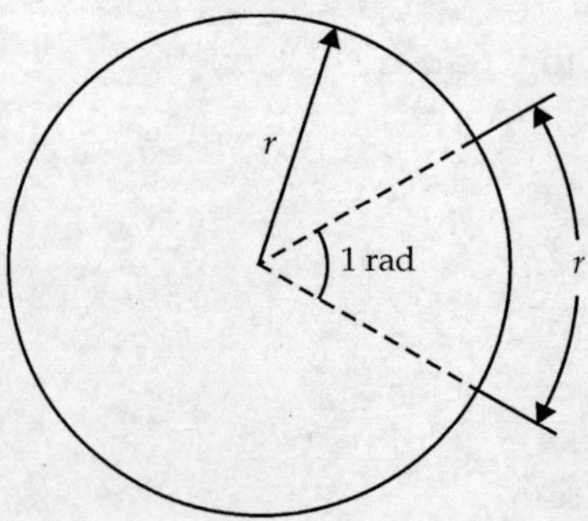

Fig. 1.7 *Radian*

Steradian is the measure of a solid angle. It is defined as the solid angle with its vertex at the centre for a sphere of radius, r which is subtended by a spherical surface area equal to the area of a square with side length r.

The differential area ds on the surface of a sphere of radius r is

$$ds = r^2 \sin\theta \, d\theta \, d\Phi \ (m^2)$$

The element of solid angle $d\Omega$ of the sphere is given by

$$d\Omega = \frac{ds}{r^2} \ (\text{Str})$$

It is illustrated in Fig. 1.8.

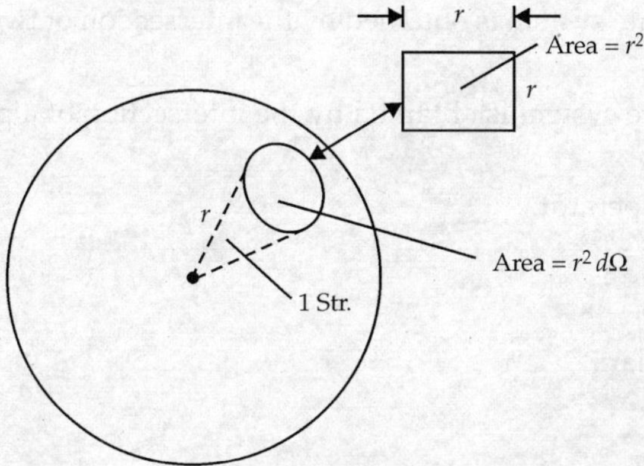

Fig. 1.8 *Steradian*

1.30 INTEGRAL THEOREMS

1.30.1 Stokes' Theorem

If a field vector like **H** and its first derivatives are continuous at all points in a region of area S bounded by a closed curve, then **Stokes' theorem** states that

$$\int_S (\nabla \times \mathbf{H}).d\mathbf{S} = \oint_L \mathbf{H}.d\mathbf{L}$$

where $d\mathbf{S}$ is along the outward normal to be the surface. This means that a surface integral is converted into line integral and vice-versa (by Stokes' theorem).

1.30.2 Divergence Theorem

If a field vector like **D** and its first derivatives are continuous at all points in a region of volume v bounded by a closed elementary surface $d\mathbf{S}$, then Divergence theorem states that

$$\oint_S \mathbf{D} \cdot d\mathbf{S} = \int_v (\nabla \cdot \mathbf{D})\, dv$$

This means that a surface integral is converted into volume integral and vice-versa by Divergence theorem.

 ## POINTS TO REMEMBER

1. Unit vector of a vector $\mathbf{A}$ is $\mathbf{a} = \dfrac{\mathbf{A}}{A}$.

2. The coordinate axes are perpendicular to each other.

3. A point in Cartesian coordinate system is obtained by the intersection of 3 planes.

4. A point in cylindrical coordinate system is obtained by the intersection of two planes and one cylindrical surface.

5. A point in spherical coordinate system is obtained by the intersection of a plane, a spherical surface and a cone.

6. Del, ∇ is a vector differential operator.

7. The unit of ∇ is $1/m$.

8. Gradient of a scalar is a vector.

9. Divergence of a vector is a scalar.

10. Curl of a vector is a vector.

11. 1 dB is defined as $10 \log_{10} \dfrac{P_1}{P_2}$.

12. 1 Neper is defined as $\log_e \left(\dfrac{V_1}{V_2} \right)$.

13. 1 Neper = 8.686 dB.

14. A determinant is represented by $\begin{vmatrix} a_1 & a_2 \\ b_1 & b_2 \end{vmatrix}$.

15. A matrix is defined as a rectangular array of numbers or functions enclosed in brackets.

16. $\cos jx = \cosh x$.

17. $\sin jx = j \sinh x$.

18. 1 degree = 0.01745 radians.

19. 1 radian = 57.29 degrees.

20. The unit of magnetic scalar potential is ampere.

21. The unit of magnetic vector potential is wb/m.

22. Reflection coefficient is a complex quantity.

23. Electric field, magnetic field, electric flux density, magnetic flux density, conduction current density, displacement current density and magnetic current density are vector quantities.

24. The unit of Poynting vector is watts/m^2.

25. Torque is a vector and has units of Newton-meter.

26. The unit of solid angle is steradian.

27. Stokes' theorem gives a relation between a closed line integral and a surface integral.

28. Divergence theorem gives a relation between a closed surface integral and a volume integral.

29. The unit of the magnetic current density is volts/m^2.

30. ∇^2 is Lapalcian scalar operator.

 ## SOLVED PROBLEMS

Problem 1.1   If a vector $\mathbf{A} = \mathbf{a}_x + 2\mathbf{a}_y + 3\mathbf{a}_z$, find its magnitude and direction.

Solution $\mathbf{A} = \mathbf{a}_x + 2\mathbf{a}_y + 3\mathbf{a}_z$

The magnitude of $\mathbf{A} = A = \sqrt{1^2 + 2^2 + 3^2} = \sqrt{1 + 4 + 9} = \sqrt{14} = 3.742$

So, $A = 3.742$

Its direction is given by unit vector

$$\mathbf{a} = \frac{\mathbf{A}}{A} = \frac{(\mathbf{a}_x + 2\mathbf{a}_y + 3\mathbf{a}_z)}{3.742}$$

So, $\mathbf{a} = 0.267\mathbf{a}_x + 0.534\mathbf{a}_y + 0.802\mathbf{a}_z.$

Problem 1.2   If $\mathbf{A} = 2\mathbf{a}_x + 5\mathbf{a}_y + 6\mathbf{a}_z$ and $\mathbf{B} = \mathbf{a}_x - 3\mathbf{a}_y + 6\mathbf{a}_z$, find $\mathbf{A} + \mathbf{B}$ and $\mathbf{A} - \mathbf{B}$.

Solution
$$\mathbf{A} + \mathbf{B} = (A_x + B_x)\,\mathbf{a}_x + (A_y + B_y)\,\mathbf{a}_y + (A_z + B_z)\,\mathbf{a}_z$$
$$= (2 + 1)\,\mathbf{a}_x + (5 - 3)\,\mathbf{a}_y + (6 + 6)\,\mathbf{a}_z$$

$$\mathbf{A} + \mathbf{B} = 3\mathbf{a}_x + 2\mathbf{a}_y + 12\mathbf{a}_z$$

$$\mathbf{A} - \mathbf{B} = (2 - 1)\,\mathbf{a}_x + (5 + 3)\,\mathbf{a}_y + (6 - 6)\,\mathbf{a}_z$$

$$\mathbf{A} - \mathbf{B} = \mathbf{a}_x + 8\mathbf{a}_y.$$

Problem 1.3   If $\mathbf{A} = \mathbf{a}_x + \mathbf{a}_y + 2\mathbf{a}_z$ and $\mathbf{B} = 2\mathbf{a}_x + \mathbf{a}_y + \mathbf{a}_z$, find $\mathbf{A} \cdot \mathbf{B}$.

Solution $\mathbf{A} \cdot \mathbf{B} = A_x B_x + A_y B_y + A_z B_z$
$$= 1.2 + 1.1 + 2.1 = 2 + 1 + 2 = 5.$$

Problem 1.4   Given $\mathbf{A} = 2\mathbf{a}_x + \mathbf{a}_y + 2\mathbf{a}_z$ and $\mathbf{B} = \mathbf{a}_x + 2\mathbf{a}_y + \mathbf{a}_z$, find out $\mathbf{A} \times \mathbf{B}$.

Solution $\mathbf{A} \times \mathbf{B} = \begin{vmatrix} \mathbf{a}_x & \mathbf{a}_y & \mathbf{a}_z \\ 2 & 1 & 2 \\ 1 & 2 & 1 \end{vmatrix}$

$$= \mathbf{a}_x\,(1-4) + \mathbf{a}_y\,(2-2) + \mathbf{a}_z\,(4-1)$$

$$\mathbf{A} \times \mathbf{B} = -3\mathbf{a}_x + 3\mathbf{a}_z.$$

Problem 1.5   For $\mathbf{A} = (1,\ 3,\ 4)$ and $\mathbf{B} = (1,\ 0,\ 2)$, find $\mathbf{A}\,.\,\mathbf{B}$.

Solution $A_x = 1, \quad A_y = 3, \quad A_z = 4$

$$B_x = 1, \quad B_y = 0, \quad B_z = 2$$

$$\mathbf{A}\,.\,\mathbf{B} = A_x\,B_x + A_y\,B_y + A_z\,B_z$$

$$= 1 \times 1 + 3 \times 0 + 4 \times 2$$

$$= 1 + 0 + 8 = 9.$$

Problem 1.6   In Cartesian coordinates, a point is described by $P\,(1,\ 2,\ 4)$. What are the orthogonal planes whose intersection gives this point?

Solution   The planes which intersect at the $P\,(1,\ 2,\ 4)$ are given by

$$x = 1$$

$$y = 2$$

and $\qquad\qquad z = 4.$

Problem 1.7   Represent the point $P\,(0.5,\ 0.2,\ 0.1)$ in terms of unit vectors along the coordinate axis.

Solution   In the point, $P\,(0.5,\ 0.2,\ 0.1)$, the coordinates are $x = 0.5$, $y = 0.2$ and $z = 0.1$. Therefore, P is represented by a vector

$$\mathbf{P} = 0.5\mathbf{a}_x + 0.2\mathbf{a}_y + 0.1\mathbf{a}_z.$$

Problem 1.8   How is a point $P\,(2.0\text{ m},\ \pi/4,\ 1.0\text{ m})$ obtained in cylindrical coordinates?

Solution   The point $P\,(2.0\text{ m},\ \pi/4,\ 1.0\text{ m})$ is obtained by the intersection of (a cylindrical surface),

$$\rho = 2.0\text{ m}$$

a plane, $\qquad\qquad \phi = \pi/4$ and

a plane, $\qquad\qquad z = 1.0\text{ m}.$

Problem 1.9   If a point in Cartesian coordinates is given by $P\,(1,\ 2,\ 3)$, express it in cylindrical coordinates?

Solution   The coordinates of P in Cartesian system are

$$x = 1,\ y = 2,\ z = 3$$

The coordinates of P in cylindrical system are given by

$$\rho = \sqrt{x^2 + y^2} = \sqrt{1^2 + 2^2} = \sqrt{5} = 2.236$$

$$\phi = \tan^{-1}\frac{y}{x} = \tan^{-1}\frac{2}{1} = \tan^{-1} 2 = 63.43^{\circ}$$

$$z = 3$$

$$P \text{ is } (2.236,\ 63.43^{\circ},\ 3).$$

Problem 1.10   If vector $\mathbf{A}$ is $4\mathbf{a}_x + 2\mathbf{a}_y + \mathbf{a}_z$, express it in the cylindrical coordinate system.

Solution $\mathbf{A} = 4\mathbf{a}_x + 2\mathbf{a}_y + \mathbf{a}_z$

We have $A_x = 4,\ A_y = 2,\ A_z = 1$

$$\phi = \tan^{-1}\frac{y}{x} = \tan^{-1}\frac{2}{4} = 26.56°$$

$$A_\rho = A_x \cos\phi + A_y \sin\phi$$

$$= 4\cos 26.56° + 2\sin 26.56°$$

That is, $A_\rho = 3.57 + 0.894 = 4.46$

$$A_\phi = -A_x \sin\phi + A_y \cos\phi$$

$$= -4 \times 0.447 + 2 \times 0.894$$

$$= -1.789 + 1.789$$

$$= 0$$

$$A_z = 1.0$$

Hence, in cylindrical coordinates,

$$\mathbf{A} = 4.46\mathbf{a}_\rho + \mathbf{a}_z.$$

Problem 1.11   How is a point $P\,(1.0\text{ m},\ 10°,\ 30°)$ obtained in spherical coordinates?

Solution $P\,(1.0\text{ m},\ 10°,\ 30°)$ is obtained by the intersection of a sphere,

$$r = 1.0\text{ m}$$

a cone, $\quad\quad\quad\quad \theta = 10°$ and

a plane, $\quad\quad\quad\quad \phi = 30°.$

Problem 1.12   If a point in Cartesian coordinates is given by $P\,(1,\ 2,\ 3)$ express it in spherical coordinates.

Solution   The coordinates of P are

$$x = 1,\ y = 2,\ z = 3$$

The coordinates of P in spherical system are given by

$$r = \sqrt{x^2 + y^2 + z^2} = \sqrt{1^2 + 2^2 + 3^2} = \sqrt{1 + 4 + 9} = \sqrt{14} = 3.741.$$

Problem 1.13   A scalar function V is given by $V = xyz^2$. Find out gradient of V.

Solution   Grad $V = \nabla V = \dfrac{\partial V}{\partial x}\mathbf{a}_x + \dfrac{\partial V}{\partial y}\mathbf{a}_y + \dfrac{\partial V}{\partial z}\mathbf{a}_z$

$$= \frac{\partial}{\partial x}(xyz^2)\,\mathbf{a}_x + \frac{\partial}{\partial y}(xyz^2)\,\mathbf{a}_y + \frac{\partial}{\partial y}(xyz^2)\,\mathbf{a}_z$$

$$\nabla V = yz^2\,\mathbf{a}_x + xz^2\,\mathbf{a}_y + 2xyz\,\mathbf{a}_z.$$

Problem 1.14   If vector $\mathbf{B} = 4xy^2\,\mathbf{a}_x + 2y^3\,\mathbf{a}_y + xyz\,\mathbf{a}_z$, find out the divergence of $\mathbf{B}$.

Solution   Div $\mathbf{B} = \nabla \cdot \mathbf{B}$

$$= \frac{\partial}{\partial x}\,B_x + \frac{\partial}{\partial y}\,B_y + \frac{\partial}{\partial z}\,B_z$$

$$= \frac{\partial}{\partial x}\,(4xy^2) + \frac{\partial}{\partial y}\,(2y^3) + \frac{\partial}{\partial z}\,(xyz)$$

$$= 4y^2 + 6y^2 + xy = 10y^2 + xy$$

Therefore, $\nabla \cdot \mathbf{B} = 10y^2 + xy.$

Problem 1.15   Given a vector $\mathbf{A} = 3x\,\mathbf{a}_x + y\,\mathbf{a}_y + 5z\,\mathbf{a}_z$, find Curl of $\mathbf{A}$.

Solution   Curl $\mathbf{A} = \nabla \times \mathbf{A}$

$$= \begin{vmatrix} \mathbf{a}_x & \mathbf{a}_y & \mathbf{a}_z \\ \dfrac{\partial}{\partial x} & \dfrac{\partial}{\partial y} & \dfrac{\partial}{\partial z} \\ 3x & y & 5z \end{vmatrix}$$

$$= \mathbf{a}_x \left[\frac{\partial}{\partial y}\,(5z) - \frac{\partial}{\partial z}\,(y) \right] + \mathbf{a}_y \left[\frac{\partial}{\partial z}\,(3x) - \frac{\partial}{\partial x}\,(5z) \right] + \mathbf{a}_z \left[\frac{\partial}{\partial x}\,(y) - \frac{\partial}{\partial y}\,(3x) \right]$$

$\nabla \times \mathbf{A} = 0.$

Problem 1.16   If the scalar potential is given by $V = x^2 - y^2 - z^2$ volts, find the Laplacian of V.

Solution $V = x^2 - y^2 - z^2$ volts

$$\nabla^2 V = \frac{\partial^2 V}{\partial x^2} + \frac{\partial^2 V}{\partial y^2} + \frac{\partial^2 V}{\partial z^2}$$

$$= 2 - 2 - 2 = -2.$$

Problem 1.17   If power gain A_p is 22, find its value in decibels.

Solution   Power gain $= 22$

A_p in dB $= 10 \log_{10} 22$

$\qquad\qquad = 13.424$ dB.

Problem 1.18   When the ratio of output and input voltages of an amplifier is 95, find its gain in dB.

Solution $A_v = $ voltage gain $= 95$

A_v in dB $= 20 \log_{10} 95$

$\qquad\qquad = 39.55$ dB.

Problem 1.19   If the power gain in a communication system is 16, what is its value in Nepers?

Solution $A_p = $ power gain $= 16$

A_p in Np $= \log_e \sqrt{A_p} = \log_e \sqrt{16} = 1.386$ Np.

Problem 1.20   If the current gain in a circuit is 34, what is its value in nepers?

Solution   Current gain $= A_I = 34$

$$A_I \text{ in Np} \quad = \log_e 34$$
$$= 3.526.$$

Problem 1.21   If a complex number A is given by $2 + j4$, find its magnitude and phase.

Solution $\qquad A = 2 + j4$

Magnitude of $A = \sqrt{2^2 + 4^2} = \sqrt{4 + 16} = \sqrt{20} = 4.472$

Phase of A, $\quad \phi = \tan^{-1} \dfrac{4}{2} = \tan^{-1} 2 = 63.43^\circ.$

Problem 1.22   If a complex number is given by $A = 1 + j3$, find its complex conjugate, magnitude of A and its phase.

Solution   The complex conjugate of A is

$$A^* = 1 - j3$$

The magnitude of $A = \sqrt{1^2 + 3^2} = \sqrt{1 + 9} = \sqrt{10} = 3.1622$

The phase of $\qquad A = \tan^{-1} \left(\dfrac{y}{x} \right) = \tan^{-1} \left(\dfrac{3}{1} \right) = 71.56^\circ.$

Problem 1.23   If the amplitude of complex number A is 5.0 and its phase is 45°, find out its real and imaginary parts.

Solution   The given complex number $A = 5.0 \angle 45^\circ$

Its magnitude, $\rho = 5.0$

Its phase, $\phi = 45^\circ$

The real part of $A = \rho \cos \phi = 5 \cos 45^\circ = 3.535$

The imaginary part of $A = \rho \sin \phi = 3.535$

A is represented as $3.535 + j3.535.$

Problem 1.24   If $A_1 = 2 + j3$ and $A_2 = 4 + j5$, find the sum of A_1 and A_2.

Solution $A_1 = 2 + j3,\ A_2 = 4 + j5$

$$A = A_1 + A_2 = 2 + j3 + 4 + j5 = 6 + j8.$$

Problem 1.25   If $A_1 = j6$ and $A_2 = 1 - j2$, find $A_1 - A_2$.

Solution $A_1 = j6,\ A_2 = 1 - j2$

$$A_1 - A_2 = j6 - 1 + j2 = -1 + j8.$$

Problem 1.26   Given two complex numbers $A = 0.4 + j5$ and $B = 2 + j3$, find the product of A and B.

Solution $\qquad\qquad A = 0.4 + j5$
$$B = 2 + j3$$

The product, $\quad AB = (0.4 + j5)\ (2 + j3)$

$$= 0.4 \times 2 + j\,0.4 \times 3 + j\,5 \times 2 + j\,5 \times j\,3$$

$$AB = -14.2 + j\,11.2.$$

Problem 1.27   Given two complex numbers $A = 10 + j6$ and $B = 2 - j3$, find the ratio of A and B.

Solution $A = 10 + j6$

$$B = 2 - j3$$

$$\frac{A}{B} = \frac{(10 + j6)}{(2 - j3)} = \frac{(10 + j6)}{(2 - j3)} \times \frac{(2 + j3)}{(2 + j3)}$$

$$= \frac{20 + j30 + j12 - 18}{4 + 9}$$

$$\frac{A}{B} = 0.153 + j3.23.$$

Problem 1.28   Given the quadratic equation $x^2 + 2x + 4 = 0$, find its roots.

Solution   For $x^2 + 2x + 4 = 0$, the roots are

$$x_{1,2} = \frac{-2 \pm \sqrt{4 - 4 \times 1 \times 4}}{2 \times 1} = \frac{-2 \pm \sqrt{-12}}{2}$$

That is,
$$x_1 = -1.0 + j1.732$$
and
$$x_2 = -1.0 - j1.732.$$

Problem 1.29   For the determinant, $\begin{vmatrix} 1 & 2 & 3 \\ 4 & 5 & 6 \\ 7 & 8 & 9 \end{vmatrix}$ find the minor of 8 and 9.

Solution   The minor of the element, 8 is $\begin{vmatrix} 1 & 3 \\ 4 & 6 \end{vmatrix}$

The minor of the element, 9 is $\begin{vmatrix} 1 & 2 \\ 4 & 5 \end{vmatrix}.$

Problem 1.30   For the determinant, $\begin{vmatrix} 1 & 3 & 2 \\ 6 & 1 & 5 \\ 7 & 9 & 8 \end{vmatrix}$ find the cofactor of 6 and 5.

Solution   The minor of 6 is $\begin{vmatrix} 3 & 2 \\ 9 & 8 \end{vmatrix}$

Its cofactor $= (-1)^{i+j} \begin{vmatrix} 3 & 2 \\ 9 & 8 \end{vmatrix}$

$$= (-1)^{2+1} \begin{vmatrix} 3 & 2 \\ 9 & 8 \end{vmatrix}$$

$$= - \begin{vmatrix} 3 & 2 \\ 9 & 8 \end{vmatrix}$$

The minor of 5 is $\begin{vmatrix} 1 & 3 \\ 7 & 9 \end{vmatrix}$

Its cofactor $= (-1)^{i+j} \begin{vmatrix} 1 & 3 \\ 7 & 9 \end{vmatrix}$

$$= (-1)^{2+3} \begin{vmatrix} 1 & 3 \\ 7 & 9 \end{vmatrix}$$

$$= \begin{vmatrix} 1 & 3 \\ 7 & 9 \end{vmatrix}.$$

Problem 1.31 Find the factorial of 4 and 6.

Solution Factorial of $4 = 4 \times 3 \times 2 \times 1 = 24$

Factorial of $6 = 6 \times 5 \times 4 \times 3 \times 2 \times 1 = 720$

$$4! = 24$$
$$6! = 720.$$

❓ OBJECTIVE QUESTIONS

1. The wavelength of an EM wave depends on its velocity. (Yes/No)

2. The velocity of an EM wave depends on the medium in which it propagates. (Yes/No)

3. The unit vectors along the Cartesian coordinate axes are not mutually perpendicular to each other. (Yes/No)

4. Radar X-band is __________ .

5. The frequency range of Radar mm-band is __________ .

6. The frequency range of UHF is __________ .

7. The wavelength range of VHF band is __________ .

8. In cylindrical coordinates, ρ has the unit of degrees. (Yes/No)

9. In cylindrical coordinates, ρ has the unit of meters. (Yes/No)

10. In spherical coordinates, θ increases in anti-clockwise direction. (Yes/No)

11. In cylindrical coordinates, ϕ increases in anti-clockwise direction. (Yes/No)

12. In spherical coordinates, θ increases in clockwise direction. (Yes/No)

13. In cylindrical coordinates, ρ is the __________ .

14. In spherical coordinates, r is __________ .

15. dv in cylindrical coordinates is __________.

16. $d\mathbf{L}$ in spherical coordinates is __________.

17. Dot product of $\mathbf{a}_z$ and $\mathbf{a}_\phi$ is __________.

18. Dot product of $\mathbf{a}_x$ and $\mathbf{a}_r$ is __________.

19. Dot product of $\mathbf{a}_x$ and $\mathbf{a}_y$ is __________.

20. In spherical coordinates, θ varies between __________.

21. In spherical coordinates, ϕ varies between __________.

22. The divergence of an incompressible fluid is __________.

23. Del is a scalar operator. (Yes/No)

24. Del has __________ units.

25. ∇^2 is a vector operator. (Yes/No)

26. ∇^2 can be operated on scalar and vector. (Yes/No)

27. A point is obtained by the intersection of __________ in spherical coordinates.

28. 1 NP is equal to __________.

29. 1 dB is defined as __________.

30. 1 NP is defined as __________.

31. 0! is __________.

32. For $x \ll 1$, $(1+x)^{-1} = $ __________.

33. For $x \ll 1$, $(1+x)^{+1} = $ __________.

34. sinh 0 is __________.

35. tanh 0 is __________.

36. cosh 0 is __________.

37. $\sin jx$ is equal to $j \sinh x$. (Yes/No)

38. $\sinh jx$ is equal to $j \sin x$. (Yes/No)

39. $\cosh jx$ is $\cos x$. (Yes/No)

40. $\tanh jx$ is $j \tan x$. (Yes/No)

41. Fresnel Integrals $C\,(\infty) = S\,(\infty)$. (Yes/No)

42. Fresnel integral, $C\,(\infty) = \sqrt{\dfrac{\pi}{8}}$. (Yes/No)

43. Error function, $\text{erf}\,(\infty) = 1$. (Yes/No)

44. Gamma function, $\Gamma\left(\dfrac{1}{2}\right)$ is __________.

45. In cylindrical coordinates, $\phi = $ constant, is a plane. (Yes/No)

46. In spherical coordinates, $\theta =$ constant, is a cone. (Yes/No)

47. $\mathbf{a}_x \cdot \mathbf{a}_\rho = \sin \phi$. (Yes/No)

48. $\mathbf{a}_x \cdot \mathbf{a}_\phi = \sin \phi$. (Yes/No)

49. $\mathbf{a}_y \cdot \mathbf{a}_\rho = 0$. (Yes/No)

50. $\mathbf{a}_z \cdot \mathbf{a}_\rho = 0$. (Yes/No)

51. $\mathbf{a}_z \cdot \mathbf{a}_\phi = 0$. (Yes/No)

52. $\mathbf{a}_y \cdot \mathbf{a}_\phi = \cos \phi$. (Yes/No)

53. The gradient of a vector does not exist. (Yes/No)

54. Divergence of a scalar does not exist. (Yes/No)

55. Curl of a scalar exists. (Yes/No)

56. Del (∇) has no units. (Yes/No)

57. Divergence of water or oil is almost zero. (Yes/No)

58. Divergence of gradient of scalar electric potential is the Laplacian of the potential. (Yes/No)

59. $j = (0,\ 1)$. (Yes/No)

60. Complex conjugate of $A = 2 + j5$ is $2 - j5$. (Yes/No)

61. A complex number $B = (x,\ y)$ is nothing but $(x + jy)$. (Yes/No)

62. $\int_0^\infty \delta(t - t_0)\, dt = 1$. (Yes/No)

63. 1 Neper = 8.686 dB. (Yes/No)

64. $\log_{10} x = 0.434 \log_{10} x$. (Yes/No)

65. $\mathbf{A} \cdot \mathbf{A} = \mathbf{A} \cdot \mathbf{A}^* = A^2$. (Yes/No)

ANSWERS

1. Yes **2.** Yes **3.** No **4.** 8-12 GHz **5.** 40-300 GHz

6. 300-3000 MHz **7.** 10 m-1 m **8.** No **9.** Yes **10.** No

11. Yes **12.** Yes **13.** Radius of the cylinder **14.** Radius of sphere

15. $\rho\, d\rho\, d\phi\, dz$ **16.** $dr\, \mathbf{a}_r + r\, d\theta\, \mathbf{a}_\theta + r \sin \theta\, d\phi\, \mathbf{a}_\phi$ **17.** Zero **18.** $\sin \theta \cos \phi$ **19.** Zero

20. 0 and π **21.** 0 and 2π **22.** Zero **23.** No **24.** $1/m$ **25.** No

26. Yes **27.** A plane, a cone and a sphere **28.** 8.686 dB

29. $10 \log_{10}\left(\dfrac{P_1}{P_2}\right)$ **30.** $\log_e\left(\dfrac{I_2}{I_1}\right)$ **31.** 1 **32.** $1 - x$ **33.** $1 + x$

34. Zero	**35.** Zero	**36.** One	**37.** Yes	**38.** Yes	**39.** Yes
40. Yes	**41.** Yes	**42.** Yes	**43.** Yes	**44.** $\sqrt{\pi}$	**45.** Yes
46. Yes	**47.** No	**48.** No	**49.** No	**50.** Yes	**51.** Yes
52. Yes	**53.** Yes	**54.** Yes	**55.** No	**56.** No	**57.** Yes
58. Yes	**59.** Yes	**60.** Yes	**61.** Yes	**62.** Yes	**63.** Yes
64. Yes	**65.** Yes.				

EXERCISE PROBLEMS

1. For the vectors $\mathbf{A} = 5\mathbf{a}_x + 3\mathbf{a}_y + 2\mathbf{a}_y$ and $\mathbf{B} = 4\mathbf{a}_y + 0.5\mathbf{a}_z$, find the magnitudes and directions.

2. If $\mathbf{A} = (2,\ 4,\ 1)$ and $\mathbf{B} = (0,\ 0.1,\ 5)$ find $\mathbf{A} + \mathbf{B},\ \mathbf{A} - \mathbf{B}$.

3. If $\mathbf{A} = 6\mathbf{a}_x + 3\mathbf{a}_z$ and $\mathbf{B} = 2\mathbf{a}_y + 5\mathbf{a}_z$ find $\mathbf{A} \cdot \mathbf{B}$.

4. When $\mathbf{A} = (2,\ 0,\ 7)$, $\mathbf{B} = (3,\ 2,\ 0)$, what is $\mathbf{A} \times \mathbf{B}$?

5. A point in Cartesian coordinates is given by $(2,\ 0,\ 4)$. Express it in cylindrical and spherical coordinate systems.

6. The electric scalar potential is given by $V = 3x + 5y^2 + 4z^3$. Find its gradient.

7. For a vector $A = 12a_x + 2a_y + 3a_z$, find the divergence and curl.

8. If a voltage gain is 22 dB, express it in Nepers.

9. Find the roots of a quadratic equation $5x^2 + 2x + 4 = 0$.

10. What are the roots of the cubic equation given by $x^3 - 2x^2 + 4x - 8 = 0$?

11. Find the value of the determinant $\begin{vmatrix} 1 & 2 & 3 \\ 2 & 1 & 3 \\ 3 & 0 & 2 \end{vmatrix}$.

12. Find the minor of '0' in Problem 1.11.

13. What is the cofactor of 4 and 3 in determinant $\begin{vmatrix} 4 & 0 & 1 \\ 5 & 2 & 6 \\ 0 & 3 & 2 \end{vmatrix}$?

14. Using the series expansion, find the value of $(1 + x)^{1/2}$ if $x = 0.01$.

15. Using the series expansion, find $\sin x$ if $x = 0.2$.

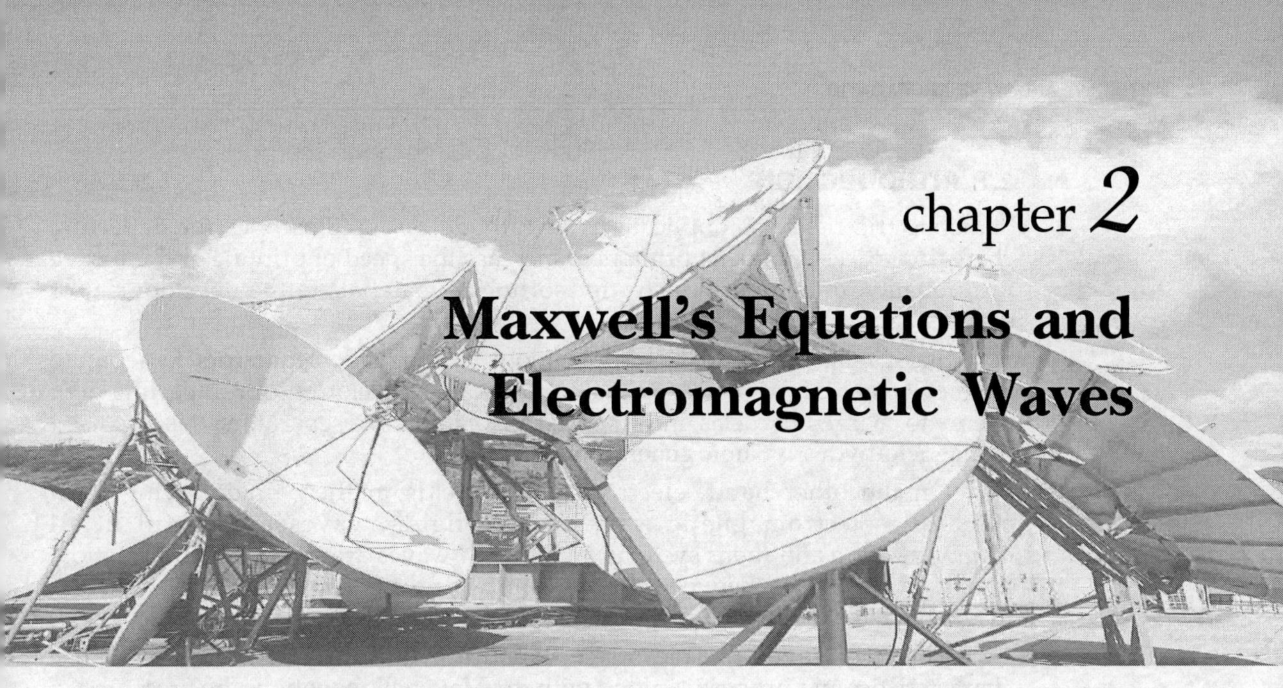

"Maxwell's equations are popularly known as electromagnetic field equations due to their importance."

CHAPTER OBJECTIVES

This chapter discusses

✦ Maxwell's equations in any media, in static, time varying and in phasor form with their complete meaning.

✦ Equation of continuity in time varying form.

✦ Different types of media.

✦ Boundary conditions in scalar and vector form.

✦ Helmholtz theorem and Lorentz Gauge condition.

✦ Objective questions and solved problems useful for class tests, final examinations and also for competitive examinations.

✦ Exercise problems to develop self problem solving skills.

2.1 INTRODUCTION

Static electric field has applications in cathode ray oscilloscopes for deflecting charged particles, in ink jet printers to increase the speed of printing and improve print quality, in mining industry for sorting of materials in the development of electrostatic voltmeters and so on.

Static magnetic field has applications in magnetic separators to separate magnetic materials from non-magnetic materials, in cyclotrons for imparting high energy to charged particles, in velocity selector and mass separators. It is also used in magneto-hydrodynamic generators.

On the other hand, electromagnetic fields in their time-varying form constitute electromagnetic waves. These fields/waves are useful in all communication and radar systems. In fact, EM waves are carriers of information, mainly in free space, between the transmitter and the receiver. The electric and magnetic fields of the EM waves are related through Maxwell's equations.

The EM waves in free space constitute uniform plane waves whose characteristics and properties along with the Maxwell's equations are presented in this chapter.

Most antenna problems can be solved with the help of Maxwell's equations and boundary conditions. In view of this Maxwell's equations are presented in detail in this chapter. Maxwell's equations are very popular and important. They are also referred to as electromagnetic field equations. In Cartesian coordinates the fields $\mathbf{E}$, $\mathbf{H}$, $\mathbf{D}$ and $\mathbf{B}$ in static form are represented as

$$\mathbf{E}\,(x,\ y,\ z)$$
$$\mathbf{H}\,(x,\ y,\ z)$$
$$\mathbf{D}\,(x,\ y,\ z)$$
and
$$\mathbf{B}\,(x,\ y,\ z).$$

In cylindrical coordinates, they are

$$\mathbf{E}\,(\rho,\ \phi,\ z)$$
$$\mathbf{H}\,(\rho,\ \phi,\ z)$$
$$\mathbf{D}\,(\rho,\ \phi,\ z)$$
and
$$\mathbf{B}\,(\rho,\ \phi,\ z)$$

and in spherical coordinates they are

$$\mathbf{E}\,(r,\ \theta,\ \phi)$$
$$\mathbf{H}\,(r,\ \theta,\ \phi)$$
$$\mathbf{D}\,(r,\ \theta,\ \phi)$$
and
$$\mathbf{B}\,(r,\ \theta,\ \phi).$$

The time-varying fields $\mathbf{E}\,(t,\ x,\ y,\ z)$, $\mathbf{H}\,(t,\ x,\ y,\ z)$ constitute electromagnetic waves which have wide applications in all the communications, radars and also in Bio-medical Engineering.

In these applications, time-varying fields are of more practical value than static electric and magnetic fields.

It may be noted that the electrostatic fields are usually produced by static charges. Magnetostatic fields are produced from the motion of electric charges with uniform velocity (direct current) or static magnetic charges (magnetic dipoles).

On the other hand, time-varying fields constituting EM waves are produced by time-varying currents. That is, any pulsating current produces radiation fields which are time-varying fields. In brief, it may be noted that:

1. Static charges produce electrostatic fields.
2. Steady currents (DC currents) produce magnetostatic fields. Static magnetic charges (magnetic dipoles) also produce magnetostatic fields.
3. Time-varying currents produce EM waves or EM fields.

2.2 EQUATION OF CONTINUITY FOR TIME-VARYING FIELDS

Equation of continuity in point form is

$$\nabla \cdot \mathbf{J} = -\dot{\rho}_v \qquad \qquad \text{...(2.1)}$$

where

$$\mathbf{J} = \text{Conduction current density (A/m}^2\text{)}$$

$$\rho_v = \text{Volume charge density (c/m}^3\text{)}, \quad \dot{\rho}_v = \frac{\partial \rho_v}{\partial t}$$

$$\nabla = \text{Vector differential operator} \left(\frac{1}{m}\right)$$

$$= a_x \frac{\partial}{\partial x} + a_y \frac{\partial}{\partial y} + a_z \frac{\partial}{\partial z}$$

Proof   Consider a closed surface enclosing a charge Q. There exists an outward flow of current given by

$$I = \oint_S \mathbf{J} \cdot d\mathbf{S} \qquad \qquad \text{...(2.2)}$$

This is known as equation of continuity in integral form.

Here I is the current flowing away through a closed surface and $d\mathbf{S}$ is the differential area on the surface whose direction is always outward, normal to the surface. As there is outward flow of current, there will be a decrease of charge the rate of which is given by $\dfrac{-dQ}{dt}$ where Q is the enclosed charge. From the principle of conservation of charge, we have

$$I = \oint_S \mathbf{J} \cdot d\mathbf{S} = \frac{-dQ}{dt} \qquad \qquad \text{...(2.3)}$$

From divergence theorem, we have

$$\oint_S \mathbf{J} \cdot d\mathbf{S} = \int_v \nabla \cdot \mathbf{J}\, dv \qquad \qquad \text{...(2.4)}$$

Hence $$\int_v (\nabla \cdot \mathbf{J})\, dv = \frac{-dQ}{dt} \qquad \qquad \text{...(2.5)}$$

By definition,
$$Q = \int_v \rho_v \, dv \qquad \qquad \text{...(2.6)}$$

where
$$\rho_v = \text{volume charge density (C/m}^3\text{)}$$

So,
$$\int_v (\nabla \cdot \mathbf{J}) \, dv = \int \frac{-\partial \rho_v}{\partial t} \, dv \qquad \qquad \text{...(2.7)}$$

$$= \int - \dot{\rho}_v \, dv \qquad \qquad \text{...(2.8)}$$

where
$$\dot{\rho}_v = \frac{\partial \rho_v}{\partial t}.$$

Two volume integrals are equal only if their integrands are equal.

So,
$$\nabla \cdot \mathbf{J} = - \dot{\rho}_v \qquad \qquad \text{...(2.9)}$$

Hence proved.

2.3 MAXWELL'S EQUATIONS FOR TIME-VARYING FIELDS

These are basically four in number.

Maxwell's equations in differential form are given by
$$\nabla \times \mathbf{H} = \dot{\mathbf{D}} + \mathbf{J} \qquad \qquad \text{...(2.10)}$$
$$\nabla \times \mathbf{E} = - \dot{\mathbf{B}} \qquad \qquad \text{...(2.11)}$$
$$\nabla \cdot \mathbf{D} = \rho_v \qquad \qquad \text{...(2.12)}$$
$$\nabla \cdot \mathbf{B} = 0 \qquad \qquad \text{...(2.13)}$$

Here
$$\mathbf{H} = \text{Magnetic field strength (A/m)}$$

$$\mathbf{D} = \text{Electric flux density (C/m}^2\text{)}$$

$$\dot{\mathbf{D}} = \frac{\partial \mathbf{D}}{\partial t} = \text{Displacement electric current density (A/m}^2\text{)}$$

$$\mathbf{J} = \text{Conduction current density (A/m}^2\text{)}$$

$$\mathbf{E} = \text{Electric field (V/m)}$$

$$\mathbf{B} = \text{Magnetic flux density wb/m}^2 \text{ or Tesla}$$

$$\dot{\mathbf{B}} = \frac{\partial \mathbf{B}}{\partial t} = \text{Time-derivative of magnetic flux density (wb/m}^2\text{-sec)}$$

$\dot{\mathbf{B}}$ is called the magnetic current density (V/m^2) or Tesla/sec

$$\rho_v = \text{Volume charge density (C/m}^3\text{)}$$

Maxwell's equations for time-varying fields in integral form are given by

$$\oint_L \mathbf{H} \cdot d\mathbf{L} = \int_S (\dot{\mathbf{D}} + \mathbf{J}) \cdot d\mathbf{S} \qquad \qquad \text{...(2.14)}$$

$$\oint_L \mathbf{E} \cdot d\mathbf{L} = -\int_S \dot{\mathbf{B}} \cdot d\mathbf{S} \qquad \qquad \text{...(2.15)}$$

$$\oint_S \mathbf{D} \cdot d\mathbf{S} = \int_v \rho_v \, dv \qquad \qquad \text{...(2.16)}$$

$$\oint_S \mathbf{B} \cdot d\mathbf{S} = 0 \qquad \qquad \text{...(2.17)}$$

Here $d\mathbf{L}$ is the differential length and $d\mathbf{S}$ is the differential area whose direction is always outward, normal to the surface.

2.3.1 Meaning of Maxwell's Equations

It is easy to understand the meaning of Maxwell's equations from their integral form.

1. The first Maxwell's equation states that the magnetomotive force around a closed path is equal to the sum of electric displacement and conduction currents through any surface bounded by the path.

2. The second equation states that the electromotive force around a closed path is equal to the negative of the time derivative of magnetic flux flowing through any surface bounded by the path. It can also be stated that the electromotive force around a closed path is equal to the inflow of magnetic current through any surface bounded by the path.

3. The third equation states that the total electric displacement flux passing through a closed surface (Gaussian Surface) is equal to the total charge inside the surface.

4. The fourth equation states that the total magnetic flux passing through any closed surface is zero.

2.3.2 Conversion of Differential Form of Maxwell's Equations to Integral Form

1. Consider the first Maxwell's equation

$$\nabla \times \mathbf{H} = \dot{\mathbf{D}} + \mathbf{J}$$

Take surface integral on both sides.

$$\int_S (\nabla \times \mathbf{H}) \cdot d\mathbf{S} = \int_S (\dot{\mathbf{D}} + \mathbf{J}) \cdot d\mathbf{S} \qquad \qquad \text{...(2.18)}$$

Applying Stokes' theorem, we can write

$$\int_S (\nabla \times \mathbf{H}) \cdot d\mathbf{S} = \oint_L \mathbf{H} \cdot d\mathbf{L} \qquad \qquad \text{...(2.19)}$$

Therefore, Equation (2.18) becomes

$$\oint_L \mathbf{H} \cdot d\mathbf{L} = \int_S (\dot{\mathbf{D}} + \mathbf{J}) \cdot d\mathbf{S} \ \text{(first law)} \qquad \qquad \text{...(2.20)}$$

2. Consider the second Maxwell's equation

$$\nabla \times \mathbf{E} = -\dot{\mathbf{B}}$$

Take surface integral on both sides

$$\int_S (\nabla \times \mathbf{E}) \cdot d\mathbf{S} = -\int_S \dot{\mathbf{B}} \cdot d\mathbf{S} \qquad \dots(2.21)$$

Applying Stoke's theorem, we get

$$\int_S (\nabla \times \mathbf{E}) \cdot d\mathbf{S} = \oint_L \mathbf{E} \cdot d\mathbf{L} \qquad \dots(2.22)$$

Therefore, Equation (2.21) becomes

$$\oint_L \mathbf{E} \cdot d\mathbf{L} = -\int_S \dot{\mathbf{B}} \cdot d\mathbf{S} \text{ (second law)} \qquad \dots(2.23)$$

3. Consider the third Maxwell's equation

$$\nabla \cdot \mathbf{D} = \rho_\upsilon$$

Take volume integral on both sides

$$\int_\upsilon (\nabla \cdot \mathbf{J}) \, d\upsilon = \int_\upsilon \rho_\upsilon \, d\upsilon \qquad \dots(2.24)$$

Applying divergence theorem, we get

$$\int_\upsilon (\nabla \cdot \mathbf{J}) \, d\upsilon = \oint_S \mathbf{D} \cdot d\mathbf{S} \qquad \dots(2.25)$$

Therefore, Equation (2.24) becomes

$$\oint_S \mathbf{D} \cdot d\mathbf{S} = \int_\upsilon \rho_\upsilon \, d\upsilon \text{ (third law)} \qquad \dots(2.26)$$

4. Consider the fourth Maxwell's equation

$$\nabla \cdot \mathbf{B} = 0$$

Take volume integral on both sides

$$\int_\upsilon \nabla \cdot \mathbf{B} \, d\upsilon = 0 \qquad \dots(2.27)$$

Applying divergence theorem, we get

$$\int_\upsilon (\nabla \cdot \mathbf{B}) \, d\upsilon = \oint_S \mathbf{B} \cdot d\mathbf{S} \qquad \dots(2.28)$$

Therefore, Equation (2.27) becomes

$$\oint_S \mathbf{B} \cdot d\mathbf{S} = 0 \text{ (fourth law)} \qquad \dots(2.29)$$

2.3.3 Maxwell's Equations for Static Fields

Maxwell's equations for static fields

$$\nabla \times \mathbf{H} = \mathbf{J} \leftrightarrow \oint_L \mathbf{H} \cdot d\mathbf{L} = \int_S \mathbf{J} \cdot d\mathbf{S} \qquad \ldots(2.30)$$

$$\nabla \times \mathbf{E} = 0 \leftrightarrow \oint_L \mathbf{E} \cdot d\mathbf{L} = 0 \qquad \ldots(2.31)$$

$$\nabla \cdot \mathbf{D} = \rho_v \leftrightarrow \oint_S \mathbf{D} \cdot d\mathbf{S} = \int_v \rho_v \, dv \qquad \ldots(2.32)$$

$$\nabla \cdot \mathbf{B} = 0 \leftrightarrow \oint_S \mathbf{B} \cdot d\mathbf{S} = 0 \qquad \ldots(2.33)$$

As the fields are static, all the field terms which have time derivatives are zero, or, $\dot{\mathbf{B}} = 0$, $\dot{\mathbf{D}} = 0$.

2.3.4 Characteristics of Free Space

Free space is characterised by the following parameters:

Relative permittivity, $\quad \epsilon_r = 1$

Relative permeability, $\quad \mu_r = 1$

Conductivity, $\quad \sigma = 0$

Conduction current density, $\quad \mathbf{J} = 0$

Volume charge density, $\quad \rho_v = 0$

Intrinsic impedance or characteristic impedance

$$\eta = 120\pi \ \text{ or } \ 377\,\Omega.$$

2.3.5 Maxwell's Equation for Free Space

$$\nabla \times \mathbf{H} = \dot{\mathbf{D}} \leftrightarrow \oint_L \mathbf{H} \cdot d\mathbf{L} = \int_S \dot{\mathbf{D}} \cdot d\mathbf{S} \qquad \ldots(2.34)$$

$$\nabla \times \mathbf{E} = -\dot{\mathbf{B}} \leftrightarrow \oint_L \mathbf{E} \cdot d\mathbf{L} = \int_S \dot{\mathbf{B}} \cdot d\mathbf{S} \qquad \ldots(2.35)$$

$$\nabla \cdot \mathbf{D} = 0 \leftrightarrow \oint_S \mathbf{D} \cdot d\mathbf{S} = 0 \qquad \ldots(2.36)$$

$$\nabla \cdot \mathbf{B} = 0 \leftrightarrow \oint_S \mathbf{B} \cdot d\mathbf{S} = 0. \qquad \ldots(2.37)$$

2.3.6 Maxwell's Equations for Static Fields in Free Space

$$\nabla \times \mathbf{H} = 0 \leftrightarrow \oint_L \mathbf{H} \cdot d\mathbf{L} = 0 \qquad \ldots(2.38)$$

$$\nabla \times \mathbf{E} = 0 \leftrightarrow \oint_L \mathbf{E} \cdot d\mathbf{L} = 0 \qquad \text{...(2.39)}$$

$$\nabla \cdot \mathbf{D} = 0 \leftrightarrow \oint_S \mathbf{D} \cdot d\mathbf{S} = 0 \qquad \text{...(2.40)}$$

$$\nabla \cdot \mathbf{B} = 0 \leftrightarrow \oint_S \mathbf{B} \cdot d\mathbf{S} = 0 \qquad \text{...(2.41)}$$

2.3.7 Proof of Maxwell's Equations

1. From Ampere's circuital law, we have

$$\nabla \times \mathbf{H} = \mathbf{J} \qquad \text{...(2.42)}$$

Take dot product on both sides

$$\nabla \cdot \nabla \times \mathbf{H} = \nabla \cdot \mathbf{J}$$

As the divergence of curl of a vector is zero,

$$\nabla \cdot \mathbf{J} = 0 \qquad \text{...(2.43)}$$

But the equation of continuity in point form says

$$\nabla \cdot \mathbf{J} = \frac{-\partial \rho_v}{\partial t} = -\dot{\rho}_v \qquad \text{...(2.44)}$$

This means that if $\nabla \times \mathbf{H} = \mathbf{J}$ is true, we get $\nabla \cdot \mathbf{J} = 0$.

As the equation of continuity is more fundamental, Ampere's circuital law should be modified. Hence

$$\nabla \times \mathbf{H} = \mathbf{J} + \mathbf{F} \qquad \text{...(2.45)}$$

Take dot product on both sides

$$\nabla \cdot \nabla \times \mathbf{H} = \nabla \cdot \mathbf{J} + \nabla \cdot \mathbf{F}$$

or $\qquad \nabla \cdot \nabla \times \mathbf{H} = 0 = \nabla \cdot \mathbf{J} + \nabla \cdot \mathbf{F} \qquad \text{...(2.46)}$

Substituting the value of $\nabla \cdot \mathbf{J}$ from the equation of continuity in Equation (2.46), we get

$$\nabla \cdot \mathbf{F} + (-\dot{\rho}_v) = 0$$

or $\qquad \nabla \cdot \mathbf{F} = \dot{\rho}_v \qquad \text{...(2.47)}$

The point form of Gauss's law says

$$\nabla \cdot \mathbf{D} = \rho_v \qquad \text{...(2.48)}$$

or $\qquad \nabla \cdot \dot{\mathbf{D}} = \dot{\rho}_v \qquad \text{...(2.49)}$

From Equations (2.47) and (2.49), we get

$$\nabla \cdot \mathbf{F} = \nabla \cdot \dot{\mathbf{D}} \qquad \text{...(2.50)}$$

The divergence of two vectors are equal only if the vectors are identical.

Hence, $\qquad \mathbf{F} = \dot{\mathbf{D}} \qquad \text{...(2.51)}$

Substituting Equation (2.50) in Equation (2.45), we get

$$\boxed{\nabla \times \mathbf{H} = \dot{\mathbf{D}} + \mathbf{J}}$$

Hence proved.

2. According to Faraday's law,

$$\text{emf} = \frac{-d\phi}{dt} \qquad \qquad \text{...(2.52)}$$

$$\phi = \text{magnetic flux, (wb)}$$

and by definition,

$$\text{emf} = \oint_L \mathbf{E} \cdot d\mathbf{L} \qquad \qquad \text{...(2.53)}$$

From Equations (2.52) and (2.53)

$$\oint_L \mathbf{E} \cdot d\mathbf{L} = \frac{-d\phi}{dt} \qquad \qquad \text{...(2.54)}$$

But

$$\phi = \oint_S \mathbf{B} \cdot d\mathbf{S} \qquad \qquad \text{...(2.55)}$$

So

$$\oint_L \mathbf{E} \cdot d\mathbf{L} = -\int_S \frac{\partial \mathbf{B}}{\partial t} \cdot d\mathbf{S} \qquad \qquad \text{...(2.56)}$$

$$= -\int_S \dot{\mathbf{B}} \cdot d\mathbf{S}, \quad \dot{\mathbf{B}} = \frac{\partial \mathbf{B}}{\partial t}$$

Applying Stokes' theorem, we get

$$\oint_L \mathbf{E} \cdot d\mathbf{L} = \int_S (\nabla \times \mathbf{E}) \cdot d\mathbf{S}$$

Hence,

$$\int_S (\nabla \times \mathbf{E}) \cdot d\mathbf{S} = \int_S -\dot{\mathbf{B}} \cdot d\mathbf{S} \qquad \qquad \text{...(2.57)}$$

Two surface integrals are equal only if their integrands are equal.

That is,

$$\boxed{\nabla \times \mathbf{E} = -\dot{\mathbf{B}}}$$

Hence proved.

3. From Gauss's law in electric field

$$\oint_S \mathbf{D} \cdot d\mathbf{S} = Q = \int_\upsilon \rho_\upsilon \, d\upsilon$$

Applying divergence theorem, we get

$$\oint_S \mathbf{D} \cdot d\mathbf{S} = \int_\upsilon (\nabla \cdot \mathbf{D}) \, d\upsilon = \int_\upsilon \rho_\upsilon \, d\upsilon$$

Two volume integrals are equal if their integrands are equal.

That is,
$$\nabla \cdot \mathbf{D} = \rho_v$$

Hence proved.

4. Gauss's law for magnetic field gives

$$\oint_S \mathbf{B} \cdot d\mathbf{S} = 0$$

Right hand side is zero as there are no isolated magnetic charges and the magnetic flux lines are a closed loop.

Applying divergence theorem, we get

$$\int_v \nabla \cdot \mathbf{B} \, dv = 0$$

or
$$\nabla \cdot \mathbf{B} = 0$$

Hence proved.

2.4 SINUSOIDAL TIME-VARYING FIELDS

In practice, electric and magnetic fields vary sinusoidally. Any periodic variation can be described in terms of sinusoidal variations with fundamental and harmonic frequencies.

The fields can be represented by

$$\tilde{\mathbf{E}} = E_m \cos \omega t$$

$$\tilde{\mathbf{E}} = E_m \sin \omega t \qquad \qquad ...(2.58)$$

where $\omega = 2\pi f$, f = frequency variation of the field, E_m is the maximum field strength.

This means that a sinusoidal time factor is attached to the field. It is also possible to represent the fields using the phasor notation. The time-varying field $\tilde{\mathbf{E}}(r, t)$ is related to phasor field $E(r)$ as

$$\tilde{\mathbf{E}}(r, t) = \text{Re}\{\mathbf{E}(r) \, e^{j\omega t}\} \qquad \qquad ...(2.59)$$

or
$$\tilde{\mathbf{E}}(r, t) = \text{Im}\{\mathbf{E}(r) \, e^{j\omega t}\} \qquad \qquad ...(2.60)$$

It may be noted that cosinusoidal variation is also considered sinusoidal in usage.

2.5 MAXWELL'S EQUATIONS IN PHASOR FORM

1. Consider the first Maxwell's equation

$$\nabla \times \tilde{\mathbf{H}} = \frac{\partial \tilde{\mathbf{D}}}{\partial t} + \tilde{\mathbf{J}} \qquad \qquad ...(2.61)$$

If
$$\tilde{\mathbf{H}} = \text{Re}\{\mathbf{H} e^{j\omega t}\}$$

$$\tilde{D} = \text{Re}\,\{\mathbf{D}e^{j\omega t}\}$$

$$\tilde{J} = \text{Re}\,\{\mathbf{J}e^{j\omega t}\}$$

Equation (2.61) becomes

$$\nabla \times \text{Re}\,\{\mathbf{H}\,e^{j\omega t}\} = \frac{\partial}{\partial t}\,\text{Re}\,\{\mathbf{D}e^{j\omega t}\} + \text{Re}\,\{\mathbf{J}e^{j\omega t}\}$$

Interchanging the operation of taking the real part, we get

$$\text{Re}\,\{\nabla \times \mathbf{H} - j\omega\mathbf{D} - \mathbf{J}\}\,e^{j\omega t} = 0$$

or
$$\nabla \times \mathbf{H} = j\omega\mathbf{D} + \mathbf{J} \qquad\qquad\qquad\qquad ...(2.62)$$

This is the first Maxwell's equation in phasor form.

2. Consider the second Maxwell's equation

$$\nabla \times \tilde{\mathbf{E}} = -\frac{\partial \tilde{\mathbf{B}}}{\partial t} \qquad\qquad\qquad\qquad ...(2.63)$$

If
$$\tilde{\mathbf{E}} = \text{Re}\,\{\mathbf{E}e^{j\omega t}\}$$

and
$$\tilde{\mathbf{B}} = \text{Re}\,\{\mathbf{B}e^{j\omega t}\}$$

Equation (2.63) becomes

$$\nabla \times \text{Re}\,\{\mathbf{E}e^{j\omega t}\} = -\frac{\partial}{\partial t}\,\text{Re}\,(\mathbf{B}e^{j\omega t})$$

or
$$\text{Re}\,[(\nabla \times \mathbf{E} + j\omega\mathbf{B})\,e^{j\omega t}] = 0$$

or
$$\nabla \times \mathbf{E} = -j\omega\mathbf{B}$$

This is the second Maxwell's equation in phasor form.

3. Similarly, considering the third Maxwell's equation,

$$\nabla \cdot \tilde{\mathbf{D}} = \rho_\upsilon$$

That is,
$$\nabla \cdot \text{Re}\,(\mathbf{D}e^{j\omega t}) = \rho_\upsilon$$

or
$$\nabla \cdot \mathbf{D} = \rho_\upsilon$$

4. Consider the fourth Maxwell's equation,

$$\nabla \cdot \tilde{\mathbf{B}} = 0$$

That is,
$$\nabla \cdot \text{Re}\,(\mathbf{B}e^{j\omega t}) = 0$$

or
$$\nabla \cdot \mathbf{B} = 0$$

In summary, Maxwell's equations in phasor form are as follows:

$$\left.\begin{aligned}
\nabla \times \mathbf{H} &= j\omega\mathbf{D} + \mathbf{J} \\
\nabla \times \mathbf{E} &= -j\omega\mathbf{B} \\
\nabla \cdot \mathbf{D} &= \rho_\upsilon \\
\nabla \cdot \mathbf{B} &= 0
\end{aligned}\right\} \qquad\qquad ...(2.64)$$

2.6 INFLUENCE OF MEDIUM ON THE FIELDS

When the sources of electric and magnetic fields exist in a medium, the medium has influence on the characteristics of the fields. The characteristics are described by the following **constitutive relations** :

$$\left.\begin{array}{l} \mathbf{D} = \in \mathbf{E} \\[2mm] \mathbf{B} = \mu \mathbf{H} \\[2mm] \mathbf{J} = \sigma \mathbf{E} \end{array}\right\} \qquad \qquad \ldots(2.65)$$

Here $\quad \in \, = \text{permittivity (F/m)}$

$\mu = \text{permeability (H/m)}$

and $\quad \sigma = \text{conductivity (mho/m) of the medium.}$

Medium can be divided into five types:

1. Homogeneous medium
2. Non-homogeneous medium
3. Isotropic medium
4. Anistropic medium
5. Source free regions.

Homogeneous medium

It is a medium for which $\in$, μ and σ are constant throughout the medium.

An example is free space.

Non-homogeneous medium

It is a medium for which $\in$, μ and σ are not constant and they are different from point to point in the medium.

An example is the Human body.

Isotropic medium

It is a medium for which $\in$ is a scalar constant.

Free space is an example.

Anistropic medium

It is a medium for which $\in$ is not a scalar constant.

For example, human body.

Source-free region

It is a medium in which there are no field sources.

2.7 SUMMARY OF MAXWELL'S EQUATIONS FOR DIFFERENT CASES

Table 2.1

Maxwell's equations in differential form (General)	Maxwell's equations in integral form (General)	Maxwell's equations in phasor form	Maxwell's equations in free space (differential form)	Maxwell's equations for static fields in differential form	Maxwell's equations in free space in integral form	Maxwell's equations in free space for static fields
$\nabla \times \mathbf{H} = \dot{\mathbf{D}} + \mathbf{J}$	$\oint_L \mathbf{H} \cdot d\mathbf{L} = \int_S (\dot{\mathbf{D}} + \mathbf{J}) \cdot d\mathbf{S}$	$\nabla \times \mathbf{H} = j\omega\mathbf{D} + \mathbf{J}$	$\nabla \times \mathbf{H} = \dot{\mathbf{D}}$	$\nabla \times \mathbf{H} = \mathbf{J}$	$\oint_L \mathbf{H} \cdot d\mathbf{L} = \int_S \dot{\mathbf{D}} \cdot d\mathbf{S}$	$\nabla \times \mathbf{H} = 0$ or $\oint_L \mathbf{H} \cdot d\mathbf{L} = 0$
$\nabla \times \mathbf{E} = -\dot{\mathbf{B}}$	$\oint_L \mathbf{E} \cdot d\mathbf{L} = -\int_S \dot{\mathbf{B}} \cdot d\mathbf{S}$	$\nabla \times \mathbf{E} = -j\omega\mathbf{B}$	$\nabla \times \mathbf{E} = -\dot{\mathbf{B}}$	$\nabla \times \mathbf{E} = 0$	$\oint_L \mathbf{E} \cdot d\mathbf{L} = -\int_S \dot{\mathbf{B}} \cdot d\mathbf{S}$	$\nabla \times \mathbf{E} = 0$ or $\oint_L \mathbf{E} \cdot d\mathbf{L} = 0$
$\nabla \cdot \mathbf{D} = \rho_v$	$\oint_S \mathbf{D} \cdot d\mathbf{S} = \int_v \rho_v \, dv$	$\nabla \cdot \mathbf{D} = \rho_v$	$\nabla \cdot \mathbf{D} = 0$	$\nabla \cdot \mathbf{D} = \rho_v$	$\oint_S \mathbf{D} \cdot d\mathbf{S} = \int_v \rho_v \, dv$	$\nabla \cdot \mathbf{D} = 0$ or $\oint_S \mathbf{D} \cdot d\mathbf{S} = 0$
$\nabla \cdot \mathbf{B} = 0$	$\oint_S \mathbf{B} \cdot d\mathbf{S} = 0$	$\nabla \cdot \mathbf{B} = 0$	$\nabla \cdot \mathbf{B} = 0$	$\nabla \cdot \mathbf{B} = 0$	$\oint_S \mathbf{B} \cdot d\mathbf{S} = 0$	$\nabla \cdot \mathbf{B} = 0$ or $\oint_S \mathbf{B} \cdot d\mathbf{S} = 0$

2.8 CONDITIONS AT A BOUNDARY SURFACE

Relations between the main fields **E, D, H, B** are expressed in terms of Maxwell's equations. These are valid at any point in a continuous medium.

As Maxwell's equations contain space derivatives, they cannot give information at points of discontinuity in the medium. However, Maxwell's equations in integral form can be used to get the information at the boundary surfaces between different media.

The electromagnetic fields that are solved using Maxwell's equations must satisfy the boundary conditions at the interface between different media.

The boundary conditions on electric and magnetic fields at any surface of discontinuity are given by:

1. The tangential component of electric field **E** is continuous across any discontinuity. That is,

$$E_{\tan 1} = E_{\tan 2}$$

Subscript tan 1 is tangential component of the field at the boundary in medium 1. Subscript tan 2 represents it in medium 2.

2. The tangential component of magnetic field **H** is continuous across any surface except at the surface of a perfect conductor. At the surface of a perfect

conductor, the tangential component of **H** is discontinuous by an amount equal to the surface current density (A/m).

$$H_{\tan 1} - H_{\tan 2} = J_s$$

3. The normal component of **B** is continuous across any discontinuity.

$$B_{n1} = B_{n2}$$

4. The normal component of **D** is continuous except at the surface which has surface charge density. At the surface where surface charge density exists, the normal component of **D** is discontinuous by an amount equal to the surface charge density.

$$D_{n1} - D_{n2} = \rho_s$$

Subscript $n1$ represents normal component of the field in medium 1 and $n2$ represents it in medium 2.

2.8.1 Proof of Boundary Conditions on *E, D, H* and *B*

To prove $\qquad E_{\tan 1} = E_{\tan 2}$ and $H_{\tan 1},\ H_{\tan 2} = J_s$

Consider the rectangular loop on the boundary of the two media (Fig. 2.1).

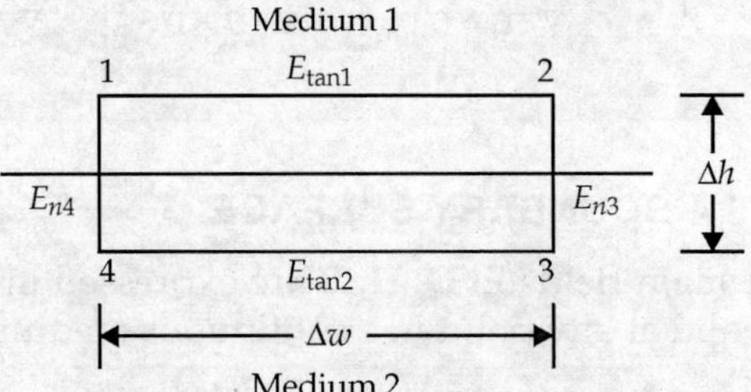

Fig. 2.1

As the electric field is a conservative, the line integral of $\mathbf{E} \cdot d\mathbf{L}$ is zero around the closed loop

or $\qquad \oint \mathbf{E} \cdot d\mathbf{L} = 0$

That is, $\quad \displaystyle\int_1^2 \mathbf{E} \cdot d\mathbf{L} + \int_2^3 \mathbf{E} \cdot d\mathbf{L} + \int_3^4 \mathbf{E} \cdot d\mathbf{L} + \int_4^1 \mathbf{E} \cdot d\mathbf{L} = 0$

and $\qquad E_{\tan 1}\, \Delta w - E_{n3}\, \Delta h - E_{\tan 2}\, \Delta w + E_{n4}\, \Delta h = 0$

As $\Delta h \to 0$, this becomes,

$$E_{\tan 1}\, \Delta w - E_{\tan 2}\, \Delta w = 0$$

or $\qquad\qquad E_{\tan 1} = E_{\tan 2} \qquad$ Hence proved.

We know that the magnetic field is not conservative. According to Ampere's circuital law, we have

$$\oint \mathbf{H} \cdot d\mathbf{L} = I$$

That is,
$$\int_1^2 \mathbf{H} \cdot d\mathbf{L} + \int_2^3 \mathbf{H} \cdot d\mathbf{L} + \int_3^4 \mathbf{H} \cdot d\mathbf{L} + \int_4^1 \mathbf{H} \cdot d\mathbf{L} = I$$

or
$$H_{\tan 1}\,\Delta w - E_{n4}\,\Delta h - E_{\tan 2}\,\Delta w + E_{n3}\,\Delta h = I$$

As $\Delta h \to 0$, this becomes,

$$H_{\tan 1}\,\Delta w - H_{\tan 2}\,\Delta w = I$$

or
$$(H_{\tan 1} - H_{\tan 2}) = \frac{I}{\Delta w} = J_s \qquad\qquad \text{Hence proved.}$$

To prove
$$D_{n1} - D_{n2} = \rho_s \quad \text{and} \quad B_{n1} = B_{n2}.$$

Consider a cylinder as in Fig. 2.2.

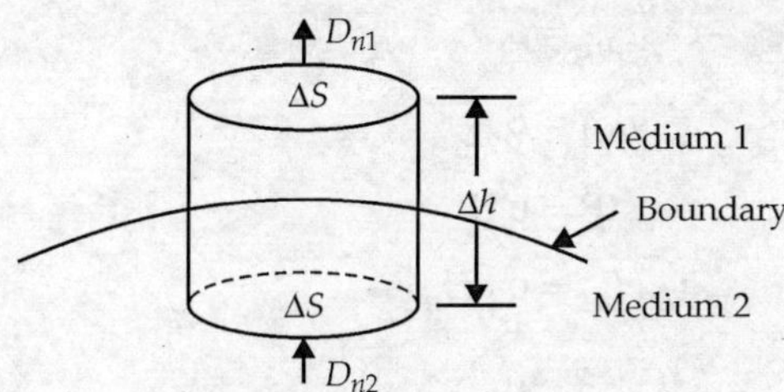

Fig. 2.2

From Gauss's law in electrostatics, we have

$$\oint_S \mathbf{D} \cdot d\mathbf{S} = Q$$

Applying this to cylindrical surfaces on the boundary spreading over medium 1 and medium 2, as $\Delta h \to 0$, we get

$$D_{n1}\,\Delta S - D_{n2}\,\Delta S = Q$$

or
$$D_{n1} - D_{n2} = \frac{Q}{\Delta S} = \rho_s$$

Thus
$$D_{n1} - D_{n2} = \rho_s \qquad\qquad \text{Hence proved.}$$

Gauss's law in magnetic field states that

$$\oint_S \mathbf{B} \cdot d\mathbf{S} = 0$$

Applying this to the cylindrical surfaces of Fig. 2.2, we get

$$B_{n1}\,\Delta S - B_{n2}\,\Delta S = 0$$

Thus
$$B_{n1} = B_{n2}$$
Hence proved.

Given the fields in one medium, it is usually required to determine fields in a second medium. This requires the knowledge of both tangential and normal components for each field. The above boundary conditions yield either tangential or normal component. The other unknown components are worked out from constitutive relations between $\mathbf{E}$ and $\mathbf{D}$, $\mathbf{H}$ and $\mathbf{B}$.

We have
$$E_{\tan 1} = E_{\tan 2}$$

But
$$\mathbf{D} = \in \mathbf{E}$$

$$\frac{D_{\tan 1}}{\in_1} = \frac{D_{\tan 2}}{\in_2}$$

Hence
$$D_{\tan 2} = \frac{\in_2}{\in_1} D_{\tan 1} = \frac{\in_{r2}}{\in_{r1}} D_{\tan 1}$$

We have
$$D_{n1} - D_{n2} = \rho_s$$

Hence
$$\in_1 E_{n1} - \in_2 E_{n2} = \rho_s$$

We have
$$B_{n1} = B_{n2}$$

But
$$\mathbf{B} = \mu \mathbf{H}$$

$$\mu_1 H_{n1} = \mu_2 H_{n2}$$

So
$$H_{n2} = \frac{\mu_2}{\mu_1} H_{n1} = \frac{\mu_{r2}}{\mu_{r1}} H_{n1}$$

We have
$$H_{\tan 1} - H_{\tan 2} = J_s$$

$$\frac{B_{\tan 1}}{\mu_1} - \frac{B_{\tan 2}}{\mu_2} = J_s.$$

2.8.2 Boundary Conditions at a Glance

$$E_{\tan 1} - E_{\tan 2} = 0$$

$$\in_1 E_{n1} - \in_2 E_{n2} = \rho_s$$

$$H_{\tan 1} - H_{\tan 2} = J_s$$

$$\mu_{r1} H_{n1} - \mu_{r2} H_{n2} = 0$$

$$B_{n1} - B_{n2} = 0$$

$$\frac{B_{\tan 1}}{\mu_1} - \frac{B_{\tan 2}}{\mu_2} = J_s$$

$$\frac{D_{\tan 1}}{\epsilon_{r1}} - \frac{D_{\tan 2}}{\epsilon_{r2}} = 0$$

$$D_{n1} - D_{n2} = \rho_s.$$

2.8.3 Boundary Conditions in Scalar Form

$$E_{t1} = E_{t2}$$

$$H_{t1} - H_{t2} = J_s$$

$$B_{n1} = B_{n2}$$

$$D_{n1} - D_{n2} = \rho_s$$

$$J_{n1} = J_{n2}$$

$$\epsilon_1 E_{n1} - \epsilon_2 E_{n2} = \rho_s$$

$$H_{n1} = \frac{\mu_2}{\mu_1} H_{n2}$$

$$\frac{B_{t1}}{\mu_1} - \frac{B_{t2}}{\mu_2} = J_s$$

$$D_{t1} = \frac{\epsilon_1}{\epsilon_2} D_{t2}$$

$$J_{t1} = \frac{\sigma_1}{\sigma_2} J_{t2}.$$

$$...(2.66)$$

2.8.4 Boundary Conditions in Vector Form

$$\mathbf{a}_n \times (\mathbf{E}_1 - \mathbf{E}_2) = 0$$

$$\mathbf{a}_n \times (\mathbf{H}_1 - \mathbf{H}_2) = \mathbf{J}_s$$

$$\mathbf{a}_n \cdot (\mathbf{B}_1 - \mathbf{B}_2) = 0$$

$$\mathbf{a}_n \cdot (\mathbf{D}_1 - \mathbf{D}_2) = \rho_s$$

$$\mathbf{a}_n \cdot (\mathbf{J}_1 - \mathbf{J}_2) = 0$$

$$\mathbf{a}_n \times \left(\frac{\mathbf{J}_1}{\sigma_1} - \frac{\mathbf{J}_2}{\sigma_2} \right) = 0$$

$$\mathbf{a}_n \cdot (\epsilon_1 \mathbf{E}_1 - \epsilon_2 \mathbf{E}_2) = \rho_s$$

$$\mathbf{a}_n \cdot \left(\frac{\mathbf{H}_1}{\mu_1} - \frac{\mathbf{H}_2}{\mu_2} \right) = 0$$

$$\mathbf{a}_n \times \left(\frac{\mathbf{B}_1}{\mu_1} - \frac{\mathbf{B}_2}{\mu_2} \right) = \mathbf{J}_s.$$

$$...(2.67)$$

2.9 TIME-VARYING POTENTIALS

It is useful to relate electric and magnetic fields in terms of their sources. However, it is often more convenient to relate potentials in terms of sources and then the fields in terms of potentials. There are two methods for doing this.

2.9.1 Heuristic Approach

In this approach, potentials already developed for static fields are generalised to obtain time-varying fields, that is, scalar electrostatic potential is given by

$$V(r) = \frac{1}{4\pi\epsilon_0} \int_v \frac{\rho(r)\,dv}{r} \qquad \qquad \text{...(2.68)}$$

and vector magnetic potential is given by

$$\mathbf{A}(r) = \frac{\mu_0}{4\pi} \int_v \frac{\mathbf{J}(r)\,dv}{r} \qquad \qquad \text{...(2.69)}$$

But the time-varying potentials are expressed in the form of

$$V(r,\ t) = \frac{1}{4\pi\epsilon_0} \int_v \frac{\rho_v(r,\ t)}{r}\,dv \qquad \qquad \text{...(2.70)}$$

and $$\mathbf{A}(\mathbf{r},\ \mathbf{t}) = \frac{\mu_0}{4\pi} \int_v \frac{\mathbf{J}(\mathbf{r},\ \mathbf{t})}{r}\,dv \qquad \qquad \text{...(2.71)}$$

These time-varying potentials are due to time-varying charge and current distributions.

The above expressions do not take care of propagation delay. To obtain far-field expressions, this delay time has to be taken into account.

Retarded Potentials

These are defined as the potentials in which a time delay or retarded time is taken into account. They are expressed as

$$V(r,\ t) = \frac{1}{4\pi\epsilon} \int_v \frac{\rho_v(r,\ t - r/v_0)}{r}\,dv,\ \text{(volt)} \qquad \qquad \text{...(2.72)}$$

$$\mathbf{A}(r,\ t) = \frac{\mu}{4\pi} \int_v \frac{\mathbf{J}(r,\ t - r/v_0)}{r}\,dv,\ \text{(wb/m)} \qquad \qquad \text{...(2.73)}$$

where retarded time or delay time is

$$t_d = (t - r/v_0)$$

(v_0 = velocity of propagation of the EM wave).

There is no change in the expressions for $\mathbf{E}$ and $\mathbf{H}$. That is,

$$\mathbf{E} = -\nabla V \qquad \qquad \text{...(2.74)}$$

and $$\mathbf{H} = \frac{1}{\mu_0}\nabla \times \mathbf{A} \qquad \qquad \text{...(2.75)}$$

2.9.2 Maxwell's Equations Approach

By definition, vector magnetic potential **A** is related to magnetic flux density as

$$\mathbf{B} \equiv \nabla \times \mathbf{A} \qquad \text{...(2.76)}$$

But the second Maxwell's equation is

$$\nabla \times \mathbf{E} = -\dot{\mathbf{B}} \qquad \text{...(2.77)}$$

From Equations (2.76) and (2.77), we have

$$\nabla \times \mathbf{E} = -\frac{\partial}{\partial t}(\nabla \times \mathbf{A}) = -\nabla \times \dot{\mathbf{A}}$$

or $\qquad \nabla \times \mathbf{E} + \nabla \times \dot{\mathbf{A}} = 0$

That is, $\qquad \nabla \times (\mathbf{E} + \dot{\mathbf{A}}) = 0 \qquad \text{...(2.78)}$

This is true only if $(\mathbf{E} + \dot{\mathbf{A}})$ is the gradient of a scalar. Therefore, setting $(\mathbf{E} + \dot{\mathbf{A}} = -\nabla V$, we get **E** as

$$\mathbf{E} = -\nabla V - \dot{\mathbf{A}} \qquad \text{...(2.79)}$$

Note that **E** is not equal to $(-\nabla V)$ for time-varying fields but it is given by Equation (2.79)

Therefore $\qquad \mathbf{E} = -\nabla V$ is valid only for static field

Now consider

$$\mathbf{B} = \nabla \times \mathbf{A}$$

Take curl on both sides,

$$\nabla \times \mathbf{B} = \nabla \times \nabla \times \mathbf{A}$$

or $\qquad \dfrac{1}{\mu_0}\nabla \times \nabla \times \mathbf{A} = \nabla \times \mathbf{H} \qquad \text{...(2.80)}$

But $\qquad \nabla \times \mathbf{H} = \dot{\mathbf{D}} + \mathbf{J} = \epsilon_0 \dot{\mathbf{E}} + \mathbf{J} \qquad \text{...(2.81)}$

From Equations (2.79) and (2.80), we get

$$\epsilon_0 \dot{\mathbf{E}} + \mathbf{J} = -\epsilon_0 \nabla V - \epsilon_0 \dot{\mathbf{A}} + \mathbf{J} \qquad \text{...(2.82)}$$

Thus Equations (2.80) to (2.81) gives

$$\frac{1}{\mu_0}\nabla \times \nabla \times A = -\epsilon_0 \nabla V - \epsilon_0 \dot{\mathbf{A}} + \mathbf{J} \qquad \text{...(2.83)}$$

But vector identity gives

$$\nabla \times \nabla \times \mathbf{A} = \nabla \nabla \cdot \mathbf{A} - \nabla^2 \mathbf{A} \qquad \text{...(2.84)}$$

That is, $\quad \nabla \nabla \cdot \mathbf{A} - \nabla^2 \mathbf{A} = -\mu_0 \epsilon_0 \nabla \dot{V} - \mu_0 \epsilon_0 \dot{\mathbf{A}} + \mu_0 \mathbf{J} \qquad \text{...(2.85)}$

The third Maxwell's equation is

$$\nabla \cdot \mathbf{E} = \rho_v / \epsilon_0 \qquad \text{...(2.86)}$$

From Equations (2.79) and (2.86), we get

$$\nabla \cdot \mathbf{E} = -\nabla \cdot \nabla V - \nabla \cdot \dot{\mathbf{A}} = \rho_v / \epsilon_0$$

or
$$\nabla^2 V + \nabla \cdot \dot{\mathbf{A}} = - \rho_v / \in_0 \qquad \text{...(2.87)}$$

We can rewrite Equation (2.85) as

$$\nabla^2 \mathbf{A} - \mu \in \dot{\mathbf{A}} = - \mu_0 \mathbf{J} + \mu_0 \in_0 \nabla \dot{V} + \nabla \nabla \cdot \mathbf{A} \qquad \text{...(2.88)}$$

The Equations (2.87) and (2.88) contain both the potentials and one of the sources. It is difficult to solve this type of equations. We should be able to get equations containing one potential and its own source to make them easier to solve. Helmholtz theorem helps in doing this.

Helmholtz Theorem

Helmholtz Theorem states that any vector field like **A** due to a finite source is uniquely specified if, and only if, its curl and the divergence are specified.

To specify **A**, we know its curl, that is,

$$\nabla \times \mathbf{A} = \mathbf{B}$$

But the $\nabla \cdot \mathbf{A}$ is specified as

$$\nabla \cdot \mathbf{A} = - \mu_0 \in_0 \dot{V} \qquad \text{...(2.89)}$$

Thus Equations (2.87) and (2.88) now become

$$\left. \begin{array}{l} \nabla^2 \mathbf{A} - \mu_0 \in_0 \ddot{\mathbf{A}} = - \mu_0 \mathbf{J} \\[2mm] \nabla^2 V - \mu_0 \in_0 \ddot{V} = - \rho_v / \in_0 \end{array} \right\} \qquad \text{...(2.90)}$$

These are the expressions between time-varying potentials and their sources. The corresponding expressions for static fields are

$$\left. \begin{array}{l} \nabla^2 \mathbf{A} = - \mu_0 \mathbf{J} \\[2mm] \nabla^2 V = - \rho_v / \in_0 \end{array} \right\} \qquad \text{...(2.91)}$$

Lorentz Gauge Condition

It is given by

$$\nabla \cdot \mathbf{A} = - \mu_0 \in_0 \dot{V}$$

Potential functions for sinusoidal fields are given by

$$\left. \begin{array}{l} \nabla^2 \mathbf{A} + \omega^2 \mu_0 \in_0 \mathbf{A} = - \mu_0 \mathbf{J} \\[2mm] \nabla^2 V + \omega^2 \mu_0 \in_0 V = - \rho_v / \in_0 \end{array} \right\} \qquad \text{...(2.92)}$$

2.10 ELECTROMAGNETIC WAVES

A wave, which is a physical phenomenon, means a recurring function of time at a point.

A **wave**, which is a physical phenomenon, means a recurring function of time at a point. In its re-occurrence, there is time delay which is proportional to the space separation between two adjacent locations.

In general, a wave is a carrier of energy or information which is a function of time as well as space.

In the context of our study a wave means **electromagnetic wave** (or, simply EM wave). Maxwell predicted the existence of EM waves and established this through his well-known Maxwell's equations. The same EM waves were investigated by Heinrich Hertz. Hertz conducted several experiments which could generate and detect EM waves. These radio waves are called Hertzian waves.

Examples of EM waves: Radio, Radar beam, TV signals and so on.

2.11 APPLICATIONS OF EM WAVES

They have a wide range of applications in all types of communication like police radio, TV, satellite, ionospheric, tropospheric, wireless, cellular, mobile communi-cations and so on as well as in all types of radars like Doppler radar, MTI radar, speed trap radar, airport surveillance radar, weather forecasting radar, remote sensing radar, ground mapping radar, IFF radar, astronomy radar, fire control radar and so on. EM waves are also used in radiation therapy, microwave ovens and so on.

The advantage in using EM waves for communication purposes is that it requires no maintenance of the medium between the transmitter and receiver. This is due to free space being the best medium for EM wave propagation.

2.12 WAVE EQUATIONS IN FREE SPACE

Wave equations in free space are given by:

$$\nabla^2 \mathbf{E} = \mu_0 \in_0 \ddot{\mathbf{E}}$$

$$\nabla^2 \mathbf{H} = \mu_0 \in_0 \ddot{\mathbf{H}}$$

where

$$\ddot{\mathbf{E}} = \frac{\partial^2 \mathbf{E}}{\partial t^2}$$

and

$$\ddot{\mathbf{H}} = \frac{\partial^2 \mathbf{H}}{\partial t^2}.$$

Proof   Free space is characterised by $\in_r = 1$, $\mu_r = 1$ or $\in \, = \, \in_0$, $\mu = \mu_0$, and $\sigma = 0$, $\rho_v = 0$ and $\mathbf{J} = 0$. Due to these characteristics of free space, Maxwell's second equation is given by

$$\nabla \times \mathbf{E} = - \dot{\mathbf{B}} \qquad \qquad \text{...(2.93)}$$

$$= - \mu_0 \dot{\mathbf{H}} \qquad \qquad [\text{as } \mathbf{B} = \mu_0 \mathbf{H}]$$

Taking curl on both sides, we get

$$\nabla \times \nabla \times \mathbf{E} = - \mu_0 \, \nabla \times \dot{\mathbf{H}} \qquad \text{...(2.94)}$$

Using standard vector identity,

$$\nabla \times \nabla \times \mathbf{E} = \nabla \nabla . \, \mathbf{E} - \nabla^2 \mathbf{E} \qquad \text{...(2.95)}$$

From the first Maxwell's equation,

$$- \mu_0 \, \nabla \times \dot{\mathbf{H}} = - \mu_0 \, \ddot{\mathbf{D}} = - \mu_0 \in_0 \ddot{\mathbf{E}} \qquad \text{...(2.96)}$$

Thus $\qquad \nabla \nabla . \, \mathbf{E} - \nabla^2 \mathbf{E} = - \mu_0 \in_0 \ddot{\mathbf{E}}$

But $\qquad\qquad \nabla . \, \mathbf{D} = \nabla . \in_0 \mathbf{E} = \in_0 \nabla . \, \mathbf{E} = 0$

So $\qquad\qquad \boxed{\nabla^2 \mathbf{E} = \in_0 \mu_0 \ddot{\mathbf{E}}} \qquad\qquad$ Hence proved.

Now consider the first Maxwell's equation

$$\nabla \times \mathbf{H} = \in_0 \dot{\mathbf{E}}$$

Taking curl on both sides

$$\nabla \times \nabla \times \mathbf{H} = \in_0 \nabla \times \dot{\mathbf{E}}$$

$$\nabla \nabla . \, \mathbf{H} - \nabla^2 \mathbf{H} = \in_0 (- \ddot{\mathbf{B}}) = - \mu_0 \in_0 \dot{\mathbf{H}}$$

Hence $\qquad\qquad \nabla^2 \mathbf{H} = \mu_0 \in_0 \ddot{\mathbf{H}} \qquad\qquad [\text{as } \nabla . \, \mathbf{H} = 0]$

Hence proved.

2.13 WAVE EQUATIONS FOR A CONDUCTING MEDIUM

The wave equations for a conducting medium ($\rho_\upsilon = 0$, $\sigma \neq 0$ and $\mathbf{J} \neq 0$) are given by

$$\nabla^2 \mathbf{E} = \mu \in \ddot{\mathbf{E}} + \mu \sigma \dot{\mathbf{E}} \qquad \text{...(2.97)}$$

and $\qquad\qquad \nabla^2 \mathbf{H} = \mu \in \ddot{\mathbf{H}} + \mu \sigma \dot{\mathbf{H}} \qquad \text{...(2.98)}$

Proof Consider the second Maxwell's equation

$$\nabla \times \mathbf{E} = - \dot{\mathbf{B}} = - \mu \dot{\mathbf{H}}$$

Taking curl on both sides

$$\nabla \times \nabla \times \mathbf{E} = - \mu \nabla \times \dot{\mathbf{H}}$$

$$\nabla \nabla . \, \mathbf{E} - \nabla^2 \mathbf{E} = - \mu \, [\in \ddot{\mathbf{E}} + \dot{\mathbf{J}}]$$

$$= - \mu \in \ddot{\mathbf{E}} - \mu \sigma \dot{\mathbf{E}}$$

Thus $\qquad\qquad \boxed{\nabla^2 \mathbf{E} = \mu \in \ddot{\mathbf{E}} + \mu \sigma \dot{\mathbf{E}}} \qquad\qquad [\text{as } \nabla . \, \mathbf{E} = 0]$

Hence proved.

Similarly, consider the first Maxwell's equation

$$\nabla \times \mathbf{H} = \dot{\mathbf{D}} + \mathbf{J} = \in \dot{\mathbf{E}} + \sigma \mathbf{E}$$

Taking curl on both sides,

$$\nabla \times \nabla \times \mathbf{H} = \in \nabla \times \dot{\mathbf{E}} + \sigma \nabla \times \mathbf{E}$$

$$\nabla\nabla \cdot \mathbf{H} - \nabla^2 \mathbf{H} = \in (-\mu\ddot{\mathbf{H}}) - \mu\sigma\dot{\mathbf{H}}$$

Thus
$$\boxed{\nabla^2 \mathbf{H} = \mu\in \ddot{\mathbf{H}} + \mu\sigma\dot{\mathbf{H}}}$$
[as $\nabla \cdot \mathbf{H} = 0$]

Hence proved.

2.14 UNIFORM PLANE WAVE EQUATION

An EM wave propagating in x direction is said to be a **uniform plane wave** if its fields **E** and **H** are independent of y and z directions.
It is defined as a wave whose electric and magnetic fields have constant amplitude over the equiphase surfaces. These waves exist only in free space at an infinite distance from the source.

A uniform plane wave propagating in x-direction has no x-components of **E** and **H**. That is, $E_x = 0$, $H_x = 0$.

The electric and magnetic fields of EM wave are always perpendicular to each other. A typical wave is shown in Fig. 2.3.

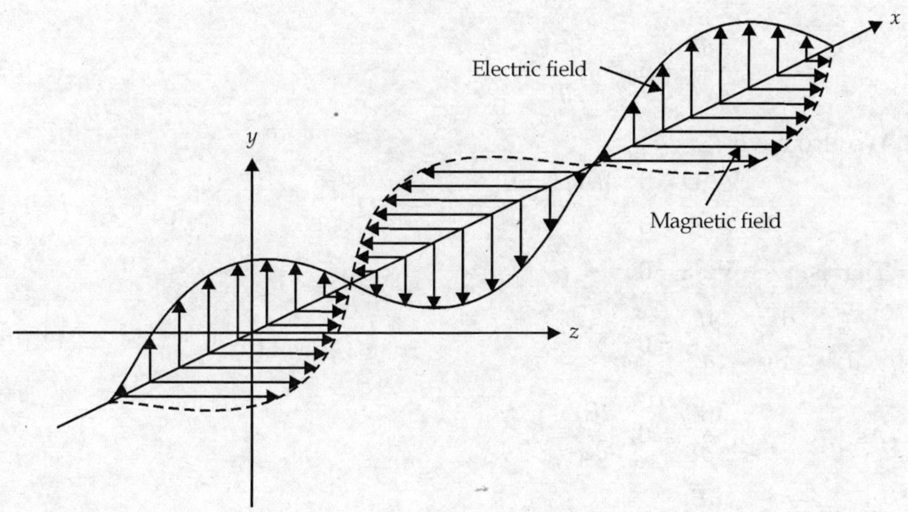

Fig. 2.3   *Electromagnetic wave*

For a uniform plane wave $E_x = 0$ and $H_x = 0$

Proof   The plane wave equation in free space is given by
$$\nabla^2 \mathbf{E} = \mu_0 \in_0 \ddot{\mathbf{E}}$$

That is,
$$\frac{\partial^2 \mathbf{E}}{\partial x^2} + \frac{\partial^2 \mathbf{E}}{\partial y^2} + \frac{\partial^2 \mathbf{E}}{\partial z^2} = \mu_0 \in_0 \frac{\partial^2 \mathbf{E}}{\partial t^2} \qquad \dots(2.99)$$

As per the definition of uniform plane wave,

$$\mathbf{E} \neq f(y) \quad \text{and} \quad \mathbf{E} \neq f(z)$$

Hence

$$\frac{\partial^2 \mathbf{E}}{\partial y^2} = 0, \quad \frac{\partial^2 \mathbf{E}}{\partial z^2} = 0$$

Equation (2.99) becomes

$$\frac{\partial^2 \mathbf{E}}{\partial x^2} = \mu_0 \in_0 \frac{\partial^2 \mathbf{E}}{\partial t^2} \tag{2.100}$$

or

$$\frac{\partial^2 E_x}{\partial x^2}\mathbf{a}_x + \frac{\partial^2 E_y}{\partial x^2}\mathbf{a}_y + \frac{\partial^2 E_z}{\partial x^2}\mathbf{a}_z = \mu_0 \in_0 \left[\frac{\partial^2 E_x}{\partial t^2}\mathbf{a}_x + \frac{\partial^2 E_y}{\partial t^2}\mathbf{a}_y + \frac{\partial^2 E_z}{\partial t^2}\mathbf{a}_z \right]$$

Equating the respective components on both sides, we get

$$\frac{\partial^2 E_x}{\partial x^2} = \mu_0 \in_0 \frac{\partial^2 E_x}{\partial t^2} \tag{2.101}$$

$$\frac{\partial^2 E_y}{\partial x^2} = \mu_0 \in_0 \frac{\partial^2 E_y}{\partial t^2} \tag{2.102}$$

$$\frac{\partial^2 E_z}{\partial x^2} = \mu_0 \in_0 \frac{\partial^2 E_z}{\partial t^2} \tag{2.103}$$

We also have

$$\nabla \cdot \mathbf{D} = 0 \quad [\text{as } \rho_\upsilon = 0]$$

or

$$\nabla \cdot \in_0 \mathbf{E} = 0$$

That is,

$$\nabla \cdot \mathbf{E} = 0$$

or

$$\frac{\partial E_x}{\partial x} + \frac{\partial E_y}{\partial y} + \frac{\partial E_z}{\partial z} = 0 \tag{2.104}$$

As

$$\frac{\partial E_y}{\partial t} = 0, \quad \frac{\partial E_z}{\partial z} = 0,$$

We have

$$\frac{\partial E_x}{\partial x} = 0 \tag{2.105}$$

Substituting Equation (2.105) in (2.101), we get

$$\frac{\partial^2 E_x}{\partial t^2} = 0 \tag{2.106}$$

This means that E_x should have one of the following solutions.

1. $E_x = 0$.

2. $E_x = $ a constant with time

3. E_x increases uniformly with time, that is, $E_x = Kt$, where K is constant.

If $E_x = a$ constant and $E_x = Kt$, it will not be a part of wave motion. Therefore,

$$E_x = 0$$

Similarly, $\qquad\qquad H_x = 0$

Hence proved.

2.15 GENERAL SOLUTION OF UNIFORM PLANE WAVE EQUATION

The wave equation in free space is

$$\nabla^2 \mathbf{E} = \mu_0 \in_0 \ddot{\mathbf{E}}$$

Applying the conditions of uniform plane wave equation, the above equation becomes

$$\nabla^2 \mathbf{E} = \frac{\partial^2 \mathbf{E}}{\partial x^2} \qquad\qquad [\text{as } \mathbf{E} \neq f(y),\ \mathbf{E} \neq f(z)]$$

Hence $\qquad \dfrac{\partial^2 \mathbf{E}}{\partial x^2} = \mu_0 \in_0 \dfrac{\partial^2 \mathbf{E}}{\partial t^2}$ $\qquad\qquad$ ...(2.107)

Equating the respective components on either side and as $E_x = 0$, we have

$$\frac{\partial^2 E_y}{\partial x^2} = \mu_0 \in_0 \frac{\partial^2 y}{\partial t^2} \qquad\qquad ...(2.108)$$

$$\frac{\partial^2 E_z}{\partial x^2} = \mu_0 \in_0 \frac{\partial^2 E_z}{\partial t^2} \qquad\qquad ...(2.109)$$

Equation (2.107) has a general solution given by

$$\mathbf{E} = f_1(x - \upsilon_0 t) + f_2(x + \upsilon_0 t) \qquad\qquad ...(2.110)$$

where f_1 and f_2 are functions of $(x - \upsilon_0 t)$ and $(x + \upsilon_0 t)$ respectively.

$$\upsilon_0 = \frac{1}{\sqrt{\mu_0 \in_0}} = \text{velocity of propagation}$$

x is the direction of propagation of the wave

$f_1(x - \upsilon_0 t)$ represents a forward wave and

$f_2(x + \upsilon_0 t)$ represents a reflected wave.

This reflected wave is present when there is a conductor which acts as a reflector. Otherwise, it is absent. As we are considering free space propagation, $\mathbf{E}$ can only be $f_1(x - \upsilon_0 t)$. That is,

$$\mathbf{E} = f(x - \upsilon_0 t) \qquad\qquad ...(2.111)$$

This is the solution of uniform plane wave equation in free space.

The behaviour is typically represented in Fig. 2.4.

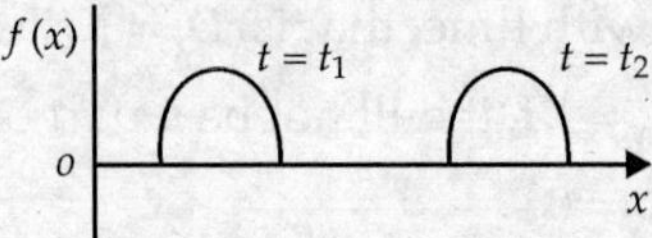

Fig. 2.4 *A wave along x-direction*

2.16 RELATION BETWEEN E AND H IN A UNIFORM PLANE WAVE

The relation between **E** and **H** is

$$\left|\frac{\mathbf{E}}{\mathbf{H}}\right| = \frac{E}{H} = 120\pi\,\Omega \approx 377\,\Omega. \qquad \text{...(2.112)}$$

Proof The first Maxwell's equation is

$$\nabla \times \mathbf{H} = \begin{vmatrix} \mathbf{a}_x & \mathbf{a}_y & \mathbf{a}_z \\ \dfrac{\partial}{\partial x} & \dfrac{\partial}{\partial y} & \dfrac{\partial}{\partial z} \\ H_x & H_y & H_z \end{vmatrix} = \dot{\mathbf{D}} = \in_0 \dot{\mathbf{E}}$$

$$= \mathbf{a}_x\left[\frac{\partial H_z}{\partial y} - \frac{\partial H_y}{\partial z}\right] + \mathbf{a}_y\left[\frac{\partial H_x}{\partial z} - \frac{\partial H_z}{\partial x}\right] + \mathbf{a}_z\left[\frac{\partial H_y}{\partial x} - \frac{\partial H_x}{\partial y}\right]$$

$$\text{...(2.113)}$$

But

$$\frac{\partial H_z}{\partial y} = 0, \ \frac{\partial H_y}{\partial z} = 0, \ H_x = 0$$

Equation (2.113) becomes

$$\nabla \times \mathbf{H} = -\frac{\partial H_z}{\partial x}\mathbf{a}_y + \frac{\partial H_y}{\partial x}\mathbf{a}_z = \in_0 \dot{\mathbf{E}}$$

Similarly,

$$\nabla \times \mathbf{E} = -\frac{\partial E_z}{\partial x}\mathbf{a}_y + \frac{\partial E_y}{\partial x}\mathbf{a}_z = -\mu_0 \dot{\mathbf{H}}$$

or

$$-\frac{\partial H_z}{\partial x}\mathbf{a}_y + \frac{\partial H_y}{\partial x}\mathbf{a}_z = \in_0 \frac{\partial E_y}{\partial t}\mathbf{a}_y + \in_0 \frac{\partial E_z}{\partial t}\mathbf{a}_z$$

and

$$-\frac{\partial E_z}{\partial x}\mathbf{a}_y + \frac{\partial E_y}{\partial x}\mathbf{a}_z = -\mu_0 \frac{\partial H_y}{\partial t}\mathbf{a}_y - \mu_0 \frac{\partial H_z}{\partial t}\mathbf{a}_z$$

Equating the respective components, we get

$$-\frac{\partial H_z}{\partial x} = \in_0 \frac{\partial E_y}{\partial t} \qquad \text{...(2.114)}$$

$$\frac{\partial H_y}{\partial x} = \in_0 \frac{\partial E_z}{\partial t} \qquad \text{...(2.115)}$$

$$\frac{\partial E_z}{\partial x} = \mu_0 \frac{\partial H_y}{\partial t} \qquad \qquad \text{...(2.116)}$$

$$\frac{\partial E_y}{\partial x} = - \mu_0 \frac{\partial H_z}{\partial t} \qquad \qquad \text{...(2.117)}$$

Writing E_y in the form of

$$E_y = f(x - v_0 t), \quad v_0 = \frac{1}{\sqrt{\mu_0 \epsilon_0}}$$

$$\frac{\partial E_y}{\partial t} = \frac{\partial f}{\partial (x - v_0 t)} \frac{\partial (x - v_0 t)}{\partial t} = - v_0 f' \qquad \text{...(2.118)}$$

where

$$f' = \frac{\partial f}{\partial (x - v_0 t)}$$

From Equations (2.114) to (2.117), we have

$$\frac{\partial H_z}{\partial x} = - \epsilon_0 \frac{\partial E_y}{\partial t} = \epsilon_0 v_0 f'$$

or

$$H_z = \int \epsilon_0 v_0 f' \, dx + A$$

As the constant A cannot be a part of wave motion, we can put $A = 0$.

So

$$H_z = \epsilon_0 v_0 \int f' \, dx$$

Since

$$\frac{\partial f}{\partial x} = f' \frac{\partial (x - v_0 f)}{\partial x} = f'$$

$$H_z = \sqrt{\frac{\epsilon_0}{\mu_0}} \int \frac{\partial f}{\partial x} \, dx$$

or

$$H_z = \sqrt{\frac{\epsilon_0}{\mu_0}} f = \sqrt{\frac{\epsilon_0}{\mu_0}} E_y$$

or

$$\frac{E_y}{H_z} = \sqrt{\frac{\mu_0}{\epsilon_0}} \qquad \qquad \text{...(2.119)}$$

Similarly, if we take

$$E_z = f(x - v_0 t)$$

$$\frac{\partial E_z}{\partial t} = - v_0 f' \qquad \qquad \text{...(2.120)}$$

From Equations (2.115) and (2.120), we have

$$\frac{\partial H_y}{\partial x} = \epsilon_0 \frac{\partial E_z}{\partial t} = - \epsilon_0 v_0 f'$$

Hence
$$H_y = -\epsilon_0 \, v_0 \int f' \, dx$$
$$= -\epsilon_0 \, v_0 \, f$$
$$= -\epsilon_0 \, v_0 \, E_z$$

or
$$\frac{E_z}{H_y} = -\sqrt{\frac{\mu_0}{\epsilon_0}} \qquad \qquad ...(2.121)$$

So
$$\frac{E}{H} = \sqrt{\frac{E_y^2 + E_z^2}{H_y^2 + H_z^2}} = \sqrt{\frac{H_z^2 \left(\dfrac{\mu_0}{\epsilon_0}\right) + H_y^2 \left(\dfrac{\mu_0}{\epsilon_0}\right)}{H_y^2 + H_z^2}}$$

Hence
$$\frac{E}{H} = \sqrt{\frac{\mu_0}{\epsilon_0}} = \sqrt{\frac{4\pi \times 10^{-7}}{\dfrac{1}{36\pi \times 10^9}}} = 120\pi\,\Omega$$

That is,
$$\frac{E}{H} = 120\pi\,\Omega$$

Hence proved.

The intrinsic impedance or characteristic impedance, η_0 is defined as
$$\eta_0 = \frac{E}{H} = 120\pi\,\Omega \qquad \qquad ...(2.122)$$

E and H of EM wave are perpendicular to each other

Proof   Consider
$$\mathbf{E} \cdot \mathbf{H} = E_x H_x + E_y H_y + E_z H_z = E_y H_y + E_z H_z \qquad [\text{as } E_x = 0, \ H_x = 0]$$

$$\frac{E_y}{H_z} = \sqrt{\frac{\mu_0}{\epsilon_0}}, \ \frac{E_z}{H_y} = -\sqrt{\frac{\mu_0}{\epsilon_0}}, \ E_x = 0, \ H_x = 0$$

We get
$$\mathbf{E} \cdot \mathbf{H} = E_x H_x + E_y H_y + E_z H_z$$
$$= \sqrt{\frac{\mu_0}{\epsilon_0}} H_y H_z - \sqrt{\frac{\mu_0}{\epsilon_0}} H_y H_z = 0$$

Therefore $\mathbf{E} \cdot \mathbf{H} = 0$ \qquad ...(2.123)

The dot product of two vectors $\mathbf{E}$ and $\mathbf{H}$ is zero only when the two vectors are perpendicular to each other. Hence proved.

2.17 WAVE EQUATIONS IN PHASOR FORM

The wave equations in free space are
$$\nabla^2 \mathbf{H} = \mu_0 \, \epsilon_0 \, \ddot{\mathbf{H}}$$

and
$$\nabla^2 \mathbf{E} = \mu_0 \, \epsilon_0 \, \ddot{\mathbf{E}}$$

If
$$\tilde{\mathbf{E}} = \mathrm{Re}\,\{\mathbf{E}e^{j\omega t}\}$$
$$\tilde{\mathbf{H}} = \mathrm{Re}\,\{\mathbf{H}e^{j\omega t}\}$$

The wave equations become
$$\nabla^2\,\mathbf{H} = -\,\omega^2\,\mu_0\in_0\,\mathbf{H}$$
$$\nabla^2\,\mathbf{E} = -\,\omega^2\,\mu_0\in_0\,\mathbf{E} \qquad \qquad ...(2.124)$$
These are wave equations in phasor form.

Note that a single time derivative of the field gives a factor of $j\omega$ and a double time derivative of the field gives a factor of $-\omega^2$.

Similarly, the wave equations in conductive medium are
$$\nabla^2\,\mathbf{E} = (-\,\omega^2\,\mu\in\, +\, j\omega\mu\sigma)\,\mathbf{E}$$
$$\nabla^2\,\mathbf{H} = (-\,\omega^2\,\mu\in\, +\, j\omega\mu\sigma)\,\mathbf{H} \qquad \qquad ...(2.125)$$

These can be represented in the following form:
$$\left.\begin{array}{c}\nabla^2\,\mathbf{E} = \gamma^2\,\mathbf{E}\\[2mm]\nabla^2\,\mathbf{H} = \gamma^2\,\mathbf{H}\end{array}\right\} \qquad \qquad ...(2.126)$$

where
$$\gamma = \text{propagation constant}\,\left(\frac{1}{m}\right)$$
$$= \sqrt{-\,\omega^2\,\mu\in\, +\, j\omega\mu\sigma}. \qquad \qquad ...(2.127)$$

2.18 WAVE PROPAGATION IN A LOSSLESS MEDIUM

The wave equation is
$$\nabla^2\,\mathbf{E} = -\,\omega^2\,\mu\in\,\mathbf{E}$$

That is,
$$\frac{\partial^2\,\mathbf{E}}{\partial x^2} = -\,\omega^2\,\mu\in\,\mathbf{E}$$

or
$$\frac{\partial^2\,\mathbf{E}}{\partial x^2} = -\,\beta^2\,\mathbf{E}$$

where
$$\beta = \omega\,\sqrt{\mu\in}\,.$$

The y-component of $\mathbf{E}$ may be written as
$$E_y = Ae^{-j\beta x} + Be^{j\beta x}$$

where A and B are arbitrary complex constants. Then
$$\tilde{E}_y\,(x,\ t) = \mathrm{Re}\,\{E_y\,(x)\,e^{j\omega t}\}$$
$$= \mathrm{Re}\,\{Ae^{j\,(\omega t - \beta x)} + Be^{j\,(\omega t + \beta x)}\} \qquad \qquad ...(2.128)$$

If A and B are real, Equation (2.128) becomes

$$\tilde{E}_y\,(x,\ t) = A \cos\,(\omega t - \beta x) + B \cos\,(\omega t + \beta x) \qquad \text{...(2.129)}$$

Equation (2.129) is the sum of two waves. They travel in opposite directions. If $A = B$, the waves combine together and form a standing wave. Such waves do not progress.

2.18.1 The Wave Velocity(υ)

It is defined as the velocity of propagation of the wave. It is also defined as

$$\upsilon \equiv \frac{\omega}{\beta} \qquad \text{...(2.130)}$$

where
$$\omega = 2\pi f = \text{angular frequency}$$
$$\beta = \text{phase constant, radians/m.}$$

Phase shift constant, β is defined as a measure of the phase shift in radians per unit length.

Wavelength of the wave, λ is defined as that distance through which the sinusoidal wave passes through a full cycle of 2π radians.

That is, $\lambda \equiv \dfrac{2\pi}{\beta}$ $\qquad$ ...(2.131)

Phase velocity, (υ_p) is defined as the velocity of some point in the sinusoidal waveform.

Intrinsic or **characteristic impedance** of a medium which has a finite value of conductivity is given by

$$\eta \equiv \sqrt{\frac{j\omega\mu}{\sigma + j\omega\epsilon}}\,. \qquad \text{...(2.132)}$$

2.19 PROPAGATION CHARACTERISTICS OF EM WAVES IN FREE SPACE

The wave equation in free space is

$$\nabla^2\,\mathbf{E} = -\,\omega^2\,\mu_0\,\epsilon_0\,\mathbf{E}$$

$$= \gamma^2\,\mathbf{E}$$

$$\gamma = \sqrt{-\,\omega^2\,\mu_0\,\epsilon_0} = j\omega\,\sqrt{\mu_0\,\epsilon_0}$$

$$= j\beta$$

The phase constant, β (rad/m)

$$\beta = \omega\,\sqrt{\mu_0\,\epsilon_0} \qquad \text{...(2.133)}$$

The propagation characteristics of EM wave in free space are:

1. Propagation constant, $\gamma = j\omega\,\sqrt{\mu_0\,\epsilon_0}$, (m^{-1})
2. Phase shift constant, $\beta = \omega\,\sqrt{\mu_0\,\epsilon_0}$, (rad/m)

3. Velocity of propagation, $\upsilon_0 = \dfrac{1}{\sqrt{\mu_0 \in_0}}$, (m/s)

4. The velocity of propagation of EM wave is the same as phase velocity, that is,

$$\upsilon_p = \frac{\omega}{\beta} = \frac{\omega}{\omega \sqrt{\mu_0 \in_0}} = \upsilon_0 \text{ (m/s)}$$

5. $\lambda = \dfrac{2\pi}{\beta} = \dfrac{2\pi}{\omega \sqrt{\mu_0 \in_0}}$, (m)

6. $\dfrac{E}{H} = 120\pi\,\Omega = \sqrt{\dfrac{\mu_0}{\in_0}}$, ($\Omega$)

7. Attenuation constant, $\alpha = 0$.

2.20 PROPAGATION CHARACTERISTICS OF EM WAVE IN A CONDUCTING MEDIUM

The wave equation in a conducting medium in phasor form is

$$\nabla^2 \mathbf{E} - \gamma^2 \mathbf{E} = 0 \qquad \qquad ...(2.134)$$

where the propagation constant, γ is

$$\gamma = j\omega\mu\,(\sigma + j\omega\in) \qquad \qquad ...(2.135)$$
$$= -\omega^2 \mu\in + j\omega\mu\sigma$$

or
$$\gamma = \alpha + j\beta \qquad \qquad ...(2.136)$$

where α is called attenuation constant, dB/m

           β is called phase constant, rad/m.

Phase constant, β is also called wave number and it is the imaginary part of propagation constant.

One solution of Equation (2.134) is

$$E(x) = E_0\, e^{-\gamma x}$$

where x is the direction of propagation and in time-varying form,

$$\tilde{\mathbf{E}}(x,\ t) = \mathrm{Re}\ \{E_0\, e^{j\omega t}\}$$
$$= \mathrm{Re}\ \{E_0\, e^{-\gamma x + \omega t}\}$$

So
$$\tilde{\mathbf{E}}(x,\ t) = e^{\alpha x}\ \mathrm{Re}\ \{E_0\, e^{\omega t - \beta x}\} \qquad \qquad ...(2.137)$$

This is the equation of EM wave propagating in x-direction and attenuated by a factor $e^{-\alpha x}$.

Attenuation constant, $\alpha(\mathrm{dB}/\mathrm{m})$ is defined as a constant which indicates the rate at which the wave amplitude reduces as it propagates from one point to another. It is the real part of propagation constant.

2.20.1 Expressions for α and β in a Conducting Medium

$$\alpha = \omega \sqrt{\frac{\mu\epsilon}{2}\left(\sqrt{1+\frac{\sigma^2}{\omega^2\epsilon^2}}-1\right)}, \quad \text{dB/m} \qquad \ldots(2.138)$$

$$\beta = \omega \sqrt{\frac{\mu\epsilon}{2}\left(\sqrt{1+\frac{\sigma^2}{\omega^2\epsilon^2}}+1\right)}, \quad \text{rad/m} \qquad \ldots(2.139)$$

Proof   From the wave equation, the propagation constant, γ is given by

$$\gamma = \sqrt{-\omega^2\mu\epsilon + j\omega\mu\sigma} = \alpha + j\beta$$

Squaring both sides, we get

$$\gamma^2 = -\omega^2\mu\epsilon + j\omega\mu\sigma$$

$$= (\alpha + j\beta)^2$$

$$= \alpha^2 - \beta^2 + 2j\alpha\beta$$

Equating real and imaginary parts,

$$\alpha^2 - \beta^2 = -\omega^2\mu\epsilon \qquad \ldots(2.140)$$

$$2\alpha\beta = \omega\mu\sigma$$

or $\qquad\qquad \alpha\beta = (\omega\mu\sigma)/2$

That is $\qquad\quad \alpha = \dfrac{\omega\mu\sigma}{2\beta} \ \text{ or } \ \beta = \dfrac{\omega\mu\sigma}{2\alpha} \qquad \ldots(2.141)$

From Equations (2.140) and (2.141), we get

$$-\omega^2\mu\epsilon = \alpha^2 - \frac{\omega^2\mu^2\sigma^2}{4\alpha^2}$$

or $\qquad -4\alpha^2\omega^2\mu\epsilon = 4\alpha^4 - \omega^2\mu^2\sigma^2$

That is, $\qquad 4\alpha^4 + 4\alpha^2\omega^2\mu\epsilon - \omega^2\mu^2\sigma^2 = 0$

Dividing by 4,

$$\alpha^4 + \alpha^2\omega^2\mu\epsilon - \frac{\omega^2\mu^2\sigma^2}{4} = 0$$

Adding and subtracting $\left(\dfrac{\omega^2\mu\epsilon}{2}\right)^2$

$$\alpha^4 + \alpha^2\omega^2\mu\epsilon + \left(\frac{\omega^2\mu\epsilon}{2}\right)^2 = \frac{\omega^2\mu^2\sigma^2}{4} + \left(\frac{\omega^2\mu\epsilon}{2}\right)^2$$

or
$$\left(\alpha^2 + \frac{\omega^2 \mu \in}{2}\right)^2 = \frac{\omega^2 \mu^2 \sigma^2}{4} + \frac{\omega^4 \mu^2 \in^2}{4}$$

or
$$\left(\alpha^2 + \frac{\omega^2 \mu \in}{2}\right)^2 = \frac{\omega^4 \mu^2 \in^2}{4}\left(1 + \frac{\sigma^2}{\omega^2 \in^2}\right)$$

Taking square root on either side, we get

$$\alpha^2 + \frac{\omega^2 \mu \in}{2} = \sqrt{\frac{\omega^4 \mu^2 \in^2}{4}\left(1 + \frac{\sigma^2}{\omega^2 \in^2}\right)}$$

or
$$\alpha^2 = \frac{\omega^2 \mu \in}{2}\sqrt{1 + \frac{\sigma^2}{\omega^2 \in^2}} - \frac{\omega^2 \mu \in}{2} \qquad \text{...(2.142)}$$

Hence
$$\alpha = \pm\, \omega \sqrt{\frac{\mu \in}{2}\left(\sqrt{1 + \frac{\sigma^2}{\omega^2 \in^2}} - 1\right)}$$

But α cannot be negative.

So
$$\alpha = \omega \sqrt{\frac{\mu \in}{2}\left(\sqrt{1 + \frac{\sigma^2}{\omega^2 \in^2}} - 1\right)}$$

Hence proved.

Now substituting Equation (2.142) in Equation (2.140), we get

$$\frac{\omega^2 \mu \in}{2}\sqrt{1 + \frac{\sigma^2}{\omega^2 \in^2}} - \frac{\omega^2 \mu \in}{2} - \beta^2 = -\omega^2 \mu \in$$

That is,
$$+\beta^2 = \frac{\omega^2 \mu \in}{2}\sqrt{1 + \frac{\sigma^2}{\omega^2 \in^2}} + \frac{\omega^2 \mu \in}{2}$$

So
$$\beta = \omega \sqrt{\frac{\mu \in}{2}\left(\sqrt{1 + \frac{\sigma^2}{\omega^2 \in^2}} + 1\right)}$$

Hence proved.

Propagation characteristics of EM wave in conducting medium are:

1. Propagation constant, $\gamma = \sqrt{-\omega^2 \mu \in + j\omega\mu\sigma}$, $\left(\dfrac{1}{m}\right)$

2. Phase shift constant, $\beta = \omega \sqrt{\dfrac{\mu \in}{2}\left(\sqrt{1 + \dfrac{\sigma^2}{\omega^2 \in^2}} + 1\right)}$, (rad/m)

3. Attenuation constant, $\alpha = \omega \sqrt{\dfrac{\mu\epsilon}{2}\left(\sqrt{1+\dfrac{\sigma^2}{\omega^2\epsilon^2}}-1\right)}$, dB/m

4. Velocity of propagation of EM wave,

$$v = f\lambda = \frac{\omega}{\beta}\frac{1}{\sqrt{\dfrac{\mu\epsilon}{2}\left(\sqrt{1+\dfrac{\sigma^2}{\omega^2\epsilon^2}}+1\right)}}\text{ m/s}$$

5. $\lambda = \dfrac{2\pi}{\beta}\,(m)$ or $\lambda = \dfrac{f}{\sqrt{\dfrac{\mu\epsilon}{2}\left(\sqrt{1+\dfrac{\sigma^2}{\omega^2\epsilon^2}}+1\right)}}$

6. Intrinsic impedance, η

$$\frac{E}{H} = \sqrt{\frac{j\omega\mu}{\sigma+j\omega\epsilon}}\ \ \Omega.$$

2.21 CONDUCTORS AND DIELECTRICS

As some media behave as good conductors at one frequency range and as good dielectrics at some other frequency range, the conventional definitions of conductors and dielectrics are not satisfactory in communication through EM waves.

The displacement current density,

$$\mathbf{J}_d = \epsilon\,\dot{\mathbf{D}} = \epsilon\,\dot{\mathbf{E}} = j\omega\epsilon\,\mathbf{E}$$

and the conduction current density,

$$\mathbf{J}_c = \sigma\mathbf{E}$$

Therefore $\qquad \dfrac{\mathbf{J}_c}{\mathbf{J}_d} = \dfrac{\sigma}{\omega\epsilon}$ $\hspace{4cm}$ ...(2.143)

If $\quad \dfrac{\sigma}{\omega\epsilon} \gg 1$, the medium is **good conductor.**

If $\quad \dfrac{\sigma}{\omega\epsilon} \ll 1$, the medium is **good dielectric.**

Dissipation factor of a dielectric material (D_f)

$$D_f \equiv \frac{\sigma}{\omega\epsilon} \hspace{4cm} ...(2.144)$$

When D_f is small for a specific type of dielectric material, the dissipation factor is practically the same as its power factor, which is equal to sin ϕ. Here,

$$\phi = \tan^{-1} D_f.$$

2.22 WAVE PROPAGATION CHARACTERISTICS IN GOOD DIELECTRICS

Attenuation constant, α is given by

$$\alpha = \omega \sqrt{\frac{\mu\epsilon}{2}\left(\sqrt{1 + \frac{\sigma^2}{\omega^2\epsilon^2}} - 1\right)} \qquad \ldots(2.145)$$

For dielectrics, $\qquad \dfrac{\sigma}{\omega\epsilon} \ll 1$

Expanding $\left(1 + \dfrac{\sigma^2}{\omega^2\epsilon^2}\right)^{1/2}$ by binomial series, higher order terms can be neglected.

Hence $\qquad \sqrt{1 + \dfrac{\sigma^2}{\omega^2\epsilon^2}} \approx \left(1 + \dfrac{1}{2\omega^2\epsilon^2}\right)$

Now Equation (2.145) becomes

$$\alpha \approx \omega \sqrt{\frac{\mu\epsilon}{2}\left[\left(1 + \frac{\sigma^2}{2\omega^2\epsilon^2}\right) - 1\right]}$$

$$= \omega \sqrt{\frac{\mu\epsilon\,\sigma^2}{4\omega^2\epsilon^2}}$$

$$= \omega \sqrt{\frac{\mu\sigma^2}{4\omega^2\epsilon}}$$

So $\qquad \alpha = \dfrac{\sigma}{2}\sqrt{\dfrac{\mu}{\epsilon}} \qquad \ldots(2.146)$

The phase shift constant β is given by

$$\beta = \omega \sqrt{\frac{\mu\epsilon}{2}\left(\sqrt{1 + \frac{\sigma^2}{\omega^2\epsilon^2}} + 1\right)} \qquad \ldots(2.147)$$

$$\approx \omega \sqrt{\frac{\mu\epsilon}{2}\left(1 + \frac{\sigma^2}{2\omega^2\epsilon^2} + 1\right)}$$

That is, $\qquad \beta \approx \omega \sqrt{\dfrac{\mu\epsilon}{2}\left(2 + \dfrac{\sigma^2}{2\omega^2\epsilon^2}\right)}$

$$= \omega \sqrt{\mu\epsilon \left(1 + \frac{\sigma^2}{4\omega^2 \epsilon^2}\right)}$$

So
$$\beta \approx \omega \sqrt{\mu\epsilon} \left(1 + \frac{\sigma^2}{8\omega^2 \epsilon^2}\right) \qquad \ldots(2.148)$$

The velocity of propagation is

$$\upsilon = \frac{\omega}{\beta} = \frac{\omega}{\omega \sqrt{\mu\epsilon} \left(1 + \frac{\sigma^2}{8\omega^2 \epsilon^2}\right)}$$

$$= \frac{1}{\sqrt{\mu\epsilon} \left(1 + \frac{\sigma^2}{8\omega^2 \epsilon^2}\right)}$$

So
$$\upsilon \approx \frac{1}{\sqrt{\mu\epsilon}} \left(1 - \frac{\sigma^2}{8\omega^2 \epsilon^2}\right) \qquad \ldots(2.149)$$

2.22.1 Intrinsic or Characteristic Impedance of a General Medium η

$$\eta = \sqrt{\frac{j\omega\mu}{\sigma + j\omega\epsilon}} = \sqrt{\frac{j\omega\mu}{j\omega\epsilon} \left(\frac{1}{1 + \frac{\sigma}{j\omega\epsilon}}\right)} = \sqrt{\frac{\mu}{\epsilon} \left(\frac{1}{1 + \frac{\sigma}{j\omega\epsilon}}\right)}$$

So
$$\eta \approx \sqrt{\frac{\mu}{\epsilon}} \left(1 + j\frac{\sigma}{2\omega\epsilon}\right) \qquad \ldots(2.150)$$

Propagation characteristics in good dielectrics are given by

1. Attenuation constant, $\alpha \approx \dfrac{\sigma}{2} \sqrt{\dfrac{\mu}{\epsilon}}$, dB/m

2. Phase shift constant, $\beta \approx \omega \sqrt{\mu\epsilon} \left(1 + \dfrac{\sigma^2}{8\omega^2 \epsilon^2}\right)$, rad/m

3. Velocity of the wave, $\upsilon = \dfrac{\omega}{\beta} \approx \dfrac{1}{\sqrt{\mu\epsilon}} \left(1 - \dfrac{\sigma^2}{8\omega^2 \epsilon^2}\right)$, m/s

4. Intrinsic impedance, $\eta \approx \sqrt{\dfrac{\mu}{\epsilon}} \left(1 + j\dfrac{\sigma}{2\omega\epsilon}\right)$, Ω

2.23 WAVE PROPAGATION CHARACTERISTICS IN GOOD CONDUCTORS

The propagation constant γ is given by

$$\gamma = \sqrt{-\omega^2 \mu\epsilon + j\omega\mu\sigma} \qquad \text{...(2.151)}$$

$$= j\omega\mu\sigma \sqrt{\left(1 + j\,\frac{\omega\epsilon}{\sigma}\right)}$$

$$\approx \sqrt{j\,\omega\mu\sigma} \qquad \left[\text{as } \frac{\omega\epsilon}{\sigma} \ll 1 \text{ for good conductors}\right]$$

$$= \sqrt{\omega\mu\sigma}\ \angle 45^\circ \qquad \text{...(2.152)}$$

That is, $\gamma = \alpha + j\beta = \sqrt{\omega\mu\sigma}\ \angle 45^\circ$

or

$$\alpha = \beta = \sqrt{\frac{\omega\mu\sigma}{2}} \qquad \text{...(2.153)}$$

Velocity of the wave in a conductor is

$$\upsilon = \frac{\omega}{\beta} = \sqrt{\frac{2\omega}{\mu\sigma}} \qquad \text{...(2.154)}$$

and the intrinsic impedance of the conductor is

$$\eta = \sqrt{\frac{j\omega\mu}{\sigma + j\omega\epsilon}} \qquad \text{...(2.155)}$$

$$= \sqrt{\frac{\mu}{\epsilon}\left(\frac{1}{1 + \dfrac{\sigma}{j\omega\epsilon}}\right)}$$

$$= \sqrt{\frac{j\omega\mu\epsilon}{\sigma\epsilon}}$$

So

$$\eta = \sqrt{\frac{j\omega\mu}{\sigma}} \qquad \text{...(2.156)}$$

Characteristics of wave propagation in good conductors are:

$$\alpha = \sqrt{\frac{\omega\mu\sigma}{2}}\ ,\ \text{dB/m}$$

$$\beta = \sqrt{\frac{\omega\mu\sigma}{2}}\ ,\ \text{rad/m}$$

$$\upsilon = \sqrt{\frac{2\omega}{\mu\sigma}}\ ,\ \text{m/s}$$

$$\eta = \sqrt{\frac{j\omega\mu}{\sigma}}\ ,\ \Omega$$

2.24 DEPTH OF PENETRATION, δ (*m*)

The **depth of penetration** is defined as that depth in which the wave attenuates to $\dfrac{1}{e}$ or approximately 37 per cent of its original amplitude. Depth of penetration is also called as skin depth. It is a measure of depth to which an EM wave can penetrate the medium.

The depth of penetration

$$\delta = \frac{1}{\alpha}, \; \alpha \text{ being attenuation constant} \qquad \qquad \text{...(2.157)}$$

The conductivity of the medium attenuates the wave during propagation. In good conductors the rate of attenuation is more at radio frequencies.

Proof $\delta = \dfrac{1}{\alpha}$:

Let the wave attenuation be represented by

$$\mathbf{E} = E_0 \, e^{-\alpha z} \qquad \qquad \text{...(2.158)}$$

z being the direction of propagation.

At $\qquad \qquad z = \delta,$

$$\mathbf{E} = E_0 \, e^{-\alpha \delta} \qquad \qquad \text{...(2.159)}$$

As per the definition of δ, we have

$$\mathbf{E} = E_0 \, e^{-1} \quad \text{at } z = \delta \qquad \qquad \text{...(2.160)}$$

From Equations (2.159) and (2.160), we have

$$E_0 \, e^{-1} = E_0 \, e^{-\alpha \delta}$$

or $\qquad \qquad \alpha \delta = 1$

or $\qquad \qquad \delta = \dfrac{1}{\alpha}$

Hence proved.

2.25 POLARISATION OF A WAVE

The **polarisation of a wave** is defined as the direction of the electric field at a given point as a function of time.

The polarisation of a composite wave is the direction of its electric field.

2.25.1 Types of Polarisations

These are of three types:

(*a*) Linear polarisation.

(*b*) Circular polarisation.

(*c*) Elliptical polarisation.

(*a*) Linear polarisation A wave is said to be linearly polarised if the electric field as a function of time remains along a straight line at some point in the medium. Linear polarisation of a wave is again of three types:

(*i*) Horizontal polarisation.

(*ii*) Vertical polarisation.

(*iii*) Theta polarisation.

If $\tilde{E}_y = 0$ and $\tilde{E}_x$ is present, when a wave travels in z-direction with $\tilde{\mathbf{E}}$ and $\tilde{\mathbf{H}}$ fields lying in xy-plane, it is said to be **x-polarised** or **horizontally polarised**. If only $\tilde{E}_y$ is present and $\tilde{E}_x = 0$, then the wave is said to be **y-polarised** or **vertically polarised**. On the other hand, if $\tilde{E}_x$ and $\tilde{E}_y$ are present and are in phase, then the wave is said to be theta polarised. This is given by

$$\theta = \left(\tan^{-1} \frac{E_y}{E_x} \right).$$

(*b*) Circular polarisation A wave is said to be circularly polarised when the electric field traces a circle. If $\tilde{E}_x$ and $\tilde{E}_y$ have equal magnitudes and a $90°$ phase difference, the locus of the resultant $\tilde{E}$ is a circle and the wave is **circularly polarised**.

Let $\mathbf{E}$ of an uniform plane wave travelling in the z-direction be represented by

$$\mathbf{E}(z) = E_c e^{-j\beta z} \qquad \ldots(2.161)$$

and in time-varying form by

$$\tilde{\mathbf{E}}(z, t) = \text{Re}\{E_c e^{-j\beta z} e^{j\omega t}\} \qquad \ldots(2.162)$$

As the wave moves in z-direction, $\tilde{E}$ and $\tilde{H}$ lie in x-y plane.

Here, $\mathbf{E}_c$ is a complex vector. That is, $\mathbf{E}_c$ can be written as

$$\mathbf{E}_c = \mathbf{E}_1 + j\,\mathbf{E}_2 \qquad \ldots(2.163)$$

where $\mathbf{E}_1$ and $\mathbf{E}_2$ are real vectors.

At some point in space, say $z = 0$, Equation (2.162) becomes

$$\tilde{\mathbf{E}}(0, t) = \text{Re}\,[(\mathbf{E}_1 + j\,\mathbf{E}_2)\, e^{j\omega t}]$$

$$= \mathbf{E}_1 \cos \omega t - \mathbf{E}_2 \sin \omega t \qquad \ldots(2.164)$$

As per the definition of circular polarisation, the electric vector at $z = 0$ is expressed as

$$\mathbf{E}_c = E_k\, \mathbf{a}_x + j E_k\, \mathbf{a}_y$$

Hence $\qquad \tilde{E}(0,\ t) = E_k \cos \omega t\, \mathbf{a}_x - E_k \sin \omega t\, \mathbf{a}_y$

$$\tilde{E}_x = E_k \cos \omega t$$

$$\tilde{E}_y = - E_k \sin \omega t$$

So $\qquad \boxed{\tilde{E}_x^2 + \tilde{E}_y^2 = E_k^2}$

This represents a circle.

(c) **Elliptical polarisation** If $\tilde{E}_x$ and $\tilde{E}_y$ are not equal in magnitude and they differ by $90°$ phase, then the tip of resultant electric vector traces an ellipse. It is said to be **elliptically polarised**. Here $\mathbf{E}_c$ can be written as

$$\mathbf{E}_c = a\mathbf{a}_x + jb\mathbf{a}_y \quad \text{where } a \text{ and } b \text{ are constants}$$

Hence, $\qquad \tilde{E}(0,\ t) = a \cos \omega t\, \mathbf{a}_x - b \sin \omega t\, \mathbf{a}_y$

and $\qquad \tilde{E}_x = a \cos \omega t$

$$\tilde{E}_y = - b \sin \omega t$$

or $\qquad \boxed{\dfrac{\tilde{E}_x^2}{a^2} + \dfrac{\tilde{E}_y^2}{b^2} = 1}$ $\qquad\qquad$ ...(2.165)

This is the equation of an ellipse and hence the wave is said to be elliptically polarised.

2.25.2 Sources of Different Polarised EM Waves

- Horizontal dipole produces horizontally polarised waves.
- Vertical dipole produces vertically polarised waves.
- Inclined dipole produces theta polarised waves.
- Circular slots produce circularly polarised waves.
- Elliptical slots produce elliptically polarised waves.

2.26 DIRECTION COSINES OF A VECTOR FIELD

The **direction cosine of a vector field** is defined as the cosine of the angle made by the vector with the required coordinate axis.

Consider a vector **E** which is arbitrarily oriented with respect to Cartesian coordinate axes. Assume **E** makes angles θ_x, θ_y and θ_z with x, y, and z-axis (Fig. 2.5). Then the component of a vector in a given direction is the projection of the vector, **E** on a line in that direction. That is,

$$E_x = \mathbf{E} \cdot \mathbf{a}_x = E \cos \theta_x \qquad\qquad ...(2.166)$$

$$E_y = \mathbf{E} \cdot \mathbf{a}_y = E \cos \theta_y \qquad\qquad ...(2.167)$$

$$E_z = \mathbf{E} \cdot \mathbf{a}_z = E \cos \theta_z \qquad\qquad ...(2.168)$$

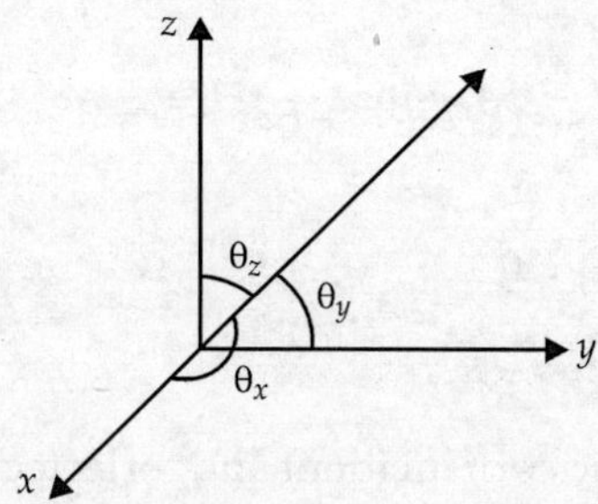

Fig. 2.5   *Arbitrarily oriented vector,* **E**

The $\cos \theta_x$, $\cos \theta_y$ and $\cos \theta_z$ are known as direction cosines of the vector along the coordinate axes.

2.27 WAVES ON A PERFECT CONDUCTOR—NORMAL INCIDENCE

1. When a wave in air is incident on a perfect conductor normally, it is entirely reflected.
2. As neither **E** nor **H** can exist in a perfect conductor, none of the energy is transmitted through it.
3. As there are no losses within a perfect conductor, no energy is absorbed in it.
4. When an EM wave travelling in one medium is incident upon a second medium, it is partially reflected and partially transmitted.

2.27.1 Total Fields of a Wave at any Point after Reflection with Normal Incidence on a Perfect Conductor

Resultant electric field

$$\tilde{\mathbf{E}}_R \,(z,\ t) = 2E_i \sin \beta z \sin \omega t \qquad \qquad ...(2.169)$$

E_i is amplitude of electric field of incident wave,

$$z = \text{direction of propagation.}$$

Resultant magnetic field,

$$\tilde{\mathbf{H}}_R \,(z,\ t) = 2H_i \cos \beta z \cos \omega t \qquad \qquad ...(2.170)$$

H_i is amplitude of magnetic field of incident wave.

Let the electric field of the incident wave be

$$E_{\text{incident}} = E_i \, e^{-j\beta z}, \quad \beta = \frac{2\pi}{\lambda} \qquad \qquad ...(2.171)$$

Then the electric field of the reflected wave is

$$E_{\text{reflected}} = E_r \, e^{j\beta z} \qquad \qquad ...(2.172)$$

The boundary condition is

$$E_{\text{tan1}} = E_{\text{tan2}} = 0 \ \text{ at } \ z = 0$$

This requires that

$$E_R = (E_i\, e^{-j\beta z} + E_r\, e^{j\beta z}) = 0$$

At $z = 0$

$$E_i + E_r = 0$$

So
$$E_i = -E_r \qquad \qquad \text{...(2.173)}$$

This means, the amplitudes of incident and reflected electric field strength are equal but with a phase reversal on reflection.

Now
$$\mathbf{E}_R\,(z) = E_i\, e^{-j\beta z} + E_r\, e^{j\beta z}$$

$$= E_i\,(e^{-j\beta z} - e^{j\beta z})$$

$$= -2j\, E_i \sin \beta z \qquad \qquad \text{...(2.174)}$$

In time-varying form

$$\widetilde{\mathbf{E}}_R\,(z, t) = \mathrm{Re}\,\{-2j\, E_i \sin \beta z e^{j\omega t}\}$$

$$\widetilde{\mathbf{E}}_R\,(x,\ t) = 2E_i \sin \beta z \sin \omega t \qquad \qquad \text{...(2.175)}$$

This obviously represents a standing wave. The variation of E_R is shown in Fig. 2.6.

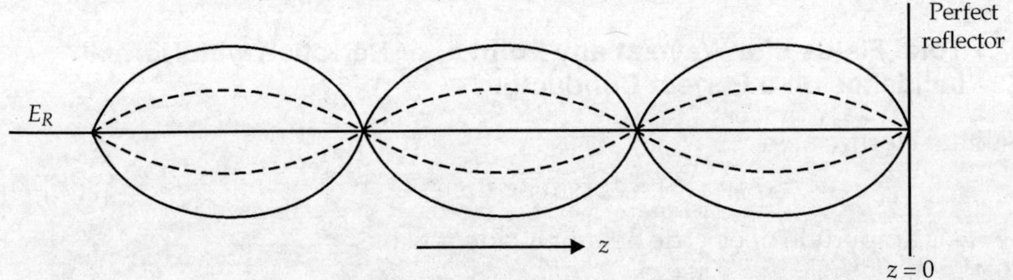

Fig. 2.6 *Variation of electric field*

Conclusions

1. The magnitude of electric field varies sinusoidally with distance from the reflecting plane.

2. $E_R = 0$ at the surface of the conductor $(z = 0)$ and also at

$$z = \frac{n\lambda}{2},\ n, = 1, 2, 3 \dots$$

3. $(E_R)_{\max} = 2E_i$.

4. E_R is maximum at $z = m\,\dfrac{\lambda}{4},\ m = 1, 3, 5, \dots$

Resultant Magnetic Field, H_R

Let us write H_R as

$$H_R(z) = H_i\, e^{-j\beta z} + H_r\, e^{j\beta z}$$

At the surface of the perfect conductor,

$$H_{\tan 1} - H_{\tan 2} = J_s$$

As J_s is not specified, we cannot use this boundary condition. Suppose $H_i = -H_r$. It leads to identical directions of incident and reflected powers. This cannot be true. Therefore, H_i and H_r should be the same at $z = 0$

or
$$H_i = +H_r$$

Hence
$$H_R(x) = H_i\, (e^{-j\beta z} + e^{j\beta z})$$
$$= 2H_i \cos \beta z$$

In time-varying form,

$$\tilde{H}_R(z,\ t) = \text{Re}\ \{2H_i \cos \beta z\, e^{j\omega t}\}$$

So
$$\tilde{H}_R(z,\ t) = 2H_i \cos \beta z \cos \omega t \qquad\qquad ...(2.176)$$

This also represents a standing wave. The variation of H_R is shown in Fig. 2.7.

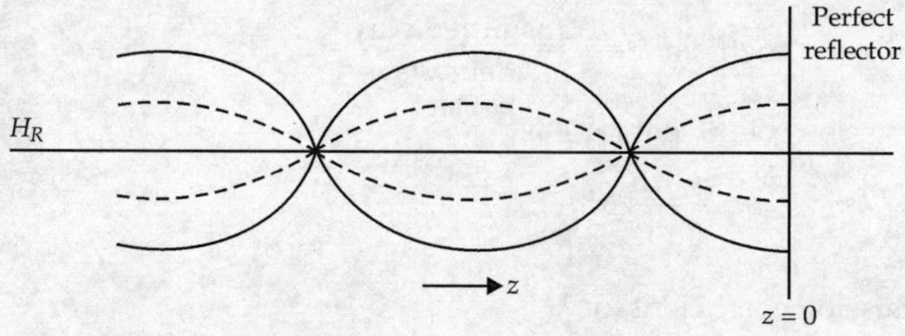

Fig. 2.7   *Variation of magnetic field*

Conclusions

1. Magnitude of H_R varies cosinusoidally with distance.

2. $H_R = 0$ at $z = m\dfrac{\lambda}{4}$, $m = 1, 3, 5, ...$

3. $(H_R)_{\max} = 2H_i$.

4. H_R is maximum at $z = n\dfrac{\lambda}{2}$, $n = 1, 2, 3, ...$ from the surface.

2.28 WAVES ON DIELECTRIC–NORMAL INCIDENCE

When an EM wave is incident normally on the surface of a dielectric reflection and transmission take place.

For perfect dielectric, $\sigma = 0$. Hence, there is no loss or no absorption of energy in it.

The reflection coefficient is defined as the ratio of reflected wave to incident wave.

That is, the reflection coefficient $\equiv \dfrac{\text{reflected wave}}{\text{incident wave}}$,

That is, Reflection coefficient for $E = \Gamma_E \equiv \dfrac{E_r}{E_i}$

Reflection coefficient for $H = \Gamma_H \equiv \dfrac{H_r}{H_i}$

where

$$E_r = \text{reflected electric field}$$
$$E_i = \text{incident electric field}$$
$$H_r = \text{reflected magnetic field}$$
$$H_i = \text{incident magnetic field}$$

Transmission coefficient is defined as the ratio of transmitted wave to incident wave. That is,

Transmission coefficient $\equiv \dfrac{\text{transmitted wave}}{\text{incident wave}}$

Transmission coefficient for E is

$$T_E \equiv \dfrac{E_t}{E_i}$$

Transmission coefficient for H is

$$T_H \equiv \dfrac{H_t}{H_i}$$

The expressions for reflection and transmission coefficients are:

$$\Gamma_E = \dfrac{E_r}{E_i} = \dfrac{\eta_2 - \eta_1}{\eta_2 + \eta_1}$$

$$\Gamma_H = \dfrac{H_r}{H_i} = \dfrac{\eta_1 - \eta_2}{\eta_1 + \eta_2}$$

$$T_E = \dfrac{E_t}{E_i} = \dfrac{2\eta_2}{\eta_1 + \eta_2}$$

$$T_H = \dfrac{H_t}{H_i} = \dfrac{2\eta_1}{\eta_1 + \eta_2}$$

where η_1 and η_2 are intrinsic impedances of medium 1 and medium 2 respectively.

Proof   Let $\in_1$, μ_1, η_1 be the permittivity, permeability and intrinsic impedance of medium 1. $\in_2$, μ_2, η_2 are those of medium 2. We know that

$$E_i = \eta_1 \, H_i \qquad \qquad \text{...(2.177)}$$

$$E_r = -\eta_1 \, H_r \qquad \qquad \text{...(2.178)}$$

$$E_t = \eta_2 \, H_t \qquad \qquad \text{...(2.179)}$$

At the boundary of a dielectric, the tangential components of E and H are continuous. That is,

$$E_i + E_r = E_t \qquad \qquad \text{...(2.180)}$$

$$H_i + H_r = H_t \qquad \qquad \text{...(2.181)}$$

Subtracting Equation (2.178) from Equation (2.177),

$$\eta_1 \, (H_i + H_r) = E_i - E_r = \eta_1 \, H_t$$

$$= \eta_1 \, H_t = (E_i - E_r)$$

$$= \frac{\eta_1}{\eta_2} \, E_t = \frac{\eta_1}{\eta_2} \, (E_i + E_r)$$

So $\qquad \qquad \eta_2 \, (E_i - E_r) = \eta_1 \, (E_i + E_r)$

or $\qquad \qquad \eta_2 \, E_i - \eta_1 \, E_i = \eta_1 \, E_r + \eta_2 \, E_r$

That is, $\qquad \qquad \dfrac{E_r}{E_i} = \Gamma_E = \dfrac{\eta_2 - \eta_1}{\eta_2 + \eta_1} \qquad \qquad \text{...(2.182)}$

Hence proved.

Now consider, $\qquad E_t = E_i + E_r$

That is, $\qquad \qquad \dfrac{E_t}{E_i} = 1 + \dfrac{E_r}{E_i}$

$$= 1 + \frac{\eta_2 - \eta_1}{\eta_2 + \eta_1}$$

$$= \frac{\eta_2 + \eta_1 + \eta_2 - \eta_1}{\eta_1 + \eta_2}$$

So $\qquad \qquad \dfrac{E_r}{E_i} = T_E = \dfrac{2\eta_2}{\eta_1 + \eta_2} \qquad \qquad \text{...(2.183)}$

Hence proved.

From the Equations (2.178) and (2.177),

$$\frac{H_r}{H_i} = -\frac{E_r}{E_i} = -\frac{\eta_2 - \eta_1}{\eta_2 + \eta_1}$$

So $\qquad \qquad \dfrac{H_r}{H_i} = \Gamma_H = \dfrac{\eta_1 - \eta_2}{\eta_1 + \eta_2} \qquad \qquad \text{...(2.184)}$

Hence proved.

Similarly, consider $H_i + H_r = H_t$

or
$$\frac{H_t}{H_i} = 1 + \frac{H_r}{H_i}$$

$$= 1 + \frac{\eta_1 - \eta_2}{\eta_1 + \eta_2}$$

So
$$\frac{H_t}{H_i} = T_H = \frac{2\eta_1}{\eta_1 + \eta_2}$$

Hence proved.

Since $\mu \approx \mu_0$ for the medium of interest,

$$\frac{E_r}{E_i} = \frac{\eta_2 - \eta_1}{\eta_2 + \eta_1} = \frac{\sqrt{\epsilon_1} - \sqrt{\epsilon_2}}{\sqrt{\epsilon_1} + \epsilon_2}$$

$$\frac{H_r}{H_i} = \frac{\eta_1 - \eta_2}{\eta_1 + \eta_2} = \frac{\sqrt{\epsilon_2} - \sqrt{\epsilon_1}}{\sqrt{\epsilon_2} + \sqrt{\epsilon_1}}$$

$$\frac{E_t}{E_i} = \frac{2\eta_2}{\eta_1 + \eta_2} = \frac{2\sqrt{\epsilon_1}}{\sqrt{\epsilon_1} + \sqrt{\epsilon_2}}$$

$$\frac{H_t}{H_i} = \frac{2\eta_1}{\eta_1 + \eta_2} = \frac{2\sqrt{\epsilon_2}}{\sqrt{\epsilon_1} + \sqrt{\epsilon_2}}.$$

2.29 REFLECTION OF WAVE FROM A GOOD CONDUCTOR WITH OBLIQUE INCIDENCE

In order to find out the reflection coefficient when a wave is incident on a good conductor obliquely, two types of polarisations are considered.

    1. Parallel polarisation   2. Perpendicular polarisation.

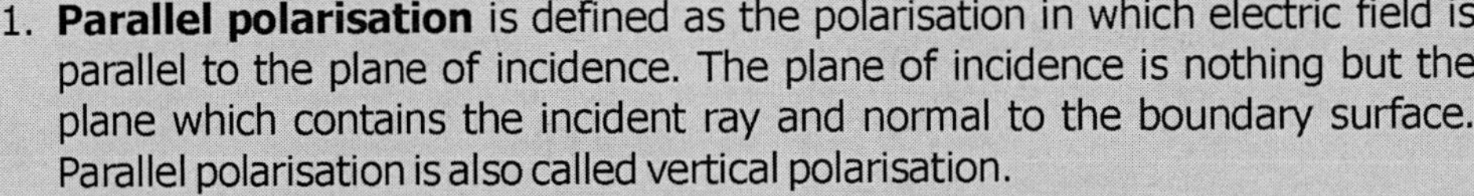

1. **Parallel polarisation** is defined as the polarisation in which electric field is parallel to the plane of incidence. The plane of incidence is nothing but the plane which contains the incident ray and normal to the boundary surface. Parallel polarisation is also called vertical polarisation.

2. **Perpendicular polarisation** is defined as the polarisation in which the electric field is perpendicular to the plane of incidence. It is also called as horizontal polarisation.

2.30 REFLECTION OF A WAVE FROM A DIELECTRIC WITH OBLIQUE INCIDENCE

Let θ_i be the angle of incidence

       θ_r be the angle of reflection and

       θ_t be the angle of transmission or refraction.

These angles are shown in Fig. 2.8.

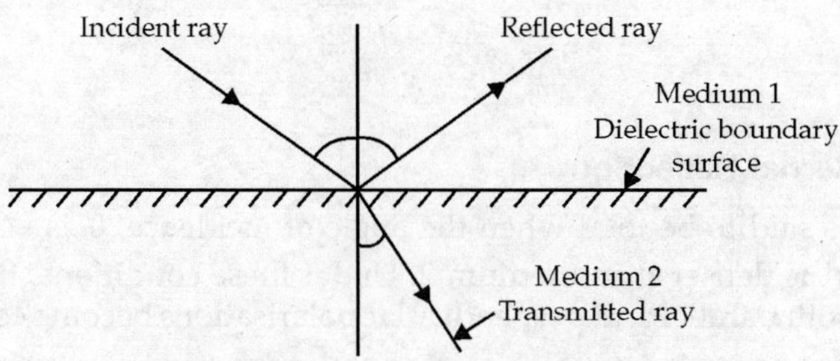

Fig. 2.8 *Reflection and transmission of a wave*

According to Snell's law, angle of incidence and angle of refraction are related by

$$\frac{\sin \theta_i}{\sin \theta_t} = \sqrt{\frac{\epsilon_2}{\epsilon_1}} \qquad \qquad ...(2.185)$$

It may also be noted that the angle of incidence is equal to the angle of reflection. That is,

$$\theta_i = \theta_r \qquad \qquad ...(2.186)$$

Using the law of conservation of energy and boundary conditions at the boundary surface, the reflection coefficient for perpendicular polarisation is given by

$$\frac{E_r}{E_i} = \Gamma_E = \frac{\cos \theta_i - \sqrt{\left(\frac{\epsilon_2}{\epsilon_1}\right) - \sin^2 \theta_i}}{\cos \theta_i + \sqrt{\left(\frac{\epsilon_2}{\epsilon_1}\right) - \sin^2 \theta_i}} \qquad \qquad ...(2.187)$$

The reflection coefficient for parallel polarisation is given by

$$\frac{E_r}{E_i} = \Gamma_E = \frac{\left(\frac{\epsilon_2}{\epsilon_1}\right)\cos \theta_i - \sqrt{\left(\frac{\epsilon_2}{\epsilon_1}\right) - \sin^2 \theta_i}}{\left(\frac{\epsilon_2}{\epsilon_1}\right)\cos \theta_i + \sqrt{\left(\frac{\epsilon_2}{\epsilon_1}\right) - \sin^2 \theta_i}} \qquad \qquad ...(2.187)$$

2.31 BREWSTER ANGLE

Brewster angle is defined as the angle of incidence at which there is no reflection.

For parallel polarisation, the Brewster angle is given by

$$\theta_b = \tan^{-1} \sqrt{\frac{\in_2}{\in_1}}.$$

2.31.1 Total Internal Reflection

The reflection is said to be total when the angle of incidence, θ_i is very high and when medium 1 is denser than medium 2. Under these conditions, the reflection coefficient for both parallel and perpendicular polarisations become complex. This happens when

$$\sin \theta_i > \sqrt{\frac{\in_2}{\in_1}}$$

It may be noted that total internal reflection does not mean that there is no field in medium 2.

That is, the field exists in a less dense medium also. Its phase progression along the boundary decreases exponentially away from it. A non-uniform plane wave has such characteristics.

The concept of total internal reflection is used in binocular optics.

2.32 POYNTING VECTOR AND FLOW OF POWER

When EM waves travel from one point to another, there will be energy flow across the surface involved.

2.32.1 Poynting Theorem

It states that the cross product of $\mathbf{E}$ and $\mathbf{H}$ at any point is a measure of the rate of energy flow per unit area at that point, that is,

$$\mathbf{P} = \mathbf{E} \times \mathbf{H} \text{ watts/m}^2$$

Poynting Vector, P is defined as
$$\mathbf{P} \equiv \mathbf{E} \times \mathbf{H} \text{ watts/ m}^2$$

If $\mathbf{E}$ and $\mathbf{H}$ are instantaneous, $\mathbf{P}$ is also instantaneous.

Proof First Maxwell's equation is
$$\nabla \times \mathbf{H} = \dot{\mathbf{D}} + \mathbf{J} = \in \dot{\mathbf{E}} + \mathbf{J}$$

or
$$\mathbf{J} = \nabla \times \mathbf{H} - \in \dot{\mathbf{E}}$$

That is,
$$\mathbf{E} . \mathbf{J} = \mathbf{E} . \nabla \times \mathbf{H} - \in \mathbf{E} . \dot{\mathbf{E}}$$

But
$$\nabla . (\mathbf{E} \times \mathbf{H}) = \mathbf{H} . \nabla \times \mathbf{E} - \mathbf{E} . \nabla \times \mathbf{H}$$

Hence
$$\mathbf{E} . \mathbf{J} = \mathbf{H} . \nabla \times \mathbf{E} - \nabla . \mathbf{E} \times \mathbf{H} - \in \mathbf{E} . \dot{\mathbf{E}}$$

As
$$\nabla \times \mathbf{E} = -\mu \dot{\mathbf{H}}, \text{ this becomes}$$
$$\mathbf{E} . \mathbf{J} = -\mu \mathbf{H} . \dot{\mathbf{H}} - \in \mathbf{E} . \dot{\mathbf{E}} - \nabla . \mathbf{E} \times \mathbf{H}$$

Here
$$\mathbf{E} \cdot \dot{\mathbf{E}} = \frac{1}{2} \frac{\partial}{\partial t} E^2$$

$$\mathbf{H} \cdot \dot{\mathbf{H}} = \frac{1}{2} \frac{\partial}{\partial t} H^2$$

So
$$\mathbf{E} \cdot \mathbf{J} = -\nabla \cdot \mathbf{E} \times \mathbf{H} - \frac{\partial}{\partial t} \left(\frac{1}{2} \mu H^2 + \frac{1}{2} \in E^2 \right)$$

Taking volume integral

$$\int_v \mathbf{E} \cdot \mathbf{J} \, dv = -\int_v (\nabla \cdot \mathbf{E} \times \mathbf{H}) \, dv - \frac{\partial}{\partial t} \int \left(\frac{1}{2} \in E^2 + \frac{1}{2} \mu H^2 \right) dv$$

By divergence theorem,

$$\int_v \nabla \cdot \mathbf{E} \times \mathbf{H} \, dv = \oint_S (\mathbf{E} \times \mathbf{H}) \cdot d\mathbf{S}$$

So
$$\int_v \mathbf{E} \cdot \mathbf{J} \, dv = -\oint_S (\mathbf{E} \times \mathbf{H}) \cdot d\mathbf{S} - \frac{\partial}{\partial t} \int_v \left(\frac{1}{2} \in E^2 + \frac{1}{2} \mu H^2 \right) dv$$

The left hand side represents energy dissipated in the volume.

The first term on the right hand side represents the rate at which the stored energy in magnetic and electric fields is changing. Negative sign indicates decrease. Therefore, by the law of conservation of energy, the rate of energy dissipation in the volume is equal to the rate at which the stored energy in static electric and magnetic fields in the volume is decreasing plus the rate at which the energy is entering the volume from outside. Therefore,

$$-\oint_S (\mathbf{E} \times \mathbf{H}) \cdot d\mathbf{S}$$

represents inward power flow, or

$$\oint_S (\mathbf{E} \times \mathbf{H}) \cdot d\mathbf{S}$$

represents outward power flow and

$$\oint_S (\mathbf{E} \times \mathbf{H}) \cdot d\mathbf{S} \text{ is in watts}$$

So $\mathbf{E} \times \mathbf{H}$ represents power flow per unit area, that is, $\mathbf{E} \times \mathbf{H}$ is in watts/m^2.

So
$$P = \mathbf{E} \times \mathbf{H}$$

Hence proved.

2.33 COMPLEX POYNTING VECTOR

It is defined as

$$\mathbf{P}_c \equiv \frac{1}{2} \mathbf{E} \times \mathbf{H}^*$$

where $\mathbf{H}^*$ is the complex conjugate of $\mathbf{H}$.

$$P_{av} = \frac{1}{2} \operatorname{Re} \{\mathbf{E} \times \mathbf{H}^*\}$$

$$P_{\text{react}} = \frac{1}{2} \operatorname{Im} (\mathbf{E} \times \mathbf{H}^*)$$

Complex Poynting vector is useful to find out average power flow. We have

$$P_{av} = \frac{1}{2} \operatorname{Re} (EH^*)$$

$$= \frac{1}{2} \operatorname{Re} (\eta HH^*)$$

If
$$H = H_m e^{j\omega t}$$

$$H^* = H_m e^{-j\omega t}$$

So
$$P_{av} = \frac{1}{2} H_m^2 \operatorname{Re} (\eta).$$

POINTS TO REMEMBER

1. Electrostatic fields are produced by static charges.

2. Steady magnetic fields are produced by steady (DC) currents.

3. Electromagnetic waves or fields are produced by time-varying currents.

4. Equation of continuity is $\nabla \cdot \mathbf{J} = -\dot{\rho}_v$.

5. Maxwell's equations in general form are:
$$\nabla \times \mathbf{H} = \dot{\mathbf{D}} + \mathbf{J}$$
$$\nabla \times \mathbf{E} = -\dot{\mathbf{B}}$$
$$\nabla \cdot \mathbf{D} = \rho_v$$
$$\nabla \cdot \mathbf{B} = 0.$$

6. The magnetic current density $\left(\dot{\mathbf{B}} = \dfrac{\partial \mathbf{B}}{\partial t} \right)$ has units of volt/m^2.

7. The propagation constant, $\gamma = \alpha + j\beta \ (m^{-1})$.

8. The attenuation constant, α has the unit of m^{-1}.

9. The phase shift constant, β has the unit of rad/m.

10. The general boundary conditions are:
$$E_{\tan 1} = E_{\tan 2}$$

$$H_{\tan 1} - H_{\tan 2} = J_s$$
$$B_{n1} = B_{n2}$$
$$D_{n1} - D_{n2} = \rho_s.$$

11. The vector magnetic potential, $\mathbf{A}$ is

$$\mathbf{A} = \frac{\mu}{4\pi} \int_{\upsilon} \frac{\mathbf{J}\,(r,\ t)}{r}\, d\upsilon \ \ (\text{wb/m}).$$

12. Lorentz gauge condition is $\nabla \cdot \mathbf{A} = -\,\mu \in \dot{\mathbf{V}}$.

13. For a uniform plane wave propagating in x-direction, $E_x = 0$ and $H_x = 0$.

14. Intrinsic impedance of free space is $120\pi\Omega$.

15. Intrinsic impedance of a medium is $\eta = \dfrac{\mathbf{E}}{\mathbf{H}}$.

16. The wave equations in a conductive medium are:
$$\nabla^2\,\mathbf{E} = \mu \in \ddot{\mathbf{E}} + \mu\sigma\dot{\mathbf{E}}$$
$$\nabla^2\,\mathbf{H} = \mu \in \ddot{\mathbf{H}} + \mu\sigma\dot{\mathbf{H}}.$$

17. The solution of uniform plane wave propagating in x-direction is
$$\mathbf{E} = f\,(x - \upsilon_0\,t).$$

18. The propagation constant is $\gamma = \sqrt{-\,\omega^2\,\mu \in +\,j\omega\mu\sigma}$.

19. The attenuation constant in free space is zero.

20. The phase constant in free space is $\beta = \omega\,\sqrt{\mu_0 \in_0}$.

21. Phase velocity in free space, $\upsilon_p = \dfrac{\omega}{\beta}$.

22. Dissipation factor, $D_f = \dfrac{\sigma}{\omega \in}$.

23. Attenuation constant in good conductors is $\alpha = \sqrt{\dfrac{\omega\mu\sigma}{2}}$.

24. Phase constant in good conductors is $\beta = \sqrt{\dfrac{\omega\mu\sigma}{2}}$.

25. In good dielectrics $\alpha = \dfrac{\sigma}{2}\,\sqrt{\dfrac{\mu}{\in}}$.

26. Depth of penetration, $\delta = \dfrac{1}{\alpha}$.

27. Poynting vector, $\mathbf{P} = \mathbf{E} \times \mathbf{H}$.

28. Complex Poynting vector, $\mathbf{P}_c = 1/2\ \mathbf{E} \times \mathbf{H}^*$.

29. Reflection coefficient $= \dfrac{\text{reflection wave}}{\text{incident wave}}$.

30. Transmission coefficient $= \dfrac{\text{transmitted wave}}{\text{incident wave}}$.

31. Brewster angle, $\theta_b = \tan^{-1} \sqrt{\dfrac{\epsilon_2}{\epsilon_1}}$.

32. Snell's law is given by $\dfrac{\sin\theta_i}{\sin\theta_t} = \sqrt{\dfrac{\epsilon_2}{\epsilon_1}}$.

 **SOLVED PROBLEMS**

Problem 2.1   Given $\mathbf{E} = 10 \sin (\omega t - \beta z)\, \mathbf{a}_y$ V/m in free space, determine $\mathbf{D}$, $\mathbf{B}$, $\mathbf{H}$.

Solution $\mathbf{E} = 10 \sin (\omega t - \beta z)\, \mathbf{a}_y$ V/m

$$\mathbf{D} = \epsilon_0\, \mathbf{E},\ \ \epsilon_0 = 8.854 \times 10^{-12} \text{ F/m}$$

So $\mathbf{D} = 10\epsilon_0 \sin (\omega t - \beta z)\, \mathbf{a}_y\ \text{C/m}^2$

Second Maxwell's equation is

$$\nabla \times \mathbf{E} = -\dot{\mathbf{B}}$$

That is $\nabla \times \mathbf{E} = \begin{vmatrix} \mathbf{a}_x & \mathbf{a}_y & \mathbf{a}_z \\ \dfrac{\partial}{\partial x} & \dfrac{\partial}{\partial y} & \dfrac{\partial}{\partial z} \\ 0 & E_y & 0 \end{vmatrix}$

or $\nabla \times \mathbf{E} = \mathbf{a}_x \left[-\dfrac{\partial}{\partial z} E_y \right] + 0 + \mathbf{a}_z \left[\dfrac{\partial}{\partial x} E_y \right]$

As $E_y = 10 \sin (\omega t - \beta z)$ V/m

$$\frac{\partial E_y}{\partial x} = 0$$

Now $\nabla \times \mathbf{E}$ becomes

$$\nabla \times \mathbf{E} = -\frac{\partial E_y}{\partial z}\, \mathbf{a}_x$$

$$= 10\beta \cos (\omega t - \beta z)\, \mathbf{a}_x$$

$$= -\frac{\partial \mathbf{B}}{\partial t}$$

So $\mathbf{B} = -\displaystyle\int 10\beta \cos (\omega t - \beta z)\, dt\, \mathbf{a}_x$

or
$$\mathbf{B} = \frac{10\beta}{\omega} \sin(\omega t - \beta z)\, \mathbf{a}_x,\ \text{wb/m}^2$$

and
$$\mathbf{H} = \frac{\mathbf{B}}{\mu_0} = \frac{-10\beta}{\mu_0\,\omega} \sin(\omega t - \beta z)\, \mathbf{a}_x,\ \text{A/m}.$$

Problem 2.2 If the electric field strength, $\mathbf{E}$ of an electromagnetic wave in free space is given by

$$\mathbf{E} = 2\cos\omega\left(t - \frac{z}{v_0}\right)\mathbf{a}_y\ \text{V/m},\ \text{find the magnetic field, } \mathbf{H}.$$

Solution We have $\dfrac{\partial \mathbf{B}}{\partial t} = -\nabla \times \mathbf{E}$

$$= -\begin{vmatrix} \mathbf{a}_x & \mathbf{a}_y & \mathbf{a}_z \\ \dfrac{\partial}{\partial x} & \dfrac{\partial}{\partial y} & \dfrac{\partial}{\partial z} \\ 0 & E_y & 0 \end{vmatrix}$$

$$= -\left[\mathbf{a}_x\left(-\frac{\partial}{\partial z}E_y\right) + \mathbf{a}_y(0) + \mathbf{a}_z\left(\frac{\partial}{\partial x}E_y\right) \right]$$

$$= \frac{\partial E_y}{\partial z}\,\mathbf{a}_x$$

$$= \frac{2\omega}{v_0}\sin\omega\left(t - \frac{z}{v_0}\right)\mathbf{a}_x$$

So
$$\mathbf{B} = \frac{2\omega}{v_0}\int \sin\omega\left(t - \frac{z}{v_0}\right)dt\,\mathbf{a}_x$$

or
$$\mathbf{B} = \frac{-2\omega}{v_0\,\omega}\cos\omega\left(t - \frac{z}{v_0}\right)\mathbf{a}_x$$

or
$$\mathbf{H} = \frac{\mathbf{B}}{\mu_0} = \frac{-2}{v_0\,\mu_0}\cos\omega\left(t - \frac{z}{v_0}\right)\mathbf{a}_x \qquad \eta = \sqrt{\frac{\mu_0}{\epsilon_0}} = 120\pi\,\Omega$$

That is,
$$\mathbf{H} = \frac{-2}{\eta_0}\cos\omega\left(t - \frac{z}{v_0}\right)\mathbf{a}_x \qquad \left[v_0 = \frac{1}{\sqrt{\mu_0\,\epsilon_0}}\right]$$

So
$$\mathbf{H} = -\frac{1}{60\pi}\cos\omega\left(t - \frac{z}{v_0}\right)\mathbf{a}_x\ \text{A/m}.$$

Problem 2.3 The parallel plates in a capacitor have an area of $4\times 10^{-4}\ \text{m}^2$ and the plates are separated by 0.4 cm. A voltage of $10\sin 10^3\,t$ volts is applied to the capacitor. Find the displacement current when the dielectric material between the plates has a relative permittivity of 4.

Solution   We have

$$\mathbf{D} = \epsilon\, \mathbf{E}$$

or

$$D = \epsilon\, E$$

The displacement current density, J_d is

$$J_d = \frac{\partial D}{\partial t} = \frac{\partial (\epsilon\, E)}{\partial t}$$

But

$$E = \frac{V}{d}$$

where

$$d = \text{plate separation} = 0.4\ \text{cm} = 0.004\ \text{m}.$$

So

$$J_d = \frac{d}{\partial t}\left(\epsilon\, \frac{V}{d}\right)$$

$$= \frac{\epsilon}{d}\, \frac{dV}{dt}$$

$$= \frac{\epsilon_0\, \epsilon_r}{d}\, \frac{dV}{dt}$$

The displacement current, I_d is

$$I_d = J_d \times A = \frac{\epsilon_0\, \epsilon_r\, A}{d}\, \frac{dV}{dt}, \quad A = \text{Area of the plate}$$

$$= C\, \frac{dV}{dt} \qquad \left[\text{as } C = \frac{\epsilon_0\, \epsilon_r\, A}{d}\right]$$

$$= \frac{4 \times 8.854 \times 10^{-12} \times 4 \times 10^{-4}}{4 \times 10^{-3}}\, \frac{dV}{dt}$$

$$= 35.42 \times 10^{-13}\, \frac{dV}{dt}$$

But

$$V = 10 \sin 10^3\, t \ \text{volt}$$

$$\frac{dV}{dt} = 10 \times 10^3 \cos 10^3\, t$$

So

$$I_d = 35.42 \cos 10^3\, t \ \text{nA}.$$

Problem 2.4   In free space, the magnetic field of an EM wave is given by

$$\mathbf{H} = 0.4\omega\epsilon_0 \cos (\omega t - 50x)\, \mathbf{a}_z \ \text{A/m}.$$

Find the electric field, $\mathbf{E}$ and displacement current density, $\dot{\mathbf{D}}$.

Solution $\mathbf{H} = 0.4\omega\epsilon_0 \cos (\omega t - 50x)\, \mathbf{a}_z \ \text{A/m}$

We have $\nabla \times \mathbf{H} = \dot{\mathbf{D}} + \mathbf{J}$

But $\mathbf{J} = 0$ for free space.

So
$$\nabla \times \mathbf{H} = \frac{\partial \mathbf{D}}{\partial t} = \in_0 \frac{\partial \mathbf{E}}{\partial t}$$

Hence
$$\in_0 \frac{\partial \mathbf{E}}{\partial t} = \begin{vmatrix} \mathbf{a}_x & \mathbf{a}_y & \mathbf{a}_z \\ \dfrac{\partial}{\partial x} & \dfrac{\partial}{\partial y} & \dfrac{\partial}{\partial z} \\ 0 & 0 & H_z \end{vmatrix}$$

$$= \mathbf{a}_x \left[\frac{\partial}{\partial y} H_z \right] + \mathbf{a}_y \left[-\frac{\partial}{\partial x} H_z \right] + [0]\, \mathbf{a}_z$$

But
$$\frac{\partial H_z}{\partial y} = 0 \qquad\qquad\qquad [\mathbf{H} \text{ is not a function of } y]$$

So
$$\in_0 \frac{\partial \mathbf{E}}{\partial t} = -\frac{\partial H_z}{\partial x}\, \mathbf{a}_y$$

$$\frac{\partial H_z}{\partial x} = \frac{\partial}{\partial x} [0.4\omega \in_0 \cos(\omega t - 50x)]$$

$$= +0.4\omega \in_0 50 \times \sin(\omega t - 50x)$$

$$= 20 \in_0 \omega \sin(\omega t - 50x)$$

That is,
$$\in_0 \frac{\partial \mathbf{E}}{\partial t} = -20 \in_0 \omega \sin(\omega t - 50x)\, \mathbf{a}_y$$

or
$$\frac{\partial \mathbf{E}}{\partial t} = -20\omega \sin(\omega t - 50x)\, \mathbf{a}_y$$

So
$$\mathbf{E} = \int -20\omega \sin(\omega t - 50x)\, dt\, \mathbf{a}_y$$

$$= \frac{-20}{\omega}\, \omega(-1) \cos(\omega t - 50x)\, \mathbf{a}_y$$

So
$$\mathbf{E} = 20 \cos(\omega t - 50x)\, \mathbf{a}_y \ \text{V/m.}$$

The displacement current density, $\mathbf{J}_d$

$$\mathbf{J}_d = \frac{\partial \mathbf{D}}{\partial t}$$

$$= \in_0 \frac{\partial \mathbf{E}}{\partial t}$$

$$= \in_0 \frac{\partial}{\partial t} (20 \cos(\omega t - 50x)\, \mathbf{a}_y$$

So
$$\mathbf{J}_d = -20 \in_0 \omega \sin(\omega t - 50x)\, \mathbf{a}_y \ \text{A/m}^2.$$

Problem 2.5   If there is a magnetic field represented by

$$\mathbf{B} = 2 \sin (\omega t - \beta x)\, \mathbf{a}_x + 2y \cos (\omega t - \beta x)\, \mathbf{a}_y$$

in a medium where $\rho_v = 0$, $\sigma = 0$ and $J = 0$, find the electric field. Assume $\epsilon_r = 1$, $\mu_r = 1$.

Solution   We have $\nabla \times \mathbf{H} = \dot{\mathbf{D}} + \mathbf{J}$

But $\qquad\qquad\qquad\qquad \mathbf{J} = 0$

Since $\qquad\qquad\qquad\quad \mathbf{D} = \epsilon_0\, \mathbf{E}, \ \ \mathbf{B} = \mu_0\, \mathbf{H}$

We can write $\quad \nabla \times \mathbf{H} = \dfrac{1}{\mu_0} \nabla \times \mathbf{B} = \epsilon_0 \dfrac{\partial \mathbf{E}}{\partial t}$

or $\qquad\qquad\qquad \nabla \times \mathbf{B} = \mu_0\, \epsilon_0 \dfrac{\partial \mathbf{E}}{\partial t}$

$$\nabla \times \mathbf{B} = \begin{vmatrix} \mathbf{a}_x & \mathbf{a}_y & \mathbf{a}_z \\[6pt] \dfrac{\partial}{\partial x} & \dfrac{\partial}{\partial y} & \dfrac{\partial}{\partial z} \\[6pt] B_x & B_y & 0 \end{vmatrix}$$

$$= \mathbf{a}_x \left[-\dfrac{\partial}{\partial z} B_y \right] + \mathbf{a}_y \left[\dfrac{\partial}{\partial z} B_x + 0 \right] + \mathbf{a}_z \left[\dfrac{\partial}{\partial x} B_y - \dfrac{\partial}{\partial y} B_x \right]$$

As B_x and B_y are independent of y and z

$$\frac{\partial B_y}{\partial z} = 0, \ \ \frac{\partial B_y}{\partial x} = 0, \ \ \frac{\partial B_x}{\partial y} = 0$$

Hence $\qquad\qquad \nabla \times \mathbf{B} = \mathbf{a}_z \left[\dfrac{\partial B_y}{\partial x} - \dfrac{\partial B_x}{\partial y} \right]$

But $\qquad\qquad \dfrac{\partial B_y}{\partial x} = \dfrac{\partial\, (2y \cos (\omega t - \beta x)}{\partial x}$

$$= + 2y\beta \sin (\omega t - \beta x)$$

$$= 2\beta y \sin (\omega t - \beta x)$$

So $\qquad\qquad \nabla \times \mathbf{B} = 2\beta y \sin (\omega t - \beta x)\, \mathbf{a}_z$

$$= \mu_0\, \epsilon_0 \dfrac{\partial \mathbf{E}}{\partial t}$$

or $\qquad\qquad \dfrac{\partial \mathbf{E}}{\partial t} = \dfrac{\mathbf{a}_z}{\mu_0\, \epsilon_0} [2\beta y \sin (\omega t - \beta x)]$

Thus $\qquad\qquad \mathbf{E} = \dfrac{\mathbf{a}_z\, 2\beta y}{\mu_0\, \epsilon_0} \int \sin (\omega t - \beta x)\, dt$

or $\qquad\qquad \mathbf{E} = \dfrac{-\, 2\beta y}{\omega \mu_0\, \epsilon_0} \cos (\omega t - \beta x)\, \mathbf{a}_z \ \text{V/m}.$

Problem 2.6 An electric field in a medium which is source free is given by

$$\mathbf{E} = 1.5 \cos (10^8 t - \beta z)\, \mathbf{a}_x \text{ V/m},$$

where E_m is the amplitude of $\mathbf{E}$, ω is the angular frequency and β is the phase constant. Obtain $\mathbf{B}$, $\mathbf{H}$, $\mathbf{D}$. Assume $\epsilon_r = 1$, $\mu_r = 1$, $\sigma = 0$.

Solution   We have

$$\nabla \times \mathbf{E} = -\frac{\partial \mathbf{B}}{\partial t}$$

or

$$\nabla \times \mathbf{E} = \begin{vmatrix} \mathbf{a}_x & \mathbf{a}_y & \mathbf{a}_z \\ \dfrac{\partial}{\partial x} & \dfrac{\partial}{\partial y} & \dfrac{\partial}{\partial z} \\ E_x & 0 & 0 \end{vmatrix}$$

$$= \mathbf{a}_x\,[0] + \mathbf{a}_y \left[\frac{\partial E_x}{\partial z}\right] - \mathbf{a}_z \left[\frac{\partial E_x}{\partial y}\right]$$

As E_x is not a function of y,

$$\frac{\partial E_x}{\partial y} = 0$$

So

$$\nabla \times \mathbf{E} = \frac{\partial E_x}{\partial z}\,\mathbf{a}_y$$

or

$$\frac{\partial \mathbf{B}}{\partial t} = -\frac{\partial E_x}{\partial z}\,\mathbf{a}_y$$

$$= -1.5\beta \sin (10^8 t - \beta z)\, \mathbf{a}_y$$

Hence

$$\mathbf{B} = -\int 1.5\beta \sin (10^8 t - \beta z)\, dt\, \mathbf{a}_y$$

or

$$\mathbf{B} = \frac{1.5\beta}{10^8} \cos (10^8 t - \beta z)\, \mathbf{a}_y \text{ wb/m}^2$$

As

$$\mathbf{B} = \mu \mathbf{H}, \quad \mu = \mu_0 = 4\pi \times 10^{-7} \text{ H/m}$$

we have

$$\mathbf{H} = \frac{\mathbf{B}}{\mu} = 12\beta \cos (10^8 t - \beta z)\, \mathbf{a}_y, \text{ mA/m}$$

But

$$\mathbf{D} = \epsilon \mathbf{E} = \epsilon_0 \mathbf{E}, \quad \epsilon_0 = 8.854 \times 10^{-12} \text{ F/m}$$

So

$$\mathbf{D} = 13.28 \cos (10^8 t - \beta z)\, \mathbf{a}_x, \text{ pC/m}^2.$$

Problem 2.7   Verify whether the following fields:

$$\mathbf{E} = 2 \sin x\, \sin t\, \mathbf{a}_y \text{ and}$$

$$\mathbf{H} = \frac{2}{\mu_0} \cos x \, \cos t \, \mathbf{a}_z$$

satisfy Maxwell's equations in free space.

Solution $\qquad \nabla \times \mathbf{H} = \dot{\mathbf{D}}$ $\hfill$ [as $\mathbf{J} = 0$]

That is, $\qquad \dfrac{-\partial H_z}{\partial x} \mathbf{a}_y = \in_0 \dfrac{\partial E_y}{\partial t} \mathbf{a}_y$

or $\qquad \dfrac{2}{\mu_0} \sin x \, \cos t = 2\in_0 \sin x \, \cos t$

That is, $\qquad \dfrac{2}{\mu_0} = 2\in_0$

or $\qquad \mu_0 \in_0 = 1$

which cannot be satisfied. Therefore, the given fields do not satisfy Maxwell's equations.

Problem 2.8 $x < 0$ defines region 1 and $x > 0$ defines region 2. Region 1 is characterised by $\mu_{r_1} = 3.0$ and region 2 is characterised by $\mu_{r_2} = 5.0$. If the magnetic field in region 1 is given by $\mathbf{H}_1 = 4.0\mathbf{a}_x + 1.5\mathbf{a}_y - 3.0\mathbf{a}_z$ A/m, find $\mathbf{H}_2$ and H_2.

Solution $\qquad \mathbf{H}_1 = 4.0\mathbf{a}_x + 1.5\mathbf{a}_y - 3.0\mathbf{a}_z,$ A/m

For the regions given,

$$\mathbf{H}_{t_1} = 1.5\mathbf{a}_y - 3.0\mathbf{a}_z,\ \text{A/m}$$

and $\qquad \mathbf{H}_{n_1} = 4.0\mathbf{a}_x$

As $\qquad \mathbf{H}_{t_1} = \mathbf{H}_{t_2}$

$$\mathbf{H}_{t_2} = 1.5\mathbf{a}_y - 3.0\mathbf{a}_z,\ \text{A/m}$$

and $\qquad \mathbf{H}_{n_2} = \dfrac{\mu_1}{\mu_2} \mathbf{H}_{n_1}$

$$= \frac{3}{5} \times 4.0\mathbf{a}_x$$

or $\qquad \mathbf{H}_{n_2} = 2.4\mathbf{a}_x$

So $\qquad \mathbf{H}_2 = \mathbf{H}_{t_2} + \mathbf{H}_{n_2}$

or $\qquad \mathbf{H}_2 = 2.4\mathbf{a}_x + 1.5\mathbf{a}_y - 3.0\mathbf{a}_z,\ \text{A/m}$

The magnitude of $\mathbf{H}_2$ is

$$H_2 = \sqrt{2.4^2 + 1.5^2 (-3.0)^2}$$

or $\qquad H_2 = 4.1243 \text{ A/m.}$

Problem 2.9 In a three-dimensional space, divided into region 1 ($x < 0$) and region 2 ($x > 0$), $\sigma_1 = \sigma_2 = 0$. $\mathbf{E}_1 = 1\mathbf{a}_x + 2\mathbf{a}_y + 3\mathbf{a}_z$. Find $\mathbf{E}_2$ and $\mathbf{D}_2$. $\in_{r1} = 1$ and $\in_{r2} = 2$.

Solution

$$E_{t_1} = 2\mathbf{a}_y + 3\mathbf{a}_z, \ \text{V/m}$$

$$E_{n_1} = \mathbf{a}_x, \quad \epsilon_{r1} = 1$$

$$D_{t_1} = \epsilon_0\,(2\mathbf{a}_y + 3\mathbf{a}_z), \ \text{C/m}^2$$

As

$$E_{t_1} = E_{t_2}$$

$$E_{t_2} = 2\mathbf{a}_y + 3\mathbf{a}_z, \ \text{V/m}$$

As

$$D_{n_1} = D_{n_2}$$

$$\epsilon_1\, E_{n_1} = \epsilon_2\, E_{n_2}$$

or

$$E_{n_2} = \frac{\epsilon_1}{\epsilon_2}\, E_{n_1} = \frac{1}{2}\, \mathbf{a}_x$$

So

$$E_2 = E_{t_2} + E_{n_2}$$

or

$$\boxed{E_2 = 0.5\mathbf{a}_x + 2\mathbf{a}_y + 3\mathbf{a}_z, \ \text{A/m}}$$

$$D_2 = \epsilon_2\, E_2$$

$$= 2\epsilon_0\,(0.5\mathbf{a}_x + 2\mathbf{a}_y + 3\mathbf{a}_z)$$

or

$$\boxed{D_2 = \epsilon_0\,(\mathbf{a}_x + 4\mathbf{a}_y + 6\mathbf{a}_z, \ \text{C/m}^2.}$$

Problem 2.10   If the retarded scalar electric potential, $V = x - v_0\,t$ and vector magnetic potential,

$$A = \left(\frac{x}{v_0} - t\right)\mathbf{a}_x, \ \text{where } v_0 \text{ is the velocity of propagation:}$$

(a) Find $\nabla \cdot \mathbf{A}$

(b) Find $\mathbf{B}, \mathbf{H}, \mathbf{E},$ and $\mathbf{D}$

(c) Show that $\nabla \cdot \mathbf{A} = -\mu_0\,\epsilon_0\,\dfrac{\partial V}{\partial t}.$

Solution (a)

$$\nabla \cdot \mathbf{A} = \frac{\partial A_x}{\partial x} + \frac{\partial A_y}{\partial y} + \frac{\partial A_z}{\partial z}$$

But

$$A_y = 0, \ A_z = 0$$

So

$$\nabla \cdot \mathbf{A} = \frac{\partial A_x}{\partial x} = \frac{\partial}{\partial x}\left(\frac{x}{v_0} - t\right) = \frac{1}{v_0}$$

or

$$\boxed{\nabla \cdot \mathbf{A} = 1/v_0}$$

(b) We have $\mathbf{B} = \nabla \times \mathbf{A}$

Here

$$A = \left(\frac{x}{v_0} - t\right)\mathbf{a}_x$$

That is,
$$A_x = \left(\frac{x}{v_0} - t\right)$$

$$\nabla \times \mathbf{A} = \begin{vmatrix} \mathbf{a}_x & \mathbf{a}_y & \mathbf{a}_z \\ \dfrac{\partial}{\partial x} & \dfrac{\partial}{\partial y} & \dfrac{\partial}{\partial z} \\ A_x & 0 & 0 \end{vmatrix}$$

$$= \mathbf{a}_x \,[0] + \mathbf{a}_y \left[\frac{\partial A_x}{\partial z}\right] + \mathbf{a}_z \left[-\frac{\partial A_x}{\partial y}\right]$$

As $A_x \neq f\,(y)$, $\dfrac{\partial A_x}{\partial y} = 0$ and A_x is also not function of z

$$\frac{\partial A_x}{\partial z} = 0$$

or
$$\nabla \times \mathbf{A} = 0$$

Hence
$$\boxed{\mathbf{B} = 0}$$

As
$$\mathbf{H} = \frac{1}{\mu_0}\,\mathbf{B}, \ \mathbf{H} = 0$$

$$\mathbf{E} = -\nabla V - \dot{\mathbf{A}}$$

$$= -\nabla V - \frac{\partial \mathbf{A}}{\partial t}$$

$$\nabla V = +\frac{\partial V}{\partial x}\,\mathbf{a}_x + \frac{\partial V}{\partial y}\,\mathbf{a}_y + \frac{\partial V}{\partial z}\,\mathbf{a}_z$$

Here
$$\frac{\partial V}{\partial y} = 0, \quad \frac{\partial V}{\partial z} = 0$$

Hence
$$\nabla V = \frac{\partial V}{\partial x}\,\mathbf{a}_x = \mathbf{a}_x$$

$$\frac{\partial \mathbf{A}}{\partial t} = \frac{\partial}{\partial t}\,(x/v_0 - t)\,\mathbf{a}_x$$

$$= -\mathbf{a}_x$$

or
$$\mathbf{E} = -\nabla V - \dot{\mathbf{A}}$$

or
$$\boxed{\mathbf{E} = -\mathbf{a}_x + \mathbf{a}_x = 0}$$

and
$$\boxed{\mathbf{D} = \epsilon_0\,\mathbf{E} = 0}$$

(c)
$$\frac{\partial V}{\partial t} = \frac{\partial}{\partial t}\,(x - v_0\,t) = -v_0$$

But
$$\nabla \cdot \mathbf{A} = \frac{1}{v_0}$$

$$\mu_0 \in_0 \frac{\partial V}{\partial t} = -v_0 \times \frac{1}{v_0^2} = -\frac{1}{v_0}$$

or
$$-\mu_0 \in_0 \frac{\partial V}{\partial t} = \frac{1}{v_0}$$

So
$$\boxed{\nabla \cdot \mathbf{A} = -\mu_0 \in_0 \frac{\partial V}{\partial t}.}$$
Hence proved.

Problem 2.11 If the electric field strength of a radio broadcast signal at a TV is given by $\mathbf{E} = 5.0 \cos(\omega t - \beta y)\,\mathbf{a}_z$, v/m, determine displacement current density. If the same field exists in a medium whose conductivity is given by 2.0×10^3 (mho)/cm, find the conduction current density.

Solution $\mathbf{E}$ at a TV in free space
$$= 5 \cos(\omega t - \beta y)\,\mathbf{a}_z \text{ v/m}$$

Electric flux density,
$$\mathbf{D} = \in_0 \mathbf{E} = 5\in_0 \cos(\omega t - \beta y)\,\mathbf{a}_z \text{ c/m}^2$$

The displacement current density
$$\mathbf{J}_d = \dot{\mathbf{D}} = \frac{\partial D}{\partial t}$$

$$= \frac{\partial}{\partial t}[5\in_0 \cos(\omega t - \beta y)\,\mathbf{a}_z]$$

That is,
$$\boxed{\mathbf{J}_d = -5\in_0 \omega \sin(\omega t - \beta y)\,\mathbf{a}_z,\ \text{A/m}^2}$$

The conduction current density,
$$\mathbf{J}_c = \sigma \mathbf{E}$$

$$\sigma = 2.0 \times 10^3 \text{ (mho)/cm}$$

$$= 2 \times 10^5 \text{ mho/m}$$

That is,
$$\mathbf{J}_c = 2 \times 10^5 \times 5 \cos(\omega t - \beta y)\,\mathbf{a}_z$$

So
$$\boxed{\mathbf{J}_c = 10^6 \cos(\omega t - \beta y)\,\mathbf{a}_z,\ \text{A/m}^2.}$$

Problem 2.12 If $\mathbf{H} = \cos(10^8 t - \beta z)\,\mathbf{a}_y$, A/m and $\mathbf{E} = 377 \cos(10^8 t - \beta z)\,\mathbf{a}_x$, v/m in free space, find the frequency, λ, phase constant β, and intrinsic impedance of the medium.

Solution $\omega = 2\pi f = 10^8$

So
$$f = \frac{10^8}{2\pi} = 15.9 \text{ MHz}$$

$$\beta = \frac{2\pi}{\lambda} = \frac{\omega}{v}$$

$$\lambda = \frac{v_0}{f} = \frac{3 \times 10^8}{15.9 \times 10^6}$$

$$= \frac{300}{15.9} = 18.86 \text{ m}$$

Hence,
$$\beta = \frac{2\pi}{\lambda} = 0.3329 \text{ rad/m}$$

$$\eta_0 = \left| \frac{\mathbf{E}}{\mathbf{H}} \right| = 377\,\Omega.$$

Problem 2.13   Find the electric flux density and volume charge density if the electric field, $\mathbf{E} = x^2\,\mathbf{a}_x + 2y^2\,\mathbf{a}_y + z^2\,\mathbf{a}_z$ v/m in a medium whose $\epsilon_r = 2$.

Solution
$$\mathbf{E} = x^2\,\mathbf{a}_x + 2y^2\,\mathbf{a}_y + z^2\,\mathbf{a}_z \text{ v/m}$$

So
$$\mathbf{D} = \epsilon\,\mathbf{E} = \epsilon_0\,\epsilon_r\,\mathbf{E}$$

$$= 8.854 \times 10^{-12} \times 2 \times (x^2\,\mathbf{a}_x + 2y^2\,\mathbf{a}_y + z^2\,\mathbf{a}_z)$$

So
$$\mathbf{D} = 17.708x^2\,\mathbf{a}_x + 35.416y^2\,\mathbf{a}_y + 17.708z^2\,\mathbf{a}_z, \text{ pc/m}^2$$

From Maxwell's equation, we have

$$\nabla \cdot \mathbf{D} = \text{Div}\,(D) = \frac{\partial}{\partial x}\,D_x + \frac{\partial}{\partial y}\,D_y + \frac{\partial}{\partial z}\,D_z = \rho_v$$

So
$$\rho_v = 35.416x + 70.832y + 35.416z, \text{ c/m}^3.$$

Problem 2.14   In free space, if a wave with a frequency of 100 MHz propagates, find the propagation constant.

Solution   Propagation constant, γ in free space is

$$\gamma = j\omega\,\sqrt{\mu_0\,\epsilon_0} = j\,\frac{\omega}{v_0} \qquad \left[\text{As } v_0 = \frac{1}{\sqrt{\mu_0\,\epsilon_0}} \right]$$

$$= j\,\frac{2\pi f}{v_0} = j\,\frac{2\pi \times 100 \times 10^6}{3 \times 10^8}$$

$$= \frac{j\,2\pi \times 10^8}{3 \times 10^8} = j\,2.094$$

So
$$\gamma = j\,2.094 \text{ 1/m.}$$

Problem 2.15   If **H** field is given by $\mathbf{H}\,(z,\ t) = 48\cos\,(10^8\,t + 40z)\,\mathbf{a}_y$, A/m, identify the amplitude, frequency and phase constant. Find the wavelength.

Solution   Amplitude of the magnetic field

$$= 48 \text{ A/m}$$

$$\omega = 10^8$$

or
$$f = \frac{10^8}{2\pi} = 15.915 \text{ MHz}$$

$$\beta = 40 \text{ rad/m}$$

Wave length,
$$\lambda = \frac{2\pi}{\beta} = \frac{2\pi}{40} = 0.157 \text{ m.}$$

Problem 2.16   When the amplitude of the magnetic field in a plane wave is 2 A/m (*a*) deter mine the magnitude of the electric field for the plane wave in free space, (*b*) determine the magnitude of the electric field when the wave propagates in a medium which is characterised by $\sigma = 0$, $\mu = \mu_0$ and $\in\ = 4\in_0$.

Solution   We have

(*a*)
$$\frac{E}{H} = \eta_0 = \sqrt{\frac{\mu_0}{\in_0}} = 120\pi\,\Omega \text{ for free space}$$

But
$$H = 2 \text{ A/m}$$
$$E = \eta H$$
$$= 120\pi \times 2 = 240\pi \text{ V/m}$$

So
$$\boxed{E = 240\pi, \ \text{V/m}}$$

(*b*)
$$\sigma = 0, \ \in_r = 4, \ \mu_r = 1$$

$$\eta = \sqrt{\frac{\mu}{\in}} = \sqrt{\frac{\mu_0}{4\in_0}} = \frac{1}{2}\sqrt{\frac{\mu_0}{\in_0}}$$

$$= \frac{1}{2} \times 120\pi = 60\pi\,\Omega$$

$$E = \eta H = 60\pi \times 2 = 120\pi \text{ V/m}$$

So
$$\boxed{E = 120\pi, \ \text{V/m.}}$$

Problem 2.17   If $\in_r = 9$, $\mu = \mu_0$, for the medium in which a wave with a frequency, $f = 0.3$ GHz is propagating, determine propagation constant and intrinsic impedance of the medium when $\sigma = 0$.

Solution $\quad \sigma = 0$

The expression for propagation constant, γ is

$$\gamma = \sqrt{j\omega\mu\,(\sigma + j\omega\in)} = j\omega\,\sqrt{\mu\in}$$

$$\mu_r = 1, \ \in_r = 9, \ \sigma = 0, \ f = 0.3 \text{ GHz}$$

$$\gamma = j\,2\pi \times 3 \times 10^8 \times \sqrt{4 \times 10^{-7} \times 9 \times 8.854 \times 10^{-12}}$$

$$= j\,6\pi \times 10^8 \times 2 \times 3 \sqrt{8.854 \times 10^{-19}}$$

$$= j\,36\pi \times 10^8 \times 10^{-9} \times 0.9409$$

So
$$\gamma = j\,10.641\ (1/\text{m})$$

Intrinsic impedance,

$$\eta = \sqrt{\frac{j\omega\mu}{\sigma + \omega\in}} = \sqrt{\frac{\mu}{\in}}$$

That is,

$$\eta = \sqrt{\frac{\mu_0}{9\in_0}} = \frac{1}{3}\sqrt{\frac{\mu_0}{\in_0}} = \frac{1}{3} \times 120\pi$$

or
$$\eta = 40\pi\,\Omega.$$

Problem 2.18   The wavelength of x-directed plane wave in a loss-less medium is 0.25 m and the velocity of propagation is 1.5×10^{10} cm/s. The wave has z-directed electric field with an amplitude equal to 10 V/m. Find the frequency and permitivity of the medium. The medium has $\mu = \mu_0$.

Solution
$$v = 1.5 \times 10^{10}\ \text{cm/sec} = 1.5 \times 10^{8}\ \text{m/s}$$

Frequency of the wave

$$f = \frac{v}{\lambda} = \frac{1.5 \times 10^{8}}{25 \times 10^{-2}} = 0.06 \times 10^{10}$$

or
$$f = 600\ \text{MHz}$$

and we have

$$v = \frac{1}{\sqrt{\mu\in}} = \frac{1}{\sqrt{\mu_0 \in_0 \in_r}}$$

$$= \frac{v_0}{\sqrt{\in_r}} = 1.5 \times 10^{8}$$

or
$$\sqrt{\in_r} = \frac{3 \times 10^{8}}{1.5 \times 10^{8}} = 2$$

So
$$\in_r = 1.141.$$

Problem 2.19   Identify frequency and phase constant when the electric field of an EM wave is given by $\mathbf{E} = 5.0 \sin (10^{8}\, t - 4.0x)\ \mathbf{a}_z$. Also find λ.

Solution
$$\mathbf{E} = 5.0 \sin (10^{8}\, t - 4.0x)\ \mathbf{a}_z$$

$$\omega = 10^{8}\ \text{or}\ 2\pi f = 10^{8}$$

So
$$f = \frac{10^{8}}{2\pi} = 15.915\ \text{MHz}$$

$$\beta = 4.0\ \text{rad/m}$$

$$\lambda = \frac{v_0}{f} = \frac{3 \times 10^{8}}{15.915 \times 10^{6}} = 18.850\ \text{m}.$$

Problem 2.20 Earth has a conductivity of $\sigma = 10^{-2}$ v/m, $\epsilon_r = 10$, $\mu_r = 2$. What are the conducting characteristics of the earth at:

(a) $f = 50$ Hz
(b) $f = 1$ kHz
(c) $f = 1$ MHz
(d) $f = 100$ MHz
(e) $f = 10$ GHz?

Solution The parameters of the earth are

$$\sigma = 10^{-2}\,\text{mho/m},\ \epsilon_r = 10,\ \mu_r = 2$$

$$\frac{\sigma}{\omega\epsilon} = \frac{10^{-2}}{2\pi f \times 10 \times \dfrac{1}{36\pi} \times 10^{-9}}$$

$$= \frac{10{-}2 \times 18}{10^{-8} \times f} = \frac{18 \times 10^6}{f}$$

(a) At $\quad f = 50$ Hz

$$\frac{\sigma}{\omega\epsilon} = \frac{18 \times 10^6}{50} = 0.36 \times 10^6 = 3.6 \times 10^5$$

This is $\gg 1$. Hence it behaves like a good conductor.

(b) At $\quad f = 1$ kHz

$$\frac{\sigma}{\omega\epsilon} = \frac{18 \times 10^6}{f} = \frac{18 \times 10^6}{10^3} = 18 \times 10^3$$

This is $\gg 1$. Hence it behaves like a good conductor.

(c) At $\quad f = 1$ MHz

$$\frac{\sigma}{\omega\epsilon} = \frac{18 \times 10^6}{10^6} = 18$$

It behaves like a moderate conductor.

(d) At $\quad f = 100$ MHz

$$\frac{\sigma}{\omega\epsilon} = \frac{18 \times 10^6}{100 \times 10^6} = 0.18$$

The earth behaves like a quasi-dielectric.

(e) At $\quad f = 10$ GHz

$$\frac{\sigma}{\omega\epsilon} = \frac{18 \times 10^6}{10 \times 10^9} = 18 \times 10^{-4}$$

That is, $\quad \dfrac{\sigma}{\omega\epsilon} \ll 1$.

The earth behaves like a good dielectric.

Problem 2.21   A medium like copper conductor which is characterised by the parameters $\sigma = 5.8 \times 10^7$ mho/m, $\in_r = 1$, $\mu_r = 1$ supports a uniform plane wave of frequency 60 Hz. Find the attenuation constant, propagation constant, intrinsic impedance, wavelength and phase velocity of wave.

Solution   Let us obtain the ratio.

$$\frac{\sigma}{\omega \in} = \frac{5.8 \times 10^7}{2\pi \times 60 \times 8.854 \times 10^{-12}} = 173 \times 10^{14}$$

This is $\gg 1$. Therefore, it is a very good conductor.

$$\text{Attenuation constant} = \alpha = \left(\frac{\omega \mu \sigma}{2}\right)^{1/2} = 117.2 \left(\frac{1}{m}\right)$$

$$\text{Phase constant} = \beta = \left(\frac{\omega \mu \sigma}{2}\right)^{1/2} = 117.2 \left(\frac{1}{m}\right)$$

$$\text{Propagation constant} = \gamma = \alpha + j\beta = 117.2 + j117.2 \left(\frac{1}{m}\right)$$

$$\lambda = \text{wavelength} = \frac{2\pi}{\beta} = \frac{2\pi}{117.2} = 0.0536 \text{ m}$$

Intrinsic impedance,

$$\eta = \sqrt{\frac{j\omega\mu}{\sigma}} = (2.022 + j2.022) \times 10^{-6} \ \Omega$$

Phase velocity of wave,
$$\upsilon = \lambda f = 0.0536 \times 60 = 3.216 \text{ m/s}$$

So
$$\alpha = 117.2 \text{ m}^{-1} \qquad \beta = 117.2 \text{ m}^{-1}$$
$$\gamma = 117.2 + j117.2 \text{ m}^{-1} \qquad \lambda = 5.36 \text{ cm}$$
$$\eta = (2.022 + j2.022) \ \mu\Omega \quad \upsilon = 3.216 \text{ m/s}.$$

Problem 2.22   Find the depth of penetration, δ of an EM wave in copper at $f = 60$ Hz and 100 MHz. For copper, $\sigma = 5.8 \times 10^7$ mho/m, $\mu_r = 1$, $\in_r = 1$.

Solution   For copper, at
$$f = 60 \text{ Hz}$$

$$\frac{\sigma}{\omega \in} = \frac{5.8 \times 10^7}{2\pi \times 60 \times 8.854 \times 10^{-12}} = 174 \times 10^{14} \gg 1.$$

Therefore, at $f = 60$ Hz, copper is very good conductor.

The depth of penetration,

$$\delta = \frac{1}{\alpha} = \sqrt{\frac{2}{\omega \mu \sigma}} = \sqrt{\frac{2}{2\pi \times 60 \times 4\pi \times 10^{-7} \times 5.8 \times 10^7}}$$

or
$$\delta = 8.53 \times 10^{-3} \text{ m}$$

At
$$f = 100 \text{ MHz}$$

$$\frac{\sigma}{\omega \in} = \frac{5.8 \times 10^7}{2\pi \times 8.854 \times 10^{-4}} = 0.10425 \times 10^{11} = 10.425 \times 10^9 \gg 1$$

Hence copper is very good conductor at
$$f = 100 \text{ MHz}$$

Now the depth of penetration,

$$\delta = \frac{1}{\alpha} = \sqrt{\frac{2}{\omega \mu \sigma}} = \sqrt{\frac{2}{2\pi \times 100 \times 10^6 \times 4\pi \times 10^{-7} \times 5.8 \times 10^7}}$$

So
$$\delta = 6.608 \times 10^{-6} \text{ m.}$$

Problem 2.23   A conduction current in a copper conductor is 10 Amperes. Find the displacement current in the conductor at 1 MHz, 100 MHz. For copper,

$$\in_r = 1, \mu_r = 1, \sigma = 5.8 \times 10^7 \text{ mho/m.}$$

Solution   We have conduction current density,
$$\mathbf{J}_c = \sigma \mathbf{E}$$

or
$$J_c = \sigma E$$

and displacement current density,

$$\mathbf{J}_d = \frac{\partial \mathbf{D}}{\partial t} = \frac{\in \partial \mathbf{E}}{\partial t}$$

$$J_d = \omega \in E$$

or
$$\frac{J_d}{J_c} = \frac{\omega \in E}{\sigma E} = \frac{\omega \in}{\sigma}$$

Thus
$$J_d = \frac{2\pi f \in}{\sigma} J_c$$

or
$$I_d = \frac{2\pi f \in}{\sigma} I_c$$

At $f = 1$ MHz

$$I_d = \frac{2\pi \times 10^6 \times \in_0 \times 10}{5.8 \times 10^7} = \frac{2\pi \times 8.854 \times 10^{-12}}{5.8}$$

or
$$I_d = 9.591 \text{ PA}$$

At
$$f = 100 \text{ MHz}$$

$$I_d = \frac{2\pi f \in}{\sigma} I_c$$

$$= \frac{2\pi \times 100 \times 10^6 \times 8.854 \times 10^{-12}}{5.8 \times 10^7} \times 10$$

$$= \frac{2\pi \times 10^9 \times 10^{-12} \times 8.854}{5.8 \times 10^7} = 9.586 \times 10^{-10}$$

or $\qquad \boxed{I_d = 959.1 \text{ PA.}}$

Problem 2.24 Verify the cosine of angle between any vectors $\mathbf{E}_1$ and $\mathbf{E}_2$ as the sum of the products of the direction cosines of these vectors.

Solution Let the vectors of $\mathbf{E}_1$ and $\mathbf{E}_2$ be

$$\mathbf{E}_1 = E_{1x}\,\mathbf{a}_x + E_{1y}\,\mathbf{a}_y + E_{1z}\,\mathbf{a}_z$$

and $\qquad \mathbf{E}_2 = E_{2x}\,\mathbf{a}_x + E_{2y}\,\mathbf{a}_y + E_{2z}\,\mathbf{a}_z$

Let $\mathbf{E}_1$ make angles θ_{1x}, θ_{1y} and θ_{1z} with x, y, z-axes and $\mathbf{E}_2$ make angles θ_{2x}, θ_{2y} and θ_{2z} with x, y, z-axes.

Let α be the angle between $\mathbf{E}_1$ and $\mathbf{E}_2$. Then

$$\mathbf{E}_1 \cdot \mathbf{E}_2 = E_{1x}E_{2x} + E_{1y}E_{2y} + E_{1z}E_{2z} = E_1 E_2 \cos\alpha \qquad \text{...(2.189)}$$

But we know

$$\left. \begin{aligned} E_{1x} &= \mathbf{E}_1 \cdot \mathbf{a}_x = E_1 \cos\theta_{1x} \\ E_{1y} &= \mathbf{E}_1 \cdot \mathbf{a}_y = E_1 \cos\theta_{1y} \\ E_{1z} &= \mathbf{E}_1 \cdot \mathbf{a}_z = E_1 \cos\theta_{1z} \end{aligned} \right\} \qquad \text{...(2.190)}$$

and

$$\left. \begin{aligned} E_{2x} &= \mathbf{E}_2 \cdot \mathbf{a}_x = E_2 \cos\theta_{2x} \\ E_{2y} &= \mathbf{E}_2 \cdot \mathbf{a}_y = E_2 \cos\theta_{2y} \\ E_{2z} &= \mathbf{E}_2 \cdot \mathbf{a}_z = E_2 \cos\theta_{2z} \end{aligned} \right\} \qquad \text{...(2.191)}$$

From Equations (2.189)–(2.191), we get

$$E_1 E_2 \cos\alpha = E_1 E_2 (\cos\theta_{1x}\,\cos\theta_{2x} + \cos\theta_{1y}\,\cos\theta_{2y} + \cos\theta_{1z}\,\cos\theta_{2z})$$

$$\therefore \qquad \cos\alpha = (\cos\theta_{1x}\cos\theta_{2x} + \cos\theta_{1y}\cos\theta_{2y} + \cos\theta_{1z}\cos\theta_{2z})$$

Hence proved.

Problem 2.25 A vessel under sea water requires a minimum signal level of 20μ V/m. What is the depth in the sea that can be reached by an aeroplane which transmits 4.0 MHz plane wave? The wave has an electric field intensity of 100 V/m. The propagation is vertical into the sea. For sea water, $\sigma = 4$ mho/m, $\mu_r = 1$, $\epsilon_r = 81$.

Solution At $\qquad f = 4.0 \text{ MHz} = 4.0 \times 10^6 \text{ Hz}$

For sea water $\quad \sigma = 4.0 \text{ } \upsilon/\text{m}$

$$\mu_r = 1$$

$$\epsilon_r = 81$$

So $\qquad \dfrac{\sigma}{\omega\epsilon} = \dfrac{4}{2\pi \times 4 \times 10^6 \times 8.854 \times 10^{-12} \times 81} = \dfrac{10^6}{2253.06} = 221.9$

This is $\qquad \gg 1.$

So sea water is a good conductor.

Intrinsic impedance of sea water,

$$\eta_2 = \sqrt{\frac{\omega\mu}{\sigma}}\ \angle 45^\circ = \sqrt{\frac{2\pi \times 10^6 \times 4 \times 4\pi \times 10^{-7}}{4}}\ \angle 45^\circ = 2.8\ \angle 45^\circ\ \Omega$$

η_1 (free space) $= 377\,\Omega$

α_1 (free space) $= 0$

$$\beta_1\ \text{(free space)} = \frac{\omega}{\upsilon} = \frac{2\pi \times 4 \times 10^6}{3 \times 10^8} = 8.377 \times 10^{-2}\ (\text{m}^{-1})$$

Transmission coefficient

$$= \frac{2\eta_2}{\eta_1 + \eta_2}$$

$$\eta_2 = 2.8\ \angle 45^\circ = 1.9798 + j\,1.9798$$

Transmission coefficient

$$= \frac{3.9596 + j\,3.9596}{377 + 1.9798 + j\,1.9798}$$

$$= \frac{5.6\ \angle 45^\circ}{378.9798 + j\,1.9798} = 14.77 \times 10^{-3}\ \angle 44.70^\circ$$

The transmitted electric field

$$= 14.77 \times 10^{-3} \times 100 = 1.477\ \text{V/m}$$

The propagation takes place in the form of $e^{-\alpha_2 x}$. Therefore, the distance at which the signal becomes $20\mu\text{V/m}$ is calculated

$$1.477\,e^{-\alpha_2 d} = 20 \times 10^{-6}$$

or
$$e^{-\alpha_2 d} = \frac{20 \times 10^{-6}}{1.477} = 13.54 \times 10^{-6}$$

or
$$-\alpha_2\, d = \ln 13.54 \times 10^{-6}$$

$$= -6 \ln 13.54$$

$$= -15.6338$$

So
$$d = \frac{+15.6338}{\alpha_2}$$

But
$$\alpha_2 = \sqrt{\frac{\omega\mu\sigma}{2}}$$

$$= \sqrt{\frac{2\pi \times 4 \times 10^6 \times 4 \times 4\pi \times 10^{-7}}{2}}$$

$$= 8\pi \times \sqrt{10^{-1}}$$

$$= 25 \times 0.316$$

$$= 7.905 \ (\text{m}^{-1})$$

Thus
$$d = \frac{15.6338}{7.905} = 1.977 \text{ m}$$

So
$$\boxed{d = 1.977 \text{ m}}$$

That is, the signal will reach the vessel if it is at a depth of 1.977 m.

Problem 2.26 The electric field of a plane wave propagating in a medium is given by

$$\mathbf{E} = 4.0 e^{-\alpha x} \cos (10^9 \ \pi t - \beta x) \ \mathbf{a}_z \ \text{V/m}$$

The medium is characterised by

$$\epsilon_r = 49, \ \mu_r = 4 \ \text{ and } \ \sigma = 4\upsilon/\text{m}$$

Find the magnetic field of the wave.

Solution Here $\omega = \pi \times 10^9$

The ratio $\dfrac{\sigma}{\omega \epsilon} = \dfrac{36 \times 4}{49} = 2.938$

This is neither $\ll 1$ nor $\gg 1$

So
$$|\eta| = \left| \sqrt{\frac{j\omega\mu}{\sigma + j\omega\epsilon}} \right| = \left| \sqrt{\frac{j\pi \times 10^9 \times 4 \times 4\pi \times 10^{-7}}{4 + j\pi \times 10^9 \times 49 \times \dfrac{1}{36\pi \times 10^9}}} \right|$$

$$= \left| \sqrt{\frac{j16\pi^2 \times 10^2 \times 36}{4 \times 36 + j49}} \right| = \frac{240\pi \ \angle 45^\circ}{\sqrt{144 + j49}}$$

$$= \frac{240\pi \ \angle 45^\circ}{(20736 + 240) \ \angle 18^\circ} = \frac{240\pi}{\sqrt{152.10}} \ \frac{\angle 45^\circ}{\angle 9.39^\circ} = 61.13 \ \angle 35.604^\circ$$

or
$$\boxed{\mathbf{H} = - 0.065 \cos (10^9 \ \pi t - \beta x - 35.604^\circ) \ \mathbf{a}_y.}$$

Problem 2.27 An elliptical polarised wave has an electric field of

$$\mathbf{E} = \sin (\omega t - \beta z) \ \mathbf{a}_x + 2 \sin (\omega t - \beta z + 75^\circ) \ \mathbf{a}_y \ \text{V/m}.$$

Find the power per unit area conveyed by the wave in free space.

Solution
$$E_x = \sin (\omega t - \beta z), \ \text{V/m}$$

$$E_y = 2 \sin (\omega t - \beta z + 75^\circ), \ \text{V/m}$$

According to Poynting theorem, we have

$$\mathbf{P}_{\text{inst}} = \mathbf{E} \times \mathbf{H}$$

So
$$P_{av} = \frac{1}{2} \frac{E^2}{\eta_0} = \frac{1}{2} \frac{(E_x^2 + E_y^2)}{\eta_0}$$

$$= \frac{1}{2}\left(\frac{1+4}{377}\right)$$

or

$$P_{av} = 0.00663 \text{ w/m}^2 = 6.63 \text{ mw/m}^2.$$

Problem 2.28   The magnetic field, **H** of a plane wave has a magnitude of 5 mA/m in a medium defined by $\in_r = 4$, $\mu_r = 1$.

Determine: (*a*) the average power flow, (*b*) the maximum energy density in the plane wave.

Solution   (*a*) We have

$$\frac{E}{H} = \sqrt{\frac{\mu}{\in}} = \sqrt{\frac{4\pi \times 10^{-7}}{4 \times 10^{-9} \times (1/36)}} = 188.4\Omega$$

So

$$E = 188.4H = 188.4 \times 5 \times 10^{-3} = 942.0 \text{ mV/m}$$

$$P_{av} = \frac{E^2}{2 \times 188.5} = \frac{942.0 \times 10^{-3} \times 942 \times 10^{-3}}{377} = 2353.75\mu \text{ w/m}^2$$

(*b*)   The maximum energy density of the wave is

$$W_E = \frac{1}{2} \in E^2 \times 2 = \in E^2 = \in_0 \in_r E^2$$

$$= 4 \times 8.854 \times 10^{-12} \times 942 \times 942 \times 10^{-6}$$

$$= 314.269 \times 10^5 \times 10^{-18}$$

$$= 31.42 \times 10^6 \times 10^{-18}$$

or

$$W_E = 31.42 \text{ PJ/m}^3.$$

Problem 2.29   A plane wave travelling in a medium of $\in_r = 1$, $\mu_r = 1$ has an electric field intensity of $100 \times \sqrt{\pi}$ V/m. Determine the energy density in magnetic field and also the total energy density.

Solution   The electric energy density is given by

$$W_E = \frac{1}{2} \in E^2 = \frac{1}{2} \in_0 \in_r E^2$$

$$= \frac{1}{2} \times 8.854 \times 10^{-12} \times 1 \times 100^2 \times \pi$$

$$= 13.9 \times 10^{-8} = 139 \times 10^{-9}$$

or

$$W_E = 139 \text{ nJ/m}^3$$

As the energy electric density is equal to that of magnetic field for a plane traveling wave,

or

$$W_H = 139 \text{ nJ/m}^3$$

So total energy density

$$W_T = 278 \text{ nJ/m}^3.$$

Problem 2.30   The conductivity of sea water $\sigma = 5\,\mho/m$, $\in_r = 80$. What is the distance an EM wave can be transmitted at 25 kHz and 25 MHz when the range corresponds to 90% of attenuation?

Solution   If the wave is moving in x-direction, we have

$$e^{-\alpha x} = 0.1$$

That is, $\ln(e^{-\alpha x}) = \ln(0.1)$

or $-\alpha x = \ln(0.1)$

$$-\alpha x = -2.30$$

or

$$x = \frac{2.30}{\alpha}$$

$$\alpha = \omega\sqrt{\frac{\mu\in}{2}\left(\sqrt{1 + \frac{\sigma^2}{\omega^2\in^2}} - 1\right)}$$

$$\alpha \text{ at } f = 25 \text{ kHz} = 0.702$$

$$\alpha \text{ at } f = 25 \text{ MHz}$$

$$\alpha = 21.96$$

$$x \text{ at } f = 25 \text{ kHz} = \frac{2.3}{0.702}$$

or

$$\boxed{x = 3.27 \text{ m}}$$

$$x \text{ at } f = 25 \text{ MHz} = \frac{2.3}{21.96} = 0.104$$

or

$$\boxed{x = 0.104 \text{ m.}}$$

Problem 2.31   A perpendicularly polarised wave is incident at angle of $\theta_i = 15°$. It is propagating from medium 1 to medium 2. Medium 1 is defined by $\in_{r_1} = 8.5$, $\mu_{r_1} = 1$, $\sigma_1 = 0$ and medium 2 is free space. If $E_i = 1.0$ mV/m, determine E_r, H_i, H_r.

Solution   The intrinsic impedance of medium 1 is given by

$$\eta_1 = \sqrt{\frac{\mu_1}{\in_1}} = \sqrt{\frac{\mu_0}{\in_0}} \times \sqrt{\frac{\mu_{r_1}}{\in_{r_1}}}$$

$$= 377 \times \sqrt{\frac{1}{8.5}}$$

or

$$\boxed{\eta_1 = 129\,\Omega}$$

As medium 2 is free space,

$$\eta_2 = 377\,\Omega$$

By Snell's law $\dfrac{\sin\theta_i}{\sin\theta_t} = \sqrt{\dfrac{\in_2}{\in_1}} = \sqrt{\dfrac{\in_0}{\in_0\, 8.5}}$

So $\qquad\qquad\qquad \theta_t = 49°$

The reflection coefficient for electric field is

$$\frac{E_r}{E_i} = \frac{\eta_2 \cos\theta_i - \eta_1 \cos\theta_i}{\eta_2 \cos\theta_i + \eta_1 \cos\theta_i} = 0.490$$

$\therefore \qquad\qquad E_r = 0.490 \times 1.0 \times 10^{-3} = 0.490 \text{ mV/m}$

$\therefore \qquad\qquad \boxed{E_r = 0.490 \text{ mV/m}}$

Now $\qquad\qquad \dfrac{E_i}{H_i} = \eta_1$

or $\qquad\qquad H_i = \dfrac{E_i}{\eta_1} = \dfrac{1.0 \times 10^{-3}}{129}$

So $\qquad\qquad \boxed{H_i = 7.75\mu \text{ A/m}}$

Similarly, $\qquad\qquad \dfrac{E_r}{H_r} = \eta_1$

or $\qquad\qquad \boxed{H_r = 3.798\mu \text{ A/m.}}$

Problem 2.32   A plane wave of frequency 2 MHz is incident normally upon a copper conductor. The wave has an electric field amplitude of $E = 2\text{ mV/m}$. Copper has $\mu_r = 1$, $\epsilon_r = 1$ and $\sigma = 5.8 \times 10^7$ mho/m. Find the average power density absorbed by copper.

Solution   Copper is a good conductor.

$$\eta = \sqrt{\frac{\mu\omega}{\sigma}} = \sqrt{\frac{4\pi \times 10^{-7} \times 2\pi \times 2 \times 10^6}{58 \times 10^6}}$$

$$= \frac{4\pi}{\sqrt{5.8}} \times 10^{-4} = \frac{4\pi}{2.40} \times 10^{-4} = 5.235 \times 10^{-4}\ \Omega$$

$$P_{av} = \frac{1}{2}\frac{E^2}{|\eta|} = \frac{\frac{1}{2} \times 4 \times 10^{-6}}{5.235 \times 10^{-4}} = 0.382 \times 10^{-2}\ \text{w/m}^2$$

or $\qquad\qquad \boxed{P_{av} = 3.82 \text{ mw/m}^2.}$

OBJECTIVE QUESTIONS

1. The first Maxwell's equation in free space

 (a) $\nabla \times \mathbf{H} = \dot{\mathbf{D}} + \mathbf{J}$ (b) $\nabla \times \mathbf{H} = \dot{\mathbf{D}}$

 (c) $\nabla \times \mathbf{H} = 0$ (d) $\nabla \times \mathbf{H} = \mathbf{J}$

2. Poynting vector is given by
   (a) $\mathbf{E} \times \mathbf{H}$
   (b) $\mathbf{E} \cdot \mathbf{H}$
   (c) $\mathbf{H} \times \mathbf{E}$
   (d) $\mathbf{H} \cdot \mathbf{E}$

3. Poynting vector gives
   (a) rate of energy flow
   (b) direction of polarisation
   (c) electric field
   (d) magnetic field

4. $\mathbf{E} \cdot \mathbf{H}$ of uniform plane wave is
   (a) EH
   (b) 0
   (c) ηE^2
   (d) None of these

5. Absolute permeability of free space is
   (a) $4\pi \times 10^{-7}\ \text{A/m}$
   (b) $4\pi \times 10^{-7}\ \text{H/m}$
   (c) $4\pi \times 10^{-7}\ \text{F/m}$
   (d) $4\pi \times 10^{-7}\ \text{H/m}^2$

6. For a static magnetic field,
   (a) $\nabla \times \mathbf{B} = \rho$
   (b) $\nabla \times \mathbf{B} = \mu J$
   (c) $\nabla \cdot \mathbf{B} = \mu_0 \mathbf{J}$
   (d) $\nabla \times \mathbf{B} = 0$

7. Electric field in free space
   (a) $\dfrac{\mathbf{D}}{\epsilon_0}$
   (b) $\dfrac{\mathbf{D}}{\mu_0}$
   (c) $\epsilon_0\, \mathbf{D}$
   (d) $\dfrac{\sigma}{\epsilon_0}$

8. For uniform plane wave in the x-direction
   (a) $E_x = 0$
   (b) $H_x = 0$
   (c) $E_x = 0$ and $H_x = 0$
   (d) None of these

9. Displacement current density is
   (a) $\mathbf{D}$
   (b) $\mathbf{J}$
   (c) $\partial \mathbf{D}/\partial t$
   (d) $\partial \mathbf{J}/\partial t$

10. Depth of penetration in free space is
    (a) α
    (b) $1/\alpha$
    (c) 0
    (d) None of these

11. Complex Poynting vector, $\mathbf{P}$ is
    (a) $P = E \times H^*$
    (b) $\mathbf{P} = \mathbf{E} \times \mathbf{H}^*$
    (c) $P = \dfrac{1}{2}\, \mathbf{E} \times \mathbf{H}^*$
    (d) None of these

12. The time-varying electric field is
    (a) $\mathbf{E} = -\nabla V$
    (b) $\mathbf{E} = -\nabla V - \dot{A}$
    (c) $\mathbf{E} = -\nabla V - \mathbf{B}$
    (d) $\mathbf{E} = -\nabla V - \mathbf{D}$

13. Uniform plane wave is
  (*a*) longitudinal in nature
  (*b*) transverse in nature
  (*c*) neither transverse nor longitudinal
  (*d*) vertically directed

14. Polarisation and direction of propagation of EM wave are one and the same. (Yes/No)

15. The direction of propagation of EM wave is obtained from
  (*a*) $\mathbf{E} \times \mathbf{H}$
  (*b*) $\mathbf{E} \cdot \mathbf{H}$
  (*c*) $\mathbf{E}$
  (*d*) $\mathbf{H}$

16. A field can exist if it satisfies
  (*a*) Gauss's law
  (*b*) Faraday's law
  (*c*) Coulomb's law
  (*d*) all Maxwell's equations

17. If $\sigma = 2.0$ mho/m, $E = 10.0$ V/m, the conduction current density is
  (*a*) 5.0 A/m^2
  (*b*) 20.0 A/m^2
  (*c*) 40.0 A/m^2
  (*d*) 20 A

18. Maxwell's equations give the relations between
  (*a*) different fields
  (*b*) different sources
  (*c*) different boundary conditions
  (*d*) None of these

19. The boundary condition on $\mathbf{E}$ is
  (*a*) $\mathbf{a}_n \times (\mathbf{E}_1 - \mathbf{E}_2) = 0$
  (*b*) $\mathbf{a}_n \cdot (\mathbf{E}_1 - \mathbf{E}_2) = 0$
  (*c*) $\mathbf{E}_1 = \mathbf{E}_2$
  (*d*) None of these

20. The boundary conditions on $\mathbf{H}$ is
  (*a*) $\mathbf{a}_n \times (\mathbf{H}_1 - \mathbf{H}_2) = \mathbf{J}_s$
  (*b*) $\mathbf{a}_n \cdot (\mathbf{H}_1 - \mathbf{H}_2) = 0$
  (*c*) $\mathbf{a}_n \times (\mathbf{H}_1 - \mathbf{H}_2) = \mathbf{J}_s$
  (*d*) $\mathbf{a}_n \cdot (\mathbf{H}_1 - \mathbf{H}_2) = 0$

21. The velocity of EM wave is
  (*a*) inversely proportional to β
  (*b*) inversely proportional to α
  (*c*) directly proportional to β
  (*d*) directly proportional to α

22. Velocity of the wave in a good conductor is
  (*a*) very small
  (*b*) very large
  (*c*) moderate
  (*d*) None of these

23. If $E = 2$ V/m, of a wave in free space, (H) is
  (*a*) $\dfrac{1}{60\pi}$ A/m
  (*b*) 60π A/m
  (*c*) 120π A/m
  (*d*) 240π A/m

24. If wet soil has $\sigma = 10^{-2}$ ℧/m, $\epsilon_r = 15$, $\mu_r = 1$, $f = 60$ Hz, it is a
  (*a*) good conductor
  (*b*) good dielectric
  (*c*) semi conductor
  (*d*) None of these

25. If wet soil has $\sigma = 10^{-2}\ \mho/m$, $\epsilon_r = 15$, $\mu_r = 1$ at 10 GHz, it is a
    - (a) good conductor
    - (b) good dielectric
    - (c) semi conductor
    - (d) semi dielectric

26. The cosine of the angle between two vectors is
    - (a) sum of the products of the directions of the two vectors
    - (b) difference of the products of the directions of the two vectors
    - (c) product of the products of the directions of the two vectors
    - (d) None of these

27. The electric field intensity $\mathbf{E}$ at a point (1, 2, 2) due to (1/9) nc located at (0, 0, 0) is
    - (a) 33 V/m
    - (b) 0.333 V/m
    - (c) 0.33 V/m
    - (d) zero

28. If $\mathbf{E}$ is a vector, then $\nabla \cdot \nabla \times \mathbf{E}$ is
    - (a) 0
    - (b) 1
    - (c) does not exist
    - (d) None of these

29. The Maxwell's equation, $\nabla \cdot \mathbf{B} = 0$ is due to
    - (a) $\mathbf{B} = \mu \mathbf{H}$
    - (b) $\mathbf{B} = \dfrac{\mathbf{H}}{\mu}$
    - (c) non-existence of a mono pole
    - (d) None of these

30. Velocity of EM wave in free space is
    - (a) independent of f
    - (b) increases with increase in f
    - (c) decreases with increase in f
    - (d) None of these

31. The direction of propagation of EM wave is given by
    - (a) the direction of $\mathbf{E}$
    - (b) the direction of $\mathbf{H}$
    - (c) the direction of $\mathbf{E} \times \mathbf{H}$
    - (d) the direction of $\mathbf{E} \cdot \mathbf{H}$

32. For uniform plane wave propagating in z-direction
    - (a) $\mathbf{E}_x = 0$
    - (b) $\mathbf{H}_x = 0$
    - (c) $\mathbf{E}_y = 0,\ \mathbf{H}_y = 0$
    - (d) $\mathbf{E}_z = 0,\ \mathbf{H}_z = 0$

33. For free space
    - (a) $\sigma = \infty$
    - (b) $\sigma = 0$
    - (c) $J \neq 0$
    - (d) None of these

34. Velocity of propagation of EM wave is
    - (a) $\sqrt{\dfrac{\epsilon_0}{\mu_0}}$
    - (b) $\dfrac{\mu_0}{\epsilon_0}$
    - (c) $\dfrac{1}{\sqrt{\mu_0 \epsilon_0}}$
    - (d) $\dfrac{\epsilon_0}{\mu_0}$

35. The electric field for time-varying potentials

(a) $\mathbf{E} = -\nabla V$

(b) $\mathbf{E} = -\nabla V - \mathbf{A}$

(c) $\mathbf{E} = \nabla V$

(d) $\mathbf{E} = -\nabla V + \mathbf{A}$

36. The intrinsic impedance of a medium whose $\sigma = 0$, $\in_r = 9$, $\mu_r = 1$ is

(a) $40\pi\,\Omega$

(b) $9\,\Omega$

(c) $120\pi\,\Omega$

(d) $60\pi\,\Omega$

37. For time-varying EM fields

(a) $\nabla \times \mathbf{H} = \mathbf{J}$

(b) $\nabla \times \mathbf{H} = \dot{\mathbf{D}} + \mathbf{J}$

(c) $\nabla \times \mathbf{E} = 0$

(d)   None of these

38. The wave length of a wave with a propagation constant $= 0.1\pi + j\,0.2\pi$ is

(a)   10 m

(b)   20 m

(c)   30 m

(d)   25 m

39. The electric field just above a conductor is always

(a)   normal to the surface

(b)   tangential to source

(c)   zero

(d) ∞

40. The normal components of $\mathbf{D}$ are

(a)   continuous across a dielectric boundary

(b)   discontinuous across a dielectric boundary

(c)   zero

(d) ∞

41. $\mathbf{E} = -\nabla V$ is valid for all types of fields.   (Yes/No)

42. The unit of vector magnetic potential is ______________.

43. $\dfrac{\partial \mathbf{B}}{\partial t}$ is magnetic current density.   (Yes/No)

44. The unit of magnetic current density is ______________.

45. Theta polarisation is called linear polarisation.   (Yes/No)

46. $\mathbf{B}$ and $\mathbf{D}$ can be related.   (Yes/No)

47. The characteristic impedance of a medium is $\sqrt{\dfrac{\in}{\mu}}$.   (Yes/No)

48. The velocity of propagation of uniform plane wave and its phase velocity are identical.

(Yes/No)

49. Brewster angle is the angle of reflection.   (Yes/No)

50. $\mathbf{P} \times \mathbf{H}$ gives average power.   (Yes/No)

51. If $\mathbf{E} = \cos(6 \times 10^7\, t - \beta z)\,\mathbf{a}_x$, β is ______________.

52. $\oint \mathbf{B} \cdot d\mathbf{S} = Q$.   (Yes/No)

53. $\nabla \cdot \mathbf{E} = - \dot{\mathbf{B}}$.   (Yes/No)

54. $\nabla \cdot \mathbf{J} = \dot{\rho}_v$.   (Yes/No)

55. $\mathbf{E} = e^{j2x}\, \mathbf{a}_y$ means that $\mathbf{E} = \sin(\omega t + 2x)\, \mathbf{a}_y$.   (Yes/No)

56. $\mathbf{E}$ and $\mathbf{H}$ in good conductors are in time phase.   (Yes/No)

57. Power density is represented by Poynting vector.   (Yes/No)

58. Complex Poynting vector is $\mathbf{E} \times \mathbf{H}^*$.   (Yes/No)

59. The units of Poynting vector are watts.   (Yes/No)

60. Depth of penetration is nothing but α.   (Yes/No)

61. In free space, $\nabla \cdot \mathbf{D} = 0$.   (Yes/No)

62. In free space, $\nabla \times \mathbf{H} = \mathbf{J}$.   (Yes/No)

63. For static fields, $\nabla \times \mathbf{H} = \mathbf{D}$.   (Yes/No)

64. The unit of $\mathbf{D}$ is wb/m^2.   (Yes/No)

65. The unit of is $\dot{\mathbf{D}}$ is A/m^2.   (Yes/No)

66. The unit of J is A/m.   (Yes/No)

67. β has the unit of radians.   (Yes/No)

68. Unit of α and β are the same.   (Yes/No)

69. Unit of propagation constant in m^{-1}.   (Yes/No)

70. Brewster angle is the same as critical angle.   (Yes/No)

ANSWERS

1. (b)	2. (a)	3. (a)	4. (b)	5. (b)	6. (b)
7. (a)	8. (c)	9. (c)	10. (b)	11. (c)	12. (b)
13. (b)	14. No	15. (a)	16. (d)	17. (b)	18. (a)
19. (a)	20. (b)	21. (a)	22. (a)	23. (a)	24. (a)
25. (b)	26. (a)	27. (c)	28. (a)	29. (c)	30. (a)
31. (c)	32. (d)	33. (b)	34. (c)	35. (a)	36. (a)
37. (b)	38. (a)	39. (a)	40. (a)	41. Yes	42. wb/m
43. Yes	44. v/m	45. Yes	46. Yes	47. No	48. Yes
49. No	50. No	51. 0.2 rad/m	52. No	53. No	54. No
55. Yes	56. No	57. Yes	58. No	59. No	60. No
61. Yes	62. No	63. No	64. No	65. Yes	66. No
67. No	68. No	69. Yes	70. No.		

 **EXERCISE PROBLEMS**

1. If $\mathbf{A}$ is vector magnetic potential, prove $\mathbf{E} = -\dfrac{\partial \mathbf{A}}{\partial t}$.

2. If $\mathbf{E} = 2.0 \sin kx \cos \omega t \; \mathbf{a}_x$ in free space, find the volume charge density.

3. If $\mathbf{E} = 10 \cos(\omega t - kz)\, a_y$ V/m, find $\mathbf{D}, \mathbf{H}, \mathbf{B}$ in free space.

4. If the electric field strength of an EM wave in free space has an amplitude of 5.0 V/m, find the magnetic field strength.

5. When the amplitude of a magnetic field in free space is 10 mA/m, find the magnitude of electric field.

6. The electric field of an EM wave is $\mathbf{E} = 50 \cos \omega \left(t - \dfrac{z}{v_0} \right) \mathbf{a}_y$, find $\mathbf{H}$.

7. Find the depth of penetration for copper at 20 MHz for copper $\sigma = 5.8 \times 10^7$ mho/m, $\mu_r = 1$.

8. What is the velocity of propagation of uniform plane wave in a medium whose $\in_r = 10$, $\mu_r = 3$?

9. The conduction current in a copper wire is 4.0 amps. Find out displacement current in the wire at 1 MHz and 10 MHz. For copper $\in_r = 1$, $\mu_r = 1$, $\sigma = 5.8 \times 10^7$ mho/m.

10. An EM wave in free space is incident normally on a dielectric whose $\in_r = 5.0$. Find the reflection and transmission coefficients.

11. If the electric field in free space is $\mathbf{E} = 2.0 \cos(\omega t - \beta z)\, \mathbf{a}_x$ V/m, find the average power flowing across a square each side of which is 2 m. The square is in z which is a constant plane.

12. If $\widetilde{\mathbf{E}} = 2 \cos(10^8 t - 20x + 40°)\, \mathbf{a}_z$, what is the phasor form of $\mathbf{E}$?

13. Derive the condition under which the electric field $\mathbf{E} = k \cos(3 \times 10^8 t - z)\, \mathbf{a}_y$ exist in a source-free dielectric medium. Here k and β are constants.

14. If the electric and magnetic fields in a medium ($\mu_r = 1$ and $\in_r = 4$) are given by

$$\mathbf{E} = 10 \cos(10^8 t - \beta z)\, \mathbf{a}_x \quad \text{and} \quad \mathbf{H} = \frac{1}{25} \cos(10^8 t - \beta z)\, \mathbf{a}_y$$

find out η, f and β.

chapter *3*

Radiation and Antennas

"An antenna is an electromagnetic radiator, a sensor, a transducer and an impedance matching device with extensive applications in all communication, Radar and in Bio-Medical systems."

CHAPTER OBJECTIVES

This chapter discusses

- ✦ Antenna fundamental parameters

- ✦ Antenna properties, characteristics and functions

- ✦ Radiation mechanism

- ✦ Basic antennas and their radiation characteristics

- ✦ Objective questions and solved problems useful for class tests, final examinations and also for competitive examinations

- ✦ Exercise problems to develop self problem solving skills

3.1 INTRODUCTION

An antenna is basically a transducer. It converts radio frequency (RF) electrical current into an electromagnetic (EM) wave of the same frequency.

It produces electric and magnetic fields, which constitute an electromagnetic field. The transmission and reception of EM energy is obtained by this field. It forms a part of the transmitter as well as the receiver circuits. Its equivalent circuit is characterised by the presence of distributed constants, namely, resistance, inductance and capacitance. The current produces a magnetic field and a charge produces an electrostatic field. These two in turn create an induction field.

When RF signal is applied to an antenna, electric and magnetic fields are produced. They are shown in Fig. 3.1. These two fields constitute the EM wave. As a result, antenna is known as a generator/radiator of EM waves and it is also a sensor of EM waves.

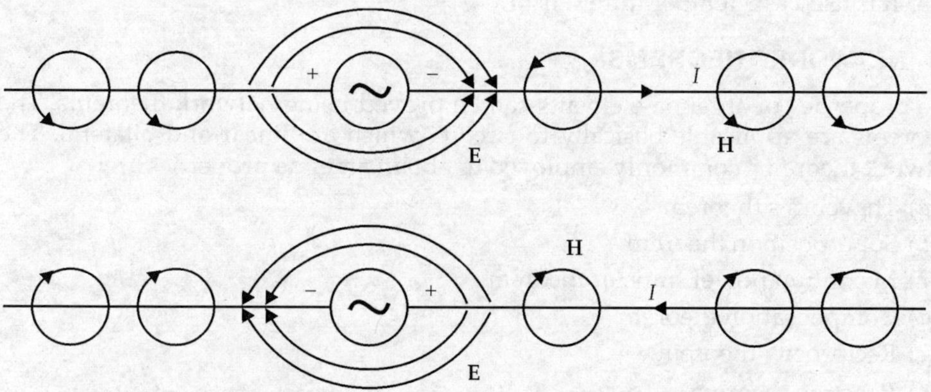

Fig. 3.1 *Electric and magnetic fields produced by RF signals*

The electric and magnetic fields of the EM wave are perpendicular to each other and hence $\mathbf{E} \cdot \mathbf{H} = 0$.

EM waves carry information signals from transmitter to receiver. There is no communication system without one type of antenna or the other. The antennas are characterised by several parameters. These parameters, antenna functions, elementary antennas and their radiation parameters are presented in this chapter.

3.2 DEFINITION OF ANTENNA

Antenna is also known as Aerial:

An antenna can be defined in the following different ways:

1. An antenna may be a piece of conducting material in the form a wire, rod or any other shape with excitation.

2. An antenna is a source or radiator of electromagnetic waves.

3. An antenna is a sensor of electromagnetic waves.

4. An antenna is a transducer.

5. An antenna is an impedance matching device.

6. An antenna is a coupler between a generator/transmission line and space or vice-versa.

3.3 FUNCTIONS OF ANTENNAS

1. It is used as a transducer. That is, it converts electrical energy into EM energy at the transmitting end and it converts EM energy back into electrical energy at the receiving end.

2. It is used as an impedance matching device. That is, it matches/couples the transmitter and free space on the transmitting side and it matches/couples free space and the receiver on the receiving side.

3. It is used to direct radiated energy in desired directions and to suppress it in unwanted directions.

4. It is used to sense the presence of electromagnetic waves.

5. It is used as a temperature sensor.

3.4 NETWORK THEOREMS

The properties of antenna elements can be proved using network theorems. The theorems are applicable basically to circuits which are linear and bilateral. The network theorems commonly employed to obtain antenna properties are:

(*a*) Thevenin's theorem

(*b*) Superposition theorem

(*c*) Maximum power transfer theorem

(*d*) Compensation theorem

(*e*) Reciprocity theorem

(*f*) Reaction theorem

(*a*) **Thevenin's theorem**   In a linear network containing one or more voltage or current sources, the current in Z_L connected between any two terminals is the same when Z_L is connected to a single voltage source. This source is equal to the open circuit voltage across the terminals of interest and its impedance is equal to the impedance of the network looking back from the terminals when all the sources are replaced by their internal impedances.

(*b*) **Superposition theorem**   In a linear and bilateral network containing more than one source, the current at any point is equal to the sum of the currents which flow when each source is considered to be present separately while the remaining sources are replaced by their internal impedances.

(*c*) **Maximum power transfer theorem**   The power transfer to an impedance connected to two terminals of a network is maximum if the impedance is the complex conjugate of the impedance seen looking back into the network from the two terminals.

(*d*) **Compensation theorem**   An impedance in a network is equal to an ideal voltage source if the generated voltage is equal to the instantaneous voltage which exists across the impedance due to flow of current in it.

(*e*) **Reciprocity theorem**   In a linear and bilateral network, if V is the applied voltage across any two terminals, and if I is current in any branch, the ratio of

V and I is the same as the ratio of V and I when the generator and ammeter are interchanged. The $\dfrac{V}{I}$ is called Transfer impedance. According to reciprocity theorem, $Z_{12} = Z_{21}$ or $Y_{12} = Y_{21}$.

(f) **Reaction theorem**

If $\mathbf{J}_1, \mathbf{M}_1$ and $\mathbf{J}_2, \mathbf{M}_2$ are two sets of sources, the **reaction theorem** is defined as

$$\langle 1, 2 \rangle = \int_{\upsilon} (\mathbf{E}_1 . \mathbf{J}_2 - \mathbf{H}_1 . \mathbf{M}_2)\, d\upsilon \text{ or } \langle 2, 1 \rangle = \int_{\upsilon} (\mathbf{E}_2 . \mathbf{J}_1 - \mathbf{H}_2 . \mathbf{M}_1)\, d\upsilon$$

$\langle 1, 2 \rangle$ represents the reaction of fields $(\mathbf{E}_1, \mathbf{H}_1)$ to sources $(\mathbf{J}_2, \mathbf{M}_2)$.
$\langle 2, 1 \rangle$ represents the reaction of fields $(\mathbf{E}_2, \mathbf{H}_2)$ to sources $(\mathbf{J}_1, \mathbf{M}_1)$.
By Reciprocity $\langle 1, 2 \rangle = \langle 2, 1 \rangle$
$\langle 1, 1 \rangle$ or $\langle 2, 2 \rangle$ are known as self-reaction. That is, they represent the reaction of the fields on their own sources.

3.5 PROPERTIES OF ANTENNA

1. It has identical impedance when used for transmitting and receiving purposes. This property is called equality of impedances.

2. It has identical directional characteristics/patterns when it is used for transmitting and receiving purposes. This property is called equality of directional patterns.

3. It has the same effective length when it is used for transmitting and receiving purposes. This property is called equality of effective lengths.

   These properties can be proved using Reciprocity theorem.

3.6 ANTENNA PARAMETERS

1. **Antenna Impedance, Z_a**   It is defined as the ratio of input voltage to input current or

$$Z_a \equiv \frac{V_i}{I_i}\, \Omega$$

Z_a is a complex quantity and it is written as

$$Z_a = R_a + j X_a$$

Here, the reactive part X_a results from fields surrounding the antenna. The resistive part, R_a is given by

$$R_a = R_l + R_r$$

R_l represents losses in the antenna. R_r is called radiation resistance.

2. **Radiation Resistance, R_r**   is defined as a fictitious or hypothetical resistance that would dissipate an amount of power equal to the radiated power.

or
$$R_r \equiv \frac{\text{power radiated}}{I_{\text{rms}}^2}$$

3. **Directional Characteristics**   These are also called radiation characteristics or radiation pattern. These are of two types:

(*a*) **Field strength pattern**   It is the variation of the absolute value of field strength as a function of θ.

$E\ Vs\ \theta$ is called Field strength pattern.

(*b*) **Power pattern**   It is the variation of radiated power with θ.

$P\ Vs\ \theta$ is called Power pattern.

An antenna radiation pattern is a three dimensional variation of the radiation field. It is a pattern drawn as a function of θ and ϕ. The pattern consists of one main lobe and a number of minor/side lobes.

Typical 3-D radiation patterns are shown in Fig. 3.2.

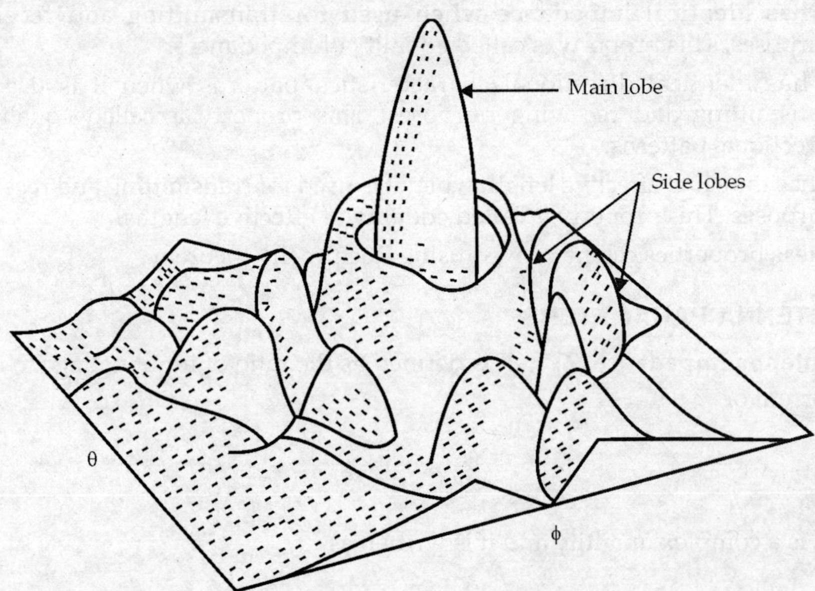

Fig. 3.2 (*a*)

4. **Effective Length of Antenna (L_{eff})**   It is used to indicate the effectiveness of the antenna as a radiator or receiver of EM energy.

L_{eff} **of Transmitting Antenna**   It is equal to the length of an equivalent linear antenna which radiates the same field strength as the actual antenna and the current is constant throughout the length of the linear antenna.

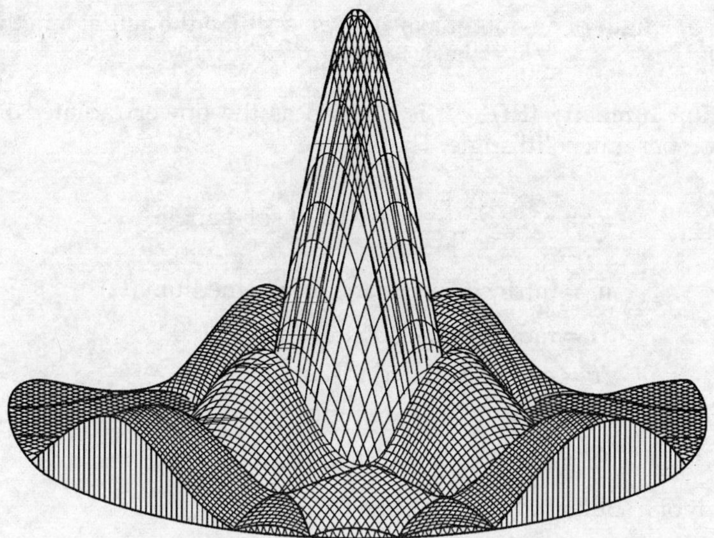

Fig. 3.2 (*b*) *Typical 3-D radiation pattern*

Refer the Fig. 3.3.

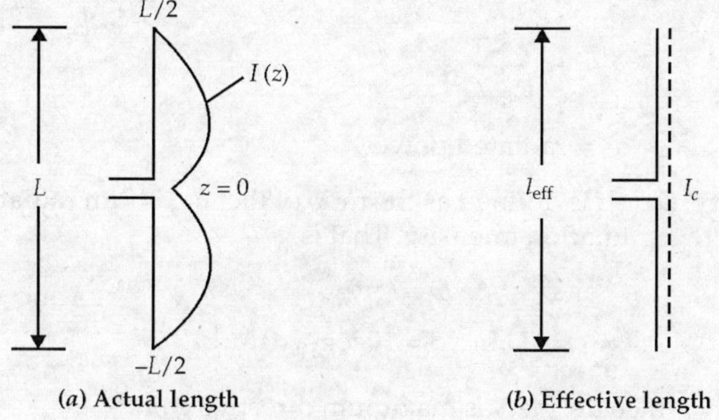

Fig. 3.3 *Definition of effective length of transmitting antenna*

L_{eff} of transmitting antenna is defined mathematically as

$$L_{\text{eff}} (Tx) = \frac{1}{I_c} \int_{-L/2}^{L/2} I(z)\, dz \ (\text{m})$$

L_{eff} of receiving antenna   It is defined as the ratio of the open circuit voltage developed at the terminals of the antenna under the received field strength, E. That is

$$L_{\text{eff}} (Rx) \equiv \frac{V_{OC}}{E} \ (\text{m})$$

Effective length of an antenna is always less than the actual length. That is, $L_{eff} < L$.

5. **Radiation Intensity (RI)**   It is defined as the power radiated in a given direction per unit solid angle. That is,

$$RI = r^2 P = \frac{r^2 E^2}{\eta_0} \text{ watts/unit solid angle}$$

Here η_0 = intrinsic impedance of the medium (Ω)

r = radius of the sphere, (m)

P = power radiated-instantaneous

E = electric field strength, (V/m)

$RI = RI\,(\theta,\ \phi)$ is a function of θ and ϕ

The unit of a solid angle is Steradian (sr).

6. **Directive Gain g_d**   It is defined as the ratio of intensity of radiation in a specified direction to the average radiation intensity. That is,

$$g_d \equiv \frac{RI}{RI_{av}} = \frac{RI}{w_r/4\pi}$$

or $$g_d = \frac{4\pi\,(RI)}{w_r}$$

w_r = radiated power.

7. **Directivity, D**   It is defined as the ratio of the maximum radiation intensity to the average radiation intensity. That is,

$$D \equiv (g_d)_{max}$$

$$D \text{ in dB} = 10 \log_{10} (g_d)_{max}$$

Directivity is also defined as maximum directive gain.

8. **Power Gain, g_p**   It is defined as the ratio of 4π times radiation intensity to the total input power. That is,

$$g_P \equiv \frac{4\pi\,(RI)}{w_t}$$

where $w_t = w_r + w_l,$

w_l = ohmic losses in the antenna.

9. **Antenna Efficiency (η)**   It is defined as the ratio of the radiated power to the input power. That is,

$$\eta \equiv \frac{w_r}{w_t} = \frac{w_r}{w_r + w_l} = \frac{g_P}{g_d}$$

10. **Effective Area**   It is defined as

$$A_e = \frac{\lambda^2}{4\pi}\, g_d \; (\text{m}^2)$$

or $$A_e \equiv \frac{w_R}{P} \; (\text{m}^2)$$

where w_R = received power (watt)

P = power flow per square meter (watts/m^2) for the incident wave.

11. **Antenna Equivalent Circuit**   It is a series R_a, L_a and C_a circuit (Fig. 3.4).

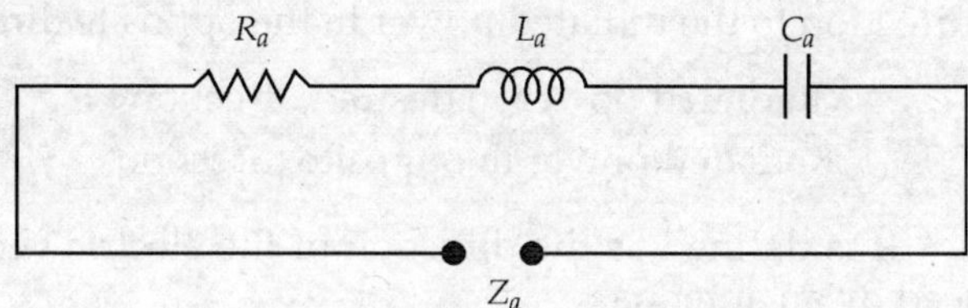

Fig. 3.4 *Antenna equivalent circuit*

The main difference between the antenna equivalent circuit and an RLC circuit is that R_a, L_a and C_a vary with frequency. As a result, the antenna conductance peak appears not at resonant frequency but at a frequency slightly away from f_r (Fig. 3.5).

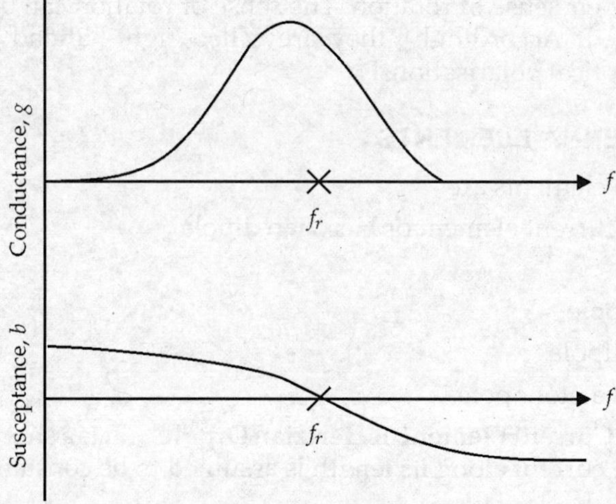

Fig. 3.5 *Typical variation of antenna conductance (g) and susceptance (b)*

The antenna impedance, $Z_a = R + j\,(X_L - X_C)$

where $$X_L = \omega L, \; X_C = \frac{1}{\omega C}$$

The corresponding admittance,

$$Y_a = \frac{1}{Z_a} = g + jb$$

Here $\quad\quad g$ = conductance

$\quad\quad\quad\quad b$ = susceptance.

12. **Antenna Bandwidth** It is defined as the range of frequencies over which the antenna maintains its characteristics and parameters, like gain, front-to-back ratio, standing wave ratio, radiation pattern, polarisation, impedance, directivity and so on, without considerable change.

13. **Front-to-Back ratio (FBR)** FBR is defined as the ratio of radiated power in the desired direction to the radiated power in the opposite direction.

That is, $\quad$ FBR $= \dfrac{\text{Radiated power in desired direction}}{\text{Radiated power in opposite direction}}$.

14. **Polarisation** It is defined as the direction of the electric vector of the EM wave produced by an antenna.

It is of three types:

 (*i*) Linear polarisation

(*ii*) Circular polarisation

(*iii*) Elliptical polarisation

Linear polarisation is again of three types, namely Horizontal, Vertical and Theta polarisations. Circular and elliptical polarisations can also be described in terms of their sense of rotation. The sense of rotation can be right-handed or left-handed. Accordingly, they are called right-handed or left-handed circular/elliptical polarisations.

3.7 BASIC ANTENNA ELEMENTS

The basic antenna elements are

1. Alternating current element or Hertzian dipole
2. Short dipole
3. Short monopole
4. Half-wave dipole
5. Quarter-wave monopole

1. **Alternating Current Element or Hertzian Dipole** It is a short linear antenna in which the current along its length is assumed to be constant.

2. **Short Dipole** It is a linear antenna whose length is less than $\dfrac{\lambda}{4}$ and the current distribution is assumed to be triangular.

3. **Short Monopole** It is a linear antenna whose length is less than $\dfrac{\lambda}{8}$ and the current distribution is assumed to be triangular.

4. **Half-wave Dipole**   It is a linear antenna whose length is $\dfrac{\lambda}{2}$ and the current distribution is assumed to be sinusoidal. It is usually centre-fed.

5. **Quarter-wave Monopole**   It is a linear antenna whose length is $\dfrac{\lambda}{4}$ and the current distribution is assumed to be sinusoidal. It is fed at one end with respect to earth.

3.8 RADIATION MECHANISM

When a transmitting antenna is excited with an alternating voltage, the initial motion is started by the balanced motion of charges in the antenna. Resonant oscillations are produced by the supplied energy. Electric and magnetic fields are generated due to sudden changes in charge. When the charges around the antenna are set in motion first, the other charges are separated from the antenna and they are also set in motion. The disturbance is spread from the antenna into space. The electric and magnetic fields so produced are perpendicular to each other. EM waves have no boundaries. The EM energy decreases as it propagates.

3.9 RADIATION FIELDS OF ALTERNATING CURRENT ELEMENT (OR OSCILLATING ELECTRIC DIPOLE)

The concept of an alternating current element, $I\,dl \cos \omega t$ is of theoretical interest. But the theory developed for this can be extended to practical antennas. The concept of retarded vector magnetic potential $\mathbf{A}$ is very useful to derive radiation fields of antenna elements including current element.

Derivation of radiation fields consists of the following steps:

1. Write expression for retarded vector magnetic potential.

2. Write expressions for the components of $\mathbf{A}$ in Cartesian coordinates.

3. Express $\mathbf{A}$ in components of spherical coordinate system.

4. Obtain the components of $\mathbf{H}$ from $\mu \mathbf{H} = \nabla \times \mathbf{A}$.

5. Obtain the components of $\mathbf{E}$ from

$$\dot{\mathbf{E}} = \frac{\partial \mathbf{E}}{dt} = \frac{1}{\in} \nabla \times \mathbf{H} \qquad \text{(as } \mathbf{J} = 0 \text{ for free space)}$$

6. $\mathbf{E} = \dfrac{1}{\in} \displaystyle\int (\nabla \times \mathbf{H})\, dt.$

7. Identify radiation and near-field terms.

An alternating current element at the origin of a spherical coordinate system is shown in Fig. 3.6.

The retarded vector magnetic potential, $\mathbf{A}\,(r,\ t)$ is given by

$$\mathbf{A}\,(r,\ t) = \frac{\mu}{4\pi} \int_{\upsilon} \frac{\mathbf{J}\,(r,\ t - r/\upsilon_0)}{r}\, d\upsilon \qquad \qquad \dots(3.1)$$

As the element is z-directed, $\mathbf{A}$ is also z-directed. $\dfrac{r}{\upsilon_0}$ is the delay time.

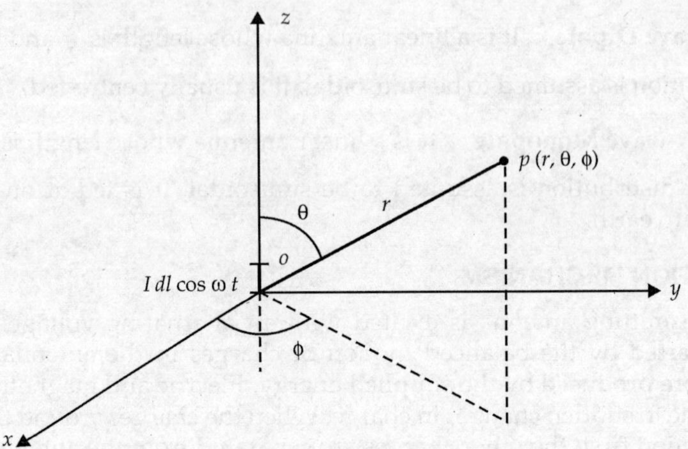

Fig. 3.6 *Alternating current element at the origin of a spherical coordinate system*

The volume integral in Equation (3.1) can be simplified by taking integration over the cross-sectional area of the element and an integration along its length. We know,

$$\oint_S \mathbf{J} \cdot d\mathbf{S} = I$$

and

$$\int_0^{dl} I \, dL = I \, dl$$

$$\mathbf{A} = A_z \, \mathbf{a}_z = \frac{\mu_0}{4\pi} \frac{I \, dl \cos \omega \, (t - r/v_0)}{r} \, \mathbf{a}_z \qquad \ldots(3.2)$$

This means **A** has only z-component and $A_x = 0$ and $A_y = 0$.

Changing Cartesian components to spherical coordinate components, we get

$$\left. \begin{aligned} A_r &= A_z \cos \theta \\ A_\theta &= - A_z \sin \theta \\ A_\phi &= 0 \end{aligned} \right\} \qquad \ldots(3.3)$$

But we know
$$\mathbf{H} = \frac{1}{\mu_0} \nabla \times \mathbf{A} \qquad \ldots(3.4)$$

So
$$H_r = \frac{1}{\mu_0} (\nabla \times \mathbf{A})_r$$

$$H_\theta = \frac{1}{\mu_0} (\nabla \times \mathbf{A})_\theta$$

and
$$H_\phi = \frac{1}{\mu_0} (\nabla \times \mathbf{A})_\phi$$

$$H_r = \frac{1}{\mu_0\, r \sin\theta} \left[\frac{\partial}{\partial\theta} (\sin\theta\, A_\phi) - \frac{\partial A_\theta}{\partial\phi} \right]$$

As $\qquad A_\phi = 0$ and $A_\theta \neq f(\phi)$, $H_r = 0$

Similarly, $\qquad H_\theta = \frac{1}{\mu_0} \left[\frac{1}{r \sin\theta} \frac{\partial A_r}{\partial\phi} - \frac{1}{r} \frac{\partial (rA_\phi)}{\partial r} \right] = 0 \qquad\qquad$ [as $A_r \neq f(\phi)$]

$$H_\phi = \frac{1}{\mu_0\, r} \left[\frac{\partial}{\partial r} (rA_\theta) - \frac{\partial A_r}{\partial r} \right] \qquad\qquad \text{...(3.5)}$$

Here $\qquad A_\theta = - A_z \sin\theta$

$$= - \frac{\mu_0}{4\pi} \frac{I\, dl \cos\omega\,(t - r/\upsilon_0)}{r} \sin\theta \qquad\qquad \text{...(3.6)}$$

$$A_r = A_z \cos\theta$$

$$= - \frac{\mu_0}{4\pi} \frac{I\, dl \cos\omega\,(t - r/\upsilon_0)}{r} \cos\theta \qquad\qquad \text{...(3.7)}$$

Substituting Equations (3.6) and (3.7) in Equation (3.5) and simplifying, we get

$$H_\phi = \frac{I\, dl \sin\theta}{4\pi} \left[\frac{-\omega \sin\omega\left(t - \dfrac{r}{\upsilon_0}\right)}{r\upsilon_0} - \frac{\cos\omega\left(t - \dfrac{r}{\upsilon_0}\right)}{r^2} \right] \qquad \text{...(3.8)}$$

From Maxwell's first equation, we have

$$\nabla \times \mathbf{H} = \dot{\mathbf{D}} = \epsilon_0\, \dot{\mathbf{E}}$$

or $\qquad\qquad \mathbf{E} = \dfrac{1}{\epsilon_0} \int (\nabla \times \mathbf{H})\, dt \qquad\qquad\qquad \text{...(3.9)}$

From Equations (3.6) and (3.9), we get E_r, E_θ and E_ϕ

But $\qquad\qquad E_\phi = 0 \qquad\qquad\qquad\qquad$ [as $\mathbf{H} = H_\phi\, \mathbf{a}_\phi$]

On simplification of Equation (3.9), we get

$$E_\theta = \frac{I\, dl \sin\theta}{4\pi\, \epsilon_0} \left[-\frac{\omega \sin\omega\, t_d}{r\upsilon_0^2} + \frac{\cos\omega\, t_d}{r^2\, \upsilon_0} + \frac{\sin\omega\, t_d}{\omega r^3} \right] \qquad \text{...(3.10)}$$

and $\qquad E_r = \dfrac{2I\, dl \cos\theta}{4\pi\epsilon_0} \left[\dfrac{\cos\omega t_d}{r^2\, \upsilon_0} + \dfrac{\sin\omega t_d}{\omega r^3} \right] \qquad\qquad \text{...(3.11)}$

where $\qquad t_d = (t - r/\upsilon_0) \qquad\qquad\qquad\qquad\qquad \text{...(3.12)}$

The **resultant field components** of an alternating current element are

$$H_\phi = \frac{I\,dl\,\sin\theta}{4\pi}\left[-\frac{\omega\sin\omega t_d}{r\upsilon_0} + \frac{\cos\omega t_d}{r^2}\right]$$

$$E_\theta = \frac{I\,dl\,\sin\theta}{4\pi\in_0}\left[-\frac{\omega\sin\omega t_d}{r\upsilon_0^2} + \frac{\cos\omega t_d}{r^2\upsilon_0} + \frac{\sin\omega t_d}{\omega r^3}\right] \qquad ...(3.13)$$

$$E_r = \frac{2I\,dl}{4\pi\in_0}\cos\theta\left[\frac{\cos\omega t_d}{r^2\upsilon_0} + \frac{\sin\omega t_d}{\omega r^3}\right]$$

$$H_r = 0, \quad H_\theta = 0, \quad E_\phi = 0.$$

3.10 RADIATED POWER AND RADIATION RESISTANCE OF CURRENT ELEMENT

The derivation of expression for radiated power consists of the following steps:

- Obtain field components as in the Section (3.8).
- Obtain expression for radiated power using Poynting vector.
- Obtain average radiated power.
- Obtain total power radiated from $P_T = \oint_S P_{av}\,ds.$

- Identify expression for radiation resistance.

Poynting vector is

$$\mathbf{P} = \mathbf{E} \times \mathbf{H} \text{ watts/m}^2$$

That is, $P_\theta = -E_r H_\phi$ $\qquad\qquad ...(3.14)$

From Equations (3.13) and (3.14), we have

$$P_\theta = -\left[\frac{2I\,dl\,\cos\theta}{4\pi\in_0}\left(\frac{\cos\omega t_d}{r^2\upsilon_0} + \frac{\sin\omega t_d}{\omega r^3}\right)\right] \times$$

$$\left[\frac{I\,dl\,\sin\theta}{4\pi}\left(-\frac{\omega\sin\omega t_d}{r\upsilon_0} + \frac{\cos\omega t_d}{r^2}\right)\right] \quad ...(3.15)$$

$$= \frac{2I^2\,dl^2\cos\theta\,\sin\theta}{16\pi^2\in_0}\left[\frac{\sin^2\omega t_d}{r^4\upsilon_0} - \frac{\cos^2\omega t_d}{r^4\upsilon_0} - \right.$$

$$\left. \frac{\sin\omega t_d\cos\omega t_d}{\omega r^5} + \frac{\omega\sin\omega t_d\cos\omega t_d}{r^3\upsilon_0}\right] \quad ...(3.16)$$

$$2 \sin \theta \, \cos \theta = \sin 2\theta$$

But
$$\sin \omega t_d \, \cos \omega t_d = \frac{\sin 2\omega t_d}{2} \qquad \qquad \text{...(3.17)}$$

$$\sin^2 \omega t_d - \cos^2 \omega t_d = \cos 2\omega t_d$$

Using these identities, Equation (3.16) becomes

$$P_\theta = \frac{I^2 \, dl^2 \sin 2\theta}{16\pi^2 \in_0} \left[-\frac{\cos 2\omega t_d}{r^4 \, v_0} - \frac{\sin 2\omega t_d}{2\omega r^5} + \frac{\omega \sin 2\omega t_d}{2r^3 \, v_0^2} \right] \qquad \text{...(3.18)}$$

P_θ in Equation (3.18) represents instantaneous power flow in θ-direction. But the average value of $\cos 2\omega t_d$ or $\sin 2\omega t_d$ over a cycle is zero. Hence, $(P_\theta)_{av} = 0$ at any value of r. This means that power in θ-direction surges back and forth.

Similarly, $P_r = E_\theta H_\phi$ $\qquad \qquad$ ...(3.19)

From Equations (3.13) and (3.18), we get

$$P_r = \frac{I^2 \, dl^2 \sin^2 \theta}{16\pi^2 \in_0} \left[\frac{\sin \omega t_d \cos \omega t_d}{\omega r^5} + \frac{\cos^2 \omega t_d}{r^4 \, v_0} - \frac{\omega \sin \omega t_d \cos \omega t_d}{r^3 \, v_0^2} \right.$$

$$\left. - \frac{\sin^2 \omega t_d}{r^4 \, v_0} - \frac{\omega \sin \omega t_d \cos \omega t_d}{r^3 \, v_0^2} + \frac{\omega^2 \sin^2 \omega t_d}{r^2 \, v_0^3} \right] \qquad \text{...(3.20)}$$

Using identities of Equation (3.17), we get

$$P_r = \frac{I^2 \, dl^2 \sin^2 \theta}{16\pi^2 \in_0} \left[\frac{\sin 2\omega t_d}{2\omega r^5} + \frac{\cos 2\omega t_d}{r^4 \, v_0} - \frac{\omega \sin 2\omega t_d}{r^3 \, v_0^2} + \frac{\omega^2 (1 - \cos 2\omega t_d)}{2r^2 \, v_0^3} \right] \qquad \text{...(3.21)}$$

It is obvious from Equation (3.21), that the average value of P_r is

$$P_{r \, (av)} = \frac{\omega^2 \, I^2 \, dl^2 \sin^2 \theta}{32\pi^2 \, r^2 \in_0 v_0^3}$$

or
$$P_{r \, (av)} = \frac{\eta_0}{2} \left(\frac{\omega I \, dl \sin \theta}{4\pi r \, v_0} \right)^2 \text{watt/m}^2 \qquad \qquad \text{...(3.22)}$$

where
$$\eta_0 = \sqrt{\frac{\mu_0}{\in_0}} = 120\pi\,\Omega$$

The total power radiated

$$P_T = \oint_{\text{surface}} P_{r \, (av)} \, ds \qquad \qquad \text{...(3.23)}$$

$P_{r\,(av)}$ is independent of ϕ and hence the element of area ds on the spherical shell is given by

$$ds = 2\pi r^2 \sin \theta \; d\theta \qquad\qquad ...(3.24)$$

Here, θ varies between 0 and π.

Now Equation (3.24) becomes

$$P_T = \int_0^\pi \frac{\eta_0}{2} \left[\frac{\omega I\, dl \sin \theta}{4\pi r v_0} \right]^2 2\pi r^2 \sin \theta \; d\theta$$

$$= \frac{\eta_0 \, \omega^2 \, I^2 \, dl^2}{16\pi v_0^2} \int_0^\pi \sin^3 \theta \; d\theta$$

But

$$\int_0^\pi \sin^3 \theta \; d\theta = \left(-\cos \theta + \frac{1}{3} \cos^3 \theta \right)_0^\pi = \frac{4}{3}$$

So

$$P_T = \frac{\eta_0 \, \omega^2 \, I^2 \, dl^2}{12\pi v_0^2} \qquad\qquad ...(3.25)$$

Here I is the peak value of current.

As

$$I = \sqrt{2}\, I_{\text{eff}}$$

$$I^2 = 2I_{\text{eff}}^2$$

Thus Equation (3.25) becomes

$$P_T = \frac{\eta_0 \, \omega^2 \, dl^2 \, I_{\text{eff}}^2}{6\pi v_0^2}$$

or

$$P_T = 80\pi^2 \left(\frac{dl}{\lambda} \right)^2 I_{\text{eff}}^2 \text{ watts} \qquad\qquad ...(3.26)$$

This is in the form of $P = I^2 R$. Hence the coefficient of I_{eff}^2 has the dimensions of resistance and it is called Radiation Resistance.

Radiation Resistance of Hertzian dipole

$$R_r = 80\pi^2 \left(\frac{dl}{\lambda} \right)^2 \Omega. \qquad\qquad ...(3.27)$$

3.11 RADIATION, INDUCTION AND ELECTROSTATIC FIELDS

The field components of current elements from Equation (3.13) are

$$H_\phi = \frac{I\, dl \sin \theta}{4\pi} \left[-\frac{\omega \sin \omega t_d}{r v_0} + \frac{\cos \omega t_d}{r^2} \right]$$

$$E_\theta = \frac{I\,dl\,\sin\theta}{4\pi\epsilon_0}\left[-\frac{\omega\sin\omega t_d}{rv_0^2} + \frac{\cos\omega t_d}{r^2\,v_0} + \frac{\sin\omega t_d}{\omega r^3}\right] \qquad ...(3.28)$$

$$E_r = \frac{2I\,dl}{4\pi\epsilon_0}\left[\frac{\cos\omega t_d}{r^2 v_0} + \frac{\sin\omega t_d}{\omega r^3}\right]$$

H_ϕ field consists of $\dfrac{1}{r}$ and $\dfrac{1}{r^2}$ terms. $\dfrac{1}{r^2}$ term dominates over $\dfrac{1}{r}$ term at points close to the current element. When r is small, $\dfrac{1}{r^2}$ term is called Induction Field.

On the other hand, $\dfrac{1}{r}$ term dominates over $\dfrac{1}{r^2}$ terms when r is large. Here $\dfrac{1}{r}$ term is called Radiation Field or distant field or far-field.

The expression for E_θ consists of three terms: $\dfrac{1}{r}, \dfrac{1}{r^2}, \dfrac{1}{r^3}$, and the expression for E_r consists of $\dfrac{1}{r^2}$ and $\dfrac{1}{r^3}$ terms. These $\dfrac{1}{r^3}$ term is called Electrostatic Field.

$\dfrac{1}{r}$ term in E and H fields is called Radiation Field

$\dfrac{1}{r^2}$ term is called Induction Field

$\dfrac{1}{r^3}$ term is called Electrostatic Field.

If the induction and radiation fields have equal amplitudes, then from Equation (3.28), we have

$$\frac{I\,dl\,\omega\sin\theta}{4\pi r v_0} = \frac{I\,dl\,\sin\theta}{4\pi r^2}$$

or
$$\frac{\omega}{r v_0} = \frac{1}{r^2} \quad \text{or} \quad r = \frac{v_0}{\omega} = \frac{\lambda}{2\pi} \approx \frac{\lambda}{6.0} \qquad ...(3.29)$$

At a distance of $r = \dfrac{\lambda}{2\pi}$, induction and radiation fields have equal amplitudes.

3.12 HERTZIAN DIPOLE

Hertzian dipole is an infinitesimal current element Idl which does not exist in real life.

(or)

Hertzian dipole is a short linear antenna which, when radiating, is assumed to carry constant current along its length.

As Hertzian dipole and alternating current elements virtually mean the same, the radiated power and radiation resistance are given by

$$P_T = 80\pi^2 \left(\frac{dl}{\lambda}\right)^2 I_{\text{eff}}^2 \text{ watts}$$

$$R_r = 80\pi^2 \left(\frac{dl}{\lambda}\right)^2 \Omega. \qquad \qquad ...(3.30)$$

3.13 DIFFERENT CURRENT DISTRIBUTIONS IN LINEAR ANTENNAS

The possible current distributions are:

1. Constant current along its length—valid in Hertzian dipole.
2. Triangular current distribution—valid in short dipole and monopole.

For a triangular current distributions

$$R_r \left(\text{short dipole, } l < \frac{\lambda}{4}\right) \approx 20\pi^2 \left(\frac{l}{\lambda}\right)^2 \Omega$$

$$R_r \left(\text{short monopole, } l < \frac{\lambda}{8}\right) \approx 10\pi^2 \left(\frac{l}{\lambda}\right)^2 \Omega \qquad \qquad ...(3.31)$$

3. Sinusoidal current distribution—valid in half-wave dipole.
4. Exact current distribution—This can be determined using the method of moment technique. However, this method is beyond the scope of this book.

3.14 RADIATION FROM HALF-WAVE DIPOLE

Radiated power by half-wave dipole, $P_T = 73.0 I_{\text{eff}}^2$

Radiation resistance of half-wave dipole, $R_r = 73\,\Omega$.

Proof Proof consists of the following steps:

- Write expressions for the assumed current distribution in the element.
- Obtain expression for vector magnetic potential, **A**.
- Obtain **H** from **A**.
- Obtain **E** from $\left(\dfrac{E}{H}\right) = \eta_0$.
- Obtain average radiated power P_{av}.
- Obtain total power radiated.
- Obtain the value of radiation resistance.

The sinusoidal current distribution is represented by Fig. 3.7.

$$I = I_m \sin \beta (H - Z) \text{ for } z > 0$$

$$= I_m \sin \beta (H + Z) \text{ for } z < 0$$

Here I_m = current maximum.

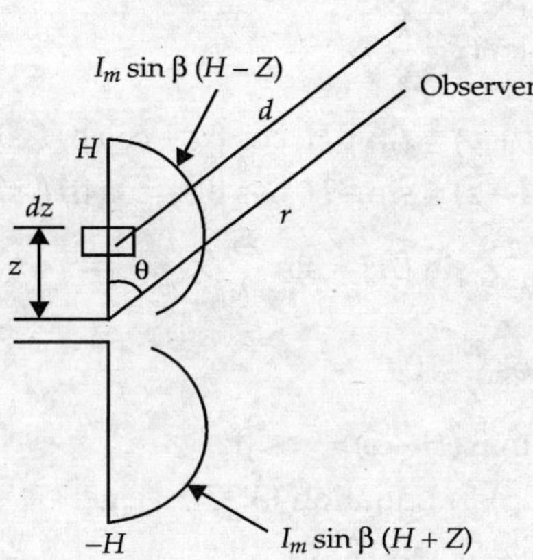

Fig. 3.7 *Dipole*

The vector potential at a point P due to the current element $I\,dz$ is given by,

$$d\mathbf{A} = dA_z\,\mathbf{a}_z = \frac{\mu_0\,Ie^{-j\beta d}\,dz}{4\pi d}\,\mathbf{a}_z \qquad \qquad ...(3.32)$$

Here d is the distance from the current element to the point P. The total vector potential at P due to all current elements is given by

$$A_z = \frac{\mu_0}{4\pi}\int_{-H}^{H}\frac{Ie^{-j\beta d}}{d}\,dz \qquad \qquad ...(3.33)$$

$$= \frac{\mu_0}{4\pi}\int_{-H}^{0}\frac{I_m \sin \beta (H+z)}{d}\,e^{-j\beta d}\,dz + \frac{\mu}{4\pi}\int_{0}^{H}\frac{I_m \sin \beta (H-z)}{d}\,e^{-j\beta d}\,dz$$

$$...(3.34)$$

It is of interest here to consider radiation fields. d in the denominator can be approximated to r. But in the numerator, d is in the phase term and it is given by

$$d = r - z \cos \theta$$

Now Equation (3.34) becomes

$$A_z = \frac{\mu_0}{4\pi}\int_{-H}^{0}\frac{I_m \sin \beta (H+z)}{r}\,e^{-j\beta (r-z \cos \theta)}\,dz$$

$$+ \frac{\mu_0}{4\pi}\int_{0}^{H}\frac{I_m \sin \beta (H-z)}{r}\,e^{-j\beta (r-z \cos \theta)}\,dz$$

$$= \frac{\mu_0 I_m e^{-j\beta r}}{4\pi r}\left[\int_{-H}^{0}\sin \beta (H+z)\,e^{j\beta z \cos \theta}\,dz + \int_{0}^{H}\sin \beta (H-z)\,e^{j\beta z \cos \theta}\,dz\right]$$

$$...(3.35)$$

For a half-wave dipole, $H = \dfrac{\lambda}{4}$

But,
$$\sin \beta\,(H+z) = \sin \beta H\,\cos \beta z + \cos \beta H\,\sin \beta z$$
$$\sin \beta\,(H-z) = \sin \beta H\,\cos \beta z - \cos \beta H\,\sin \beta z$$

As
$$\beta = \frac{2\pi}{\lambda}, \quad \sin \beta H = \sin \frac{2\pi}{\lambda} \times \frac{\lambda}{4} = 1,$$

$$\cos \beta H = \cos \frac{\pi}{2} = 0$$

So $\quad \sin \beta\,(H+z) = \sin \beta\,(H-z) = \cos \beta z \qquad\qquad \ldots(3.36)$

Putting Equation (3.36) in Equation (3.35), we get

$$A_z = \frac{\mu_0\, I_m\, e^{-j\beta r}}{4\pi r}\left[\int_{-H}^{0} \cos \beta\, z e^{+j\beta z \cos \theta} + \int_{0}^{H} \cos \beta\, z e^{+j\beta z \cos \theta}\, dz \right]$$

$$\ldots(3.37)$$

But $\displaystyle \int_{-H}^{0} \cos \beta z e^{+j\beta z \cos \theta}\, dz = \int_{0}^{H} \cos \beta z e^{-j\beta z \cos \theta}\, dz$

$$A_z = \frac{I_m\, \mu_0}{4\pi r}\, e^{-j\beta r}\left[\int_{0}^{\lambda/4} \cos \beta z\, (e^{j\beta z \cos \theta} + e^{-j\beta z \cos \theta})\, dz \right]$$

$$= \frac{I_m\, \mu_0}{4\pi r}\, e^{-j\beta r}\left[\int_{0}^{\lambda/4} \cos \{\beta z\,(1 + \cos \theta)\} + \cos \{\beta z\,(1 - \cos \theta)\} \right] dz$$

$$= \frac{I_m\, \mu_0}{4\pi r}\, e^{-j\beta r}\left[\frac{\sin \{\beta z\,(1 + \cos \theta)\}}{\beta\,(1 + \cos \theta)} + \frac{\sin \{\beta z\,(1 - \cos \theta)\}}{\beta\,(1 - \cos \theta)} \right]_{0}^{\lambda/4}$$

$$= \frac{\mu_0\, I_m}{4\pi \beta r}\, e^{-j\beta r}\left[\frac{(1 - \cos \theta)\,\cos \left(\dfrac{\pi}{2} \cos \theta\right) + (1 + \cos \theta)\,\cos \left(\dfrac{\pi}{2} \cos \theta\right)}{\sin^2 \theta} \right]$$

$$A_z = \frac{\mu_0\, I_m}{2\pi \beta r}\, e^{-j\beta r}\left[\frac{\cos \dfrac{\pi}{2} \cos \theta}{\sin^2 \theta} \right] \qquad\qquad \ldots(3.38)$$

But we have

$$\mu_0\, H_\phi = \frac{1}{r}\left[\frac{\partial}{\partial r}\,(r A_\theta) - \frac{\partial}{\partial \theta}\, A_r \right]$$

$$= \frac{1}{r}\left[\frac{\partial}{\partial r}\, r\,(- A_z \sin \theta) - \frac{\partial}{\partial \theta}\,(A_z \cos \theta) \right]$$

$$\mu_0 \, H_\phi = - \sin \theta \, \frac{\partial A_z}{\partial r} \qquad \qquad ...(3.39)$$

From Equations (3.38) and (3.39), we have

$$\mu_0 \, H_\phi = - \frac{\partial}{\partial r} \left(\frac{j\mu_0 \, I_m \, e^{-j\beta r}}{2\pi\beta r} \; \frac{\cos\left(\dfrac{\pi}{2}\cos\theta\right)}{\sin^2\theta} \right) \sin\theta$$

$$H_\phi = \frac{j I_m \, e^{-j\beta r}}{2\pi r} \; \frac{\cos\left(\dfrac{\pi}{2}\cos\theta\right)}{\sin\theta} \qquad \qquad ...(3.40)$$

We also know that $E_\theta = \eta_0 \, H_\phi, \quad \eta_0 = 120\pi\,\Omega$

$$E_\theta = \frac{j\,120\pi \, I_m \, e^{j\beta r}}{2\pi r} \left[\frac{\cos\left(\dfrac{\pi}{2}\cos\theta\right)}{\sin\theta} \right]$$

$$= \frac{j\,60 I_m \, e^{-j\beta r}}{r} \left[\frac{\cos\left(\dfrac{\pi}{2}\cos\theta\right)}{\sin\theta} \right] \qquad \qquad ...(3.41)$$

The magnitude of E for the radiation field is

$$E_\theta = \frac{60 I_m}{r} \left[\frac{\cos\left(\dfrac{\pi}{2}\cos\theta\right)}{\sin\theta} \right] \text{V/m} \qquad \qquad ...(3.42)$$

E_θ and H_ϕ are in time phase. Hence the maximum value of Poynting vector is

$$P_{\max} = (E_\theta)_{\max} \; (H_\phi)_{\max}$$

$$= \frac{60 I_m}{r} \left[\frac{\cos\left(\dfrac{\pi}{2}\cos\theta\right)}{\sin\theta} \right] \times \frac{I_m}{2\pi r} \left(\frac{\cos\left(\dfrac{\pi}{2}\cos\theta\right)}{\sin\theta} \right)$$

$$= \frac{30 I_m^2}{\pi r^2} \left[\frac{\cos^2\left(\dfrac{\pi}{2}\cos\theta\right)}{\sin^2\theta} \right] \qquad \qquad ...(3.43)$$

The average value of Poynting vector is one half of the peak value.

So
$$P_{av} = \frac{15 I_m^2}{\pi r^2} \left[\frac{\cos^2\left(\dfrac{\pi}{2}\cos\theta\right)}{\sin^2\theta} \right]$$

or
$$P_{av} = \frac{\eta_0 I_m^2}{8\pi^2 r^2} \left[\frac{\cos^2\left(\dfrac{\pi}{2}\cos\theta\right)}{\sin^2\theta} \right] \qquad \ldots(3.44)$$

Therefore, total power radiated through a spherical surface by half wave dipole is

$$P_T = \oint P_{av}\, ds = \frac{\eta_0 I_m^2}{8\pi r^2} \int_0^{\pi} \frac{\cos^2\left(\dfrac{\pi}{2}\cos\theta\right)}{\sin^2\theta}\, 2\pi r^2 \sin\theta\, d\theta$$

$$= \frac{\eta_0 I_m^2}{4\pi} \int_0^{\pi} \frac{\cos^2\left(\dfrac{\pi}{2}\cos\theta\right)}{\sin\theta}\, d\theta \qquad \ldots(3.45)$$

But the numerical evaluation of the integral $\displaystyle\int_0^{\pi} \frac{\cos^2\left(\dfrac{\pi}{2}\cos\theta\right)}{\sin\theta}\, d\theta$ by Simpson's or the Trapezoidal rule gives a value of 1.218.

So
$$P_T = \frac{\eta_0 I_m^2}{4\pi}$$

$$= \frac{120\pi I_m^2}{4\pi} \times 1.218 = 36.54 I_m^2 \qquad \ldots(3.46)$$

As $I_m = \sqrt{2} I_{\text{eff}}$, Equation (3.46) becomes

$$P_T = 36.54 \times 2 \times I_{\text{eff}}^2$$

or
$$P_T = 73.08\Omega\ I_{\text{eff}}^2,\ \text{watts} \qquad \ldots(3.47)$$

The coefficient of I_{eff}^2 is the radiation resistance. That is,

$$R_r = 73.08\Omega. \qquad \ldots(3.48)$$

3.15 RADIATION FROM QUARTER-WAVE MONOPOLE

Radiated power of quarter-wave monopole, $P_T = 36.5 I_{\text{eff}}^2$ watts Radiation resistance, $R_r = 36.5\Omega$.

Proof   Consider Fig. 3.8 in which monopole with current distribution is shown.

Obtain P_{av} exactly as described in half-wave dipole. That is, from Equation (3.44), we have

$$P_{av} = \frac{\eta_0 I_m^2}{8\pi^2 r^2}\left[\frac{\cos^2\frac{\pi}{2}\cos\theta}{\sin^2\theta}\right] \qquad \dots(3.49)$$

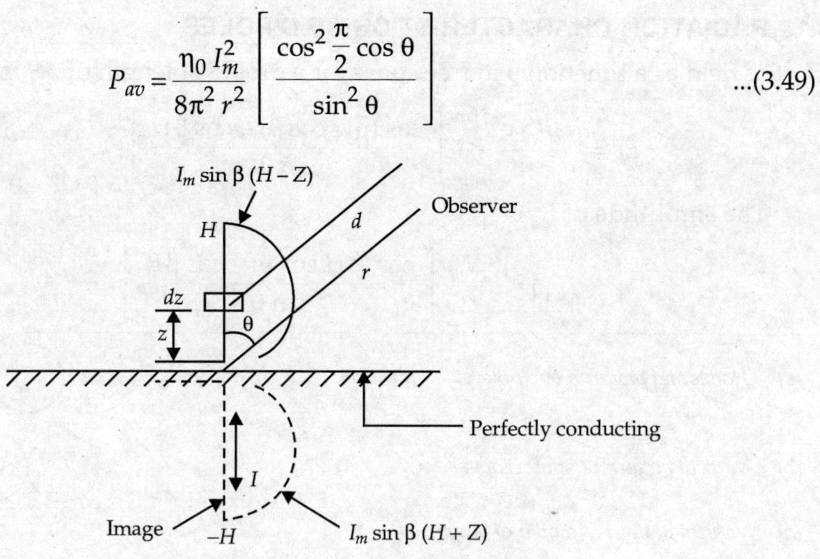

Fig. 3.8 *Monopole with current distribution*

As the monopole is fed with a perfectly conducting plane at one end, it radiates only through a hemi-spherical surface. Therefore, the total radiated power is

$$P_T = \oint P_{av}\,ds$$

$$= \frac{\eta_0 I_m^2}{8\pi r^2}\int_0^{\pi/2}\frac{\cos^2\left(\dfrac{\pi}{2}\cos\theta\right)}{\sin^2\theta}\,2\pi r^2 \sin\theta\,d\theta$$

$$= \frac{\eta_0 I_m^2}{4\pi}\int_0^{\pi/2}\frac{\cos^2\left(\dfrac{\pi}{2}\cos\theta\right)}{\sin^2\theta}\,d\theta$$

Numerical evaluation of the integral $\displaystyle\int_0^{\pi/2}\frac{\cos^2\left(\dfrac{\pi}{2}\cos\theta\right)}{\sin\theta}\,d\theta$ by Simpson's or the Trapezoidal rule gives a value of 0.609.

So $\qquad\qquad P_T = \dfrac{\eta_0 I_m^2}{4\pi}\times 0.609$

$$= 18.27 I_m^2$$

As $\qquad\qquad I_m = \sqrt{2}\,I_{\text{eff}}$

$$P_T = 36.54 I_{\text{eff}}^2 \text{ watts} \qquad \dots(3.50)$$

The Radiation resistance, $R_r = 36.54\,\Omega$ $\qquad\qquad \dots(3.51)$

3.16 RADIATION CHARACTERISTICS OF DIPOLES

Electric field as a function θ in free space for a dipole of length of $2H$ is given by

$$E_\theta = \frac{j60I_m\, e^{-j\beta r}}{r}\left[\frac{\cos(\beta H \cos\theta) - \cos\beta H}{\sin\theta}\right] \qquad \ldots(3.52)$$

The amplitude of E_θ is

$$|E_\theta| = \frac{60I_m}{r}\left[\frac{\cos(\beta H \cos\theta) - \cos\beta H}{\sin\theta}\right]$$

(a)   Horizontal patterns of dipole of length = 2H

(b)   Vertical pattern of short diople

(c)   Vertical pattern of diople of length = $\dfrac{\lambda}{2}$

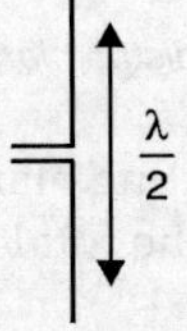

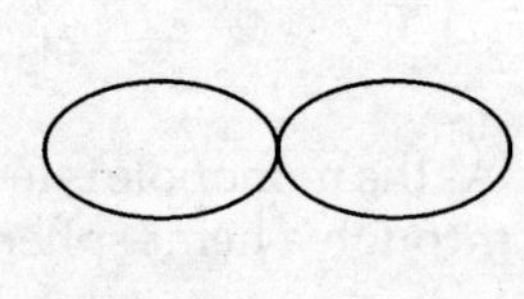

(d)   Vertical pattern of diople of length = λ

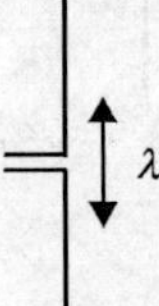

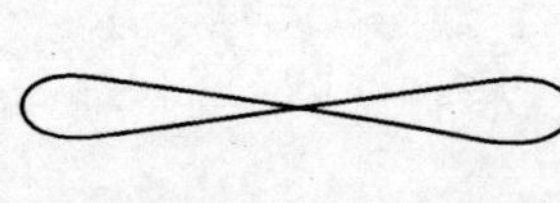

(e)   Vertical patterns of diople of lengths = $3\,\dfrac{\lambda}{2}$

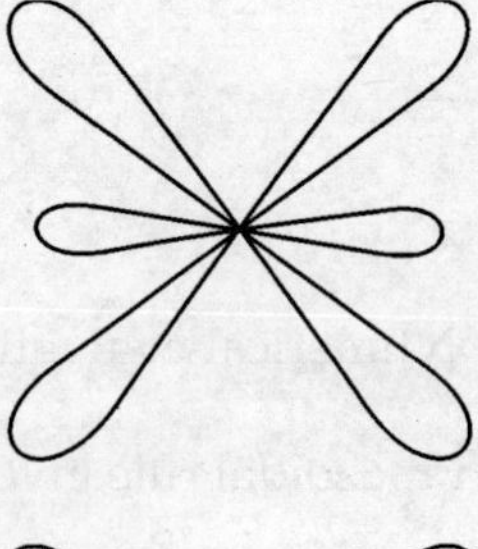

(f)   Vertical pattern of diople of length = 2λ

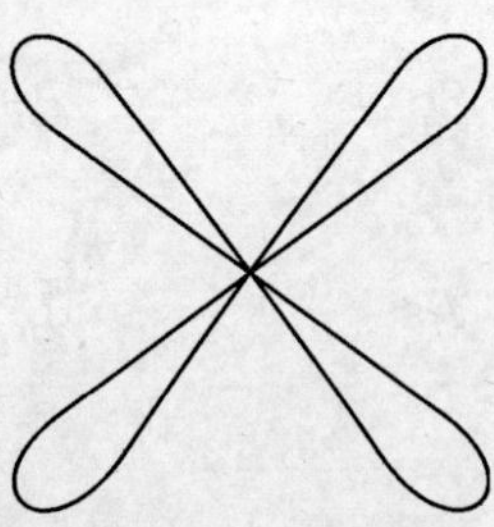

Fig. 3.9   *Radiation patterns of dipole*

The normalised (E_θ) is

$$|E_\theta|_n = \dfrac{\dfrac{60 I_m}{r}\left[\dfrac{\cos(\beta H \cos\theta) - \cos\beta H}{\sin\theta}\right]}{\dfrac{60 I_m}{r}}$$

So
$$|E_\theta|_n = \left[\dfrac{\cos(\beta H \cos\theta) - \cos\beta H}{\sin\theta}\right] \qquad\qquad ...(3.53)$$

The radiation patterns are the variation of $|E_\theta|_n$ with θ. These patterns for different lengths of dipole are shown in Fig. 3.9.

POINTS TO REMEMBER

1. Radiation intensity, $RI = \dfrac{r^2 E^2}{\eta_0}$ watts/unit solid angle.

2. Directive gain, $g_d = \dfrac{4\pi \times (RI)}{w_r}$.

3. Directivity, $D = (g_d)_{\max}$.

4. Power gain, $g_p = \dfrac{4\pi \times (RI)}{w_t}$.

5. Antenna efficiency, $\eta = \dfrac{g_p}{g_d}$.

6. Effective area, $A_e = \dfrac{\lambda^2}{4\pi} g_d$ or $A_e = \dfrac{\text{received power}}{\text{power flow of incident waves}}$.

7. Far-field is represented by $\dfrac{1}{r}$ field term.

8. Induction field is represented by $\dfrac{1}{r^2}$ field term.

9. Radiation resistance of Hertzian dipole is $80\pi^2 \left(\dfrac{dl}{\lambda}\right)^2 \Omega$.

10. Electrostatic field is represented by $\dfrac{1}{r^3}$.

11. The far-field and induction field have equal magnitudes at $r = \dfrac{\lambda}{2\pi}$.

12. Radiation resistance of half-wave dipole is 73Ω.

13. Radiation resistance of quarter-wave monopole is $36.5\,\Omega$.

14. Horizontal pattern of vertical dipole is a circle.

15. Radiated power flow of a vertical dipole is in the radial direction.

 ## SOLVED PROBLEMS

Problem 3.1 Find the radiation resistance of a Hertzian dipole of length $\dfrac{\lambda}{40},\ \dfrac{\lambda}{60},\ \dfrac{\lambda}{80}$.

Solution The radiation resistance of Hertzian dipole of length dl is

$$R_r = 80\pi^2\left(\frac{dl}{\lambda}\right)^2\ \Omega$$

If $dl = \dfrac{\lambda}{40}$ 　　 $R_r = 80\pi^2\left(\dfrac{\lambda}{40}\times\dfrac{1}{\lambda}\right)^2$

or 　　　　　　　$R_r = 0.493\,\Omega$

If $dl = \dfrac{\lambda}{60}$ 　　 $R_r = 80\pi^2\,\dfrac{1}{60^2}$

or 　　　　　　　$R_r = 0.219\,\Omega$

If $dl = \dfrac{\lambda}{80}$ 　　 $R_r = 80\pi^2\,\dfrac{1}{80^2}$

or 　　　　　　　$R_r = 0.123\,\Omega.$

Problem 3.2 Find the directivity of a current element, $I\,dl$.

Solution From Equation (3.10), the amplitude of electric far-field is

$$E = \frac{I\,dl\,\sin\theta}{4\pi\epsilon_0}\times\frac{\omega}{r\upsilon_0^2}$$

$$= \frac{\omega I\,dl\,\sin\theta}{4\pi\epsilon_0\,r\upsilon_0^2}$$

$$= \frac{2\pi f\,I\,dl\,\sin\theta}{4\pi\epsilon_0\,r\,\dfrac{1}{\mu_0\,\epsilon_0}} \qquad\qquad \left[\text{as } \omega = 2\pi f,\ \upsilon_0 = \frac{1}{\sqrt{\mu_0\,\epsilon_0}}\right]$$

$$= \frac{I\,dl\,\sin\theta}{2r}\times f\,\sqrt{\mu_0}\,\epsilon_0\times\frac{\sqrt{\mu_0\,\epsilon_0}}{\epsilon_0}$$

But 　　　　　$\eta_0 = \sqrt{\dfrac{\mu_0}{\epsilon_0}},\quad \lambda = \dfrac{\upsilon_0}{f}$

or
$$\frac{1}{\lambda} = \frac{f}{v_0} = f\,\sqrt{\mu_0\,\epsilon_0}$$

So
$$E = \frac{60I\,dl\,\sin\theta}{\lambda r}$$

Maximum radiation occurs at

$$\theta = \frac{\pi}{2}$$

or
$$E_{\max} = \frac{60I\,dl}{\lambda r} \qquad\qquad ...(3.54)$$

The radiated power of current element is

$$P_r = 80\pi^2 \left(\frac{dl}{\lambda}\right)^2 I^2 \text{ watts}$$

If P_r is assumed to be 1 watt, then

$$I = \frac{\lambda}{\sqrt{80}\,\pi\,dl}\text{ amp} \qquad\qquad ...(3.55)$$

From Equations (3.54) and (3.55), we get

$$E_{(\max)} = \frac{60}{r\,\sqrt{80}}\text{ V/m} \qquad\qquad ...(3.56)$$

The maximum radiation intensity is given by

$$RI = \frac{r E_{(\max)}^2}{\eta_0}$$

$$= \frac{r^2}{120\pi}\,\frac{60^2}{r^2 \times 80}$$

or
$$RI = \frac{3}{8\pi}$$

The maximum directive gain, g_d (max)

$$g_d\,(\max) = \frac{4\pi\,(RI)}{P_r} \qquad\qquad \text{[as } P_r = 1\text{ watt]}$$

As $P_r = 1$ watt
$$= \frac{4\pi\,(RI)}{P_r}$$

$$= 4\pi \times \frac{3}{8\pi}$$

$$= \frac{3}{2} = 1.5$$

The directivity of current element

$$D = g_d\,(\text{max}) = 1.5$$

or

$$D_{\text{in dB}} = 10\log_{10} 1.5 = 1.76\ \text{dB.}$$

Problem 3.3 Find the directivity of a half-wave dipole.

Solution For a half-wave dipole, from Equation (3.42)

$$E_{(\text{max})} = \frac{60I}{r}$$

But

$$P_r = 73I^2\ \text{watts}$$

For

$$P_r = 1\ \text{w}$$

$$I = \frac{1}{\sqrt{73}}$$

or

$$E_{(\text{max})} = \frac{60}{r} \times \frac{1}{\sqrt{73}}$$

$$g_{d\,(\text{max})} = \frac{4\pi\,(RI)}{P_r}$$

$$= 4\pi\,(RI) \qquad\qquad [\text{as } P_r = 1\ \text{watt}]$$

$$= 4\pi \times \frac{r^2 E^2}{\eta_0} \qquad\qquad \left[\text{as } RI = r^2\frac{E^2}{\eta_0}\right]$$

$$= \frac{4\pi \times r^2}{\eta_0}\,\frac{60^2}{r^2}\,\frac{1}{73}$$

$$= \frac{4\pi \times 60 \times 60}{120\pi}\,\frac{1}{73}$$

$$= \frac{120}{73} = 1.644$$

So

$$g_d\,(\text{max}) = D = 1.644.$$

Problem 3.4 An antenna whose radiation resistance is $300\,\Omega$ operates at a frequency of 1 GHz and with a current of 3 amperes. Find the radiated power.

Solution Radiated power,

$$P_r = I^2 R_r$$

$$= 3^2 \times 300$$

$$= 9 \times 300$$

So

$$P_r = 2700\ \text{watts.}$$

Problem 3.5   What is the effective area of a half-wave dipole operating at 500 MHz?

Solution   The effective area of an antenna is

$$A_e = \frac{\lambda^2}{4\pi}\, g_d$$

As $f = 500$ MHz

$$\lambda = \frac{3 \times 10^8}{500 \times 10^6}$$

$$= \frac{3}{5} = 0.6 \text{ m}$$

Directivity of a half-wave dipole is
$$(g_d)_{max} = D = 1.644$$

So $$A_e = \frac{0.6^2}{4\pi} \times 1.644$$

or $$A_e = 0.047 \text{ m}^2.$$

Problem 3.6   Find the effective area of a Hertzian dipole operating at 100 MHz.

Solution   As $f = 100$ MHz, $\lambda = \dfrac{3 \times 10^8}{100 \times 10^6} = 3$ m

Directivity of Hertzian dipole, $D = 1.5$
$$A_e = \text{effective area}$$

$$= \frac{\lambda^2}{4\pi} = \frac{3^2 \times 1.5}{4\pi} = 1.07 \text{ m}^2$$

or $$A_e = 1.07 \text{ m}^2.$$

 **OBJECTIVE QUESTIONS**

1. An antenna is a transducer.   (Yes/No)

2. An antenna is a sensor of EM waves.   (Yes/No)

3. An antenna acts as an impedance matching device.   (Yes/No)

4. Effective length of a wire antenna is always greater than the actual length.   (Yes/No)

5. Directive gain = Power gain for an antenna.   (Yes/No)

6. The units of radiation intensity are _______________.

7. Directivity is _______________.

8. Efficiency of an antenna is _______________.

9. Efficiency of an antenna in terms of directive and power gains is _______________.

10. Effective area is _______________.

11. The radiation fields are nothing but far-fields.   (Yes/No)

12. The far-field is indicated by the presence of _______________.

13. The induction field is indicated by the presence of _______________.

14. The electrostatic field is indicated by the presence of _______________.

15. The radiation resistance of an isolated half-wave dipole is _______________.

16. The radiation resistance of a quarter-wave monopole is _______________.

17. The current distribution in a half-wave dipole is _______________.

18. The current distribution in an alternating current element is _______________.

19. The current distribution in very short dipoles is _______________.

20. The radiation pattern of vertical and horizontal dipoles are identical.   (Yes/No)

21. The directivity of current element is _______________.

22. The directivity of half-wave dipole is _______________.

23. The patterns of half-wave dipole and quarter-wave monopole are identical.   (Yes/No)

24. If a current element is x-directed, vector magnetic potential is _______________.

25. Radiation resistance of short monopole is _______________.

26. Radiation resistance of short dipole is _______________.

27. The radiated fields of z-directed half-wave dipole consists of E_θ, E_r, H_θ, terms.   (Yes/No)

28. The radiated fields of z-directed dipole consists of only E_θ, E_r and H_ϕ.   (Yes/No)

29. At LF and VLF, polarisation often used is_______________.

30. dB_i means _______________.

31. dB_m means power gain in dB _______________

32. If the signal level is 1 mW, power gain is
      (a)   0 dBm   (b)   1 dBm
      (c) 10^{-3} dBm   (d)   10 dBm

33. Marconi antenna has a physical length of
      (a) $\lambda/4$ (b) $\lambda/2$
      (c) $3\lambda/2$ (d) λ

34. For a 300Ω antenna operating with 5 A of current, the radiated power is

    (*a*)   7500 W   (*b*)   750 W

    (*c*)   75 W   (*d*)   1500 W

35. Effective area of antenna is a function frequency.   (Yes/No)

36. Antenna used in mobile communications is _______________.

37. If a current element is z-directed, vector magnetic potential is _______________.

38. If vector magnetic potential has only A_z, E_ϕ is _______________.

39. Radiation resistance of current element is _______________.

40. Radiation resistance of quarter-wave monopole is _______________.

41. Directional pattern of a short dipole in the horizontal plane is a _______________.

42. Directional pattern of a horizontal half-wave centre fed dipole is _______________.

43. Effective length of a dipole is always _______________ than the actual length.

44. The directivity in dB of half-wave dipole is _______________.

45. The directivity in dB of current element is _______________.

46. Effective area of a Hertzian dipole operating at 100 MHz is _______________.

ANSWERS

1. Yes   **2.** Yes   **3.** Yes   **4.** No   **5.** No

6. Watts/unit solid angle   **7.** Maximum directive gain   **8.** $\dfrac{w_r}{(w_r + w_l)}$ **9.** g_p/g_d

10. $\dfrac{\lambda^2}{4\pi} g_d$ **11.** Yes   **12.** $\dfrac{1}{r}$ term   **13.** $\dfrac{1}{r^2}$ term   **14.** $\dfrac{1}{r^3}$ term   **15.** 73Ω

16. 36.5Ω **17.** Sinusoidal   **18.** Constant   **19.** Triangular   **20.** No   **21.** 1.5

22. 1.64   **23.** No   **24.** x-directed   **25.** $100 \left(\dfrac{1}{\lambda}\right)^2$ **26.** $200 \left(\dfrac{1}{\lambda}\right)^2$ **27.** No

28. Yes   **29.** Vertical

30. Power gain of the antenna in dB relative to isotropic antenna

31. Compared to 1 mW   **32.** (*a*)   **33.** (*a*)   **34.** (*a*)   **35.** Yes

36. Whip antenna   **37.** z-directed   **38.** Zero   **39.** $80\pi^2 \left(\dfrac{dl}{\lambda}\right)^2 \Omega$

40. 36.5Ω **41.** Circle   **42.** Figure of eight   **43.** Less   **44.** 2.15

45. 1.64   **46.** 1.07 m^2.

 **EXERCISE PROBLEMS**

1. The field amplitude due to half-wave dipole at 10 km is 0.1 V/m. It operates at 100 MHz. Find the dipole length and its radiated power.

2. What is the length of a half-wave dipole at frequencies of 10 MHz, 50 MHz and 100 MHz?

3. Find the maximum effective area of an antenna at a frequency of 2 GHz when the directivity is 100.

4. Obtain the gain of an antenna whose area is 12 m^2 and operating at a frequency of 6 GHz.

5. Find the radiated power of an antenna if a current of 10 amp exists and its radiation resistance is 32.0 Ω.

6. What is the radiation resistance of an antenna if it radiates a power of 120 W and the current in it is 10 amp.

7. Find the directivity, efficiency and effective area of an antenna if its $R_r = 80\,\Omega$, $R_l = 10\,\Omega$. The power gain is 10 dB and antenna operates at a frequency of 100 MHz.

8. If the transmitting power is 10 kW, find the power density at distances of 10 km, 50 km and 100 km, assuming that the radiator is isotropic.

9. If the current element is z-directed, find the far-field components of H.

10. Derive an expression for distant field θ-field component of **E** for a dipole of length L.

11. (a) Find the current required to radiate power of 50 W at 60 MHz from a 0.1λ Hertzian dipole.

    (b) Determine the radiation resistance in the element.

12. Find the radiation efficiency of a Hertzian dipole of length 0.03λ at a frequency of 100 MHz if the loss resistance is 0.01 Ω.

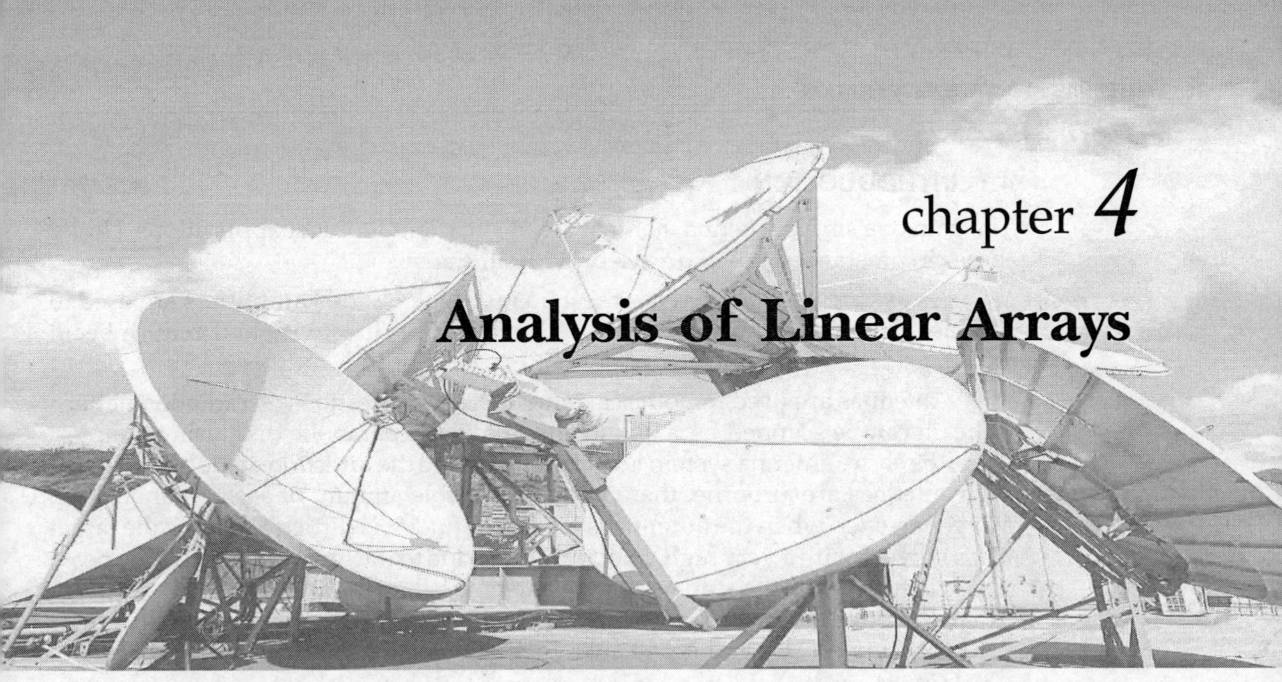

chapter 4

Analysis of Linear Arrays

"Array antennas are similar to cascaded electronic amplifiers in increasing gain."

CHAPTER OBJECTIVES

This chapter discusses

- ✦ The analysis of the arrays
- ✦ The directional behaviour of dipoles and monopoles
- ✦ The pattern characteristics of uniform and linear arrays
- ✦ Pattern calculations and effects of the earth
- ✦ Impedance matching techniques
- ✦ Transmission loss between transmitter and receiver
- ✦ Antenna signal to noise ratio
- ✦ Objective questions and solved problems useful for class tests, final examinations and also for competitive examinations
- ✦ Exercise problems to develop self problem solving skills

4.1 INTRODUCTION

The gain of a single antenna element is not sufficient for most applications. Under these circumstances, use of arrays provides the answer.

Arrays are used to increase the gain just like cascaded amplifiers are used to increase the gain. Arrays are also used to increase directivity and reduce beam width.

Antennas are used in both scan and unscan applications. A radiation pattern or beam can be scanned by a single antenna like a parabolic dish using a motor. Here, the entire antenna system is rotated to change the direction of the beam. When such antennas are airborne, there is a considerable amount of aerodynamic drag. Moreover, even when the antenna is on the ground, it is difficult to track the target if it is moving with a very high velocity. This is due to the limited speed of the motor which in turn limits the scan rate of the antenna. When the antenna directs its beam in a direction to catch the target, the target will be in a different direction due to its velocity being greater and it will always be out of sight.

In such situations, array antennas are more useful as it is possible to scan the beam from the arrays electronically. The electronic scanning can be with either phase control or frequency control. Here, the antenna is fixed and only the beam is rotated. The scan rate can be as high as a fraction of a microsecond with digital phase shifters.

In view of the above facts, array antennas are extremely useful in both airborne as well as ground-based applications. There is no aerodynamic drag as there is no movement of the antennas.

Moreover, arrays provide a greater number of parameters and offer better flexibility for the designer. In this chapter, analysis of array antennas is presented in view of their significance.

4.2 DIRECTIONAL CHARACTERISTICS OF DIPOLE ANTENNAS

The directional characteristics are nothing but the radiation patterns of antennas. They indicate the distribution of radiation power in free space in different angular regions.

In other words, the radiation pattern of an antenna is a graphical representation of radiation as a function of direction. The radiation patterns of an antenna are of two types :

1. Field strength pattern
2. Power pattern.

The power pattern is proportional to the square of the field strength pattern. The patterns are plotted either in polar coordinates or in linear coordinates. The patterns represent far-field variation. They are presented in the form of the variation of absolute normalised field strength in dB as a function of θ.

A three dimensional pattern plotted as a function of θ and ϕ gives the complete information. Sometimes, cross-sections of patterns in horizontal and vertical planes are presented.

Horizontal pattern   It is a pattern obtained for $\theta = 90^\circ$.

Vertical pattern   It is a pattern obtained for ϕ = constant.

Sometimes, patterns are shown in E-plane and H-plane.

E-plane   It is a plane passing through the antenna in the direction of the maximum beam and parallel to the E-field.

H-plane   It is a plane passing through the antenna in the direction of the maximum beam and parallel to the H-field.

Principal planes   The E-plane and H-plane are known as the principal planes.

E and H-plane patterns are known as principal plane patterns.

The radiation characteristics are also sometimes shown separately for theta and phi polarisations.

Theta polarisation   It is the same as vertical polarisation. An example is a vertical dipole.

Phi polarisation   It is the same as horizontal polarisation. An example is a horizontal loop.

4.3 RADIATION PATTERN OF AN ALTERNATING CURRENT ELEMENT

The alternating current element is also called **oscillating dipole** or **elementary dipole** or **current element**.

The radiation field of a z-directed current element is given by

$$E = \frac{60\pi I\, dL}{\lambda r} \sin\theta, \ \text{V/m} \qquad\qquad ...(4.1)$$

where
$$I\, dL = \text{current element}$$
$$dL = \text{element length}$$
$$I = \text{constant current in the element}$$
$$r = \text{far-field distance}$$
$$\lambda = \text{operating wavelength}$$
$$\theta = \text{angle between dipole axis and the line of far-field point}$$

Let the above field be represented by
$$E = E_m \sin\theta$$

where
$$E_m = \frac{60\pi I\, dL}{\lambda r}$$

The normalised field

$$E_n = \frac{E}{E_m} = \sin\theta$$

The horizontal pattern of elementary dipole is shown in Fig. 4.1. This is obtained for $\theta = 90^\circ$.

It is evident from the above expression that the field is independent of ϕ.

The vertical pattern of vertical current element is shown in Fig. 4.2.

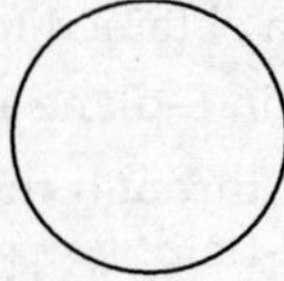

Fig. 4.1 *Horizontal pattern of elementary dipole*

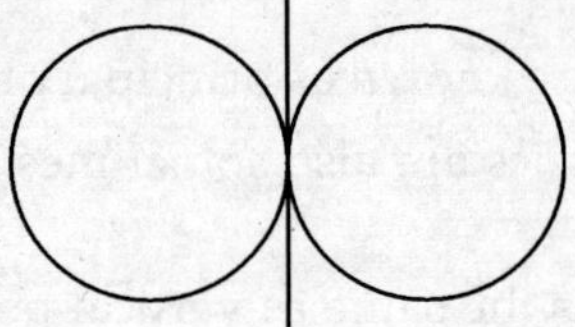

Fig. 4.2 *Vertical pattern of vertical current element*

When the current element is horizontal, the vertical pattern is shown in Fig. 4.3.

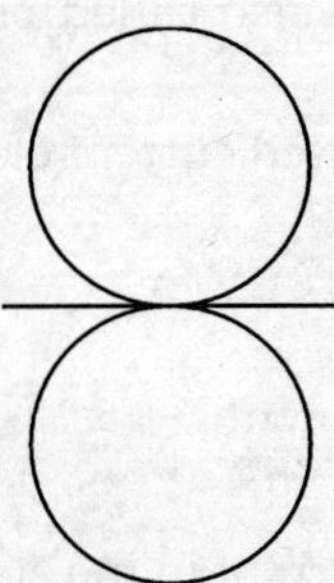

Fig. 4.3 *Vertical pattern of horizontal current element*

4.4 RADIATION PATTERN EXPRESSIONS OF CENTRE-FED VERTICAL DIPOLES OF FINITE LENGTH

The magnitude of the radiation field of a vertical dipole of length 'L' is given by

$$E = \frac{60 I_m}{r}\left[\frac{\cos\left(\dfrac{\beta L \cos\theta}{2}\right) - \cos\left(\dfrac{\beta L}{2}\right)}{\sin\theta}\right] \qquad \ldots(4.2)$$

The normalised radiation field is

$$E_n = \frac{\left[\cos\left(\dfrac{\beta L}{2}\cos\theta\right) - \cos\dfrac{\beta L}{2}\right]}{\sin\theta} \qquad \ldots(4.3)$$

For half-wave dipole, E_n is

$$E_n = \frac{\cos\left(\beta\,\dfrac{\lambda}{4}\cos\theta\right) - \cos\left(\beta\,\dfrac{\lambda}{4}\right)}{\sin\theta}$$

$$= \frac{\cos\left(\dfrac{2\pi}{\lambda}\cdot\dfrac{\lambda}{4}\cos\theta\right) - \cos\left(\dfrac{2\pi}{\lambda}\cdot\dfrac{\lambda}{4}\right)}{\sin\theta}$$

So

$$E_n = \frac{\cos\left(\dfrac{\pi}{2}\cos\theta\right)}{\sin\theta}$$

The horizontal pattern of a dipole of any length is represented by

$$E_n = \left.\frac{\left[\cos\left(\dfrac{\beta L}{2}\cos\theta\right) - \cos\left(\dfrac{\beta L}{2}\right)\right]}{\sin\theta}\right|_{\theta=90^\circ}$$

$$= \text{constant}$$

The horizontal pattern of a dipole is therefore a circle.

4.5 RADIATION PATTERNS OF CENTRE-FED VERTICAL DIPOLES

The variation of E_n with θ for different lengths of dipoles gives the vertical patterns. They are shown in Fig. 4.4.

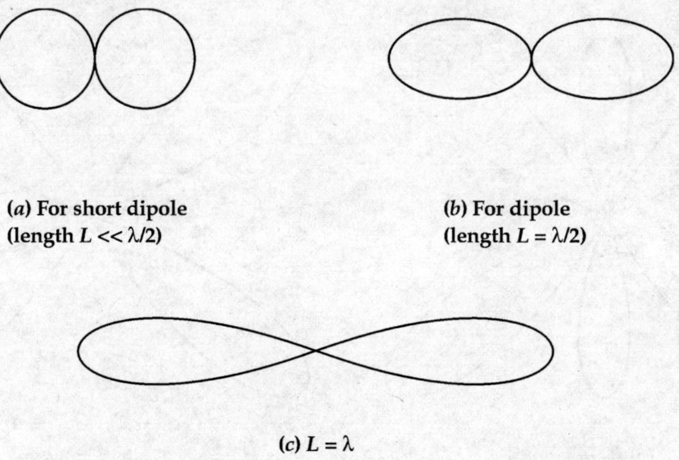

Fig. 4.4   *Vertical patterns of centre-fed dipoles of different lengths*

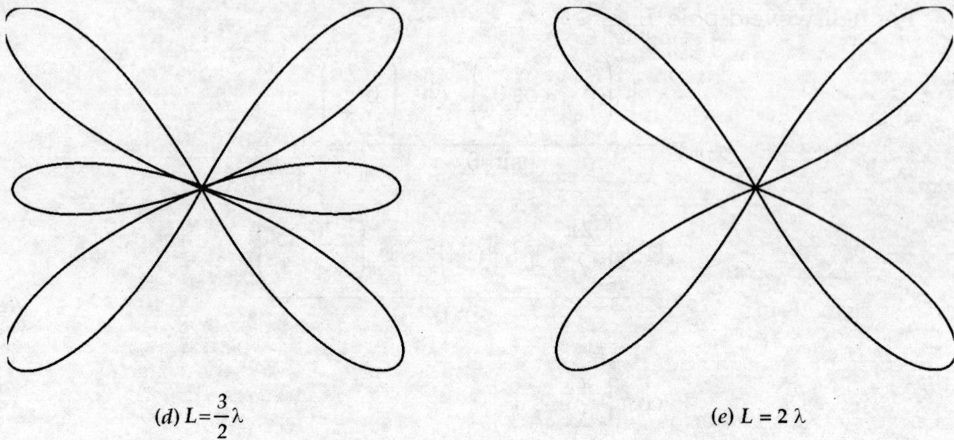

$$(d)\; L = \frac{3}{2}\lambda \qquad\qquad\qquad (e)\; L = 2\,\lambda$$

Fig. 4.4   *Vertical patterns of centre-fed dipoles of different lengths*

4.6 RADIATION PATTERNS OF CENTRE-FED HORIZONTAL DIPOLES OF FINITE LENGTH

The radiation pattern for different lengths of centre-fed horizontal dipole are shown in Fig. 4.5.

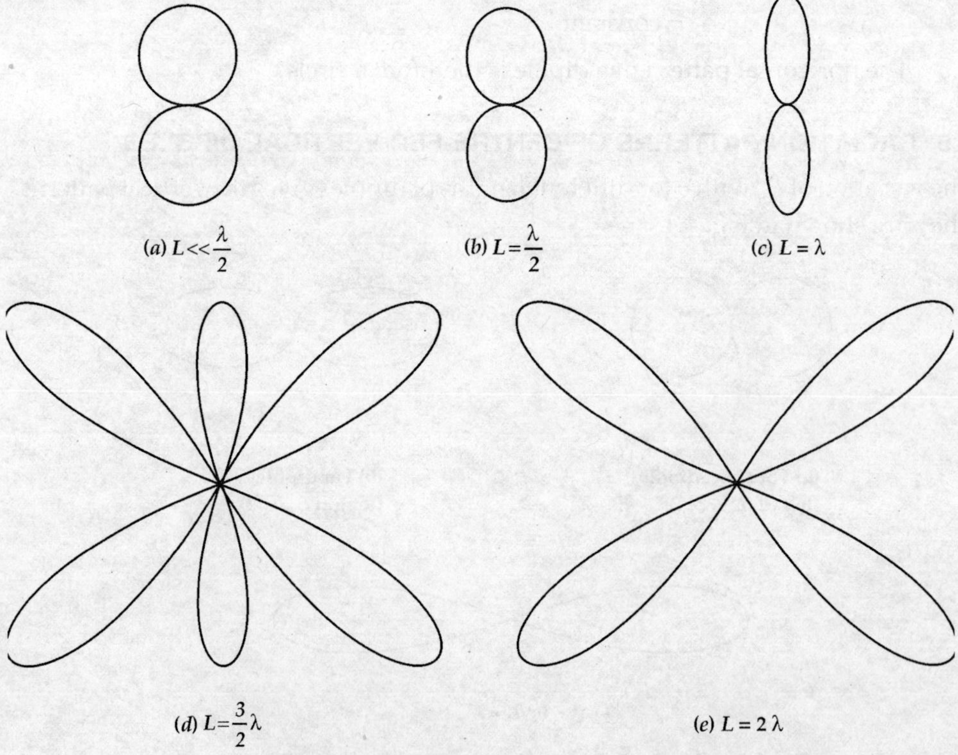

$$(a)\; L \ll \frac{\lambda}{2} \qquad\qquad (b)\; L = \frac{\lambda}{2} \qquad\qquad (c)\; L = \lambda$$

$$(d)\; L = \frac{3}{2}\lambda \qquad\qquad\qquad (e)\; L = 2\,\lambda$$

Fig. 4.5   *Patterns of horizontal dipoles of finite length*

4.7 RADIATION PATTERNS OF VERTICAL MONOPOLES

The radiation patterns of vertical monopoles of different lengths are the same as those of vertical centre-fed dipoles. However, as the monopole is fed with conducting ground plane, the length of the monopole is just half of the dipole. Here, the image forms the second half. The patterns of monopoles are shown in Fig. 4.6.

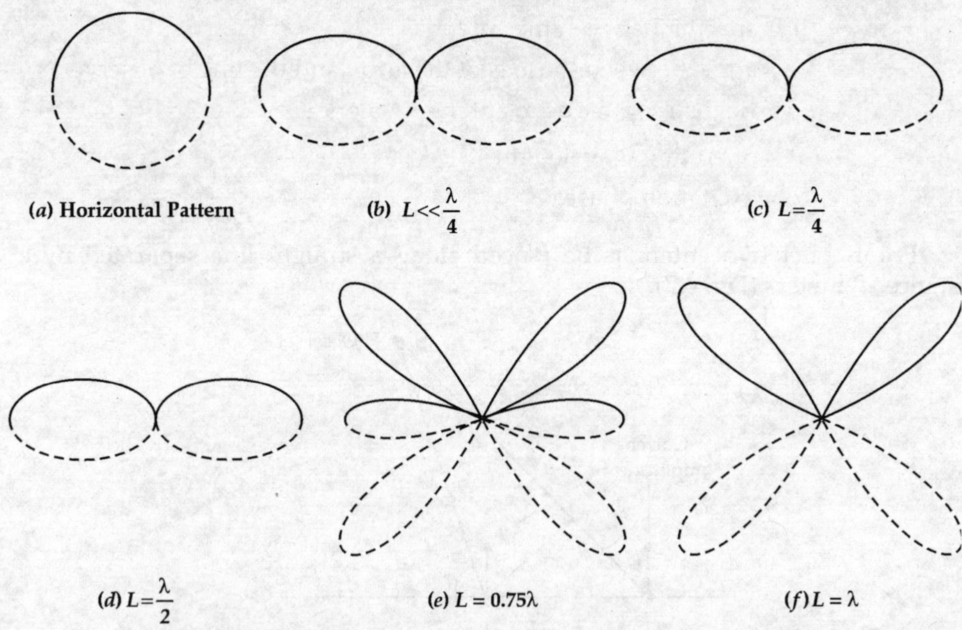

Fig. 4.6 *Vertical radiation patterns of monopoles of different lengths*

The above patterns of monopoles are obtained from the expression

$$E = E_\theta = \frac{60I}{r} \left[\frac{\cos \beta \dfrac{L}{2} - \cos \left(\beta \dfrac{L}{2} \cos \theta \right)}{\sin \theta} \right] \qquad ...(4.4)$$

Here $\dfrac{L}{2}$ = length of the monopole.

4.8 TWO-ELEMENT UNIFORM ARRAY

An **array of radiators** is defined as a system of antennas which are similar or non-similar and either similarly oriented or differently oriented. Arrays are used to increase directivity and gain.

However, in this book we consider only arrays of similar antennas with similar orientation.

Expression for resultant radiation pattern of two-element array:

It is given by

$$E_R = 2E_A \cos\left(\frac{\pi d \cos \phi}{\lambda} + \frac{\alpha_e}{2}\right) \qquad \text{...(4.5)}$$

Here d = spacing between the antennas

ϕ = angle between the axis of the array and the line of observer

E_A = field strength due to antenna A alone

λ = operating wavelength

α_e = excitation phase.

Proof   Let two antennas be placed along a straight line separated by a distance 'd' meters (Fig. 4.7).

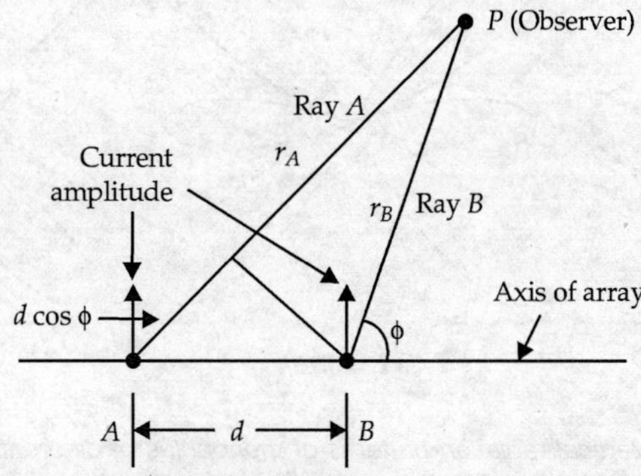

Fig. 4.7   *Two-element uniform array*

If the point P is far away from the array, Ray A and Ray B can be assumed to be parallel. Hence in Fig. 4.7, the path difference between the two ray paths is

$$r_A - r_B = d \cos \phi \qquad \text{...(4.6)}$$

or
$$r_B = r_A - d \cos \phi \qquad \text{...(4.7)}$$

This exact expression must be used in the phase term of the field. But in the magnitude term of the field, we can use the approximation, $r_A \approx r_B$.

Now the resultant phase difference due to spacing of the antennas is given by

Phase difference, α_d = wave number × path difference

Phase difference, $\alpha_d = \beta \times d \cos \phi$

$$= \beta d \cos \phi \qquad \text{...(4.8)}$$

If the excitation phase difference is α_e, the total phase difference is

$$\psi = \beta d \cos \phi + \alpha_e \qquad \text{...(4.9)}$$

Here, α_e is the phase angle by which current, I_B in antenna B leads the current I_A in antenna A.

The resultant field in phasor form when the two antennas are uniformly excited is given by

$$E_R = E_A \left(1 + e^{j\psi}\right) \qquad\qquad ...(4.10)$$

The magnitude of total field strength, when $I_A = I_B$

$$|E_R| = E = |E_A (1 + e^{j\psi})|$$

$$= E_A (1 + \cos\psi + j\sin\psi)$$

$$= E_A \sqrt{(1 + \cos\psi)^2 + \sin^2\psi}$$

$$= E_A \sqrt{1 + \cos^2\psi + 2\cos\psi + \sin^2\psi}$$

$$= E_A \sqrt{1 + \cos^2\psi + \sin^2\psi + 2\cos\psi}$$

$$= E_A \sqrt{2 + 2\cos\psi} \qquad\qquad (\text{as } \cos^2\psi + \sin^2\psi = 1)$$

$$= E_A \sqrt{2(1 + \cos\psi)}$$

But $\qquad (1 + \cos\psi) = 2\cos^2\dfrac{\psi}{2} \qquad\qquad\qquad ...(4.11)$

$$E = E_A \sqrt{2 \times 2\cos^2\dfrac{\psi}{2}}$$

$$= 2E_A \cos\dfrac{\psi}{2}$$

or $\qquad E = 2E_A \cos\left(\dfrac{\beta d\cos\phi}{2} + \dfrac{\alpha_e}{2}\right)$

$$E = 2E_A \cos\left(\dfrac{\pi d\cos\phi}{\lambda} + \dfrac{\alpha_e}{2}\right) \qquad \left(\because \beta = \dfrac{2\pi}{\lambda}\right) \qquad ...(4.12)$$

In practical applications, two-element array is rarely used. Mostly, arrays with more number of elements are used to get high directivity and gain and to have control over more parameters like spacing, current magnitude, phases and antenna configurations.

4.9 UNIFORM LINEAR ARRAYS

A **uniform linear array** is an array where the elements are spaced and excited equally along a straight line.

These arrays are suitable for production of narrow radiation beams. These types of radiation patterns are required for point to point communication at higher frequencies. These are also used in high angular resolution radars.

4.10 FIELD STRENGTH OF A UNIFORM LINEAR ARRAY

The normalised field strength of a uniform linear array is

$$E = \left| \frac{\sin \dfrac{N\psi}{2}}{\sin \dfrac{\psi}{2}} \right| \qquad \qquad ...(4.13)$$

Here N = number of elements in the array

$\psi = \beta d \cos \phi + \alpha_e$

β = wave number = $2\pi/\lambda$

d = spacing between the elements

ϕ = angle between the axis of the array and line of observer

α_e = excitation phase (progressive phase shift).

Proof   Consider a uniform linear array of non-directional elements of Fig. 4.8.

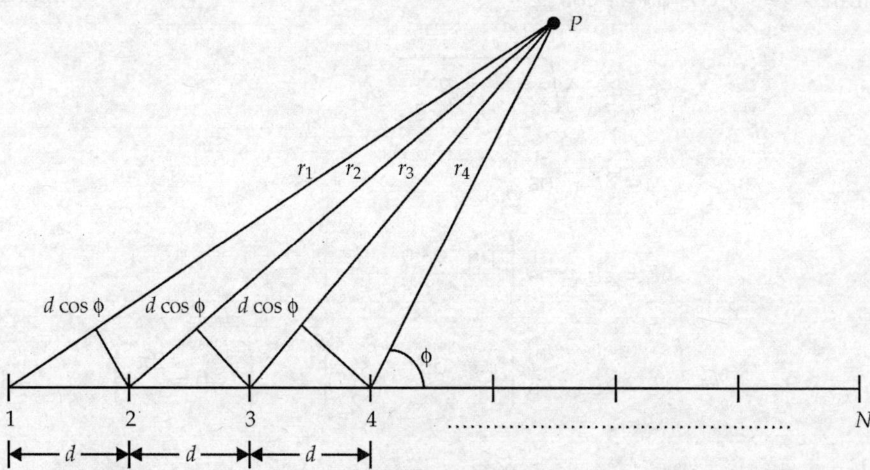

Fig. 4.8 *Uniform linear array*

Assume that N = number of antennas in the array

d = spacing between antenna elements

$r_1, r_2, r_3, ..., r_N$ = ray paths of 1, 2, 3, ..., N^{th} antenna to P.

Let P be a point far away, λ be the operating wavelength.

If the field of antenna 1 is E_1,

the field of antenna 2 is $E_1 e^{j\psi}$ and

the field of antenna 3 is $E_1 e^{j2\psi}$

Similarly, the field of antenna, N is $E_1 e^{j(N-1)\psi}$.

The total field is the vectorial sum of all the fields.

That is,
$$E_R = E_1 (1 + e^{j\psi} + e^{j2\psi} + \ldots + e^{j(N-1)\psi}) \qquad \ldots(4.14)$$

where
$$\psi = \beta d \cos \phi + \alpha_e$$

$$\alpha_e = \text{progressive phase shift between the antennas.}$$

Equation (4.14) is in geometric progression.

Multiplying both sides of Equation (4.14) by $e^{j\psi}$, we get
$$E_R\, e^{j\psi} = E_1 (e^{j\psi} + e^{j2\psi} + \ldots + e^{jN\psi}) \qquad \ldots(4.15)$$

Subtracting Equation (4.14) from Equation (4.15), we get
$$E_R (e^{j\psi} - 1) = E_1 (e^{jN\psi} - 1)$$

or
$$\frac{E_R}{E_1} = \frac{(e^{jN\psi} - 1)}{(e^{j\psi} - 1)}$$

The normalised magnitude of field strength
$$\left| \frac{E_R}{E_1} \right| = E = \left| \frac{(e^{jN\psi} - 1)}{(e^{j\psi} - 1)} \right| \qquad \ldots(4.16)$$

$$
\begin{aligned}
|e^{j\psi} - 1| &= |\cos \psi + j \sin \psi - 1| \\
&= \sqrt{(\cos \psi - 1)^2 + \sin^2 \psi} \\
&= \sqrt{1 + \cos^2 \psi - 2 \cos \psi + \sin^2 \psi} \\
&= \sqrt{2(1 - \cos \psi)} \\
&= \sqrt{2}\ \sqrt{(1 - \cos \psi)}
\end{aligned}
$$

or
$$|(e^{j\psi} - 1)| = 2 \sin \frac{\psi}{2} \qquad \ldots(4.17)$$

Similarly,
$$|(e^{jN\psi} - 1)| = 2 \sin \frac{N\psi}{2} \qquad \ldots(4.18)$$

$$E = \left| \frac{\sin \dfrac{N\psi}{2}}{\sin \dfrac{\psi}{2}} \right| \qquad \ldots(4.19)$$

The normalised field strength, E is plotted as a function of ψ. A typical variation is shown in Fig. 4.9 for $\alpha_e = 0$.

Salient features of uniform linear array:

1. The maximum value of normalised field strength, $E = \left| \dfrac{E_R}{E_1} \right| = N$.

2. The maximum value occurs at $\psi = 0$.

3. The maximum value, N at $\psi = 0$ is called **principal maximum** of the array.

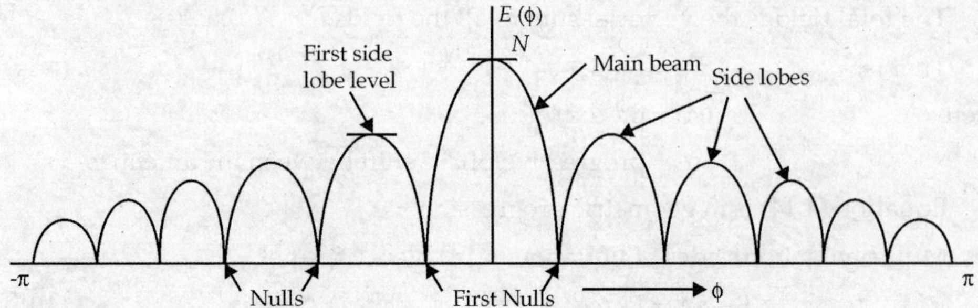

Fig. 4.9 *Variation of E with ϕ*

4. The minimum value of E occurs at $\dfrac{N\psi}{2} = \pm\, K\pi,\ K = 1,\ 2,\ 3,\ \dots$ These **minimas** are called **Nulls**.

5. Secondary maximum occurs approximately between the nulls. These secondary maximas occur when the numerator of Equation (4.19) becomes maximum.

   That is, secondary maxima occur at

$$\frac{N\psi}{2} = \pm\,(2m+1)\,\frac{\pi}{2}, \qquad\qquad m = 1,\ 2,\ 3,\ \dots$$

6. The first secondary maximum is called the **first side lobe level (SLL)**.

7. The ratio of the first secondary maximum to the principal maximum is called the **side lobe ratio (SLR)**.

8. The angular difference between the first nulls on either side of the main beam is called **Null-to-Null beam width**.

9. The variation of radiated power P (proportional to E^2) with ϕ is called **power pattern**.

10. The angular width between three dB points of the main beam is called **half-power beam width**.

11. The first side lobe ratio of uniform linear array is 0.212 (or -13.47 dB).

12. The Null-to-Null beam width of broadside array $= \dfrac{2\lambda}{Nd}$.

13. The Null-to-Null beam width of end-fire array $= 2\sqrt{\dfrac{2\lambda}{Nd}}$.

14. The Null-to-Null beam width of end fire array is greater than that of broadside array.

15. The half power beam width is approximately half of Null-to-Null beam width.

16. As $Nd = $ length of the array, Null-to-Null beam width of broadside array

$$\text{B.W.F.N.} = \frac{2\lambda}{l} = \frac{2}{l/\lambda}$$

where $\qquad l = $ length of the array

So $\quad$ H.P.B.W. $= \dfrac{1}{2} \times$ B.W.F.N. $= \dfrac{1}{l/\lambda} = \dfrac{114.6}{l/\lambda}$, deg.

17. Null-to-Null beam width of end-fire array

$$2 \sqrt{\dfrac{2\lambda}{l}} = 2 \sqrt{\dfrac{2}{l/\lambda}}$$

or $\qquad$ H.P.B.W. $= \dfrac{1}{2} \times$ B.W.F.N. $= \sqrt{\dfrac{2}{l/\lambda}} = \dfrac{57.3}{l/\lambda}$ deg.

18. The directivity of broadside array, $D = 2\,(l/\lambda)$.
19. The directivity of end-fire array, $D = 4\,(l/\lambda)$.

4.11 FIRST SIDE LOBE RATIO (SLR)

First Side Lobe Ratio (SLR) is defined as the ratio of the first side lobe level to the main beam level. That is

$$\text{SLR} \equiv \frac{\text{first side lobe level}}{\text{main side lobe level}}$$

SLR for uniform array is –13.5 dB.

Proof$\quad$From Equation (4.19), secondary maxima occurs approximately at the centre between the nulls. That is, they occur at

$$\frac{N\psi}{2} = \pm(2m+1)\,\frac{\pi}{2}, \ m = 1, 2, 3, \dots \qquad \dots(4.20)$$

Hence, the first side lobe maximum occurs at

$$\frac{N\psi}{2} = \pm(2+1)\,\frac{\pi}{2} = \frac{3\pi}{2} \qquad \dots(4.21)$$

Substituting Equation (4.21) in Equation (4.19), we get

$$E = \left| \frac{\sin\left(\dfrac{3\pi}{2}\right)}{\sin\left(\dfrac{3\pi}{2N}\right)} \right| = \left| \frac{1}{\sin\left(\dfrac{3\pi}{2N}\right)} \right|$$

For large values of N, $\dfrac{3\pi}{2N}$ is very small.

So $\qquad \left| \dfrac{1}{\sin\left(\dfrac{3\pi}{2N}\right)} \right| \approx \dfrac{1}{\left(\dfrac{3\pi}{2N}\right)}$

That is, $\qquad E = \dfrac{2N}{3\pi}$

Therefore, the amplitude of the first secondary lobe is $\left(\dfrac{2N}{3\pi}\right)$.

The amplitude of main lobe or principal maximum is N.

So
$$\text{SLR} = \frac{\dfrac{2N}{3\pi}}{N}$$

or
$$\text{SLR} = \frac{2}{3\pi} = 0.212$$

Hence
$$(\text{SLR})_{\text{dB}} = 20 \log_{10}(0.212) = -13.47 \text{ dB}. \qquad \qquad \text{...(4.22)}$$

4.12 BROADSIDE AND END-FIRE ARRAYS

Broadside Array is defined as an array of elements for which radiation maximum (or main beam) occurs perpendicular to the axis of the array, that is at $\phi = 90°$

We have
$$\psi = \beta d \cos \phi + \alpha_e$$

We know maximum of E occurs at $\psi = 0$ and for broadside array it occurs at $\phi = 90°$.

That is,
$$0 = \beta d \cos(90) + \alpha_e$$

$$\boxed{\alpha_e = 0 \text{ is the condition for broadside array}}$$

The Null-to-Null beam width for broadside array is

$$\text{B.W.} = \frac{2\lambda}{Nd} \qquad \qquad \text{...(4.23)}$$

Proof   We have

$$E = \left| \frac{\sin \dfrac{N\psi}{2}}{\sin \dfrac{\psi}{2}} \right|$$

From this, it is evident that the nulls appear at

$$\frac{N\psi}{2} = \pm K\pi, \quad K = 1, \ 2, \ 3, \ ...$$

First Null occurs at $(K = 1)$

$$\frac{N\psi_1}{2} = \pm \pi$$

or
$$\psi_1 = \frac{2\pi}{N} \qquad \qquad \text{...(4.24)}$$

Now consider
$$\psi = \beta d \cos \phi + \alpha_e$$
$$= \beta d \cos \phi \qquad\qquad (\text{as } \alpha_e = 0)$$

First Null occurs at $\psi = \psi_1$. That is,

$$\psi = \psi_1 = \beta d \cos \left(\frac{\pi}{2} + \Delta\phi\right) \qquad\qquad ...(4.25)$$

This is clear from Fig. 4.10.

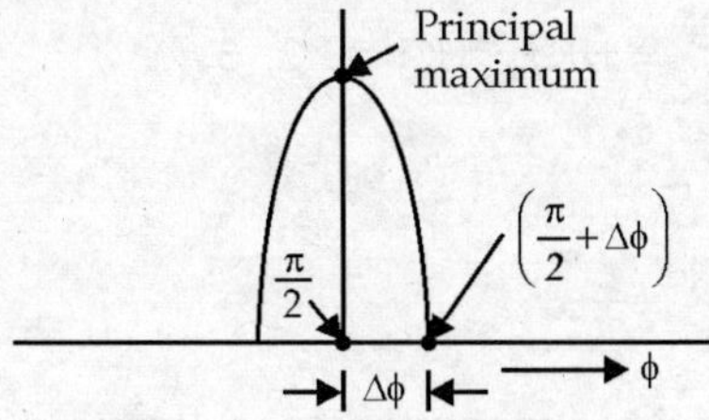

Fig. 4.10   *Radiation beam of broadside array*

From Equations (4.24) and (4.25), we have

$$\frac{2\pi}{N} = \beta d \cos \left(\frac{\pi}{2} + \Delta\phi\right)$$

$$\approx \beta d \, \Delta\phi$$

or
$$\Delta\phi = \frac{2\pi}{\beta d N}$$

But
$$\text{B.W.} = 2\Delta\phi = \frac{2 \times 2\pi}{\dfrac{2\pi}{\lambda} Nd}$$

$$\text{B.W.} = \frac{2\lambda}{Nd} \text{ for broadside array}$$

End-fire Array is defined as an array of elements for which radiation maximum (or main beam) occurs along the axis of the array, that is, $\phi = 0°$ 

Consider
$$\psi = \beta d \cos \phi + \alpha_e$$

Beam maximum is at $\psi = 0$ and also at $\phi = 0$

or
$$0 = \beta d \cos (0) + \alpha_e$$

$$\alpha_e = -\beta d \text{ is the condition for end-fire array}$$

Null-to-Null beam width of end-fire array is

$$\text{B.W.} = 2\sqrt{\frac{2\lambda}{Nd}} \qquad\qquad ...(4.26)$$

Proof   Consider

$$E = \left| \frac{\sin \dfrac{N\psi}{2}}{\sin \dfrac{\psi}{2}} \right|$$

From this, it is evident that Nulls occur at

$$\frac{N\psi}{2} = \pm K\pi, \quad K = 1,\ 2,\ 3, \ldots$$

First Null occurs at $(K = 1)$

$$\frac{N\psi}{2} = \pm \pi$$

Consider the negative sign for convenience, that is,

$$\psi_1 = -\frac{2\pi}{N} \qquad \ldots(4.27)$$

The corresponding ψ_1 from

$$\psi = \beta d \cos \phi + \alpha_e \text{ is}$$
$$\psi_1 = \beta d \cos \Delta\phi - \beta d \qquad \ldots(4.28)$$

This is clear from Fig. 4.11.

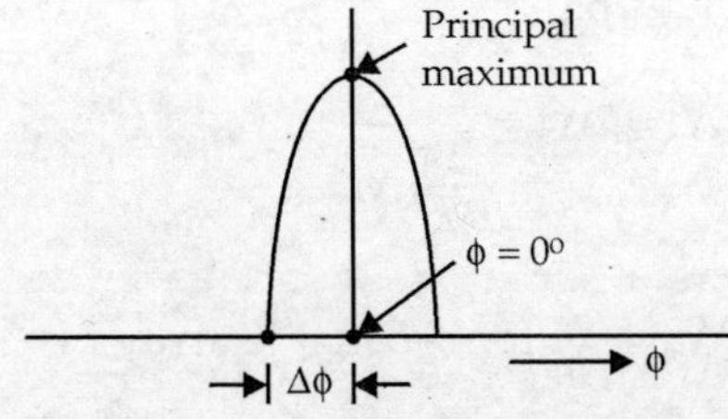

Fig. 4.11   *Radiation beam of end-fire array*

From Equations (4.27) and (4.28), we get

$$-\frac{2\pi}{N} = \beta d \cos \Delta\phi - \beta d$$

$$= \beta d (\cos \Delta\phi - 1) \qquad \ldots(4.29)$$

But $\qquad \cos \Delta\phi = 1 - \dfrac{\Delta\phi^2}{2!} + \dfrac{\Delta\phi^4}{4!} + \ldots$

For small $\Delta\phi$,

$$\cos \Delta\phi \approx 1 - \frac{\Delta\phi^2}{2} \qquad \ldots(4.30)$$

From Equations (4.29) and (4.30), we get

$$-\frac{2\pi}{N} = \beta d \left[1 - \frac{\Delta\phi^2}{2} - 1 \right]$$

$$= -\frac{\Delta\phi^2}{2} \beta d$$

or

$$\Delta\phi^2 = \frac{4\pi}{\beta d N} = \frac{4\pi}{\dfrac{2\pi}{\lambda} Nd} = \frac{2\lambda}{Nd}$$

That is,

$$\Delta\phi = \sqrt{\frac{2\lambda}{Nd}}$$

$$\boxed{B.W = 2\Delta\phi = 2\sqrt{\frac{2\lambda}{Nd}}} \qquad\qquad ...(4.31)$$

4.13 PATTERNS OF ARRAY OF NON-ISOTROPIC RADIATORS

As mentioned in the preceding sections, true isotropic nature of radiators does not exist. Almost all the practical elements are non-isotropic. In view of this, it is required to know the radiation pattern of non-isotropic elements.

Consider Fig. 4.12 in which an array of two non-isotropic radiators are shown.

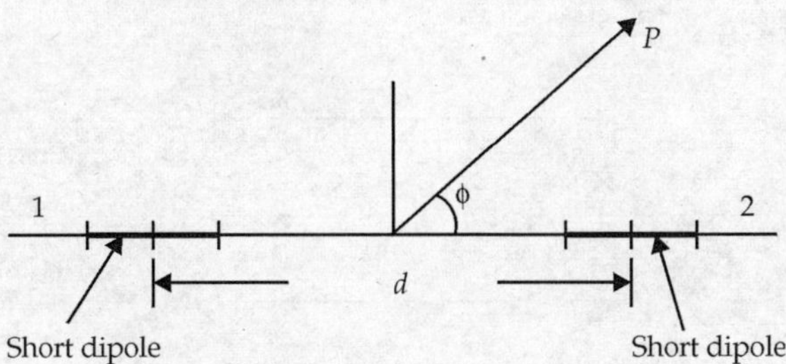

Fig. 4.12 *Two-element array of short dipoles*

The field pattern of an isolated short dipole

$$E_1 = E_0 \sin \phi \qquad\qquad ...(4.32)$$

As derived in the preceding Section (4.7) the field of a two element array is

$$E = 2E_1 \cos \psi/2 \qquad\qquad ...(4.33)$$

Here

$$\psi = \beta d \cos \phi + \alpha_e$$

From Equations (4.32) and (4.33), we have

$$E = 2E_0 \sin \phi \cos \psi/2 \qquad\qquad ...(4.34)$$

The maximum value of E is $2E_0$

or

$$(E)_{\max} = E_m = 2E_0$$

The normalised value of E is

$$\frac{E}{E_m} = \frac{E}{2E_0} = \sin \phi \; \cos \frac{\psi}{2} \qquad \text{...(4.35)}$$

or $\qquad\qquad\qquad E_n = \sin \phi \; \cos \psi/2 \qquad\qquad\qquad \text{...(4.36)}$

That is, $\qquad\qquad E_n$ = pattern of individual element × pattern of
array of two elements

The element pattern of an isotropic radiator is unity.

4.14 MULTIPLICATION OF PATTERNS

Principle of pattern multiplication states that the radiation pattern of an array is the product of the pattern of the individual antenna with the array pattern. The array pattern is a function of the location of the antennas in the array and their relative complex excitation amplitudes.

Advantage of method of multiplication. It helps to sketch the radiation pattern of array antennas rapidly from the simple product of element pattern and array pattern.

Disadvantage The principle is applicable only for arrays containing identical elements.

Example 4.1 Consider an array of four elements (isotropic or non-directional) of Fig. 4.13.

Fig. 4.13 *Array of four elements*

The elements are spaced at $\lambda/2$. The elements 1 and 2 are considered as one unit, 3 and 4 are considered as one unit as in Fig. 4.13. Since the elements are identical, both the units have the same radiation pattern.

The unit pattern is the pattern of two elements spaced at $\lambda/2$. This is given in Fig. 4.14.

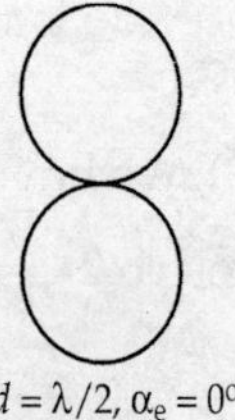

Fig. 4.14 *Pattern of two isotropic elements spaced $\lambda/2$ and excitation phase $\alpha_e = 0$*

The units represented by A and B are separated by λ. These two units are considered to be one unit whose radiation pattern is shown in Fig. 4.15.

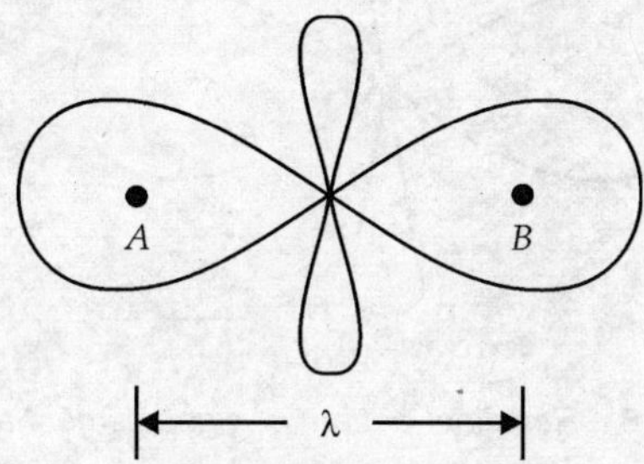

Fig. 4.15 *Pattern of two elements separated by λ and excitation phase, $\alpha_e = 0^\circ$*

The resultant pattern is given by the product of unit pattern ($\lambda/2$ spacing) of 1 and 2 elements or 3 and 4 elements and a group pattern (λ spacing) of A and B. That is, the product of Fig. 4.14 and Fig. 4.15 as shown in Fig. 4.16.

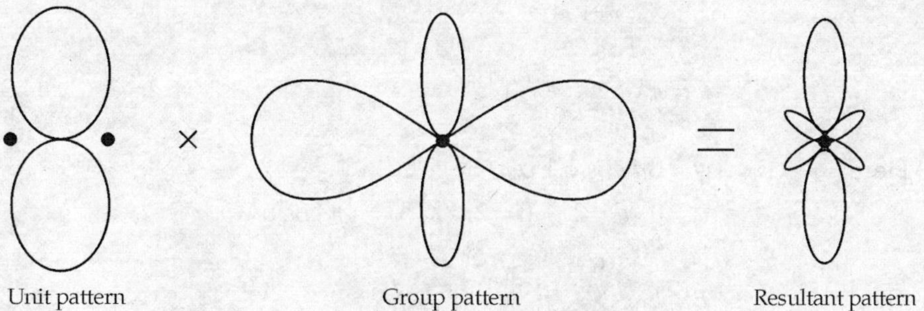

Fig. 4.16 *Resultant pattern of four isotropic elements spaced at $\lambda/2$*

Example 4.2   Consider an array of eight elements spaced at $\lambda/2$ (Fig. 4.17).

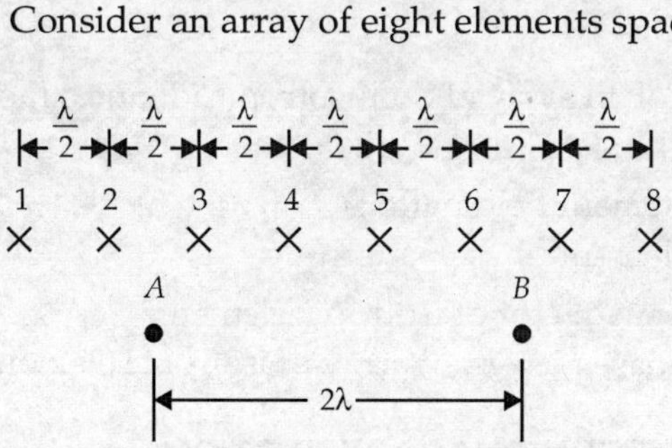

Fig. 4.17 *Array of eight elements*

Centre of the first four elements and last four elements are marked as A and B. The unit pattern is pattern of four elements and group pattern is the pattern of two elements spaced at 2λ. The resultant pattern is again the product of unit pattern and group pattern (Fig. 4.18).

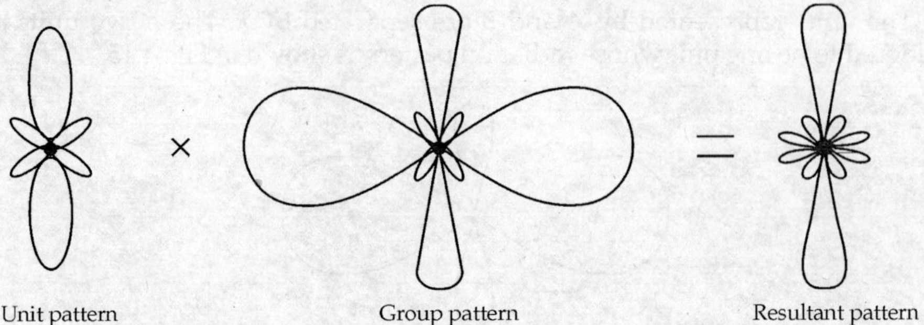

Fig. 4.18 *Resultant pattern of eight-element array*

The resultant pattern of eight element array is also obtained analytically. That is,

$$E = E_0 \left(1 + e^{j\psi} + \ldots + e^{j(n-1)\psi}\right)$$

$$= E_0 \sum_{n=1}^{N} e^{j(n-1)\psi} \qquad \ldots(4.37)$$

or

$$E = E_0 \sum_{n=1}^{8} e^{j(n-1)\psi} \qquad \ldots(4.38)$$

Here

$$\psi = \beta d \cos\phi + \alpha_e$$

$$\alpha_e = 0, \ d = 2\lambda, \ \beta = 2\pi/\lambda, \ \psi = 4\pi \cos\phi$$

So

$$E = E_0 \sum_{n=1}^{8} e^{j(n-1)4\pi \cos\phi}. \qquad \ldots(4.39)$$

4.15 GENERALISED EXPRESSION OF PRINCIPLE OF PATTERN MULTIPLICATION

The resultant pattern of an array of non-isotropic identical radiators is given by

$$E = f(\theta, \phi) \ F(\theta, \phi) \times (f_P(\theta, \phi) + F_P(\theta, \phi)$$

where

$f(\theta, \phi) = $ element field pattern

$f_P(\theta, \phi) = $ element phase pattern

$F(\theta, \phi) = $ array factor of isotropic elements

$F_P(\theta, \phi) = $ phase pattern of array of isotropic elements.

4.16 RADIATION PATTERN CHARACTERISTICS

The radiation pattern of any general antenna is bi-directional or multi-directional. It contains a main lobe and adjacent minor or side lobes. The presence of side lobes is undesirable and they create electromagnetic interference (EMI) in the radar receivers.

The EMI creates false targets in the radar display. The presence of EMI is also prone to jamming. The width of the main lobe is required to be small. But the general tendency is that if the side lobe levels are reduced, the beam width increases and vice-versa.

Ideally, there should be no side lobes as there will be waste of power in undesired directions of radiation and reduced gain in the desired directions.

It is possible to reduce the side lobes by appropriate tapered amplitude excitations to the array elements.

It is possible to reduce the beam width by increasing the array length and the number of elements in the array. By increasing the array length, directivity can be increased.

4.17 BINOMIAL ARRAYS

Binomial array is an array whose elements are excited according to the current levels determined by the binomial coefficient, nc_r, n being the number of element in the array.

The **advantage** of the binomial array is that there will be **no side lobes** in the resultant pattern of the array. The **disadvantage** is that the **beam width of the main lobe is large** which is undesirable. The directivity is small and high excitation levels are required for the centre elements of large arrays.

For the arrays of two to nine elements, the relative amplitude levels are tabulated in Table 4.1.

Table 4.1   Excitation levels of binomial array for n = 2 to 9

Number of elements	Relative amplitude
$n = 2$	1, 1
3	1, 2, 1
4	1, 3, 3, 1
5	1, 4, 6, 4, 1
6	1, 5, 10, 10, 5, 1
7	1, 6, 15, 20, 15, 6, 1
8	1, 7, 21, 35, 35, 21, 7, 1
9	1, 8, 28, 56, 70, 56, 28, 8, 1

The excitation levels for any number of radiating elements can be found from Pascal's triangle. In this triangle, each internal integer is the sum of the above adjacent integers leaving the side integers. This is shown in Table 4.2.

Table 4.2   Pascal's triangle

								1							
							1		1						
						1		2		1					
					1		3		3		1				
				1		4		6		4		1			
			1		5		10		10		5		1		
		1		6		15		20		15		6		1	
	1		7		21		35		35		21		7		1
1		8		28		56		70		56		28		8		1

4.18 EFFECT OF EARTH ON VERTICAL PATTERNS

When an antenna is not isolated and is close to the earth, earth has considerable effect on the patterns. This is significant due to the flow of current in the earth which modifies the patterns. The magnitude and phase of these currents depend on frequency, conductivity and permittivity of the earth. In fact, earth is a perfect conductor at low and medium frequencies. On the other hand, metals are perfect conductors at radio frequencies.

In Fig. 4.19, one vertical antenna and one horizontal antenna are shown. Direct and reflected rays, charges, image charges and currents are also shown.

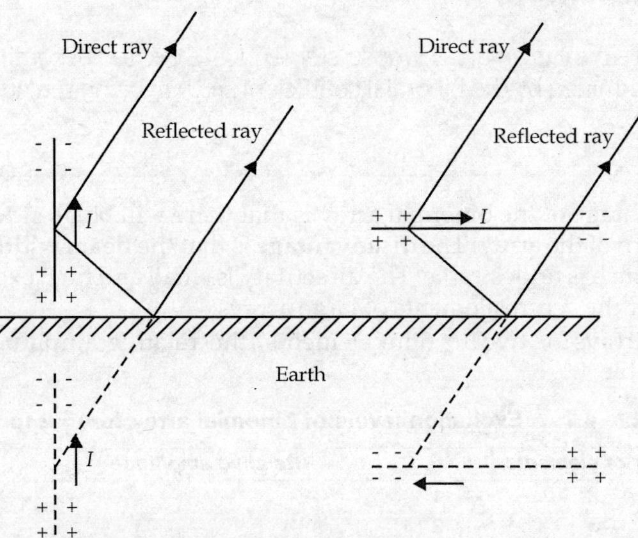

Fig. 4.19 *Actual and image antennas*

The actual and image charges have opposite polarity. In the case of vertical antennas, currents are in the same direction. But in horizontal antennas, they have opposite directions.

Boundary conditions at the surface of a perfect conductor   The tangential component of *E* and the normal component of *H* are zero. This means **E** is normal and **H** is tangential at the surface. The field of such antennas is the resultant of actual and image antennas. The image principle can be used to obtain the resultant field.

Radiation patterns of vertical antennas above the earth   When a vertical antenna is above the earth, its image is created by the earth. That is, the earth is replaced by an image antenna. If antenna height is $\lambda/2$ from the earth, its image height is λ from the actual antenna as in Fig. 4.20.

The resultant field pattern (Fig. 4.20) is the product of vertical pattern of the antenna and the vertical pattern of two uniformly excited non-directional radiators spaced at λ.

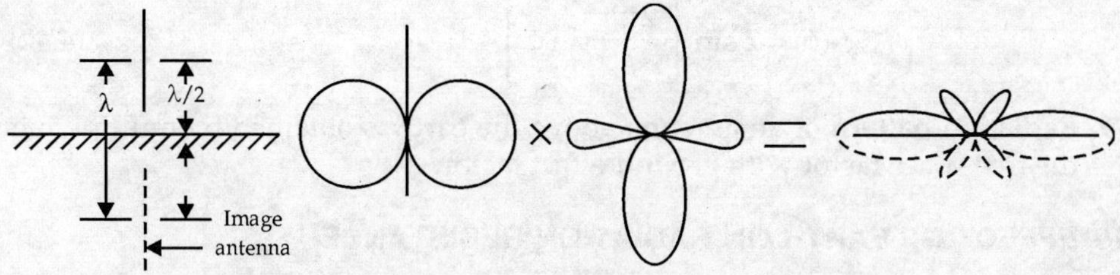

Fig. 4.20 *Vertical pattern of vertical antenna above the earth*

Radiation patterns of horizontal antennas above the earth   For a horizontal antenna, the vertical pattern is the product of the vertical pattern of the antenna and the vertical pattern of two uniformly excited non-directional radiators spaced at λ. The resultant pattern appears as shown in Fig. 4.21.

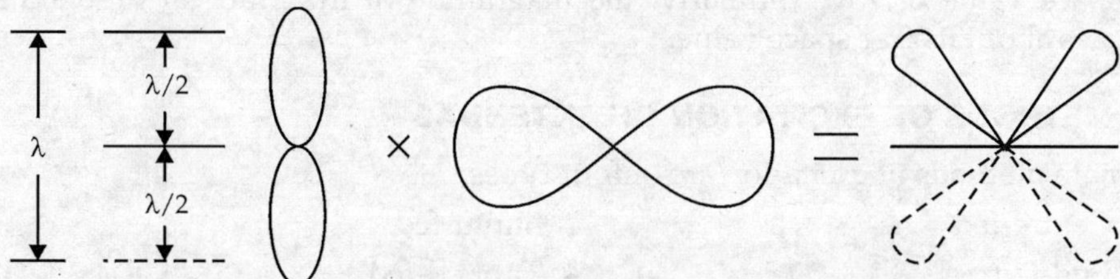

Fig. 4.21 *Vertical pattern of horizontal antenna above the earth*

Summary of the effects of earth on vertical radiation patterns:

1. The radiation pattern becomes that of a two-element array as earth introduces a second element called image antenna.
2. The directivity is increased.
3. The effect of earth on vertical antennas is different from that on horizontal antennas.
4. The effect of earth on the radiation pattern depends on the length of antennas also.
5. It depends on type of earth.
6. The effect of earth depends on frequency, permittivity and conductivity of the earth.
7. For a vertical antenna, the ground effect factor of a perfectly conducting earth is

$$f(\phi) = 2\cos\left[\frac{2\pi h}{\lambda}\sin\phi\right] \qquad \qquad ...(4.41)$$

where h = height of antenna

ϕ = angle of elevation above the horizontal.

8. For a horizontal antenna in the plane perpendicular to the axis of the antenna, the ground factor is

$$f(\phi) = 2 \sin \left[\frac{2\pi h}{\lambda} \sin \phi \right] \qquad \qquad ...(4.42)$$

9. Radiation pattern of an antenna above the earth is obtained by multiplying the free space factor with ground effect factor.

4.19 EFFECT OF EARTH ON RADIATION RESISTANCE

The theoretical radiation resistance of a thin isolated half-wave dipole is 73Ω. But its value lies between 65 and 73Ω if its thickness is finite. Moreover, antennas are always erected above the ground and hence free space conditions are not applicable.

The effect of earth on input impedance of antennas is taken care of by replacing the earth by its image. In fact, the input impedance of a theoretical half-wave dipole varies with the height above the perfectly conducting ground plane. It consists of resistive and reactive components. The magnitude of resistance oscillates about the free space value of 73Ω. Similarly, the magnitude of the practical dipole also oscillates about its free space value.

4.20 METHODS OF EXCITATION OF ANTENNAS

The main methods of excitation are of four types:

1. Centre-fed
2. Shunt-fed
3. End-fed
4. Tapped-fed

1. **Centre-fed** Here a dipole is fed at the centre by an open-wire transmission line.

2. **Shunt-fed** This is also called Delta-match excitation. This provides good impedance match and the standing waves are eliminated by choosing proper dimensions. The feeding is done through solid dielectric low impedance lines.

3. **End-fed** This is the simplest method of excitation. Antenna is end-fed with a single vertical wire. But it also radiates which is undesirable. Moreover, standing waves exist due to improper impedance matching between the antenna and the line. This results in high VSWR.

4. **Tapped fed** Horizontal antenna is tapped by a vertical wire which results in lower impedance point. This reduces radiation by the wire. It has a travelling wave current distribution. Standing waves are reduced as the mismatch is reduced.

4.21 IMPEDANCE MATCHING TECHNIQUES

Impedance matching can be done by using

1. Stubs
2. Folded dipoles
3. Baluns

1. **Stubs** A stub is a piece of transmission line one end of which is connected to the transmission line for matching with the load impedance. Its purpose is to tune out the reactance. The second end of the stub can be open circuited or short circuited. Short circuited stub lines are usually used to avoid radiation losses.

The method of stub matching of a pair of transmission lines is shown in Fig. 4.22.

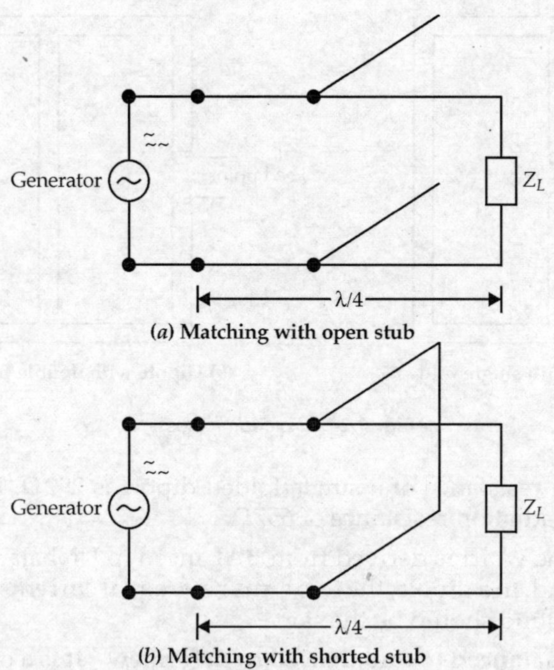

(a) Matching with open stub

(b) Matching with shorted stub

Fig. 4.22 *Stub matching of a pair of transmission lines*

The matching of a coaxial line is shown in Fig. 4.23.

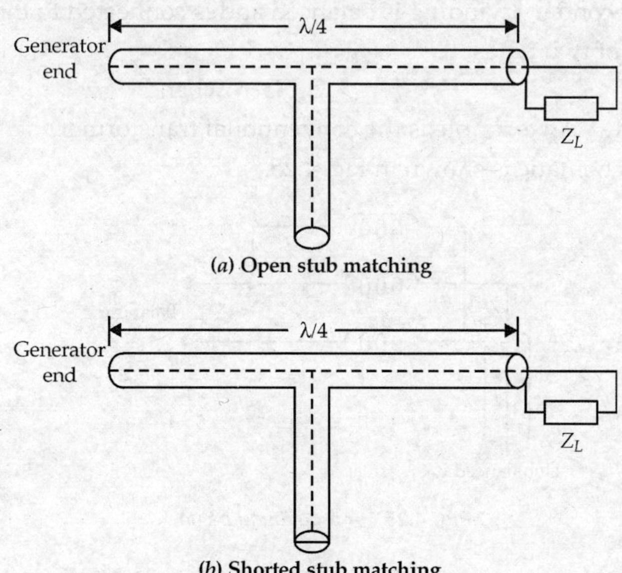

(a) Open stub matching

(b) Shorted stub matching

Fig. 4.23 *Stub matching of coaxial transmission lines*

2. **Folded dipoles**   These are used to increase the impedance. Typical folded dipoles are shown in Fig. 4.24.

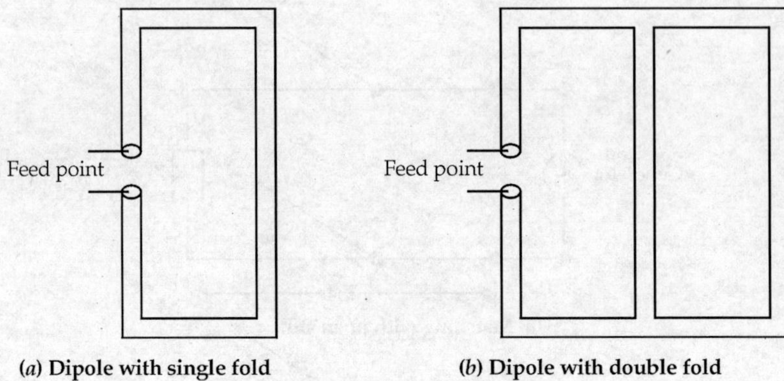

Fig. 4.24   *Folded dipoles*

The radiation resistance of a single folded dipole is $292\,\Omega$. A folded double dipole has a radiation resistance of $657\,\Omega$.

3. **BALUN**   The word is derived from **BAL**anced to **UN**balanced. A dipole is a balanced load. In a dipole, the two arms have equal currents. The impedance of arms to ground should be the same.

A balun is a balanced to unbalanced transformer   It is a circuit element. It is usually connected between a balanced line and an unbalanced line or an antenna. As a coaxial line is unbalanced, a balun is connected between a coaxial line and a dipole. Transmission lines are used as baluns at high frequencies and centre tapped transformers are used as baluns at lower frequencies. The primary is unbalanced. Centre tapped secondary winding is balanced and is connected to the antenna.

Baluns are of two types:

1. Broadband   2. Narrowband

1. **Broadband**   An example is the conventional transformer.

A typical schematic is shown in Fig. 4.25.

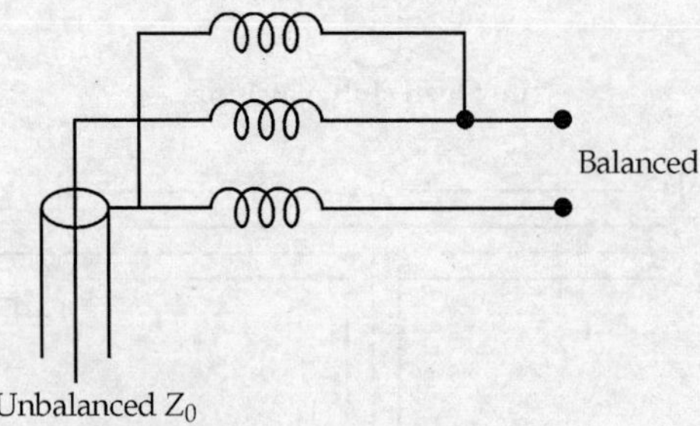

Fig. 4.25   *A broadband balun*

In this figure, a coaxial cable of $75\,\Omega$ or $50\,\Omega$ is used to feed a dipole. Its impedance transfer ratio is $1 : 1$.

2. **Narrowband Balun**   An example is bazooka or sleeve or choke.

A dipole is connected to the inner conductor of a coaxial line with a sleeve around it (Fig. 4.26).

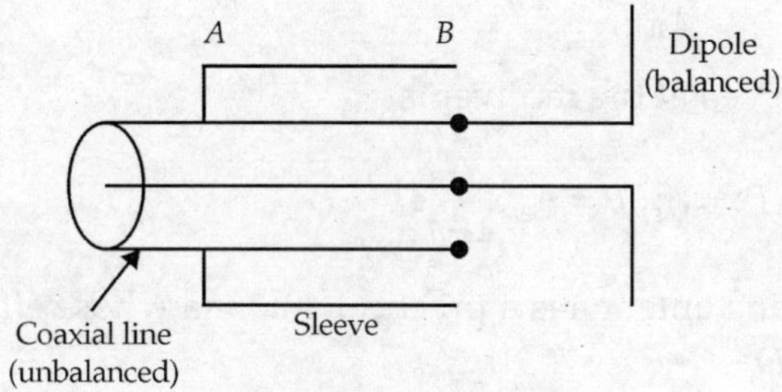

Fig. 4.26 *Bazooka balun*

4.22 TRANSMISSION LOSS BETWEEN TRANSMITTING AND RECEIVING ANTENNAS (FRIIS FORMULA)

Transmission loss is a measure of the ratio of radiated power to the received power. This is usually expressed in decibels.

Transmission loss in a communication system is defined as the loss which occurs due to the propagation medium between the transmitting and receiving antennas.

It is also defined as the **ratio of power radiated by the transmitting antenna to the resultant power which would be available from an equivalent loss-less receiving antenna**. This loss depends on the distance between the transmitting and receiving antennas and frequency.

The actual **transmission loss**, L_a is defined as

$$L_a \equiv \frac{W_T}{W_R} = 10 \log \left\{ \left(\frac{4\pi d}{\lambda} \right)^2 \frac{1}{G_T G_R} \right\} \qquad \dots(4.47)$$

The basic transmission loss, L_b is defined as

$$L_b \equiv 10 \log_{10} \left(\frac{4\pi d}{\lambda} \right)^2 \quad [G_T = G_R = 1] \qquad \dots(4.48)$$

where
$\qquad W_T$ = radiated power

$\qquad W_R$ = received power

$\qquad d$ = distance between transmitting and receiving antennas

$\qquad \lambda$ = operating wavelength

$\qquad G_T = TX$ antenna gain

$\qquad G_R = RX$ antenna gain.

Proof If W_T is the power radiated by the isotropic antenna, the power available per unit area at the receiving antenna is

$$P = \frac{W_T}{4\pi d^2} \qquad \qquad ...(4.49)$$

The power received at the receiver is

$$P_r = A_{er}\, P = A_{er}\, \frac{W_T}{4\pi d^2} \qquad \qquad ...(4.50)$$

If the transmitting antenna is a practical antenna whose effective area is A_{et}, then the directivity, D_t

$$D_t = \frac{4\pi}{\lambda^2}\, A_{et} \qquad \qquad ...(4.51)$$

Hence power received at the receiver is

$$W_R = P_r\, D_t$$

$$= A_{er}\, \frac{W_T}{4\pi d^2} \times \frac{4\pi}{\lambda^2}\, A_{et}$$

$$\frac{W_R}{W_T} = \frac{A_{er}\, A_{et}}{\lambda^2\, d^2} \qquad \qquad ...(4.52)$$

But $\qquad \qquad A_{er} = \dfrac{\lambda^2}{4\pi}\, G_R \qquad \qquad ...(4.53)$

and $\qquad \qquad A_{et} = \dfrac{\lambda^2}{4\pi}\, G_T \qquad \qquad ...(4.54)$

From Equation (4.52) and Equation (4.54), we get

$$\frac{W_R}{W_T} = \frac{\lambda^2}{4\pi} \times \frac{\lambda^2}{4\pi} \times \frac{G_R\, G_T}{\lambda^2\, d^2}$$

$$= \frac{G_R\, G_T\, \lambda^2}{(4\pi d)^2}$$

or $\qquad \qquad \dfrac{W_T}{W_R} = \left(\dfrac{4\pi d}{\lambda}\right)^2 \dfrac{1}{G_R\, G_T}$

$$\boxed{\text{Transmission loss in dB} = 10 \log\left[\left(\frac{4\pi d}{\lambda}\right)^2 \frac{1}{G_R\, G_T}\right]} \qquad \qquad ...(4.55)$$

Here, $\left(\dfrac{\lambda}{4\pi d}\right)^2$ is called **free space loss factor or basic loss factor**.

Equation (4.55) is known as **FRIIS formula**. This formula relates the power received, W_R and the radiated power W_T.

4.23 ANTENNA TEMPERATURE AND SIGNAL-TO-NOISE RATIO

The **antenna temperature** (T_A) is defined as the temperature of far or distant regions of space and near surroundings which are coupled to the antenna through radiation resistance.

It may be noted that the antenna temperature has no relation with the physical temperature of the antenna. It depends on the temperature of the regions to which the antenna is radiating. In this connection, a receiving antenna is regarded as a remote sensing, temperature measuring device.

Expression for S/N   According to Nyquist relation, the noise power available from a resistor R is given by

$$P_a = k\,T\,B \qquad \qquad ...(4.56)$$

where
$$k = \text{Boltzmann's constant} = 1.37 \times 10^{-23}\ \text{joules/}^\circ\text{K}$$

$$T = \text{absolute temperature, } ^\circ\text{K}$$

$$B = \text{bandwidth, Hz}$$

From Equation (4.56), it is obvious that the noise power is independent of R and is directly proportional to T.

The maximum power available from any source whose resistance is R, is

$$P_a = \frac{V^2}{4R} \qquad \qquad ...(4.57)$$

The thermal noise voltage across R at $T\,^\circ$K is given by

$$V = 2\sqrt{KTBR} \qquad \qquad ...(4.58)$$

If the receiver to which the antenna is connected has a gain of G, signal and source noise are amplified. Moreover, noise is also added by the receiver. As a result, input signal-to-noise ratio is decreased at the output.

If S_A is the input signal produced by the receiving antenna at a temperature of T_A, the output signal power S_0 from an amplifier of power gain G is given by $S_A\,G$. On the other hand, the output noise power (N_0), is the sum of the amplified antenna noise and the receiver noise P_N.

So
$$N_0 = KT_A\,BG + P_N$$

or
$$N_0 = KT_A\,BG + KT_e\,BG \qquad \qquad ...(4.59)$$

Here T_e is the effective noise temperature of the receiver network. The output signal-to-noise ratio

$$\frac{S_0}{N_0} = \frac{S_A\,G}{(T_A + T_e)\,KBG}$$

$$\frac{S_0}{N_0} = \frac{S_A}{(T_A + T_e)\,KB} \qquad \qquad ...(4.60)$$

The effective temperature, T_e is related to Noise figure, F_N.

Noise figure, F_N is defined as the ratio of actual noise power output (if connected at input to a source at standard temperature $T_0 = (290°K)$ to the noise power output that would exist for the same input, if the network is noiseless. That is

$$F_N = \frac{KT_0\,BG + P_N}{KT_0\,BG} = \frac{KT_0\,BG + KT_e\,BG}{KT_0\,BG}$$

$$F_N = 1 + \frac{T_e}{T_0} \qquad\qquad ...(4.61)$$

Effective noise temperature (T_e) is defined as the temperature at the input of the network that would account for the noise ΔN at the output of the network.

$$T_e = T_0\,(F_N - 1). \qquad\qquad ...(4.62)$$

POINTS TO REMEMBER

1. Horizontal pattern is a pattern obtained for $\theta = 90°$.

2. Vertical pattern is a pattern obtained for $\phi = $ constant.

3. The principal planes are E and H-planes.

4. Vertical pattern of horizontal current element is a figure of eight.

5. Vertical pattern of vertical current element is a dumbbell.

6. Array is a group of elements.

7. Arrays are used to increase gain and directivity.

8. The resultant pattern of a two-element array is $E_R = 2E_A \left(\dfrac{\pi d \cos \phi}{\lambda} + \dfrac{\alpha_e}{2} \right)$.

9. The normalised field strength of a uniform linear array is $E = \left| \dfrac{\sin \dfrac{N\psi}{2}}{\sin \dfrac{\psi}{2}} \right|$.

10. Side Lobe Ratio (SLR) $= \dfrac{\text{first side lobe level}}{\text{main lobe level}}$.

11. For broadside array, $\alpha_e = 0$.

12. For end-fire array, $\alpha_e = -\beta d$.

13. SLR of a uniform linear array is -13.5 dB.

14. BW for broadside array is $\dfrac{2\lambda}{Nd}$.

15. BW for end-fire array is $2\sqrt{\dfrac{2\lambda}{Nd}}$.

16. In binomial arrays, the excitation levels are found from nc_r.

17. SLR is zero for a binomial array.

18. Stub is a piece of transmission line used for impedance matching.

19. Balun means BALanced to UNbalanced.

20. FRIIS formula is $L_a = 10 \log \left(\dfrac{4\pi d}{\lambda} \right)^2 \dfrac{1}{G_T G_R}$.

21. Free space loss factor is $\left(\dfrac{\lambda}{4\pi d} \right)^2$.

22. Noise figure is given by $F_N = 1 + \dfrac{T_e}{T_0}$.

23. Antenna signal-to-noise ratio is $\dfrac{S_0}{N_0} = \dfrac{S_A}{(T_A + T_e)\, KB}$.

24. Effective noise temperature is $T_e = T_0\,(F_N - 1)$.

 ## SOLVED PROBLEMS

Problem 4.1   Find out Null-to-Null beam width of a broadside array:

(a) when array length $= 10\lambda$ and number of elements $= 20$

(b) when array length $= 50\lambda$ and number of elements $= 100$

(c) when array length $= 20\lambda$ and number of elements $= 50$.

Solution   (a) Array length,

$$l = 10\lambda, \ N = 20$$

$$d = \frac{10\lambda}{20} = \frac{1}{2}\lambda$$

$$\text{B.W.} = \frac{2\lambda}{Nd} = \frac{2\lambda}{20 \times \dfrac{\lambda}{2}} = \frac{4}{20} = 0.2 \text{ radians}$$

(b) $l = 50\lambda, \ N = 100$

$$\therefore \quad d = \frac{50}{100}\lambda = \frac{\lambda}{2}$$

$$\text{B.W.} = \frac{2\lambda}{Nd} = \frac{2\lambda}{100 \times \dfrac{\lambda}{2}} = 0.04 \text{ radians}$$

(c) $l = 20\lambda, \; N = 50$

$\therefore \qquad d = \dfrac{20}{50}\lambda = \dfrac{2}{5}\lambda$

$$\text{B.W.} = \frac{2\lambda}{Nd} = \frac{2\lambda}{50 \times \dfrac{2}{5}\lambda} = \frac{1}{10} = 0.1 \text{ radians.}$$

Problem 4.2   Find the Null-to-Null beam width of end-fire array:

(a) when the array length, $l = 10\lambda$ and $N = 20$

(b) $1 = 50\lambda$ and $N = 100$

(c) $1 = 10\lambda, \; N = 50.$

Solution   (a) $l = 10\lambda, \; N = 20$

$$\text{B.W.} = 2\sqrt{\frac{2\lambda}{Nd}} = 2\sqrt{\frac{2\lambda}{20 \times (\lambda/2)}} = 0.8944 \text{ radians}$$

(b) $l = 50\lambda, \; N = 100$

$$\text{B.W.} = 2\sqrt{\frac{2\lambda}{Nd}} = 2\sqrt{\frac{2\lambda}{100 \times \dfrac{\lambda}{2}}} = 0.4 \text{ radians}$$

(c) $l = 20\lambda, \; N = 50$

$$\text{B.W.} = 2\sqrt{\frac{2\lambda}{Nd}} = 2\sqrt{\frac{2\lambda}{50 \times \dfrac{2}{5}\lambda}} = 0.632 \text{ radians.}$$

Problem 4.3   If an array of isotropic radiators is operated at a frequency of 6 GHz and is required to produce a broadside beam, find Null-to-Null beam width if the array length is 10 m. Also find the directivity.

Solution   Frequency of operation,

$$f = 6 \text{ GHz}$$

$$\lambda = \frac{3 \times 10^8}{6 \times 10^9} = 0.05 \text{ m}$$

Array length, $l = 10 \text{ m}$

Null-to-Null beam width

$$\text{B.W.F.N.} = \frac{2}{(l/\lambda)} = \frac{2}{10} \times 0.05$$

$$\boxed{\text{B.W.F.N.} = \frac{0.1}{10} = 0.01 \text{ radians}}$$

Directivity, $D = 2\left(\dfrac{l}{\lambda}\right)$

$$= 2 \times \frac{10}{0.05}$$

$$= \frac{20}{5} \times 10^2 = 400$$

$$D = 400.$$

Problem 4.4   A uniform linear array is required to produce an end-fire beam when it is operated at a frequency of 10 GHz. It contains 50 radiators and are spaced at 0.5λ. Find the progressive phase shift required to produce the end-fire beam. Find the array length.

Solution   Frequency $\quad = 10\,\text{GHz}$

$$\text{Wavelength} \quad = \frac{3 \times 10^8}{10 \times 10^9} = 0.03\,\text{m}$$

Number of radiators, $N = 50$

Element spacing, $\quad d = 0.5\lambda$

So the progressive phase shift,

$$\alpha_e = -\beta d$$

$$= \frac{2\pi}{\lambda} \times 0.5\lambda$$

$$\alpha_e = \pi \text{ radians}$$

Array length, $\quad l = Nd$

$$= 50 \times 0.5\lambda = 25\lambda = 25 \times 0.03$$

$$l = 0.75\,\text{m.}$$

Problem 4.5   An array contains 100 isotropic radiators with an inter element spacing of 0.5λ. It is required to produce broadside and end-fire beams

(a) Find Null-to-Null beam width and half-power beam width in degrees.

(b) Also find the directivity of both forms of arrays.

Solution   (a) $\quad N = 100,\ d = 0.5\lambda$

Array length, $\quad l = Nd = 100 \times 0.5 = 50\lambda$

$$\text{B.W.F.N.} = \frac{114.6}{(l/\lambda)} = \frac{114.6}{50\lambda/\lambda} = \frac{114.6}{50} = 2.292^\circ$$

$$\text{H.P.B.W.} = \frac{1}{2} \times \text{B.W.F.N.} = \frac{573}{(l/\lambda)} = 11.46^\circ$$

(b) Directivity of broad side array $= D = 2 \left(\frac{l}{\lambda}\right) = 2 \times 50\,\frac{\lambda}{\lambda} = 100$

and directivity of end-fire array,

$$D = 4 \left(\frac{l}{\lambda}\right) = 4 \times \left(\frac{50\lambda}{\lambda}\right) = 200.$$

Problem 4.6 Obtain the resultant pattern of an array of two directional (but point sources) short collinear dipoles of Fig. 4.27.

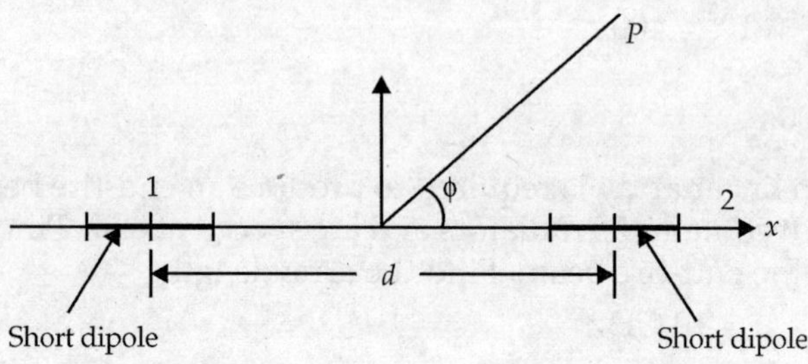

Fig. 4.27 *Array of short collinear dipoles*

Solution Normalised individual element pattern is given by

$$E_1 = \sin \phi$$

Normalised array factor,

$$E_a = \cos \psi/2, \quad \psi = \beta d \cos \phi + \alpha_e$$

If $\qquad \alpha_e = 0, \ d = \lambda/2$

normalised resultant pattern is given by

$$E = \text{element pattern} \times \text{array factor}$$
$$= \sin \phi \ \cos \psi/2$$

$$E = \sin \phi \cos \left(\frac{\pi}{2} \cos \psi \right)$$

The resultant pattern obtained from multiplication of pattern is shown in Fig. 4.28.

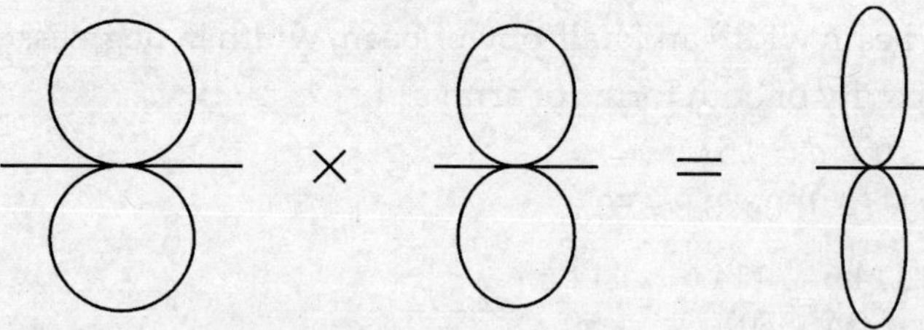

Fig. 4.28 *Resultant pattern of the array*

Problem 4.7 Obtain the resultant pattern of two short vertical dipoles of Fig. 4.29.

Solution Element pattern is

$$E_1 = E_2 = \cos \phi$$

Array factor $\qquad E_a = \cos \left(\frac{\pi}{2} \cos \phi \right), \ \text{for } \alpha_e = 0, \ d = \lambda/2$

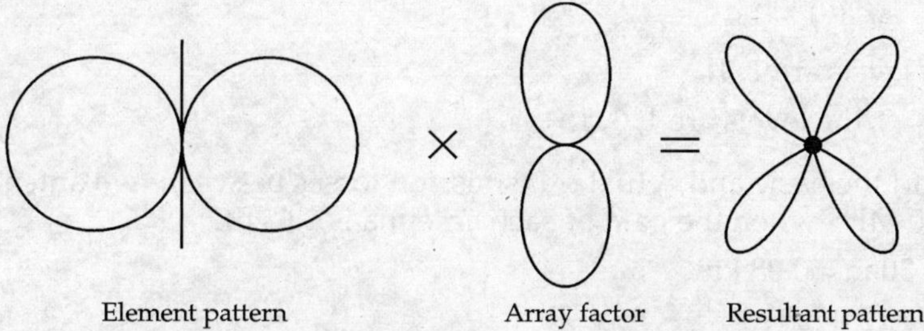

Fig. 4.29 *Array of two short vertical dipoles*

The resultant pattern is given by

$$E = \cos \phi \; \cos \left(\frac{\pi}{2} \cos \phi \right)$$

The pattern is given in Fig. 4.30.

Element pattern $\times$ Array factor $=$ Resultant pattern

Fig. 4.30 *Pattern of two short vertical dipoles*

Problem 4.8 Obtain the pattern of a two-element array fed $180°$ out of phase (end-fire) and spaced at $d = \lambda/2$.

Solution The unit pattern is 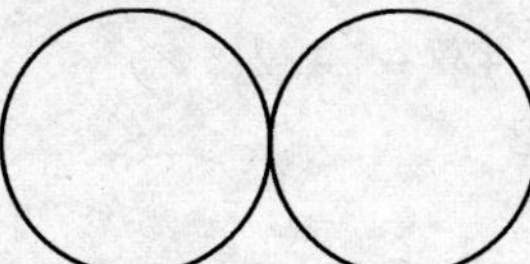

The array factor when $\alpha_e = 180°$, $d = \lambda/2$ is

The resultant pattern is

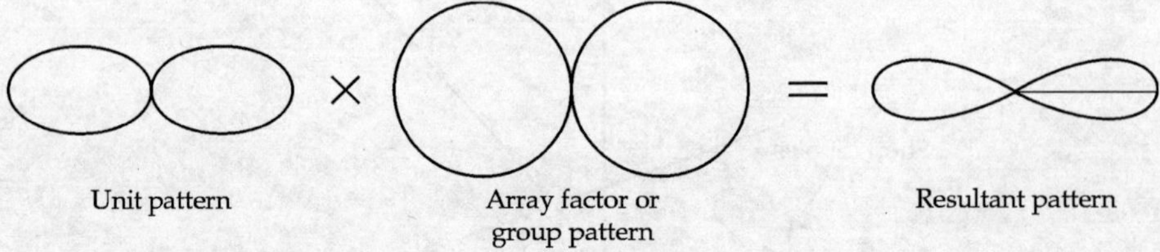

Fig. 4.31 *Pattern of two-element array for $\alpha_e = 180°$ and $d = \lambda/2$*

Problem 4.9   Find the relative excitation levels of a binomial array of 2 and 3 elements.

Solution
$$nc_r = \frac{n!}{r!\,(n-r)!}$$

where
$$r = 0,\ 1,\ 2,\ 3,\ ...,\ (n-1)$$
and the length of the array is n half wavelengths long.

If $n = 2$,
$$1c_0 = \frac{1!}{0!\,(1)!} = 1$$

$$1c_1 = \frac{1!}{1!\,(0)!} = 1$$

The excitation levels are 1, 1.

Similarly, if $n = 3$, the levels are 1, 2, 1.

Problem 4.10   Find the basic and actual transmission losses between two antennas separated by 30 m operating at 10 MHz when the gain of each antenna is 1.65 dB.

Solution
$$d = 30\ \text{m} = 0.03\ \text{km}$$
$$f = 10\ \text{MHz}$$

$$\lambda = \frac{3 \times 10^8}{10 \times 10^6} = 30\ \text{m}$$

$$G_T = 1.65\ \text{dB}$$
$$G_R = 1.65\ \text{dB}.$$

Basic transmission loss,

$$L_b = 10 \log_{10} \left(\frac{4\pi d}{\lambda} \right)^2$$

$$= 20\,[\log_{10} 4 + \log_{10} \pi + \log_{10} d - \log_{10} \lambda]$$
$$= 20\,[0.602 + 0.497 + 1.477 - 1.477]$$
$$= 20\,[1.099] = 21.98\ \text{dB}$$

Actual transmission loss,
$$L_a = L_b - G_T - G_R = +18.7\ \text{dB}$$

$$L_b = 21.98 \text{ dB}$$
$$L_a = 18.68 \text{ dB.}$$

Problem 4.11 Find the basic transmission loss between a ground-based antenna and airborne antenna when the distance between the antennas are 1.6, 16, 160 and 320 km at frequencies equal to:

(a) $f = 0.3$ GHz (b) $f = 3$ GHz.

Solution (a) $f = 0.3$ GHz $= 300 \times 10^6$ Hz

$$\lambda = \frac{3 \times 10^8}{3 \times 10^8} = 1.0 \text{ m}$$

(i) $\qquad d = 1.6$ km

$\qquad L_b =$ basic transmission loss

$$= 10 \log_{10} \left(\frac{4\pi d}{\lambda} \right)^2 = 20 \log_{10} \left(\frac{4\pi d}{\lambda} \right)$$

$$\left(\frac{4\pi d}{\lambda} \right) = 20{,}106.2$$

$$L_b = 20 \log_{10} (20{,}106.2) = 86.07 \text{ dB}$$

(ii) At $\qquad d = 16{,}000$ m

$$4\pi d = 2{,}01{,}062$$

$$L_b = 20 \log_{10} (2{,}01{,}062) = 106.1 \text{ dB}$$

(iii) $\qquad d = 160$ km $= 160 \times 10^3$ m

$$4\pi d = 20{,}10{,}620$$

$$L_b = 126.06 \text{ dB.}$$

(iv) $\qquad d = 320$ km $= 32 \times 10^4$ m

$$4\pi d = 40{,}21{,}238.6$$

$$L_b = 132.06 \text{ dB.}$$

(b) $\qquad f = 3$ GHz $= 3 \times 10^9$ Hz

$$\lambda = \frac{3 \times 10^8}{3 \times 10^9} = 0.1 \text{ m.}$$

(i) $\qquad d = 1.6$ km

$$\left(\frac{4\pi d}{\lambda} \right) = 2{,}01{,}061.9$$

$$L_b = 106.06 \text{ dB.}$$

(*ii*)
$$d = 16 \text{ km} = 16{,}000 \text{ m}$$

$$\left(\frac{4\pi d}{\lambda}\right) = 20{,}10{,}619.3$$

$$L_b = 126.06 \text{ dB.}$$

(*iii*)
$$d = 160 \text{ km} = 160 \times 10^3 \text{ m}$$

$$\left(\frac{4\pi d}{\lambda}\right) = 2{,}01{,}06{,}192.98$$

$$L_b = 146.06 \text{ dB.}$$

(*iv*)
$$d = 320 \text{ km} = 320 \times 10^3 \text{ m}$$

$$\left(\frac{4\pi d}{\lambda}\right) = 4{,}03{,}22{,}385.97$$

$$L_b = 152.087 \text{ dB.}$$

Problem 4.12 Find the actual transmission loss between two antennas separated by 1.6 km, 16 km, 160 km, 320 km at the frequencies (*a*) $f = 0.3$ GHz, (*b*) $f = 3$ GHz, when the gain of the transmitting and receiving antenna are the same and is equal to 10 dB.

Solution (*a*) $f = 0.3$ GHz

(*i*) $d = 1.6$ km

From the previous problem,

$$L_b = 86.07 \text{ dB}$$

Actual loss, $L_a = L_b - G_T - G_R$

$$L_a = 86.07 - 10 - 10 = 66.07 \text{ dB.}$$

(*ii*) $d = 16$ km

$$L_b = 106.1$$

$\therefore$ $L_a = 106.1 - 10 - 10 = 86.1 \text{ dB.}$

(*iii*) $d = 160$ km

$$L_b = 126.06 \text{ dB}$$

$$L_a = 126.06 - 10 - 10 = 106.06 \text{ dB.}$$

(*iv*) $d = 320$ km

$$L_a = 132.06 \text{ dB}$$

$$L_a = 132.06 - 10 - 10 = 112.06 \text{ dB.}$$

(*b*) $f = 3$ GHz

(*i*) $d = 1.6$ km

$$L_b = 106.06 \text{ dB}$$

$$L_a = 106.06 - 10 - 10 = 86.06 \text{ dB.}$$

(*ii*)
$$d = 16 \text{ km}$$
$$L_b = 126.06 \text{ dB}$$
$$L_a = 126.06 - 10 - 10 = 106.06 \text{ dB.}$$

(*iii*)
$$d = 160 \text{ km}$$
$$L_b = 146.06 \text{ dB}$$
$$L_a = 146.06 - 10 - 10 = 126.06 \text{ dB.}$$

(*iv*)
$$d = 320 \text{ km}$$
$$L_b = 152.087 \text{ dB}$$
$$L_a = 152.087 - 10 - 10 = 132.087 \text{ dB.}$$

Problem 4.13   Two dipoles of gain 1.64 each are used for transmitting and receiving purposes. They are separated by a distance of 10 m. The radiated power by the transmitting antenna is 15 W at a frequency of 60 MHz. Determine the receiving power.

Solution
$$W_T = 15 \text{ W}$$
$$f = 60 \text{ MHz}$$
$$d = 10 \text{ m}$$

$$\lambda = \frac{3 \times 10^8}{60 \times 10^6} = 5 \text{ m}$$

The receiving power,

$$W_R = W_T\, G_T\, G_R\, \frac{\lambda^2}{(4\pi d)^2}$$

$$= \frac{(15)\ (1.64)\ (1.64)\ (5)^2}{16\pi^2 \times 10^2}$$

$$= \frac{15 \times 1.96 \times 25}{15791.36}$$

$$= \frac{1008.6}{15791.36}$$

$$= 0.06387 \text{ W}$$

$$W_R = 63.87 \text{ mW.}$$

OBJECTIVE QUESTIONS

1. Gain of isotropic radiator is _______________.

2. Effective area of isotropic radiator is _______________.

3. Gain of infinitesimal dipole is _______________.

4. Effective area of infinitesimal dipole is _______________.

5. Effective area of optimum horn whose mouth area is A, is ________________.

6. Theta polarisation is synonymous with vertical polarisation.   (Yes/No)

7. Phi polarisation is synonymous with horizontal polarisation.   (Yes/No)

8. The directional characteristics of centre-fed 2λ dipole is________________.

9. For broadside array, the excitation phase should be ________________.

10. For end-fire array, the progressive phase shift should be equal to ________________.

11. The effect of earth on vertical patterns is that the ________________ increases and the radiation is in the upper hemisphere.

12. Binomial array is a uniform linear array.   (Yes/No)

13. The first side lobe level of a uniform linear array is ________________.

14. The side lobe level of a binomial array is ________________.

15. The beam width of a binomial array is greater than that of a uniform linear array.   (Yes/No)

16. The radiation resistance of a dipole due to earth effect is ________________.

17. The radiation patterns of terminated and unterminated antennas are the same.   (Yes/No)

18. Stub is ________________.

19. BALUN means ________________.

20. Basic transmission loss between two antennas depends on
    (*a*)   frequency   (*b*)   distance
    (*c*)   frequency and distance   (*d*)   gain of antennas.

21. Actual transmission loss between two antennas depends on ________________.

22. Antenna temperature is the temperature of the antenna.   (Yes/No)

23. The relation between effective noise temperature and noise figure is ________________.

24. When side lobe level increases, beam width also increases in general.   (Yes/No)

25. Half-power beam width in terms of Null-to-Null beam width is ________________.

26. The disadvantage of side lobes is ________________.

27. The number of sources required in a uniform array of 10 elements is ________________.

28. The number of sources required in a binomial array of 11 elements is ________________.

29. Q of an antenna made for a frequency of 15 MHz having a band width of 4 MHz is ________________.

30. The expression for Q of an antenna in terms of resonant frequency and bandwidth is ________________.

31. The radiation resistance of an antenna which is radiating 10 kW and is fed at 10 amps is ________________.

32. Directivity of a broadside array of 10λ length is ________________.

33. Directivity of end-fire array of 10λ length is ________________.

34. B.W.F.N. of broadside array of length 10λ is ______________.

35. HPBW of end-fire array of length 10λ is ______________.

36. Antenna wires are associated with ______________ and magnetic fields.

37. In a dipole antenna, the two ends are at equal ______________ relative to the mid-point.

38. The directive gain is ______________ as the antenna length increases.

39. Grounded vertical antenna of length $\lambda/4$ is an ______________ antenna which radiates equally in all directions.

40. In an end-fire array, there exists no radiation at ______________ to the axis of the array.

41. Antenna radiates maximum when it is at ______________.

42. An electric field is developed ______________ the ends of an antenna.

43. The magnetic field is developed ______________ an antenna.

44. Hertz antenna operates on ______________ of its fundamental frequency.

45. Receiving antennas in radio receivers normally are ______________.

46. The directivity of an antenna is determined by the beam width.   (Yes/No)

47. A dummy antenna is a non-radiating antenna.   (Yes/No)

48. A dummy antenna is used for tuning of preliminary transmitter and for approximating the output power of the transmitter.   (Yes/No)

49. The input impedance at the base of a $\lambda/8$ Marconi antenna is ______________.

50. If radiation resistance is low, antenna efficiency is ______________.

51. Field strength is improved with a good ground screen.   (Yes/No)

52. Marconi antenna is an ungrounded antenna.   (Yes/No)

53. Resonant length of a dipole is ______________.

54. Resonant length of a practical dipole is always ______________.

55. Actual resonant length of a dipole depends on its thickness.   (Yes/No)

56. In coaxial cables or waveguides, transmission loss per unit length is independent of distance.   (Yes/No)

57. In coaxial cables or waveguides, transmission loss per unit length increases with frequency.   (Yes/No)

58. In radio wave system or microwave links, transmission loss per unit length decreases with increasing distance.   (Yes/No)

59. Actual transmission loss is less than the basic transmission loss between transmitting and receiving antenna.   (Yes/No)

60. The received power in a communication system is inversely proportional to the square of frequency. (Yes/No)

61. The number of secondary lobes depend on the number of nulls in the resultant pattern. (Yes/No)

62. The number of nulls in the resultant pattern are the sum of the nulls in the individual pattern and array pattern. (Yes/No)

63. An ungrounded antenna near the ground acts as ______________.

64. The standard reference antenna for directive gain is the ______________.

65. Top loading is sometimes used with an antenna in order to increase its ______________.

ANSWERS

1. One **2.** $\dfrac{\lambda^2}{4\pi}$ **3.** 1.5 **4.** $\dfrac{1.5\lambda^2}{4\pi}$ **5.** 0.81λ **6.** Yes

7. Yes **8.** 4 directional **9.** Zero **10.** $-\beta d$ **11.** Directivity

12. No **13.** -13.5 dB **14.** Zero **15.** Yes

16. Oscillates between 65 and 73 Ω **17.** Yes

18. Impedance matching device **19.** BALanaced to UNbalanced **20.** (c)

21. f, d and gains of antenna **22.** No **23.** $T_e = T_0\,(F_N - 1)$ **24.** No

25. $\text{HPBW} = \dfrac{1}{2} \times \text{BWFN}$ **26.** waste power in undesired directions **27.** One

28. 6 **29.** 3.75 **30.** $Q = \dfrac{f_r}{\text{B.W.}}$ **31.** 100 Ω **32.** 20 **33.** 40

34. 0.2 **35.** 0.894 **36.** Electric **37.** Potential **38.** Increased

39. Omni directional **40.** Right angle **41.** Resonance **42.** Between

43. Around **44.** Harmonics **45.** ferrite rod and loop antenna **46.** Yes

47. Yes **48.** Yes **49.** $(8 - j500)$ ohm **50.** Low **51.** Yes

52. No **53.** 0.5λ **54.** Less than $\lambda/2$ **55.** Yes **56.** Yes

57. Yes **58.** Yes **59.** No **60.** Yes **61.** Yes **62.** Yes

63. an antenna array **64.** Isotropic antenna **65.** Band width

 ## EXERCISE PROBLEMS

1. Obtain radiation pattern of a binomial array of 5 elements. Compare with that of uniform arrays of 5 elements.

2. Draw the typical radiation pattern of an array of 5 elements with edge excitation given by 1, 0, 0, 0, 1.

3. The noise figure of an antenna is 0.5 dB at a temperature of 30°C. Calculate its equivalent temperature.

4. Find the Null-to-Null beam width of a broadside uniform linear array of length 20λ when the elements are spaced at 0.4λ.

5. What is the Null-to-Null beam width of an end-fire array of length 50λ when the elements are uniformly spaced at 0.5λ?

6. Find the current excitation of a binomial array of 15 elements.

7. A transmitting antenna has a gain of 20 dB and the receiving antenna has a gain of 50 dB in a satellite communication system. The satellite is at a height of 36,000 km. Find the basic and actual transmission losses.

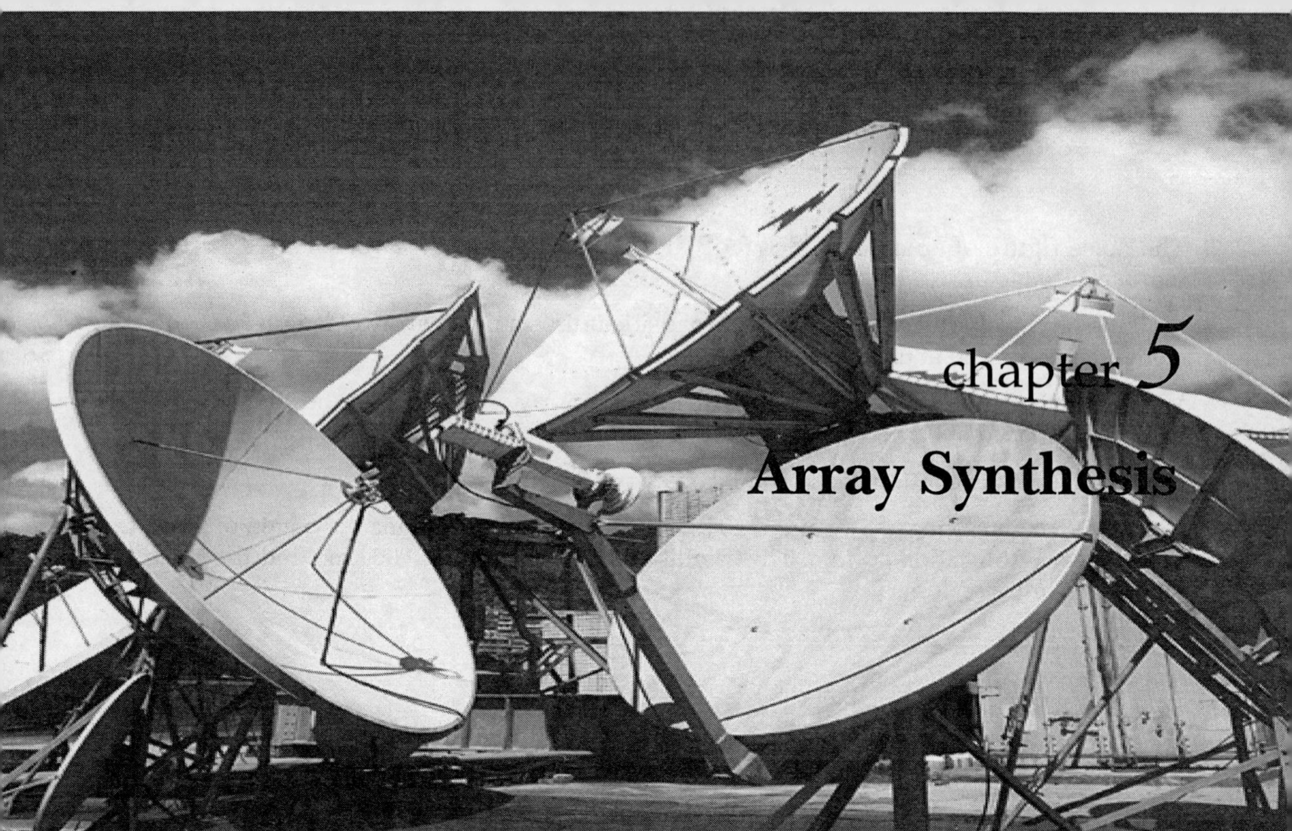

chapter *5*

Array Synthesis

"Array antennas can produce desired shape of electric field of EM wave like electronic circuits produce any shape of signal wave form."

CHAPTER OBJECTIVES

This chapter discusses

✦ The design of arrays using the best techniques

✦ Design of array using standard amplitude distributions

✦ Objective questions and solved problems useful for class tests, final examinations and also for competitive examinations

✦ Exercise problems to develop self problem solving skills

5.1 INTRODUCTION

Antenna analysis determines the radiation pattern for a given input distribution. **Antenna synthesis** is the inverse process. In antenna array synthesis, input or source distribution is determined for a specified radiation pattern.

Array synthesis may be done by the following methods:
1. Source amplitude distribution control method.
2. Source phase distribution control method.
3. Element space control method.
4. Any combination of the above.

However, in this book, array synthesis is done using the first method, that is, source amplitude distribution control method. It is required to find out a physically realisable source distribution for fixed spacing and phase distributions. In general, it is required to design an array to produce a pattern with narrow beam width, low side lobes and decaying minor lobes. The main task is to find out the antenna configuration, its geometrical dimensions and excitation or input or source distribution. The designed array should give a specified radiation pattern. At times, it is required to produce a shaped pattern in the entire visible region. Then the method is called beam shaping.

In general, pattern synthesis is applied to:

* continuous line source and

* discrete linear array.

Continuous line source   It is defined as an array of a large number of elements placed along a line of finite length. It means that there is no spacing between the elements. It is purely a theoretical case and does not exist practically. The patterns of a continuous line source are in the form of radiation integral. The amplitude distribution for a continuous line source is a function of only one coordinate. In the pattern calculations, the question of element pattern does not arise. A typical geometry of line source is shown in Fig. 5.1.

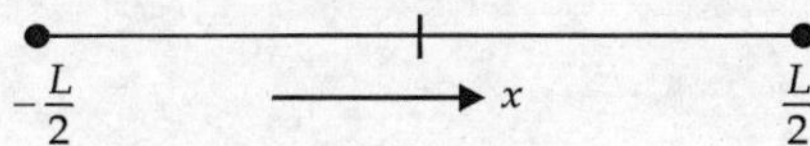

Fig. 5.1   *Line source*

Discrete linear array   It is a group of elements arranged along a line with a finite spacing between them. The radiation patterns of these arrays are in the form of finite summation. The amplitude distributions for the arrays are also discrete.

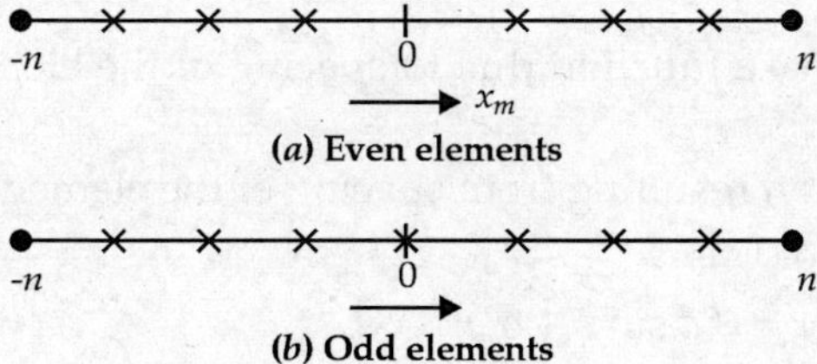

Fig. 5.2   *Linear array of even and odd number of elements*

They exist at the locations of the elements. The amplitude distribution obtained for a continuous line source can be used to determine the excitation levels of the linear array. A typical linear array is shown in Fig. 5.2.

5.2 SYNTHESIS METHODS

The following array synthesis techniques are presented in this book:

1. Schelkunoff polynomial method.
2. Fourier transform method.
3. Woodward-Lawson method.
4. Dolph-Tchebyscheff or Chebyshev method.
5. Taylor's method.
6. Laplace transform method.
7. Some standard amplitude distributions.

5.3 SCHELKUNOFF POLYNOMIAL METHOD

This is a method of design of an array which produces a pattern with the nulls in specified directions. For the design of an array, the number of nulls and their locations are specified. From the knowledge of nulls and their locations, the number of radiating elements and their excitation levels are obtained.

Design methodology   The array factor of a N-element array is given by

$$E = \sum_{n=1}^{N} a_n e^{j\,(n-1)\,(kd\,\sin\phi + \alpha_e)} \qquad \text{...(5.1)}$$

where a_n is the excitation level of the nth element,

d is the spacing between adjacent elements,

$$k = \frac{2\pi}{\lambda},$$

α_e is phase excitation,

ϕ is the angle between the line of observer and broadside to the array.

Equation (5.1) can be rewritten as

$$E = \sum_{n=1}^{N} a_n e^{j\,(n-1)\,\psi} \qquad \text{...(5.2)}$$

Here, ψ is the phase function due to spacing of the elements and excitation phase.

The phase function resulting from spacing of the elements is $kd\,\sin\phi$.

The expression for ψ is

$$\psi = (kd\,\sin\phi + \alpha_e) \qquad \text{...(5.3)}$$

If $\qquad x = e^{j\psi} \qquad \text{...(5.4)}$

$$= 1 \angle \psi$$

Equation (5.1) becomes

$$E = \sum_{n=1}^{N} a_n x^{n-1}$$

$$= a_1 + a_2 x + a_3 x^2 + \dots + a_N x^{N-1} \qquad \dots(5.5)$$

This polynomial has $(N-1)$ roots. Expressing the polynomial Equation (5.5) as a product of $(N-1)$ terms, we can write

$$E = a_n (x - x_1)(x - x_2) \dots (x - x_{N-1}) \qquad \dots(5.6)$$

Here, $x_1, x_2, \dots, x_{N-1}$ are the roots of the polynomial. These roots may be complex or real. The magnitude of **E** is given by

$$|E| = E = |a_n| \, |x - x_1| \, |x - x_2| \dots |x - x_{N-1}| \qquad \dots(5.7)$$

From Equations (5.3) and (5.4), we have

$$x = 1 \angle \psi = 1 \angle (kd \sin \phi + \alpha_e) \qquad \dots(5.8)$$

The magnitude of x is unity and its phase depends on λ, d, ϕ and α_e.

Case I   For $\alpha_e = 0$ and $d = \dfrac{\lambda}{8}$

$$\psi = \frac{2\pi}{\lambda} \times \frac{\lambda}{8} \sin \phi + \alpha_e = \frac{\pi}{4} \sin \phi$$

For these values, the variation of x is plotted in Fig. 5.3.

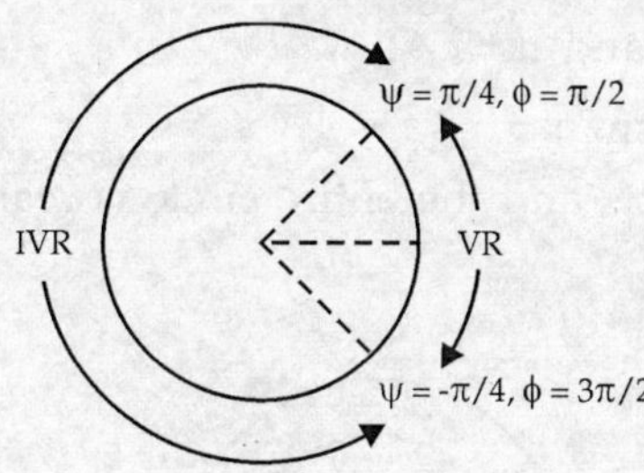

Fig. 5.3   *Visible and invisible regions for $\alpha_e = 0$, $d = \lambda/8$*

For

$$\psi = \pi/4, \ \phi = \pi/2$$
$$\psi = -\pi/4, \ \phi = 3\pi/2$$

In the range $-\pi/4 \le \psi \le \pi/4$, ϕ exists and it is realisable.

In the range beyond $-\pi/4 \le \psi \le \pi/4$, ϕ does not exist and it is not realisable.

Visible Region (VR) is defined as the region which contains the realisable part of the circle.

Invisible Region (IVR) is defined as the region which contains the unrealisable part of the circle.

Both these regions depend on the element spacing and its relative position on the circle by the progressive phase shift of the elements.

Case II   For $\alpha_e = 0$ and $d = \lambda/4$

$$\psi = kd \sin \phi = \frac{2\pi}{\lambda} \times \frac{\lambda}{4} \sin \phi$$

$$= \frac{\pi}{2} \sin \phi$$

For $-\dfrac{\pi}{2} \leq \psi \leq \dfrac{\pi}{2},$ ϕ exists and it is realisable.

For the range beyond $-\dfrac{\pi}{2} \leq \psi \leq \dfrac{\pi}{2},$ ϕ does not exist and it is not realisable.

These are shown in Fig. 5.4.

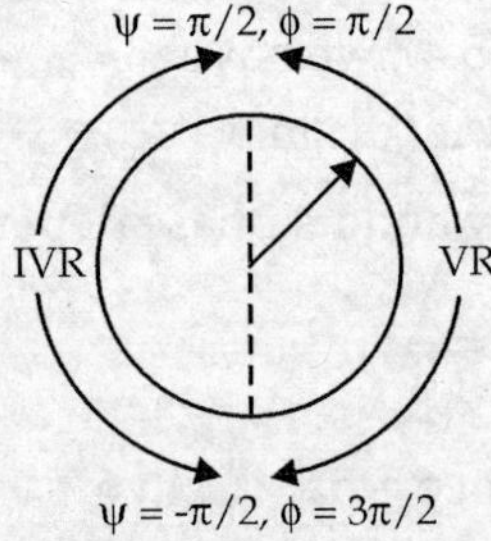

Fig. 5.4   *Visible and invisible regions for $\alpha_e = 0,\ d = \lambda/4$*

Case III   For $\alpha_e = 0$ and $d = \lambda/2$

$$\psi = \pi \sin \phi$$

For $-\pi \leq \psi \leq \pi,$ ϕ exists and the entire circle is realisable. This is shown in Fig. 5.5.

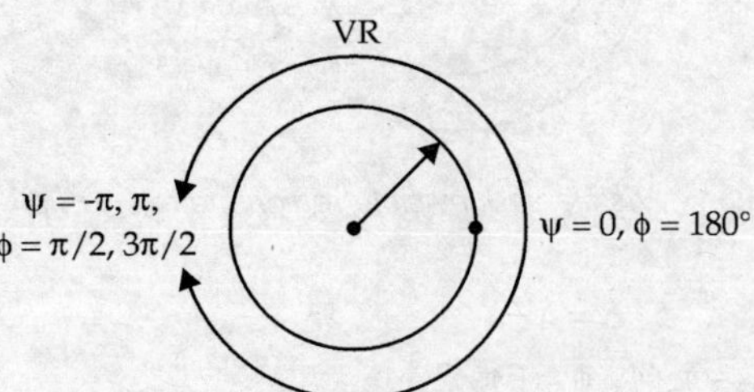

Fig. 5.5   *Visible region for $\alpha_e = 0,\ d = \lambda/2$*

5.4 FOURIER TRANSFORM METHOD

Fourier Transform of $f(x)$ is defined as

$$F(\omega) = \frac{1}{\sqrt{2\pi}} \int_{-\infty}^{\infty} f(x) e^{-j\omega x}\, dx$$

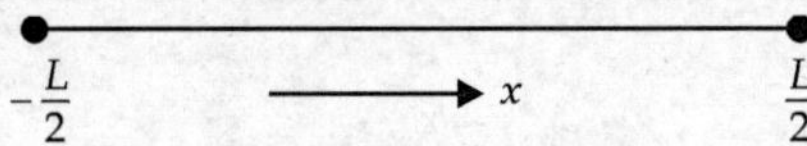

Inverse Fourier Transform of $F(\omega)$ is defined as Inverse Fourier Transform of

$$F(\omega) \equiv f(x) = \frac{1}{\sqrt{2\pi}} \int\limits_{-\infty}^{\infty} f(\omega) e^{j\omega x} \, d\omega$$

Since $F(\omega)$ and $f(x)$ are related, Fourier Transform pair can be used to relate far-field and amplitude distribution of the array. As a result, this method is used to design the excitation distribution of either a continuous line source or a discrete array for a specified radiation pattern.

5.5 LINE SOURCE DESIGN BY FOURIER TRANSFORM METHOD

Line source is defined as a continuous distribution of current along a line segment. A typical line source is shown in Fig. 5.6.

Fig. 5.6 *Line source*

Here the aim is to design amplitude distribution for a specified radiation pattern.

Let L be the length of the line source. Then the normalised space factor is given by

$$E(\phi) = \int\limits_{-L/2}^{L/2} A(x) \, e^{j\,(k\sin\phi + \alpha_e)\,x} \, dx$$

$$= \int\limits_{-L/2}^{L/2} A(x) \, e^{j\,yx} \, dx \qquad\qquad ...(5.9)$$

Here x is a point on the line source.

$$y = k\sin\phi + \alpha_e$$

or

$$\phi = \sin^{-1}\left(\frac{y - \alpha_e}{k}\right) \qquad\qquad ...(5.10)$$

α_e = excitation phase

$A(x)$ = desired amplitude distribution

$E(\phi)$ = desired radiation pattern

Although the amplitude distribution extends from $-L/2$ to $L/2$ only, the limits can be extended to infinity in Equation (5.9). Hence, using the concept of Fourier Transform, the field expression is written as

$$E(\phi) = \frac{1}{\sqrt{2\pi}} \int\limits_{-\infty}^{\infty} A(x) \, e^{j\,yx} \, dx \qquad\qquad ...(5.11)$$

Equation (5.11) represents Fourier Transform and it relates amplitude distribution $A(x)$ of a continuous source to its radiation pattern, $E(\phi)$.

Therefore, its corresponding Transform pair is given by

$$A(x) = \frac{1}{\sqrt{2\pi}} \int_{-\infty}^{\infty} E(y)\, e^{-jxy}\, dy$$

$$\text{or}$$

$$A(x) = \frac{1}{\sqrt{2\pi}} \int_{-\infty}^{\infty} E(\phi)\, e^{-jxy}\, dy \qquad \qquad ...(5.12)$$

Here $E(\phi)$ represents specified radiation pattern and $A(x)$ is the required amplitude distribution.

For a line source, the normalised amplitude distribution is given by

$$A(x) = \int_{-\infty}^{\infty} E(\phi)\, e^{-jxy}\, dy \qquad \qquad ...(5.13)$$

for $\qquad\qquad -L/2 \le x \le L/2$

$$= 0 \ \text{ elsewhere.}$$

5.6 DESIGN OF LINEAR ARRAY BY FOURIER TRANSFORM METHOD

Let the desired pattern array factor of a linear array be given by

$$E(\phi) = E(y) = \sum_{m=-n}^{n} A_m(x)\, e^{jmy} \qquad \qquad ...(5.14)$$

Here $\qquad y = kd\,\sin\phi\ (5.15)$

$\qquad A_m(x) = $ excitation coefficients to be designed

$\qquad N = 2n + 1 = $ number of elements

$\qquad d = $ spacing between the elements

The location of elements for odd number of elements is given by

$$X_m = md, \ m = 0, \ \pm 1, \ \pm 2, \ ..., \ \pm n \qquad \qquad ...(5.16)$$

The normalised excitation coefficients or levels of the array are obtained by Fourier formula:

$$A_m(x) = \int_{-\pi/2}^{\pi/2} E(\phi)\, e^{-jmy}\, dy, \ \ -n \le m \le n \qquad \qquad ...(5.17)$$

When the number of elements are even ($N = 2n$), the desired field pattern is given by

$$E(\phi) = E(y) = \sum_{m=-n}^{-1} A_m(x)\, e^{j\,[(2m+1)/2]\,y} + \sum_{m=+1}^{n} A_m(x)\, e^{j\,[(2m-1)/2]\,y}$$

$$...(5.18)$$

The locations of the elements are at

$$
\left.\begin{aligned}
x_m &= \left(\frac{2m-1}{2}\right) d, \quad 1 \le m \le n \\[2ex]
x_m &= \left(\frac{2m+1}{2}\right) d, \quad -n \le m \le -1
\end{aligned}\right\} \qquad ...(5.19)
$$

The normalised excitation coefficients of the array are again obtained by the Fourier formula given by

$$
A_m(x) = \int_{-\pi/2}^{\pi/2} E(\phi)\, e^{-j\,[(2m+1)/2]\,y}\, dy, \quad -n \le m \le -1 \qquad ...(5.20)
$$

$$
A_m(x) = \int_{-\pi/2}^{\pi/2} E(\phi)\, e^{-j\,[(2m-1)/2]\,y}\, dy, \quad 1 \le m \le n \qquad ...(5.21)
$$

When the array is symmetric, it is sufficient to calculate $A_m(x)$ from either Equation (5.20) or Equation (5.21).

When the number of elements in the array are odd ($N = 2n + 1$), the normalised excitation coefficients are determined from

$$
A_m(x) = \int_{-\pi/2}^{\pi/2} E(\phi)\, e^{-jmy}\, dy, \quad -n \le m \le n \qquad ...(5.22)
$$

$d = \dfrac{\lambda}{2}$ is usual spacing in many applications. If $d < \dfrac{\lambda}{2}$, super directive arrays result which are undesirable and impractical. If $d > \dfrac{\lambda}{2}$, the patterns contain grating lobes which are again undesirable.

5.7 LINEAR ARRAY DESIGN BY WOODWARD-LAWSON METHOD

Linear array antenna is defined as an array antenna in which the centres of the radiating elements lie along a straight line.

Woodward method is very useful for beam shaping. This method consists of sampling the desired pattern at the locations of the elements of the array. By this method, it is possible to find the excitation function for a continuous line source and excitation levels for a discrete array.

This method is simple and elegant. But there is no control over the side lobe levels in the trade off region of the pattern. Moreover, this method deals with the design of field patterns. The specified pattern may be represented either graphically or analytically.

Let the desired radiation pattern be represented by

$$E(\phi) = \sum_{m=-n}^{n} B_m \frac{\sin\left[\dfrac{N}{2} kd\,(\sin\phi - \sin\phi_m)\right]}{N\sin\left[\dfrac{1}{2} kd\,(\sin\phi - \sin\phi_m)\right]} \qquad \ldots(5.23)$$

Here, $\quad B_m = $ excitation coefficient

$\qquad N = $ number of elements in the array

$\qquad \phi = $ angle between the line of observer from broadside

$$k = \frac{2\pi}{\lambda}$$

These coefficients at the sampled points are obtained from the values of the desired pattern. Hence

$$B_m = E\,(\phi = \phi_m)_d$$

Here $\qquad \phi_m = \sin^{-1}\left[m\,\frac{\lambda}{Nd}\right]$

The normalised amplitude level of each element of the array is, therefore, obtained from

$$A_m(x) = \frac{1}{N} \sum_{m=-n}^{n} B_m\, e^{-jkx_m \sin\phi_m} \qquad \ldots(5.24)$$

where x_m is the location of nth radiating element in the array.

5.8 DOLPH-CHEBYCHEV METHOD (TSCHEBYSCHEFF DISTRIBUTION)

This method is basically a compromise between uniform and binomial arrays. It provides a means of determination of suitable polynomials which give the excitation coefficients to obtain satisfactory patterns.

In fact, the Dolph-Tschebyscheff method yields a radiation pattern containing one main beam and side lobes with the same level. This is done by finding the spacing of nulls.

The Dolph-Tschebyscheff array design for zero side lobes reduces to the binomial design. The excitation coefficients become identical in this case.

Tschebyscheff polynomial is defined as

$$T_m(x) \equiv \cos(m\cos^{-1} x),\; -1 < x < 1 \qquad \ldots(5.25)$$

and
$$T_m(x) \equiv \cosh(m\cosh^{-1} x),\; |x| > 1$$

Derivation of different polynomials   Rewriting Equation (5.25) as

$$T_m(x) = \cos(m\Delta), \quad \Delta = \cos^{-1} x \qquad \ldots(5.26)$$

If $m = 0$, $\qquad$ $T_0 (x) = \cos (0) = 1$ $\qquad$ ...(5.27)

$m = 1$, $\qquad$ $T_1 (x) = \cos (\Delta) = x$ $\qquad$ ...(5.28)

$m = 2$, $\qquad$ $T_2 (x) = \cos (2\Delta) = 2 \cos^2 \Delta - 1$

$$= 2x^2 - 1 \qquad ...(5.29)$$

$m = 3$, $\qquad$ $T_3 (x) = \cos (3\Delta) = 4\cos^3 \Delta - 3 \cos \Delta$

$$= 4x^3 - 3x \qquad ...(5.30)$$

$m = 4$, $\qquad$ $T_4 (x) = \cos (4\Delta) = 2 \cos^2 2\Delta - 1$

$$= 2 (2 \cos^2 \Delta - 1)^2 - 1$$

$$= 2 [4 \cos^4 \Delta - 4 \cos^2 \Delta + 1] - 1$$

or $\qquad$ $T_4 (x) = 8 \cos^4 \Delta - 8 \cos^2 \Delta + 2 - 1$

That is, $\qquad$ $T_4 (x) = 8x^4 - 8x^2 + 1$ $\qquad$ ...(5.31)

Similarly $T_5 (x)$, $T_6 (x)$ can be found. The polynomial can be generalised using the recursion formula. That is,

$$T_m (x) = 2x \, T_n (x) - T_{n-1} (x) \qquad ...(5.32)$$

here $m = n + 1$

If $m = 4$, n becomes 3

$$T_4 (x) = 2x \, T_3 (x) - T_2 (x)$$

$$= 2x (4x^3 - 3x) - 2x^2 + 1$$

$$= 8x^4 - 6x^2 - 2x^2 + 1$$

So $\qquad$ $T_4 (x) = 8x^4 - 8x^2 + 1$

If $m = 5$, n becomes 4

$$T_5 (x) = 2x \, T_4 (x) - T_3 (x)$$

$$= 2x (8x^4 - 8x^2 + 1) - 4x^3 + 3x$$

$$= 16x^5 - 16x^3 + 2x - 4x^3 + 3x$$

So $\qquad$ $T_5 (x) = 16x^5 - 20x^3 + 5x$ $\qquad$ ...(5.33)

If $m = 6$, n becomes 5

$$T_6 (x) = 2x \, T_5 (x) - T_4 (x)$$

$$= 2x (16x^5 - 20x^3 + 5x) - (8x^4 - 8x^2 + 1)$$

or $\qquad$ $T_6 (x) = 32x^6 - 48x^4 + 18x^2 - 1$ $\qquad$ ...(5.34)

The polynomials corresponding to $m = 0$ to 6 are grouped below for convenient reference.

$$T_0\,(x) = 1$$
$$T_1\,(x) = x$$
$$T_2\,(x) = 2x^2 - 1$$
$$T_3\,(x) = 4x^3 - 3x$$
$$T_4\,(x) = 8x^4 - 8x^2 + 1 \qquad \qquad ...(5.35)$$
$$T_5\,(x) = 16x^5 - 20x^3 + 5x$$
$$T_6\,(x) = 32x^6 - 48x^4 + 18x^2 - 1$$

The general characteristics of the polynomial $T_m\,(x)$ as a function of x is shown in Fig. 5.7.

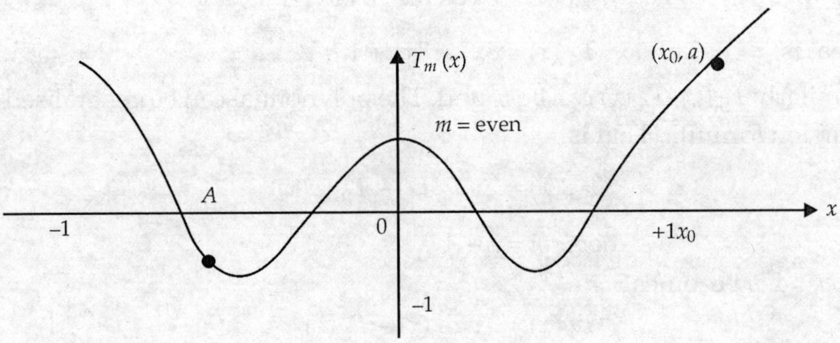

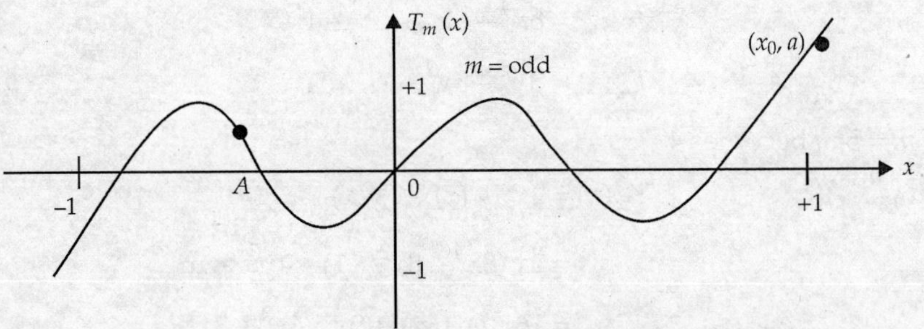

Fig. 5.7 *Tschebyscheff polynomials for different values of m*

Properties of the polynomials

1. $T_m\,(x)$ increases continuously beyond $|x| = 1$.

2. The polynomials $T_m\,(x)$ crosses the axis m times between -1 and $+1$.

3. The polynomials oscillate between -1 and $+1$.

4. If x varies from point A to a point x_0 and back, then $T_m\,(x)$ traces a pattern containing a main lobe, many side lobes or minor lobes.

5. The minor lobes have equal amplitude.

6. The minor lobes are below the main beam level by a value equal to $\left(\dfrac{1}{a}\right)$. This ratio is called as **side lobe ratio**.

7. The side lobe ratio $\left(\dfrac{1}{a}\right)$ depends on the choice of x_0.

8. The side lobes exist in $|x| < 1$.

9. The main lobe exists for $|x| > 1$.

10. The roots determine the nulls if $\cos(m\Delta) = 0$.

11. The nulls exists if $\cos(m\Delta) = 0$

    That is, $\qquad\qquad T_m(x) = \cos(m\Delta) = 0$

    This is possible if $\qquad \Delta_i = \dfrac{(2i-1)\,\pi}{2m}, \; i = 1, 2, ..., m.$

5.9 DETERMINATION OF DOLPH-CHEBYCHEV AMPLITUDE DISTRIBUTION

Procedure

1. Let a be the acceptable side lobe ratio.

   That is, $\qquad a = \dfrac{\text{main lobe maximum or level}}{\text{side lobe maximum or level}}$

   Then calculate SLL in dB.

   That is, SLL in dB $= 20\log_{10}(a)$. $\qquad\qquad\qquad$ ...(5.36)

2. Tschebyscheff Polynomial $T_m(x)$ having the same degree as the array polynomial is chosen.

   If n is the number of elements in the array, the degree of array polynomial is $(n-1)$ and the degree of the polynomial is m, then

   $$T_m(x_0) = T_{n-1}(x_0) \qquad\qquad\qquad ...(5.37)$$

   Here, $\qquad\qquad m = n - 1.$

3. Equate $T_m(x_0)$ and a. That is

   $$T_m(x_0) = T_{n-1}(x_0) = a. \qquad\qquad\qquad ...(5.38)$$

4. Determine x_0 from

   $$x_0 = \frac{1}{2}\,[\{a + \sqrt{a^2 - 1}\}^{1/m} + \{a - \sqrt{a^2 - 1}\}^{1/m}] \qquad\qquad ...(5.39)$$

5. Find E_R.

   As $a > 1$, x_0 is greater than 1. But x lies in the range of $-1 \le x \le 1$. This is a peculiar situation which is overcome by using a change of scale. That is,

$$z = \frac{x}{x_0}$$

As the definition, $T_m(x) = \cos(m \cos^{-1} x)$ is satisfied, we get,

$$z = \cos \Delta = \cos \frac{\psi}{2} \qquad \qquad ...(5.40)$$

As z lies in $-1 \le z \le 1$, the pattern polynomial for even elements becomes

$$E_R = \sum_{m=0}^{N-1} A_m \cos \frac{(2m+1)\,\psi}{2}$$

$$= A_0 \cos \frac{\psi}{2} + A_1 \cos \frac{3\psi}{2} + A_2 \cos \frac{5\psi}{2} + ...$$

$$= A_0 z + A_1 [4z^3 - 3z] + A_2 [16z^5 - 20z^3 + 5z] + ... \qquad ...(5.41)$$

and the pattern polynomial for odd elements becomes

$$E_R = \sum_{m=0}^{N} A_m \cos(2m\psi/2)$$

$$= A_0 + A_1 \cos 2\frac{\psi}{2} + A_2 \cos 4\frac{\psi}{2} + ...$$

$$= A_0 + A_1 [2z^2 - 1] + A_2 [8z^4 - 8z^2 + 1] + A_3 [32z^6 - 48z^4 + 18z^2 - 1] + ...$$

$$...(5.42)$$

6. Calculate the coefficients, A_0, A_1, A_2 etc. by equating array polynomial and Tschebyscheff polynomial. That is,

$$E_R = T_{N-1}(x). \qquad \qquad ...(5.43)$$

7. These give the required excitation coefficients or amplitude distribution for the specified side lobe ratio of a.

Resultant patterns of arrays with even and odd number of elements Assume that A_0, A_1, A_2, ..., A_m are current amplitudes of the elements. Let them be spaced at d. Origin is at the centre of the array. Consider broadside arrays of even and odd number of elements (Fig. 5.8).

Broadside array is defined as a linear or planar array antenna whose maximum radiation is along the perpendicular to the line or plane of the array.

The resultant far-field of the array of even number of isotropic elements is given by

$$E_R = 2A_0 \cos \frac{\psi}{2} + 2A_1 \cos \frac{3\psi}{2} + ... + 2A_m \cos \left(\frac{n-1}{2}\right)\psi \qquad ...(5.44)$$

Here $\qquad n = 2N$

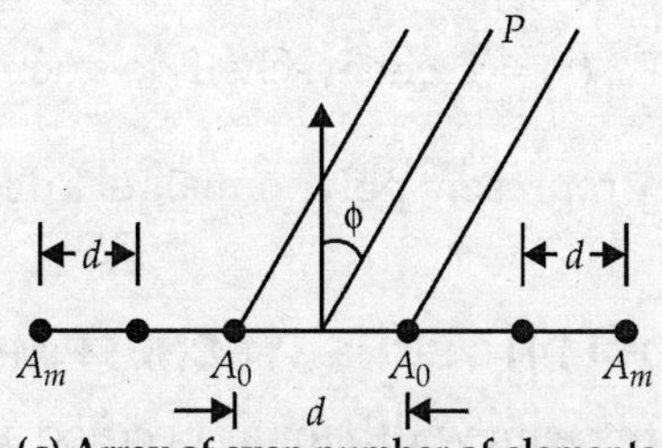

(a) Array of even number of elements

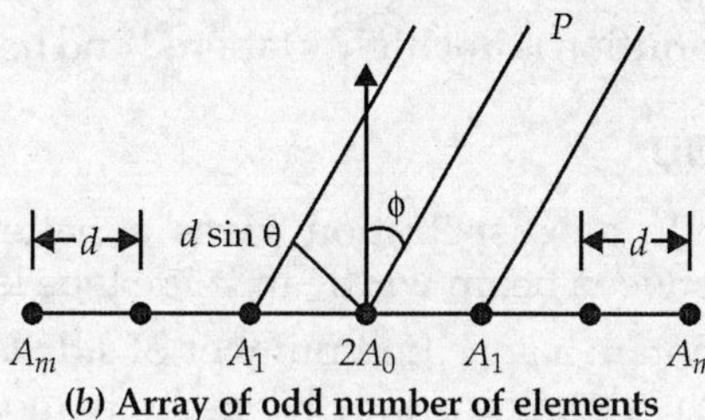

(b) Array of odd number of elements

Fig. 5.8 *Broadside arrays*

$$\psi = k \times \text{path difference}, \quad k = \frac{2\pi}{\lambda}$$

or
$$\psi = \frac{2\pi}{\lambda} d \sin \phi \qquad \qquad \text{...(5.45)}$$

$$n = 2, 4, 6, \ldots 2\,(m+1), \quad m = 0, 1, 2, 3, \ldots \qquad \text{...(5.46)}$$

or
$$(n-1) = 2\,(m+1) - 1 = 2m + 1$$

That is,
$$\left(\frac{n-1}{2}\right) = (2m+1)/2$$

So
$$E_R = 2 \sum_{m=0}^{N-1} A_m \cos\left(\frac{2m+1}{2}\right)\psi \qquad \text{...(5.47)}$$

For odd number of isotropic elements in the array, the resultant pattern is given by

$$E_R = 2A_0 + 2A_1 \cos\psi + 2A_2 \cos 2\psi + \ldots + 2A_m \cos\left(\frac{n-1}{2}\right)\psi$$

$$\text{...(5.48)}$$

Here
$$n = 1, 3, 5, \ldots, (2m+1), \quad m = 0, 1, 2, \ldots$$

That is,
$$\left(\frac{n-1}{2}\right) = \frac{2m+1-1}{2} = m$$

So
$$E_R = 2 \sum_{m=0}^{N} A_m \cos\left(\frac{n-1}{2}\right)\psi \qquad \text{...(5.49)}$$

$$\cos \frac{m\psi}{2} = T_m(x) = \text{Tschebyscheff polynomial}$$

This indicates that they represent polynomials of a degree equal to the number of elements minus one.

5.10 ADVANTAGES OF DOLPH-TSCHEBYSCHEFF METHOD

- It provides a minimum beam width for a specified side lobe level.
- It provides pattern which contains side lobes of equal level.
- The amplitude distribution is not highly tapered and hence it is more practical.

5.11 TAYLOR'S METHOD

Taylor's design of amplitude distribution gives a pattern which exhibits an optimum compromise between beam width and side lobe level.

Taylor's patterns contain a specified number of side lobes, close to the main lobe, at equal level and the remaining side lobes decay monotonically. However, practically, the closest few side lobes decay slightly. This decay is a function of space over which these side lobes are required to be at the same level. If this space increases, the rate of decay of the closest side lobes decreases.

Design procedure   The expression for the desired pattern as given by Taylor is

$$E(u) = \cosh(\pi A)\,\frac{\sin(u)}{u}\,\prod_{n=1}^{\overline{n}-1}\left[\frac{1 - \dfrac{(u)^2}{\sigma^2\pi^2\left\{A^2 + \left(n - \dfrac{1}{2}\right)^2\right\}}}{\left\{1 - \dfrac{(u)^2}{(\pi n)^2}\right\}}\right] \qquad \ldots(5.50)$$

Here $\overline{n}$ is an integer which divides the radiation pattern into a uniform side lobe region surrounding the main beam and the region of decaying side lobes.

$u = L/\lambda \sin \phi$

$L = $ array length

$\phi = $ angle measured from maximum radiation

$A = $ an adjustable real parameter having the property that $\cosh(\pi A)$ is the side lobe ratio

$$\sigma = \frac{\overline{n}}{\left[A^2 + \left(\overline{n} - \dfrac{1}{2}\right)^2\right]^{1/2}} \qquad \ldots(5.51)$$

Using Equation (5.50), the amplitude distribution of the array is found. Applying Woodward's method, the aperture distribution $A(x)$ is given by

$$A(x) = \sum_{n=-\infty}^{\infty} a_n\, e^{-jn\pi x} \qquad \qquad \text{...(5.52)}$$

where $x = X/(L/2)$, X being a variable point on the array.

The pattern $E(u)$ related to $A(x)$ is given by

$$E(u) = \int_{-1}^{1} A(x)\, e^{jux}\, dx \qquad \qquad \text{...(5.53)}$$

From Equations (5.52) and (5.53), we get,

$$E(u) = \sum_{n=1}^{\infty} a_n \frac{\sin(u - n\pi)}{(u - n\pi)} \qquad \qquad \text{...(5.54)}$$

This gives $\qquad a_n = E(u)\big|_{u = n\pi} \qquad \qquad \text{...(5.55)}$

Therefore, the expression for $E(u)$ reduces to

$$E(u) = \sum_{n=1}^{\infty} E(n\pi) \frac{\sin(u - n\pi)}{(u - n\pi)} \qquad \qquad \text{...(5.56)}$$

The aperture distribution is obtained in the form of

$$A(x) = a_0 + \sum_{n=1}^{\infty} 2a_n \cos(n\pi x)$$

or

$$A(x) = E(0) + 2 \sum_{n=1}^{\infty} [E(n\pi) \cos(n\pi x)] \qquad \qquad \text{...(5.57)}$$

and $\qquad E(n\pi) = 0 \ \text{ for } \ n \geq \bar{n}.$

5.12 LAPLACE TRANSFORM METHOD

Laplace Transform of a function, $f(t)$ is defined as

$$F(s) \equiv \int_{0}^{\infty} e^{-st} f(t)\, dt$$

where $f(t)$ is defined for all $t \geq 0$

$\quad F(s)$ is known as Laplace Transform of $f(t)$ and it is a function of s.

$\quad s$ is called complex frequency given by $s = \sigma + j\omega.$ If $\sigma = 0, s = j\omega.$

$\quad$ The Inverse Laplace Transform is defined as:

$$F(s) \equiv f(t) = \frac{1}{2\pi j} \int_{-j\infty}^{j\infty} F(s)\, e^{st}\, dt$$

The Laplace Transform pair can also be used to represent the far-field pattern in terms of source distribution.

The field function $E\,(\phi)$ and the source amplitude function $A\,(x)$ are related by Transform pair. That is,

$$E\,(\phi) = E\,(y) = \int_0^\infty A\,(x)\,e^{jkd\,\sin\phi}\,dx$$

$$= \int_0^\infty A\,(x)\,e^{jyx}\,dx \qquad \qquad ...(5.58)$$

and
$$A\,(x) = \frac{1}{2\pi j}\int_{-j\infty}^{j\infty} E\,(\phi)\,e^{-jyx}\,dy \qquad \qquad ...(5.59)$$

where x is a point on the array axis

$$y = kd\,\sin\phi,\ \ k = \frac{2\pi}{\lambda}$$

$A\,(x) = $ amplitude distribution

For a specified radiation pattern, $E\,(\phi)$, $A\,(x)$ is found.

5.13 STANDARD AMPLITUDE DISTRIBUTIONS

The common tapered **amplitude distributions** for linear arrays are:

1. Interferometer
2. Uniform
3. Circular
4. Parabolic
5. Cosinusoidal
6. Raised cosinusoidal
7. Triangular.

1. **Two-element Interferometer**   The excitation function is represented by

$$A\,(x) = \delta\left(x + \frac{l}{2}\right) + \delta\left(x - \frac{l}{2}\right) \qquad \qquad ...(5.60)$$

This is shown in Fig. 5.9.

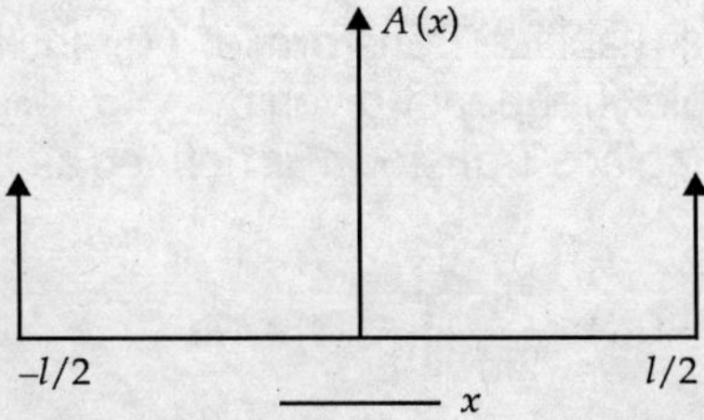

Fig. 5.9 *Two-element interferometer*

The normalised radiation pattern is given by

$$E(u) = \cos\left(\frac{\pi l u}{\lambda}\right) \qquad \ldots(5.61)$$

Here
$$u = \sin\theta$$
$$l = \text{length of the array}$$

A typical radiation pattern is shown in Fig. 5.10.

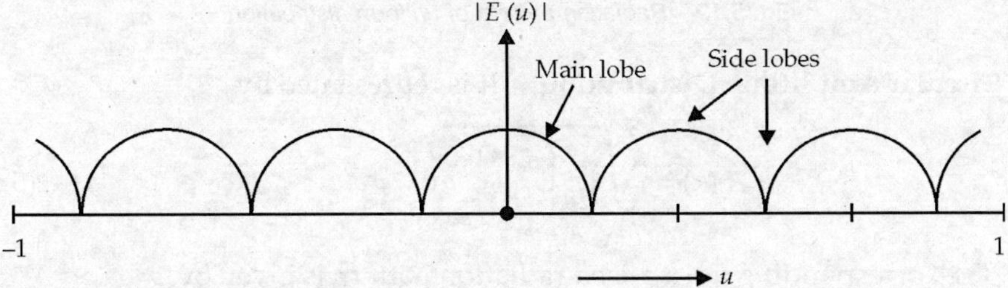

Fig. 5.10 *Radiation pattern of two-element interferometer*

The largest side lobe level is 0 dB.

2. **Uniform Distribution**   This is shown in Fig. 5.11.

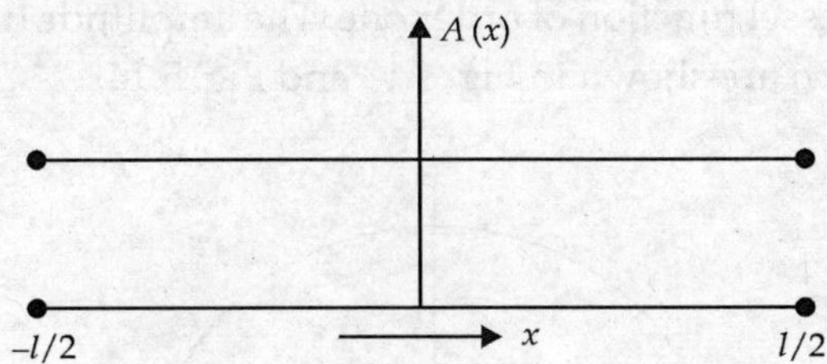

Fig. 5.11 *Uniform amplitude distribution*

It is represented by

$$A(x) = C = \text{constant} \qquad \ldots(5.62)$$

The corresponding normalised radiation pattern is given by

$$E(u) = \frac{\sin\left(\dfrac{\pi l u}{\lambda}\right)}{\left(\dfrac{\pi l u}{\lambda}\right)} \qquad \ldots(5.63)$$

The pattern is shown in Fig. 5.12.

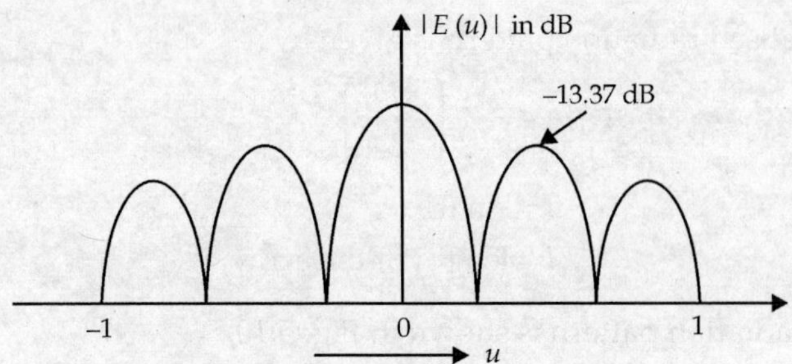

Fig. 5.12 *Radiation pattern of uniform distribution*

3. **Circular Amplitude Distribution**   It is represented by

$$A(x) = \sqrt{1 - \left(\frac{2x}{l}\right)^2} \qquad \qquad \text{...(5.64)}$$

The corresponding normalised radiation pattern is given by

$$E(u) = \frac{J_1\left(\dfrac{K\,lu}{2}\right)}{\left(\dfrac{K\,lu}{2}\right)} \qquad \qquad \text{...(5.65)}$$

Here, $J_1(x)$ = Bessel function of order one. The amplitude distribution and the radiation pattern are shown in Fig. 5.13 and Fig. 5.14.

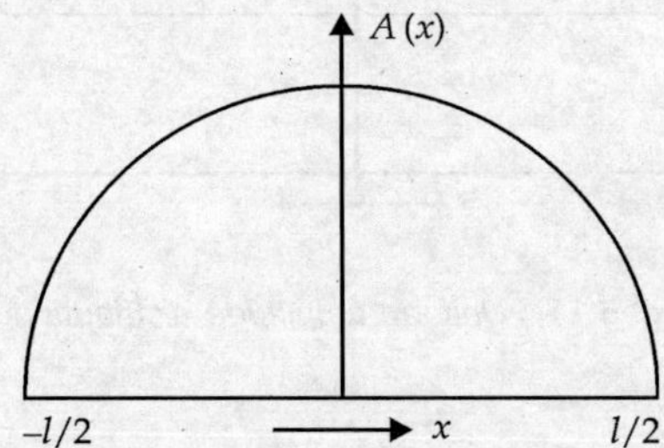

Fig. 5.13 *Circular amplitude distribution*

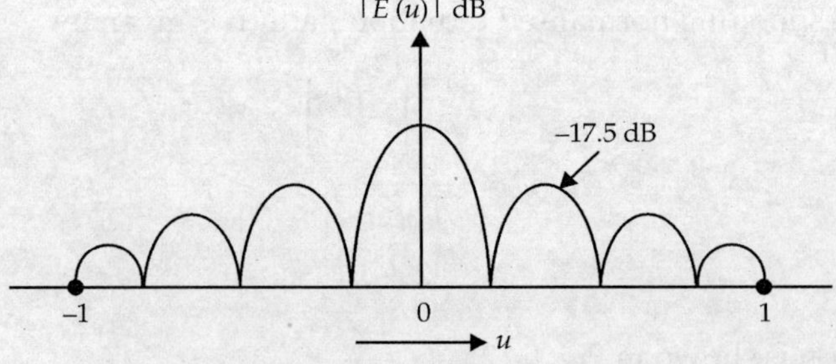

Fig. 5.14 *Radiation pattern of circular amplitude distribution*

4. Parabolic Amplitude Distribution   It is represented by

$$A(x) = \left[1 - \left(\frac{2x}{l} \right)^2 \right]$$

...(5.66)

Its variation with x is shown in Fig. 5.15.

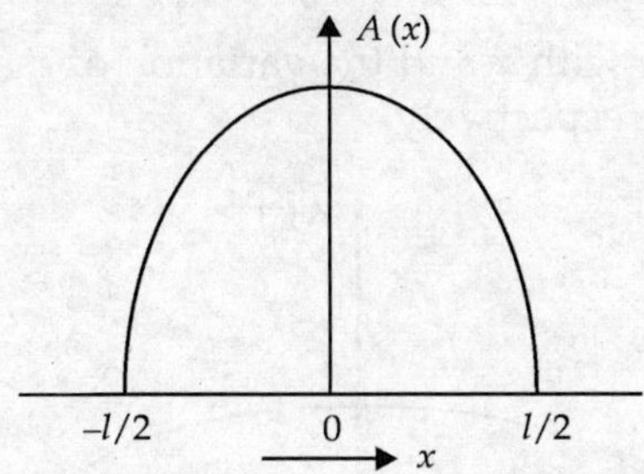

Fig. 5.15   *Parabolic amplitude distribution*

The radiation pattern for parabolic distribution is given by

$$E(u) = \sum_{n=1}^{N} \left[1 - \left(\frac{2x_n}{l} \right)^2 \right] e^{\,j\frac{\pi L}{\lambda} u x_n}$$

...(5.67)

where x_n is the location of the nth element, $u = \sin \phi$.

The corresponding radiation pattern is given in Fig. 5.16.

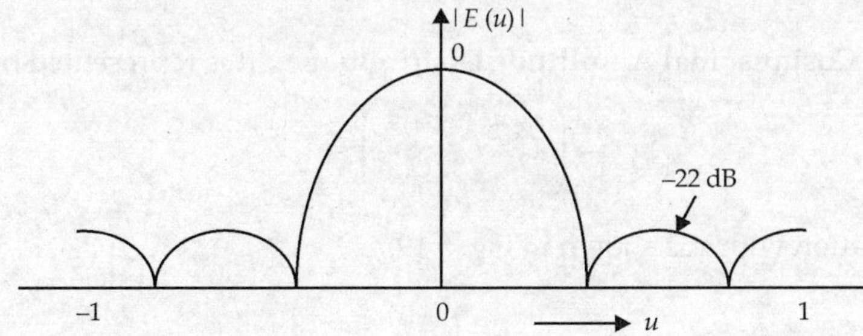

Fig. 5.16   *Radiation pattern for parabolic amplitude distribution*

5. Cosinusoidal Amplitude Distribution   It is represented by

$$A(x) = \cos \left(\frac{\pi x}{l} \right)$$

...(5.68)

The corresponding radiation pattern is given by

$$E(u) = \frac{\pi}{4}\left\{ \frac{\sin\left(\dfrac{\pi l}{\lambda}\right)\left(u - \dfrac{\lambda}{2l}\right)}{\left(\dfrac{\pi l}{\lambda}\right)\left(u - \dfrac{\lambda}{2l}\right)} + \frac{\sin\left(\dfrac{\pi l}{\lambda}\right)\left(u + \dfrac{\lambda}{2l}\right)}{\left(\dfrac{\pi l}{\lambda}\right)\left(u + \dfrac{\lambda}{2l}\right)} \right\} \qquad ...(5.69)$$

The variation of $A(x)$ with x and the variation of $E(u)$ with u are shown in Figs. (5.17) and (5.18) respectively.

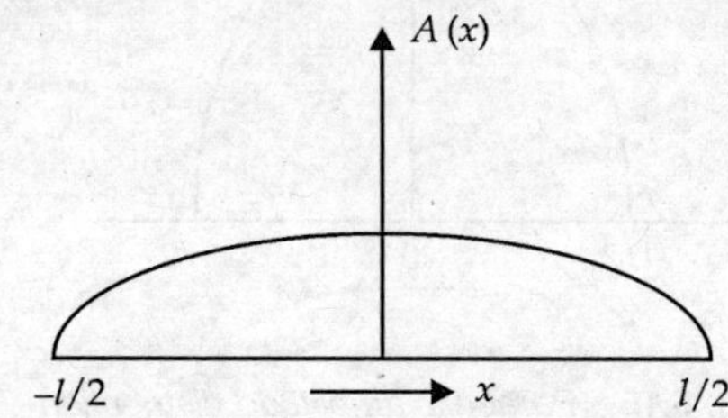

Fig. 5.17 *Cosinusoidal amplitude distribution*

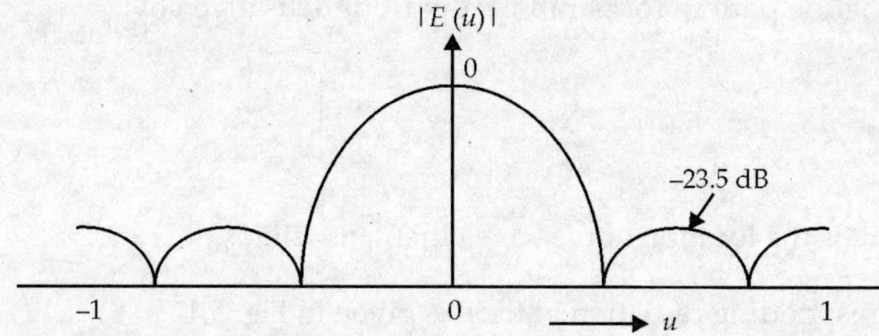

Fig. 5.18 *Radiation pattern for cosinusoidal amplitude distribution*

6. **Raised Cosinusoidal Amplitude Distribution**   It is represented by

$$A(x) = \left(1 + \cos\frac{2\pi x}{l}\right) \qquad ...(5.70)$$

Its variation with x is shown in Fig. 5.19.

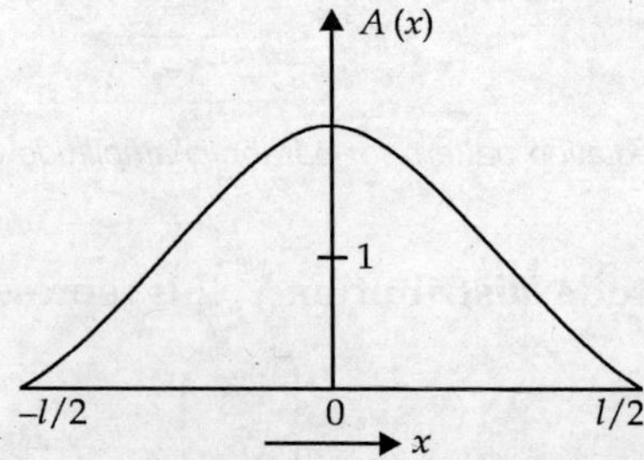

Fig. 5.19 *Raised cosinusoidal distribution*

The radiation pattern is given by

$$E\left(u\right) = \sum_{n=1}^{N} \left[1 + \cos\left(\frac{2\pi x_n}{l}\right)\right] e^{\,j\frac{\pi L}{\lambda} u x_n}$$

The corresponding radiation pattern is shown in Fig. 5.20.

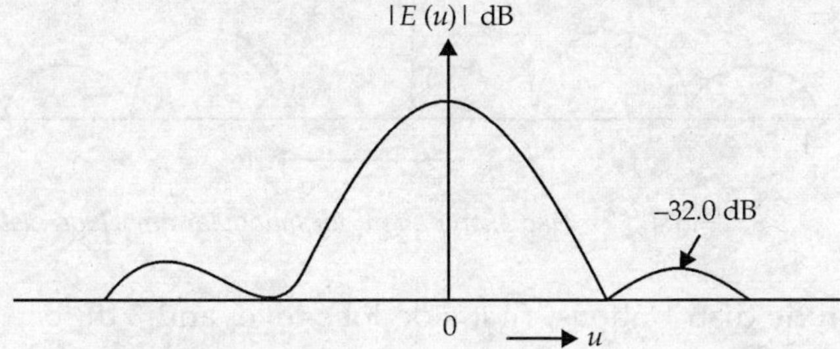

Fig. 5.20   *Radiation pattern for raised cosinusoidal amplitude distribution*

7. **Triangular Amplitude Distribution**   It is represented by

$$A\left(x\right) = \text{tri}\left(l\right) \qquad\qquad ...(5.71)$$

$$A\left(x\right) = \left(1 + \frac{2x}{l}\right) \text{ for } -\frac{l}{2} \leq x \leq 0$$

$$= \left(1 - \frac{2x}{l}\right) \text{ for } 0 \leq x \leq \frac{l}{2}$$

The variation of triangular distribution is shown in Fig. (5.21).

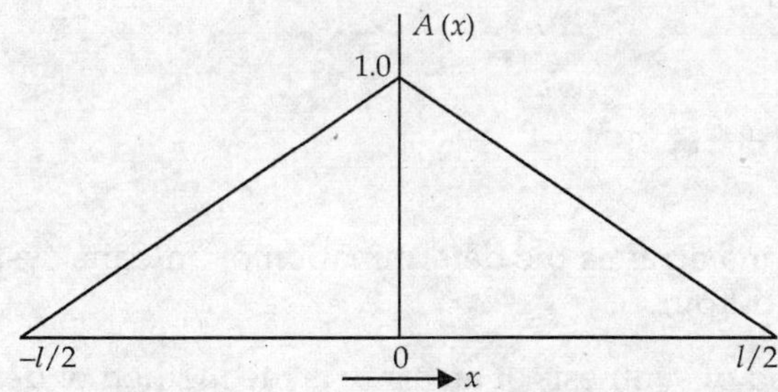

Fig. 5.21   *Triangular amplitude distribution*

The corresponding normalised radiation pattern is given by

$$E\left(u\right) = \left[\frac{\sin\left(\dfrac{\pi l u}{2\lambda}\right)}{\left(\dfrac{\pi l u}{2\lambda}\right)} \right]^2 \qquad\qquad ...(5.72)$$

The radiation pattern is shown in Fig. 5.22.

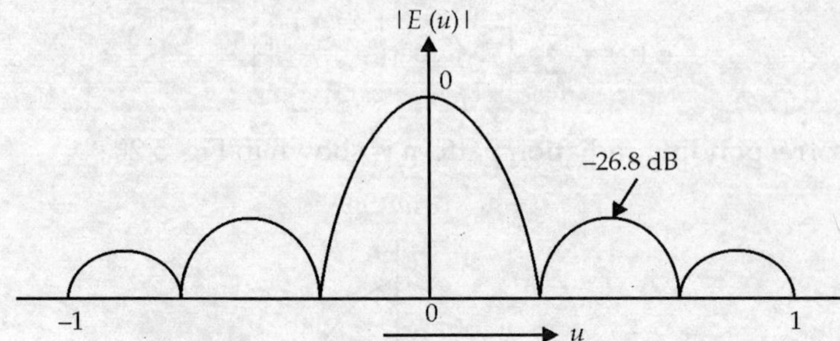

Fig. 5.22   *Radiation pattern for triangular amplitude distribution*

The amplitude distributions, first side lobe level and 3 dB beam width are tabulated in Table 5.1.

Table 5.1   Characteristics of common amplitude distributions

Amplitude distribution	Side lobe level (dB)	3 dB B.W(λ/l)
Uniform $A(x) = 1$	− 13.47	0.88
Circular $A(x) = (1 - x^2)^{1/2}$	− 17.5	1.27
Parabolic $A(x) = (1 - x^2)$	− 22.0	1.16
Cosine $A(x) = \cos \pi x/l$	− 23.5	1.5
Raised cosine $A(x) = 1 + \cos \pi x/l$	− 32.0	2.0
Triangular $A(x) = \left(1 + \dfrac{2x}{l}\right) - \dfrac{l}{2} \le x \le 0$ $A(x) = \left(1 - \dfrac{2x}{l}\right)$ for $0 \le x \le l/2$	− 26.8	1.27

 **POINTS TO REMEMBER**

1. Synthesis or design of an antenna array is the determination of antenna system details for a given input and for a required output.

2. Schelkunoff polynomial method of synthesis of an array is the method of designing an array to produce a pattern with specified nulls.

3. Visible region is a region of space which contains the realisable part of the circle.

4. Invisible region is a region of space which contains the unrealisable part of the circle.

5. Fourier Transform method is a method of array design to produce desired patterns.

6. Woodward method is a method of array design often used to obtain shaped beams.

7. Chebychev method of array design gives a radiation pattern containing side lobes of equal height.

8. Taylor's method of array design gives a radiation pattern containing two regions of side lobes. One region contains side lobes of specified equal height. The second region contains decaying side lobes.

9. Raised cosine type of amplitude distribution yields a pattern with the first side lobe level of -32.0 dB.

 ## SOLVED PROBLEMS

Problem 5.1 If the array factor of a linear array has zeros at $\phi = 90°, 180°, 270°$ and the elements are spaced at $\dfrac{\lambda}{4}$, design the array.

Solution
$$d = \frac{\lambda}{4}, \quad \alpha_e = 0, \quad kd = \frac{2\pi}{\lambda} \times \frac{\lambda}{4} = \frac{\pi}{2}$$

If zero is at $\phi = 90°$ or $\dfrac{\pi}{2}$, then

$$x = e^{j\psi} = e^{j(kd \sin \phi + 0)}$$
$$= e^{j\pi/2} = j$$

If zero is at $\phi = 180°$ or π
$$x = e^{j(kd \sin (180))} = 1$$

If zero is at $\phi = 270°$ or $\dfrac{3\pi}{2}$,

$$x = -j$$

The normalised array factor is
$$E = (x - x_1)\ (x - x_2)\ (x - x_3)$$
or
$$E = (x - j)\ (x - 1)\ (x + j)$$
or
$$E = -1 + x - x^2 + x^3$$

So the excitation coefficients are $(-1,\ 1,\ -1,\ 1)$.

The number of elements required in the array are 4.

Problem 5.2 Design a line source to obtain a radiation pattern given by

$$E(\phi) = 1 \quad \text{for} \quad 45° \le \phi \le 135° \quad \text{and}$$

$$E(\phi) = 0 \quad \text{outside this angular region.}$$

Solution The amplitude distribution for a specified radiation pattern is given by

$$A(x) = \frac{1}{2\pi} \int\limits_{-\infty}^{\infty} E(\phi)\, e^{-jxy}\, dy$$

Here $E(\phi) = 1$ for $\quad 45° \le \phi \le 135°$

and $y = k \sin \phi$ for $\qquad \phi = 45°, \; y = \dfrac{k}{\sqrt{2}}$

for $\qquad\qquad\qquad \phi = 135°, \; y = -\dfrac{k}{\sqrt{2}}$

$$A(x) = \frac{1}{\sqrt{2\pi}} \int\limits_{-k/\sqrt{2}}^{k/\sqrt{2}} e^{-jxy} \, dy$$

That is,
$$A(x) = \frac{k}{\pi\sqrt{2}} \left[\frac{\sin\left(\dfrac{kx}{\sqrt{2}}\right)}{\left(\dfrac{kx}{\sqrt{2}}\right)} \right]$$

The normalised amplitude distribution is

$$A_n(x) = \frac{\sin(kx/\sqrt{2})}{kx/\sqrt{2}}.$$

Problem 5.3   Find the excitation coefficients of a four-element broadside Dolph- Tchebyscheff array which produces a radiation pattern with its first side lobe level of $a = 9.0$ below the main beam. Assume the spacing of the elements to be $\lambda/2$.

Solution   The required side lobe ratio is

$$a = 9.0$$

So $\qquad$ SLL in dB $= 20 \log(a)$

$$= 20 \log(9.0)$$

$$= 19.08 \text{ dB}$$

The number of elements in the array, $n = 4$.

The degree of the polynomial,

$$m = n - 1 = 4 - 1 = 3$$

We know that

$$T_m(x_0) = T_{n-1}(x_0) = T_3(x_0)$$

$$= a = 9.0$$

That is, $\qquad 4x_0^3 - 3x_0 = T_3(x_0) = 9.0$

Adding $6x_0^2$ in both sides, we get

$$4x_0^3 + 6x_0^2 - 3x_0 = 6x_0^2 + 9.0$$

$$4x_0^3 - 6x_0^2 + 6x_0^2 + 6x_0 - 9x_0 - 9.0 = 0$$

$$2x_0^2(2x_0 - 3) + (6x_0^2 - 9x_0) + (6x_0 - 9) = 0$$

$$2x_0^2(2x_0 - 3) + 3x_0(2x_0 - 3) + 3(2x_0 - 3) = 0$$

$$(2x_0 - 3)\ (2x_0^2 + 3x_0 + 3) = 0$$

Therefore,
$$(2x_0 - 3) = 0$$

and
$$(2x_0^2 + 3x_0 + 3) = 0$$

or
$$x_0 = 3/2 = 1.5$$

As the number of elements in the array are even, the pattern is given by

$$E_R = \sum_{m=0}^{N-1} A_m \cos\left[(2m+1)\,\psi/2\right]$$

$$= A_0\,z + A_1\,(4z^3 - 3z) = T_3\,(x)$$

or
$$A_0\,z + A_1\,(4z^3 - 3z) = 4x^3 - 3x$$

$$\text{where } z = \frac{x}{x_0}$$

$$A_0\left(\frac{x}{x_0}\right) + A_1\left[\left(\frac{x}{x_0}\right)^2 - 3\left(\frac{x}{x_0}\right)\right] = 4x^3 - 3x$$

Equating the respective terms, we get

$$x\left[\frac{A_0 - 3A_1}{x_0}\right] = -3x$$

$$\frac{A_0 - 3A_1}{1.5} = 3$$

or
$$A_0 - 3A_1 = -4.5$$

Also we have

$$4A_1\left(\frac{x}{x_0}\right)^3 = 4x^3$$

or
$$4A_1\left(\frac{x^3}{(1.5)^3}\right) = 4x^3$$

$$A_1 = (1.5)^3 = 3.375$$

and
$$A_0 = -4.5 + 3A_1$$

$$= -4.5 + 3 \times 3.375$$

or
$$A_0 = 5.625$$

Normalising the coefficients to a maximum of 1, we get the array excitations as

$$\boxed{0.6 \quad 1 \quad 1 \quad 0.6}$$

OBJECTIVE QUESTIONS

1. The extent of visible region can be controlled by the spacing between elements.   (Yes/No)

2. The relative position on the circle is controlled by the progressive phase excitation of the elements.   (Yes/No)

3. The Schelkunoff polynomial method is useful to design an array of elements which produces a pattern with nulls in the desired directions.   (Yes/No)

4. Fourier Transform method of array synthesis is more accurate if the array length is very small.   (Yes/No)

5. The Schelkunoff polynomial method of array design gives symmetrical excitation coefficients.   (Yes/No)

6. Fourier Transform design method is applicable only to discrete arrays.   (Yes/No)

7. Continuous line sources and discrete arrays can be designed by Fourier Transform method.   (Yes/No)

8. Any type of radiation beam can be designed by Fourier Transform method.   (Yes/No)

9. Beam shaping can be done by Woodward method.   (Yes/No)

10. Only odd element arrays can be designed by Woodward method.   (Yes/No)

11. Grating lobes are extremely useful.   (Yes/No)

12. Dolph-Tschebyscheff method results in a radiation pattern with a maximum band width.   (Yes/No)

13. Dolph-Chebychev method yields a pattern which contains side lobes of unequal level.   (Yes/No)

14. The amplitude taper is very high in Dolph-Chebychev method.   (Yes/No)

15. If impulses or large peaks first appear in the nth derivative of the excitation, SLL is of the order of ________________.

16. 3 dB beam width is greater than the Null-to-Null beam width in a typical radiation pattern.   (Yes/No)

17. The last side lobe level is usually higher than that of the first side lobe level in several cases.   (Yes/No)

18. The side lobe level of triangular distribution for the array is higher than that of uniform linear array.   (Yes/No)

19. The 3 dB beam width of triangular excitation distribution is smaller than that of uniform linear array.   (Yes/No)

20. If the amplitude distribution is highly tapered towards the end of the array, the side lobe level becomes small.   (Yes/No)

21. The side lobe level in the pattern of cosinusoidal amplitude distribution is higher than that of uniform array.   (Yes/No)

22. Radiation pattern remains the same whether there is an element or not in the centre of array.
(Yes/No)

23. Collinear array means _______________.

24. The parasitic antenna element derives power from _______________.

25. A point source is a radiator which _______________.

26. Radiation pattern can be controlled by amplitude distribution only.   (Yes/No)

27. Radiation pattern can be controlled by phase control only.   (Yes/No)

28. For a binomial array the directivity is _______________.

29. The beam width of a binomial array is _______________ compared to that of uniform linear array.

30. The advantages of Dolph-Tschebyscheff distribution is that _______________ a minimum beam width for a given side lobe level.

31. The side lobes of the patterns of Dolph-Tschebyscheff amplitude distribution are _______________.

32. The gain in super directive arrays is _______________.

33. If the side lobe level below the main lobe is 19.1 dB, the side lobe level is _______________.

34. The efficiency of super directive antenna is _______________.

35. The power gain of super directive gain antenna is _______________.

36. Beam shaping can be done by Fourier Transform method.   (Yes/No)

37. In Taylor's method of array design, σ is called _______________.

38. In Taylor's method of array design, the parameter A is _______________.

39. In Taylor's method of array design, scaling factor is defined as _______________.

40. In an array pattern, the number of nulls are influenced by the number of elements in the array.
(Yes/No)

41. The space factor of an array is _______________ of a similar array of non-directive or isotropic elements.

42. A linear array with certain spacing between the elements can be represented by a polynomial.
(Yes/No)

43. There exists additional phase in an array space factor without introducing phase from excitation.   (Yes/No)

44. The space factor of a linear array of n elements is the product of $(n-1)$ virtual couplets with their null points at the zeros of E.   (Yes/No)

45. Tschebyscheff polynomials are defined as _______________.

46. The invisible region is reduced by the spacing of the elements.   (Yes/No)

47. The common spacing of radiating elements in linear arrays is _______________.

48. Super directive characteristics are obtained when the spacing of the radiating elements is
____

49. Grating lobes are often produced when the spacing of the radiating elements is _______________.

50. The side lobe level of a triangular amplitude distribution is less than that of uniform.
(Yes/No)

51. Parabolic amplitude distribution is better than the circular distribution as far as side lobe levels are concerned.
(Yes/No)

52. In two-element interferometer, the pattern consists of equal radiation lobes.　(Yes/No)

53. The range of visible region depends on _______________.

54. The degree of Tschebyscheff polynomial is equal to the number of elements minus one.
(Yes/No)

ANSWERS

1. Yes	**2.** Yes	**3.** Yes	**4.** No	**5.** Yes	**6.** No
7. Yes	**8.** Yes	**9.** Yes	**10.** Yes	**11.** No	**12.** No
13. No	**14.** No	**15.** $-10n$ dB	**16.** No	**17.** No	**18.** No
19. Yes	**20.** Yes	**21.** No	**22.** No		

23. An array in which the antennas are arranged co-axially

24. Driven element by radiation　　　　**25.** Has no volume　　　　**26.** No

27. No　　　**28.** Small　　　**29.** High　　　**30.** A minimum

31. At equal level　　　**32.** High　　　**33.** 9.0　　　**34.** Small　　　**35.** Small

36. Yes　　　**37.** Scaling factor

38. A constant which is related to maximum desired side lobe level

39. $\sigma = \dfrac{\bar{n}}{\left(A^2 + \left(\bar{n} - \dfrac{1}{2}\right)^2\right)^{1/2}}$　　　**40.** Yes　　　**41.** The radiation pattern of　　　**42.** Yes

43. Yes　　　**44.** Yes

45. $T_m(x) = \cos(m \cos^{-1} x),\ -1 < x < 1$　　$T_m(x) = \cosh(m \cosh^{-1} x),\ |x| > 1$　　　**46.** Yes

47. $\dfrac{\lambda}{2}$　　　**48.** $< \dfrac{\lambda}{2}$　　　**49.** $> \dfrac{\lambda}{2}$　　　**50.** Yes　　　**51.** Yes　　　**52.** Yes

53. Elements spacing, excitation phase and frequency　　　**54.** Yes.

 ## EXERCISE PROBLEMS

1. Find the nulls in the radiation pattern of uniform array of 5 elements.

2. Find out an expression for the relative field strength pattern of a four-element broadside array with $\lambda/2$ spacing.

3. Obtain the nulls of the pattern of an end-fire array with the elements having spacing of 0.25λ. Also determine the expression for the field strength and current levels of the elements.

4. Find the excitation coefficients of a three-element broadside Chebychev array which produces a radiation pattern with SLR = 30 dB. The spacing of the elements is 0.5λ.

5. Find the normalised excitation levels of a ten-element array for triangular amplitude distribution. The elements are spaced at $\lambda/2$.

6. What are the normalised excitation levels of a twenty-element linear array when the array is excited with parabolic distribution? The elements are spaced at 0.25λ.

7. For a five-element array, determine current levels for circular distribution if the elements are spaced at 0.5λ.

8. Obtain the required amplitude distribution for a flat beam over $-\dfrac{\pi}{2} \le \phi \le \dfrac{\pi}{2}$ using Fourier Transform method.

9. Design excitation coefficients of an array of twenty-one elements to obtain a radiation pattern given by

$$E\,(\phi) = 1 \ \text{ for } \ 45^\circ \le \phi \le 135^\circ$$
$$E\,(\phi) = 0 \ \text{ outside this angular range.}$$

10. Obtain Tschebyschev polynomials corresponding to $m = 7,\ 8,\ 9$.

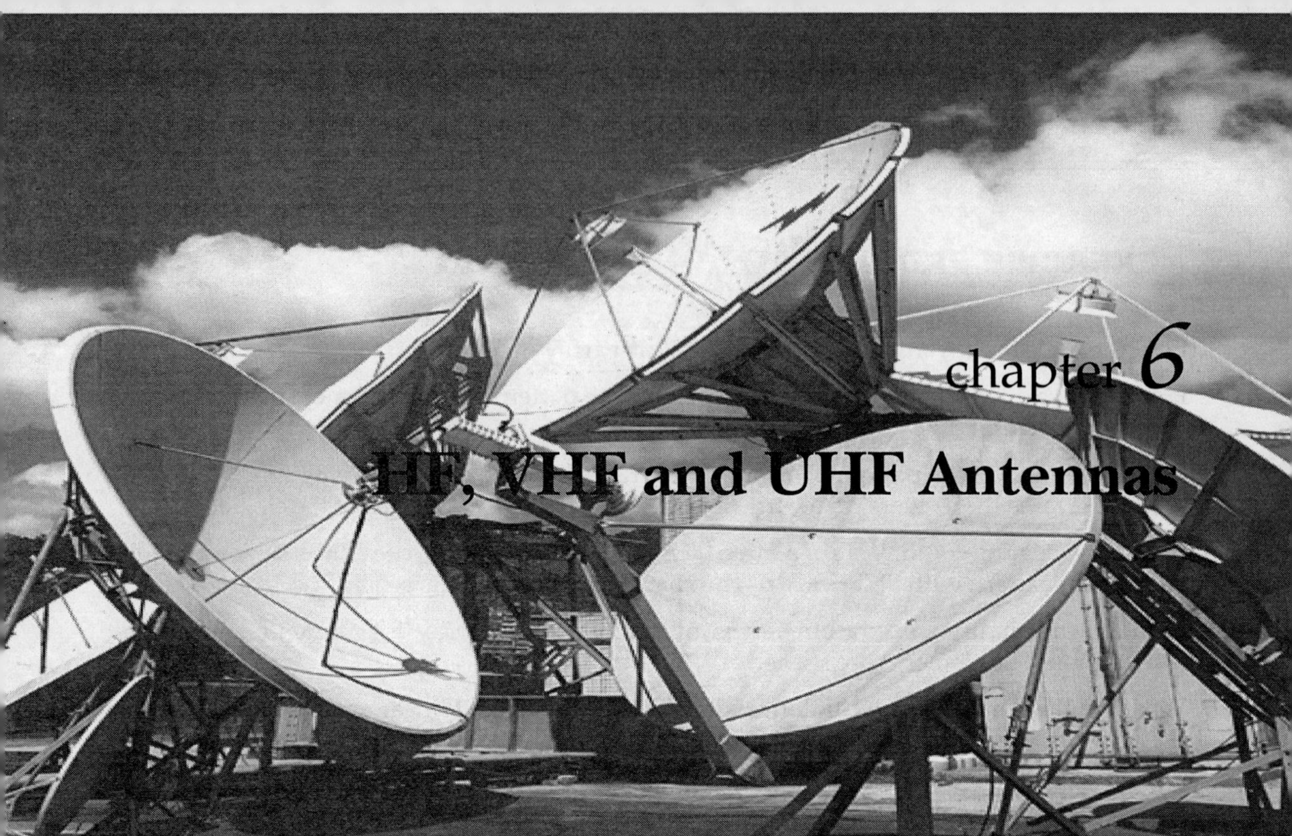

"The size of antennas is small at high frequencies and it is large at low frequencies."

CHAPTER OBJECTIVES

This chapter discusses

✦ The design, construction, application and performance parameters of all types of HF, VHF and UHF antennas

✦ Merits and demerits of antennas

✦ Objective questions and solved problems useful for class tests, final examinations and also for competitive examinations

✦ Exercise problems to develop self problem solving skills

6.1 INTRODUCTION

The HF, VHF and UHF spectrum is between 3 MHz and 1 GHz. The range above 1 GHz is usually branded as the Microwave range. The main difference between the lower frequencies and HF, VHF and UHF spectrum is in the operating wavelength. It may be noted that the frequency and the wavelength are inversely proportional to each other. As a result, at low frequency, the wavelength is large and vice-versa. Each application makes use of any one of these ranges of frequency. For instance, a 6-meter amateur band lies in the lower end of the VHF range. It may be noted that the antenna bandwidth is a function of the ratio of its length to diameter. Hence, broadbanding of an antenna in VHF and UHF is easy.

It must be noted that the concept of HF, VHF and UHF antennas has limited validity as, in principle all forms of antennas can be used at all bands. The main limitation lies in the size of the antenna. When the antenna becomes larger, it is difficult to handle and it is also difficult to install.

Large array antennas, even in HF, VHF and UHF ranges, have high "windsail area". These antennas are therefore subjected to a lot of wind force. It is always advisable to install such antennas with a helper and with the help of hosts and other tools.

The impedance matching of VHF and UHF antennas is essential and suitable BALUNS are selected for the purpose.

An antenna is an essential device in all communications and radar systems. It acts as a transducer, impedance-matching device, radiator and receiver of electromagnetic waves. With a well-designed antenna, it is possible to have communication from one point to any other point on the entire globe. With a badly-designed antenna, it is not possible to send signals even beyond the premises of the transmitter.

There are many types of antennas. The choice of antenna depends upon the frequency of operation, polarisation, gain requirements and application.

Antennas are classified into different categories.

First Classification

In this classification, antennas are classified into:

(*a*) Isotropic radiators

(*b*) Directional antennas

(*c*) Omni-directional antennas.

6.2 ISOTROPIC RADIATORS

An **isotropic radiator** is defined as a hypothetical element which radiates equally in all directions.

Examples

1. A point source
2. A star.

It is an ideal antenna but it is not realisable practically. It is useful as a reference antenna for determining directive properties of practical antennas.

If P_i is the input power to a loss-less isotropic radiator, the power density is

$$P_D = \frac{P_i}{4\pi r^2} \; \text{W/m}^2 \qquad\qquad ...(6.1)$$

Here, $\qquad\qquad r = $ radius of a sphere

6.3 DIRECTIONAL ANTENNAS

These are the antennas which radiate or receive electromagnetic waves more effectively in some directions than in others.

Examples

1. Dipoles   2. Horns

3. Paraboloids and so on.

6.4 OMNI-DIRECTIONAL ANTENNA

It is defined as an antenna which has a non-directional pattern in azimuth and has a directional pattern in elevation. An omni-directional pattern is a special type of directional pattern.

Examples

1. A circular loop antenna. Omni-directional refers only to the horizontal plane. In this plane, the pattern is a circle.

2. Vertical Hertz antenna

3. Marconi antenna

4. Quarter-wave monopole

The radiation patterns of uni-directional, omni-directional antennas are shown in Fig. 6.1.

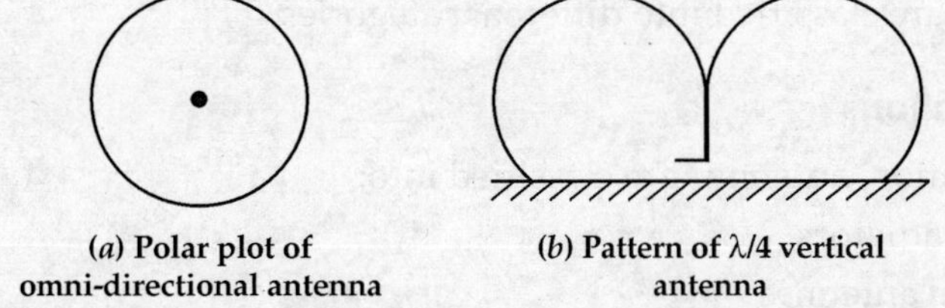

(*a*) Polar plot of
omni-directional antenna

(*b*) Pattern of $\lambda/4$ vertical
antenna

Fig. 6.1 *Polar plots of omni-directional and uni-directional antennas*

Second classification

The antennas are also classified into:

1. Resonant antennas and   2. Non-resonant antennas.

6.5 RESONANT ANTENNAS

The features of these antennas are:

1. The length of a resonant antenna is in exact multiples of $\dfrac{\lambda}{2}$.

2. These antennas are open at both ends.

3. These are not terminated in any resistance.

4. They are used at a fixed frequency.

5. In these antennas, forward/incident and backward/reflected waves exist.

6. A standing wave exists in these antennas.

7. The radiation patterns of these antennas are multi-directional.

8. The current distribution in resonant antennas is shown in Fig. 6.2.

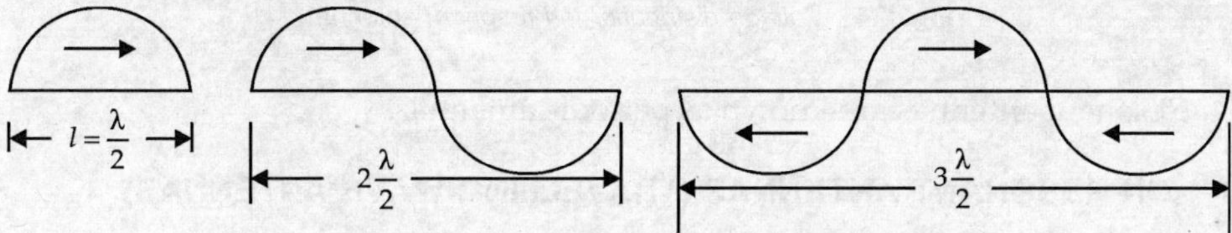

Fig. 6.2 *Current distribution on resonant dipole*

9. Radiation patterns of different resonant dipoles are shown in Fig. 6.3.

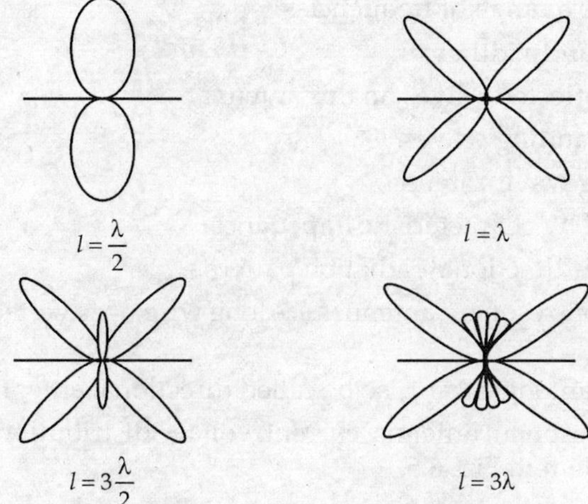

Fig. 6.3 *Radiation patterns of resonant dipoles*

10. The half-wave antenna has distributed inductance and capacitance and it acts like a resonant circuit.

11. The voltage and current on it are not in phase.

12. The voltage distribution on half-wave dipole is shown in Fig. 6.4.

13. The length of a resonant antenna is found from

$$l = \frac{\upsilon_0}{f} \times F$$

Here, F = velocity factor

(The velocity factor of wire compared to air, $F \approx 0.95$)

v_0 = velocity of propagation.

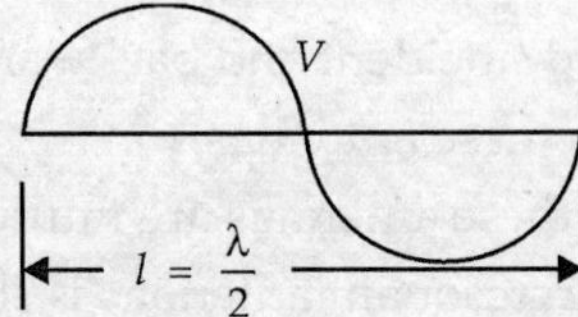

Fig. 6.4 *Voltage distribution on resonant antenna*

14. Resonant antennas are known as periodic antennas.

6.6 NON-RESONANT ANTENNAS (TRAVELLING WAVE ANTENNAS)

The features of these antennas are:

1. The length of a non-resonant antenna is other than in multiples of $\dfrac{\lambda}{2}$.

2. At one end of the antenna, it is excited and the other end is terminated.

3. It operates over a range of frequencies.

4. It has a wide bandwidth.

5. There are no reflected waves on this antenna.

6. There are no standing waves.

7. It is a travelling wave antenna.

8. It is terminated by characteristic impedance.

9. It produces uni-directional radiation patterns.

10. Examples of non-resonant antennas are long wire, V, inverted V and Rhombic antennas.

11. Non-resonant antennas can also be called directional antennas.

12. A typical non-resonant antenna, current, voltage distribution and its radiation pattern are shown in Fig. 6.5.

13. Non-resonant antennas are known as aperiodic antennas.

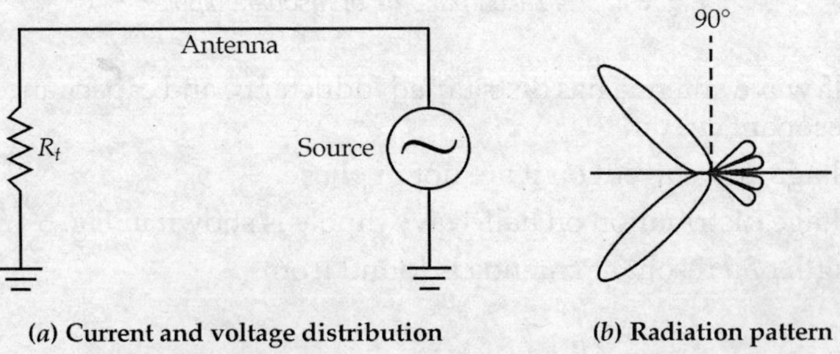

(*a*) Current and voltage distribution
on non-resonant antenna

(*b*) Radiation pattern

Fig. 6.5 *Non-resonant antenna*

14. The pattern expression of a travelling wave antenna is given by

$$E = \frac{30 I_m \sin\theta}{r\,(1 - \cos\theta)} \{2 - 2\cos[kL\,(1 - \cos\theta)]\}^{1/2} \qquad ...(6.2)$$

where I_m = maximum current in the element

r = distance from the source, $k = \dfrac{2\pi}{\lambda}$

L = length of the element.

15. The pattern is not symmetric about $\theta = 90°$.

Third Classification

This is in terms of:

1. Standing wave antennas.   2. Travelling wave antennas.

It is evident from the previous section that a standing wave antenna is nothing but a resonant antenna. Similarly, travelling wave antenna is nothing but a non-resonant antenna.

Standing wave is defined as a wave in which the ratio of the instantaneous value of any component of the wave at one point of that at any other point does not vary with time.

Travelling wave is defined as a wave whose frequency component have exponential variation of amplitude and linear variation of phase with distance.

Fourth Classification

This is on the basis of frequency range over which the antenna can be used. These are LF, HF, VHF, UHF and microwave frequency antennas.

The range of frequency and applications are shown in Table 6.1.

Table 6.1   Frequency ranges and applications

S. No.	Band name	Frequency range	Typical Applications
1.	VLF	(3-30 kHz)	Telegraphy
2.	LF	(30-300 kHz)	Marine and navigational aids
3.	MF	(300 kHz-3 MHz)	AM broadcast, navigation
4.	HF	(3-30 MHz)	Aircraft radio, short wave broadcast
5.	VHF	(30-300 MHz)	FM, television, radar and so on
6.	UHF	(300 MHz-3 GHz)	Radar, television, short distance communication
7.	Microwave	(3-30 GHz)	Radar, satellite communication and so on
8.	EHF	(30-300 GHz)	Experimental purposes

6.7 LF ANTENNAS

These antennas operate at low frequencies (75 – 160 meter band). Examples of these antennas are:

1. Inductance loaded vertical antennas (Fig. 6.6).

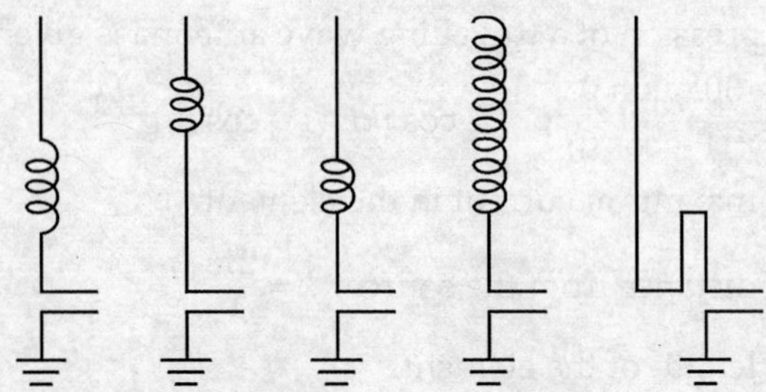

Fig. 6.6 *Inductance loaded vertical antennas*

2. Inductance loaded horizontal dipoles (Fig. 6.7).

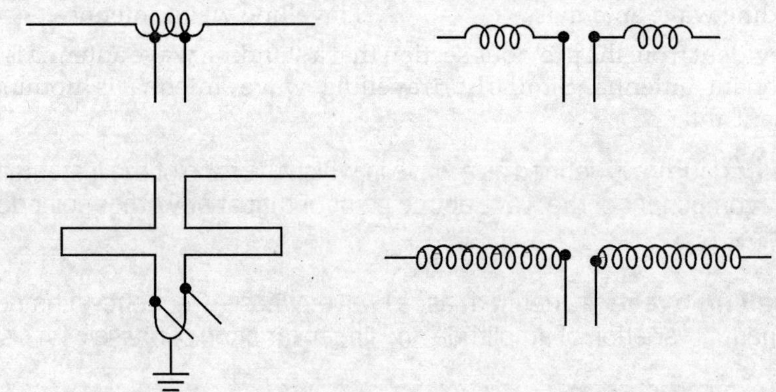

Fig. 6.7 *Inductance loaded horizontal dipoles*

3. Tower antenna (Fig. 6.8).
4. Inverted-*L* antenna. This is shown in Fig. 6.9.

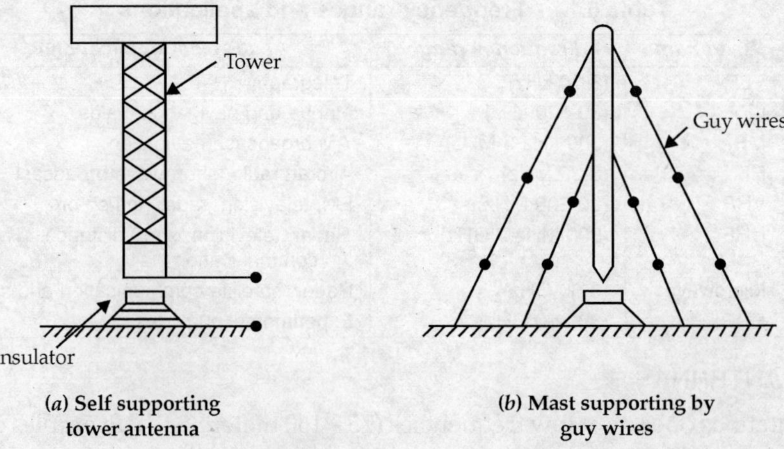

Fig. 6.8 *Tower antennas*

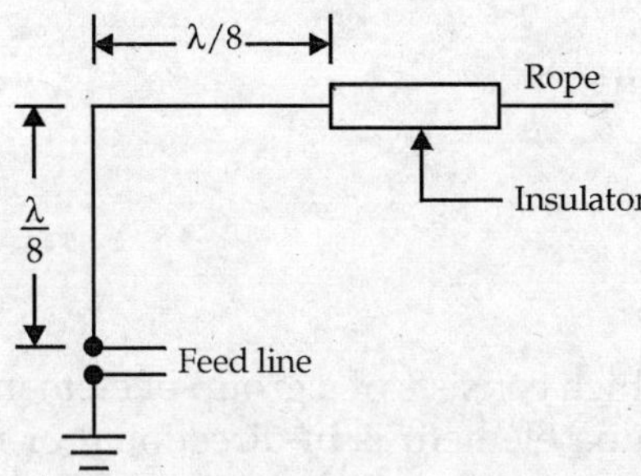

Fig. 6.9   *Inverted-L section quarter-wave antenna*

5. A short vertical monopole with top capacitance. (Fig. 6.10).

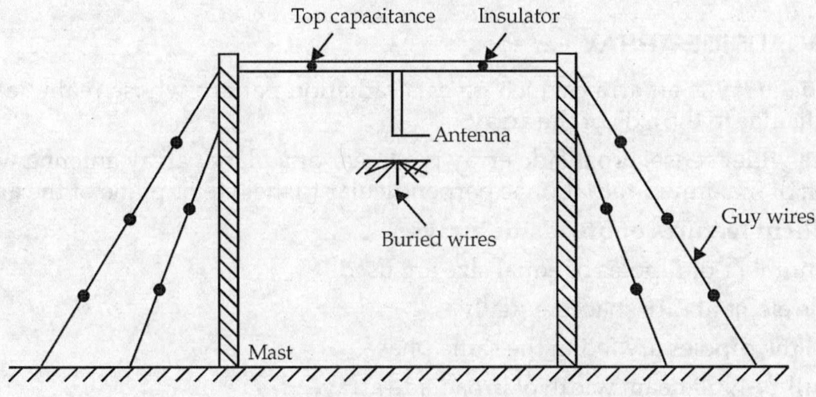

Fig. 6.10   *Monopole*

6.8 ANTENNAS FOR HF, VHF, UHF

In these frequency ranges, the following antennas are used. They are described in detail.

1. Dipole arrays : Broadside and end-fire arrays.
2. Folded dipole.
3. *V* antennas.
4. Inverted *V* antennas.
5. Rhombic antennas.
6. Yagi-Uda antennas.
7. Log-periodic antennas.
8. Loop antennas.
9. Helical antennas.
10. Whip antennas.
11. Ferrite rod antennas.

12. Turnstile antennas.
13. Super turnstile antennas.
14. Discone antennas.
15. Notch antennas.

6.9 DIPOLE ARRAYS

An array antenna is one which consists of a group of elements arranged linearly or in a plane. When an antenna element is bi-directional or multi-directional in its radiation characteristics, an array of such elements yields a uni-directional pattern. An array is said to be linear if the elements are arranged along a straight line with equal spacing. Arrays are divided into:

1. Broadside array

2. End-fire array.

6.10 BROADSIDE ARRAY

Broadside array is an array which gives a radiation pattern whose main beam is perpendicular to the axis of the array.

In a wider sense, broadside array is a linear or a planar array antenna whose direction of maximum radiation is perpendicular to the line or plane of the array.

Salient features of broadside arrays

1. A number of dipoles of equal size are used.
2. The elements are spaced equally.
3. All the dipoles are fed in the same phase.
4. Null-to-Null beam width of broadside array.

Beam width between first Nulls

$$= \frac{2\lambda}{Nd} \qquad \qquad ...(6.3)$$

where λ = wavelength

N = number of elements

d = spacing between the elements.

5. The length of the broadside array can be 2 to 10λ.

6. Typical spacing between the elements vary from $\frac{\lambda}{2}$ to λ.

7. The number of elements to be used depends on the beam width requirement, cost and space available.

8. A broadside array is often used along with a reflector antenna. The back lobe is now reflected forward and adds to the forward lobe.

9. When a broadside array is used with a reflector, it is possible to improve its gain and directivity and the broadside array becomes uni-directional.

10. This array is often used in overseas broadcast systems.

11. It is used for LF, MF, HF and higher band of frequencies. A typical broadside array is shown in Fig. 6.11.

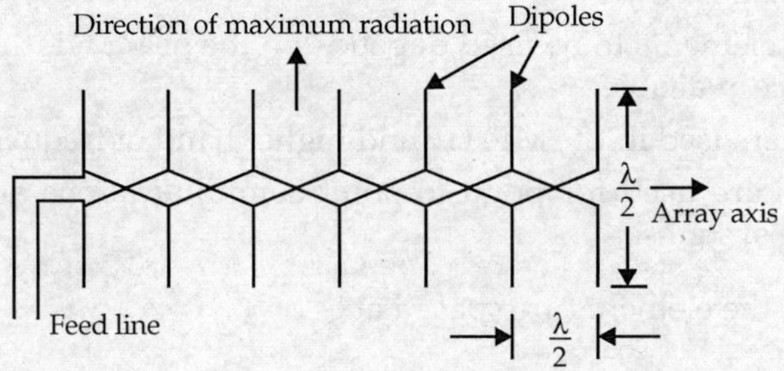

Fig. 6.11 *Broadside array*

12. A typical radiation pattern of a broadside array is shown in Fig. 6.12.

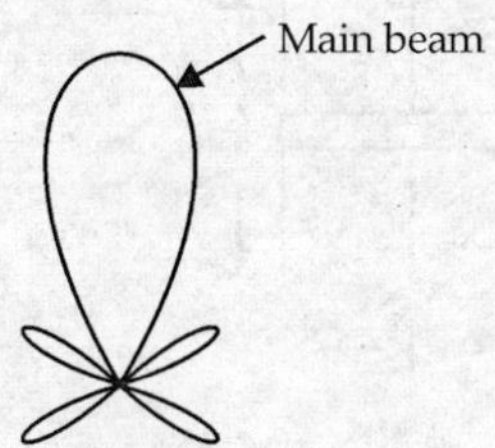

Fig. 6.12 *Radiation pattern of broadside array*

6.11 END-FIRE ARRAY

End-fire array is an array which gives a radiation pattern whose main beam is along the axis of the array.

In a wider sense, end-fire array is a linear or planar antenna whose direction of maximum radiation is along the line or in the plane of the array.

Salient features of end-fire array

1. A number of dipoles or elements of equal size are used.

2. The elements are equally spaced.

3. The elements are fed with different phases.

4. The additional phase for each element is given by

$$\alpha_{N-1} = (N-1)\,kd\,\cos\phi$$

where $k = \dfrac{2\pi}{\lambda}$, d = spacing, ϕ is the angle between the line of observation and axis of the array.

5. Null-to-Null to beam width of an end-fire array is

$$\text{B.W.F.N.} = 2\sqrt{\dfrac{2\lambda}{Nd}} \qquad\qquad \text{...(6.4)}$$

6. In this, the pattern is uni-directional.

7. The physical arrangement of the elements in the end-fire array is the same as that of broadside array.

8. The number elements to be used depends on the beam width requirements, cost and space available.

9. These are often used in LF, MF, HF and higher band of frequencies.

10. These arrays are used for point-to-point communications and in overseas broadcasting systems.

11. In this array, the elements are spaced at $\dfrac{\lambda}{4}$ or $3\dfrac{\lambda}{4}$.

A typical end-fire structure is shown in Fig. 6.13 and its radiation pattern is shown in Fig. 6.14.

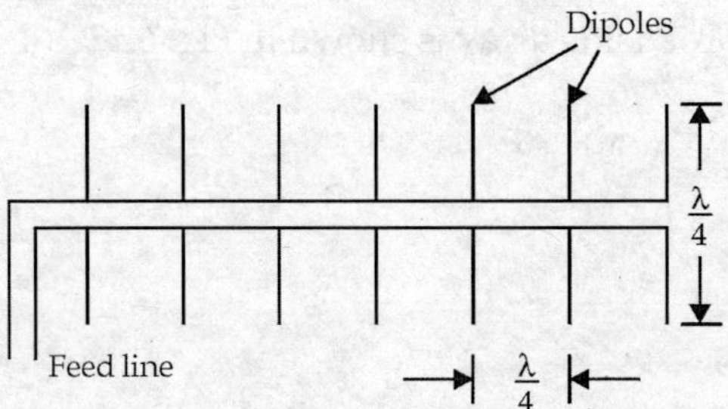

Fig. 6.13 *End-fire array*   **Fig. 6.14** *Radiation pattern*

6.12 FOLDED DIPOLE

It is an antenna composed of two or more parallel and closely spaced dipole antennas connected together at their ends with one of the dipole antennas being centre fed.

The folded dipole antenna is shown in Fig. 6.15.

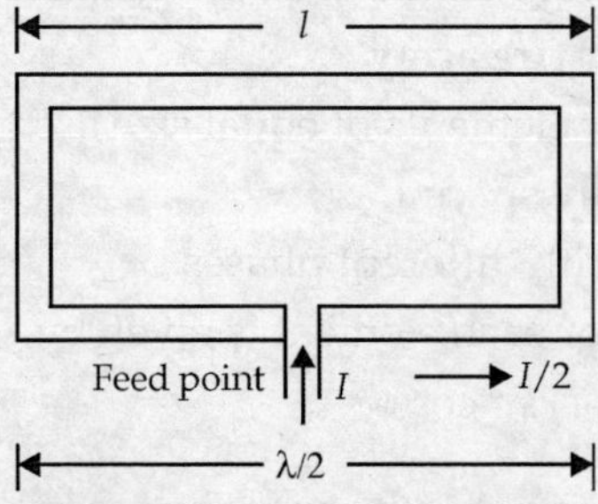

Fig. 6.15 *Folded dipole antenna*

Salient features of folded dipole

1. It is a single antenna, but consists of two elements.

2. The first is fed directly and the second is inductively coupled at the ends.

3. Its radiation pattern is the same as that of a straight dipole.

4. If the current fed is I, then the current in each arm is $\dfrac{I}{2}$, provided the two arms have the same dimensions. If it is a straight dipole, the total current I flows.

5. When the same power is applied, only half of the current flows in the first arm. Therefore, the input impedance is four times that of the straight dipole. That is,

$$R_r = 4 \times 73 = 292\,\Omega.$$

6. If the diameters of the two arms of folded dipole are different, impedance transformation of 1.5 to 25 is achievable.

7. The spacing between the arms is very small and is of the order of $\dfrac{\lambda}{100}$.

8. Folded dipole is used in Yagi-Uda antenna as an active element.

9. It has the advantages of high input impedance, greater band width, ease and low cost of construction with better impedance-matching characteristics.

A typical radiation pattern of a folded dipole is shown in Fig. 6.16.

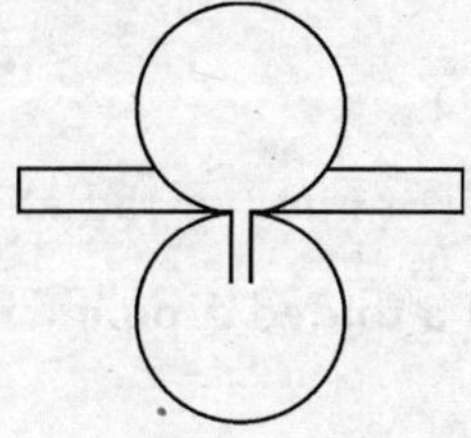

Fig. 6.16 *Radiation pattern of folded dipole*

Impedance of the folded dipole   It is given by

$$Z = 292\,\Omega$$

Proof   The equivalent diagram of the folded dipole of Fig. 6.15 is shown in Fig. 6.17.

When a voltage V is applied to the folded dipole, it is divided equally in each arm of the dipole. That is, the voltage in each dipole is $\dfrac{V}{2}$. Hence, we have

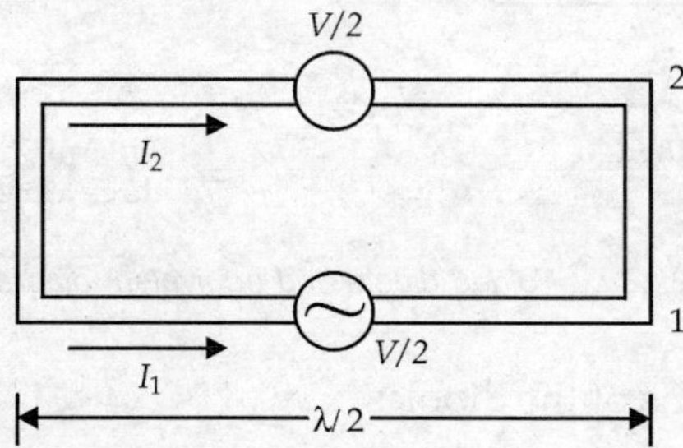

Fig. 6.17 *Equivalent diagram of $(\lambda/2)$ folded dipole*

$$\frac{V}{2} = I_1 Z_{11} + I_2 Z_{12}$$

Here, $I_1 =$ current in (1) and I_2 is current in (2)

$Z_{11} =$ self impedance of dipole (1)

$Z_{12} =$ mutual impedance between (1) and (2)

For equal dimensions of the dipoles

$$I_1 = I_2 = I$$

So $$\frac{V}{2} = I (Z_{11} + Z_{12})$$

As the two dipoles are very close and the spacing is very small,

$$Z_{11} = Z_{12}$$

or $$\frac{V}{2} = I (2Z_{11})$$

or $$Z = \frac{V}{I} = 4Z_{11}$$

$$= 4 \times 73 = 292\,\Omega$$

$$R_i = Z_r = \text{terminal or input or radiation resistance} = 292\,\Omega$$

The radiation resistance of a folded dipole with 3 arms It is given by

$$R_r = 657\,\Omega.$$

Proof The three armed folded dipole is shown in Fig. 6.18 (*a*) and its equivalent diagram is shown in Fig. 6.18 (*b*).

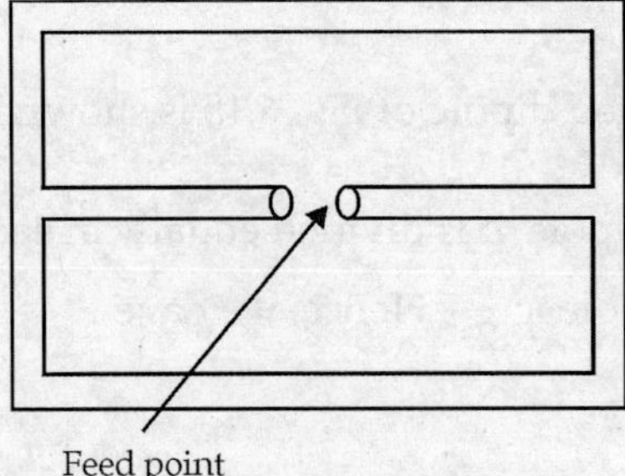

(*a*) **Three armed folded dipole**

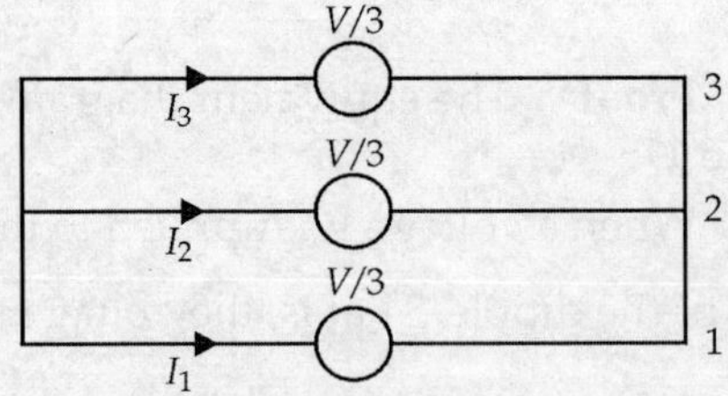

(*b*) **Equivalent diagram of three armed folded dipole**

Fig. 6.18 *Folded dipole and equivalent diagram*

For equal dimensions of the dipoles,

$$I_1 = I_2 = I_3 = I$$

So
$$\frac{V}{3} = I\,(3Z_{11})$$

or
$$\frac{V}{I} = 3 \times 3Z_{11} = 9z_{11}$$

$$= 9 \times 73 = 657\,\Omega$$

That is,
$$Z_i = R_r = 657\,\Omega$$

or, in general
$$R_r = Z_i = K^2 \times 73\,\Omega$$

where
$$K = \text{number of arms.}$$

The impedance of the dipole depends on

1. spacing between dipoles and
2. radius of the dipoles.

Case 1   If the radii of the dipoles are r_1 and r_2, then

$$Z_i = 73 \left(1 + \frac{r_2}{r_1}\right)^2 \Omega \qquad \qquad ...(6.5)$$

Case 2   If the radii of the dipoles are r_1 and r_2 and d is the spacing between the elements, then

$$Z_i = 73 \left[1 + \frac{\log \dfrac{d}{r_1}}{\log \dfrac{d}{r_2}}\right]^2 \qquad \qquad ...(6.6)$$

$$R_r = Z_i = 73 \times b$$

where b = impedance transformation ratio which is given by

$$b = \left[1 + \frac{\log \dfrac{d}{r_1}}{\log \dfrac{d}{r_2}}\right]^2 \qquad \qquad ...(6.7)$$

In folded dipole, dimension of one element can be changed to obtain desired resistance (Fig. 6.19).

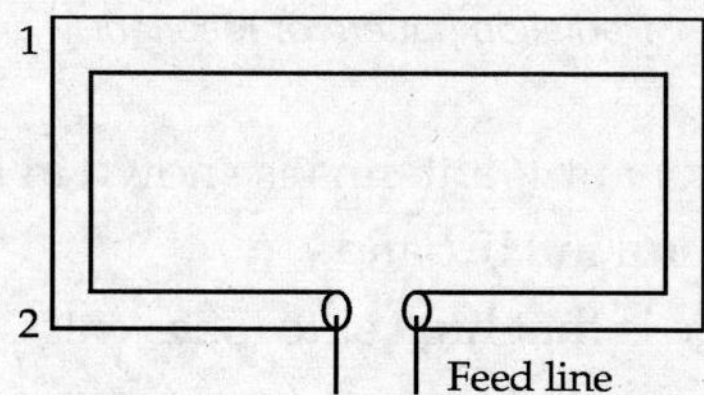

Fig. 6.19   *Folded dipole with different dimensions of arms*

If the diameter of the arm 2 is larger than that of 1, the impedance is reduced. If the diameter of the arm 2 is smaller than that of 1, the impedance is increased.

6.13 *V* ANTENNA

It is an antenna in which the conductors are arranged in *V* shape. It is balanced-fed at the apex and the included angle, length and elevation are chosen to obtain the desired directional properties.

The structure of a *V* antenna is shown in Fig. 6.20.

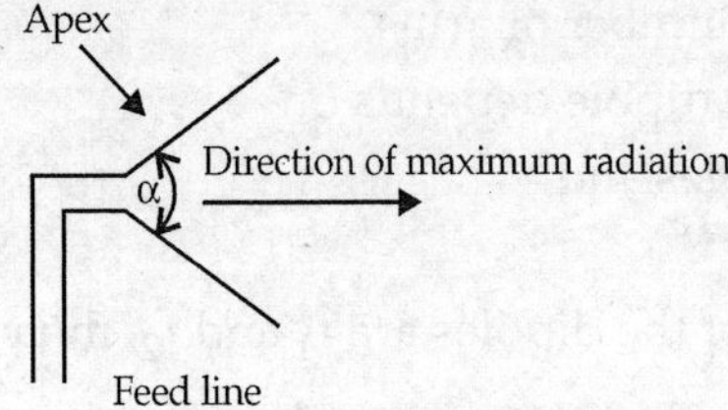

Fig. 6.20 *V antenna*

Salient features of *V*-antenna

1. It consists of two long wire antennas arranged in the form of *V* and it is fed at the apex.
2. The excitation to each wire is out of phase.
3. It offers greater gain and directivity when length of each leg or wire is increased.
4. Its radiation pattern is bi-directional.
5. *V* antennas are of two types: the first one is resonant and the second one is non-resonant.
6. The pattern of resonant *V* antenna is shown in Fig. 6.21.

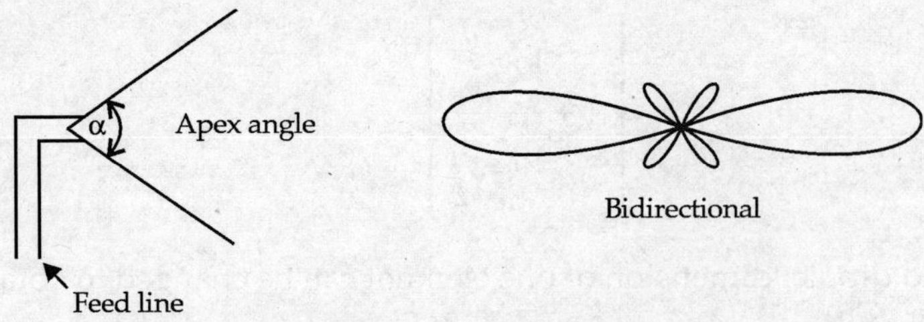

Fig. 6.21 *Radiation pattern of resonant V antenna*

7. The pattern of non-resonant *V* antenna is shown in Fig. 6.22.
8. These antennas are useful in HF band.
9. The main disadvantage is that high side lobes exist.
10. The apex angle ranges between $36°$ to $72°$ for *V* antennas of 8λ to 2λ length.

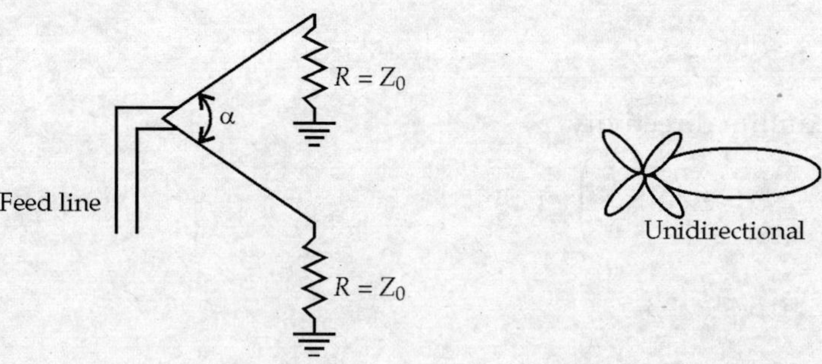

Fig. 6.22   *Non-resonant V antenna with radiation pattern*

11. It is easy to construct and they are cheap.

12. Using *V* antennas, end-fire and broadside antennas can be easily constructed. These are shown in Figs. 6.23 and 6.24 respectively.

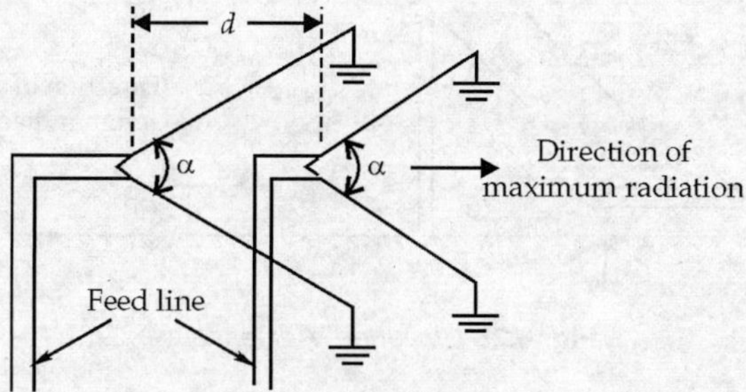

Fig. 6.23   *End-fire array of V antennas*

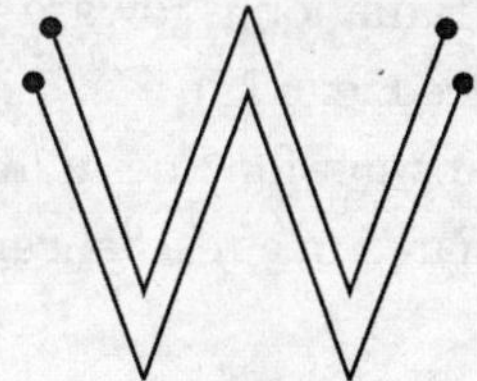

Fig. 6.24   *Broadside array of V antennas*

13. The optimum included angle, α is

$$\alpha = -149.3 \left(\frac{l}{\lambda} \right)^3 + 603.4 \left(\frac{l}{\lambda} \right)^2 - 809.5 \left(\frac{l}{\lambda} \right) + 443.6 \qquad ...(6.8)$$

for $\qquad 0.5 \le \dfrac{l}{\lambda} \le 1.5$

and $\qquad \alpha = 13.39 \left(\frac{l}{\lambda} \right)^2 - 78.27 \left(\frac{l}{\lambda} \right) + 169.77 \qquad ...(6.9)$

for $\qquad 1.5 \le \dfrac{l}{\lambda} \le 3.$

14. The maximum directivity

$$D = 2.94 \left(\dfrac{l}{\lambda} \right) + 1.15 \qquad\qquad\qquad ...(6.10)$$

for $\qquad 0.5 \le \dfrac{l}{\lambda} \le 3.$

6.14 INVERTED-*V* ANTENNA

It is an antenna in which the conductors are arranged in the shape of an inverted-*V*.

Its typical structure is shown in Fig. 6.25.

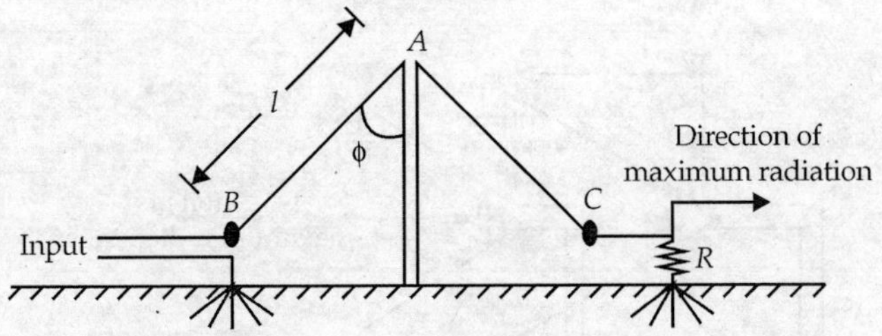

Fig. 6.25   *Inverted V antenna*

Salient features of inverted-*V* antenna

1. It is a travelling antenna.
2. The direction of maximum radiation is towards the terminated end.
3. Input is given at point *B*. (See Fig. 6.25).
4. Feeding is done through transmission lines with respect to radial earth wires.
5. Antenna wire at *C* is terminated in a resistance.
6. The angle φ is the tilt angle.
7. The terminating resistance is about $400\,\Omega$.
8. Angle of main lobe corresponds to $\dfrac{l}{\lambda}$.
9. Gain is a function of angle of tilt, leg lengths and terminating resistance.
10. These antennas are useful in HF band.
11. They have considerable band width.
12. The length of the leg equal to λ is used.
13. The main disadvantage is that it has high side lobes in its radiation pattern.
14. Inverted *V* antenna with its image looks like a rhombus.
15. It is used up to 60 MHz for receiving purposes.

16. Ground and surface waves are best received in upper HF band.

17. Arrays can be used for high gain.

18. It is not preferred for transmitting purposes.

19. Its radiation pattern contains high side lobes.

6.15 RHOMBIC ANTENNA

This is an antenna which is in the shape of a rhombus. It is usually terminated in a resistance. The side of the rhombus, the angle between the sides, the elevation, termination and height above the earth are chosen to obtain the desired radiation characteristics. A typical Rhombic antenna and radiation pattern are shown in Fig. 6.26.

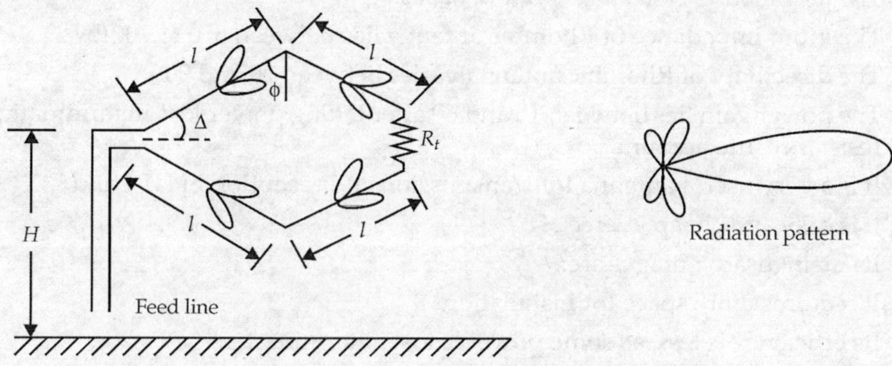

Fig. 6.26   *Rhombic antenna and radiation pattern*

Salient features of Rhombic antenna

1. It is a long wire antenna and consists four non-resonant wires.

2. It provides greater directivity than *V* antenna.

3. Its band width is high.

4. It is a HF non-resonant antenna.

5. It is very useful for point-to-point communications.

6. It is a travelling wave antenna and there are no reflections.

7. It also finds wide applications where the angle of elevation of the main lobe (measured from the plane of the antenna to the radiation axis) is less than 30°.

8. At elevation angle above 30°, the gain is very low for practical applications.

9. The directivity of each wire is

$$D\,(\theta) = \frac{60I}{r} \sin\theta \left[\frac{\sin\left[\dfrac{\pi l}{\lambda}(1 - \cos\theta) \right]}{(1 - \cos\theta)} \right] \qquad ...(6.11)$$

where I = the magnitude of the current in element

θ = the polar angle

λ = wavelength

l = length of the radiator

r = the distance from the radiator to the elevation point.

The total directivity of the Rhombic antenna is the vector sum of directivity of each wire.

10. The length of equal radiators vary from 2 to 8λ.

11. The tilt angle, ϕ varies between 40° and 75°.

12. ϕ is determined from leg length.

13. The terminating resistance is about $800\,\Omega$.

14. The input impedance of Rhombic antenna lies between 650 to $700\,\Omega$.

15. The directivity of Rhombic antenna varies between 20 and 90.

16. The power gain lies between 15 and 60 after taking power loss in terminating resistance into account.

17. It is a very useful antenna for transmission and reception in HF band.

18. It is easy and cheap to erect.

19. Its main disadvantages are:

(*i*) It requires more space for installation.

(*ii*) Its efficiency is less, as some power is lost in termination.

20. Its radiation pattern in a vertical plane is shown in Fig. 6.27.

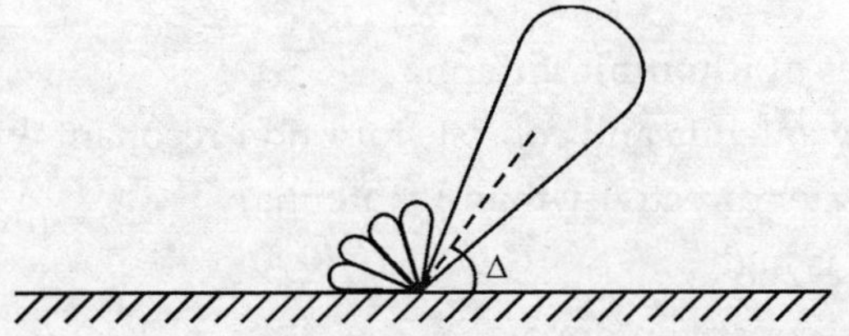

Fig. 6.27   *Radiation pattern of Rhombic antenna in vertical plane*

In the design of Rhombic antennas the maximum point of the main lobe of the radiation pattern is aligned with the desired angle of elevation. The angle of elevation is also called angle of radiation.

The **design parameters** of Rhombic antenna are:

1. Rhombic height, H

2. Angle of elevation, ϕ

3. Wire length, l.

The design equations are

$$H = \frac{\lambda}{4 \sin \Delta}$$

...(6.12)

$$\Delta = \text{elevation angle}$$

Δ is complement of tilt angle, ϕ

$$(\sin \phi = \cos \Delta)$$

$$l = \frac{\lambda}{2 \cos^2 \phi} = \frac{\lambda}{2 \sin^2 \Delta} \qquad \ldots(6.13)$$

Alignment design equations.

Tilt angle, $\phi = 90° -$ elevation angle

$$= 90° - \Delta \qquad \ldots(6.14)$$

Rhombic height,

$$H = \frac{\lambda}{4 \sin \Delta} = \frac{\lambda}{4 \cos \phi} \qquad \ldots(6.15)$$

Wire length, $\quad l = \dfrac{\lambda}{2 \sin^2 \Delta} \times K$

$$= \frac{\lambda}{2 \cos^2 \phi} \times K \qquad \ldots(6.16)$$

where $\qquad K = 0.74$

The results are given in Table 6.2 for quick reference.

Table 6.2

Frequency, f = 30 MHz, λ = 10 m					
Angle of elevation, Δ	Tilt angle ϕ	Rhombic height H in λ	H in meters	Wire length, l in λ	Wire length in meters
10°	80°	1.439	14.39	16.58	165.8
15°	75°	0.966	9.66	7.46	74.6
20°	70°	0.730	7.30	4.27	42.7
25°	65°	0.591	5.91	2.79	27.9
30°	60°	0.500	5.00	2.00	20.0
35°	55°	0.435	4.35	1.52	15.2
40°	50°	0.320	3.90	1.21	12.1

6.16 YAGI-UDA ANTENNA

This antenna was developed by Prof. Yagi and Prof. Uda. It is an array antenna which consists of one active element and a few parasitic elements. The active element consists of a folded dipole whose length is $\lambda/2$. The parasitic elements consist of one reflector and a few directors. The length of the reflector is greater than $\lambda/2$. It is located behind the active element. The length of each director is less than $\lambda/2$ and they are placed in front of the active element. The spacing between each element is not identical and hence it can be considered as a non-linear array. The number of directions in the antenna depends on the gain requirements. The impedance of the active element is resistive. The impedance of the reflector is inductive. The impedances of the directors are capacitive. A typical structure of Yagi-Uda antenna is shown in Fig. 6.28.

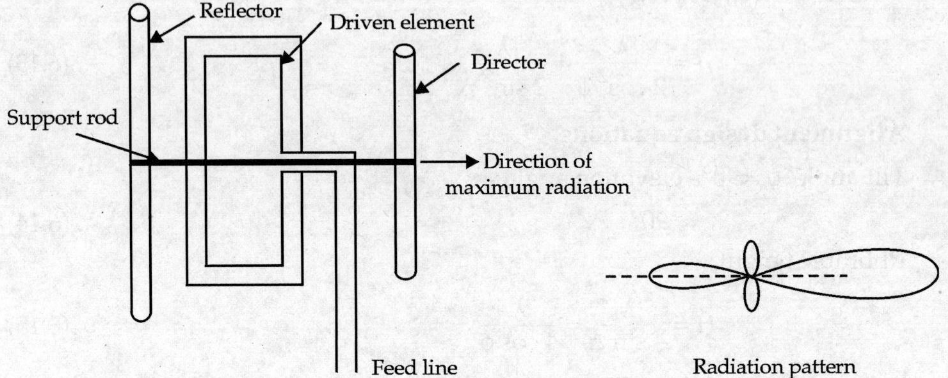

Fig. 6.28 *Yagi-Uda antenna and radiation pattern*

Salient features of Yagi-Uda antenna

1. It consists of a driven element, a reflector and one or more directors.

2. Driven element is usually a folded dipole which is excited. Director is a straight conductor placed in front of the driven element towards transmitter. Reflector is also a straight conductor placed behind the driven element.

3. Directors and reflector are called parasitic elements.

4. The length of the folded dipole is about $\frac{\lambda}{2}$ and it is at resonance. Length of the director is less than $\frac{\lambda}{2}$ and length of the reflector is greater than $\frac{\lambda}{2}$.

5. The optical equivalent of Yagi-Uda antenna is shown in Fig. 6.29.

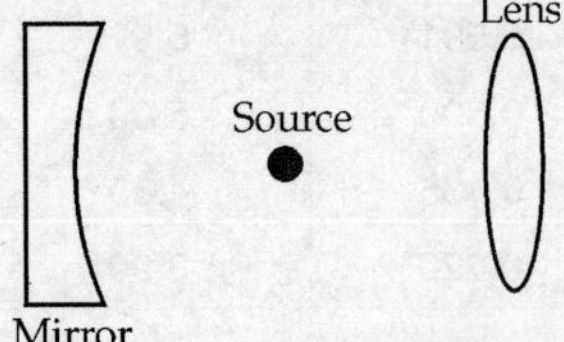

Fig. 6.29 *Optical equivalent of Yagi-Uda antenna*

6. Its radiation pattern is almost uni-directional and gives a gain of about 7 dB.

7. It is used as a transmitting antenna at HF and used for TV reception at VHF.

8. Back lobe can be reduced by bringing the elements closer. This reduces the input impedance of the antenna and hence there will be a mismatch.

9. The effect of parasitic elements depends on their distance and tuning. In other words, the effect depends on the magnitude and phase of the current induced in them.

10. Reflector resonates at a lower frequency and director resonates at a higher frequency compared to that of a driven element.

11. Folded dipole is used to obtain high impedance for proper matching between transmitter and free space.

12. More directors can be used to increase the gain. In this case, directors can be of equal length or decreasing slightly, away from the driven element. But adding too many directors will change the impedance.

13. It is relatively broadband because of the use of folded dipole.

14. Although it is compact, its gain is not high.

15. The purpose of reflector and directors is to increase the gain but they load the driven element.

16. The mutual impedance of the antenna depends on the spacing and the length of the elements.

17. Highest gain is obtained when the reflector is slightly greater than $\dfrac{\lambda}{2}$ in length and spaced at $\dfrac{\lambda}{4}$ from the driven element and when the length of the director is about 10% less than $\dfrac{\lambda}{2}$ with an optimal spacing of about $\dfrac{\lambda}{3}$.

18. It is possible to produce circular polarisation, when two Yagi-Uda antennas are placed across at right angles on the same boom, when the driven elements are fed in phase quadrature. The driven elements can be fed in phase by displacing one array by $\dfrac{\lambda}{4}$ along the boom with respect to the other.

19. The directors, whose lengths are shorter than the driven element, are characterised by capacitive reactance at the resonant frequency of the driven element.

20. The current which flows in the directors is the leading current.

21. The reflectors whose lengths are longer than the driven element are characterised by inductive reactance at the resonant frequency of the driven element.

22. Yagi-Uda antenna has exceptional sensitivity.

23. It has a good front-to-back ratio.

24. Its band width is limited.

Design Parameters   The design parameters of six-element Yagi-Uda antenna which gives a directivity of about 12 dB_i at the centre of a band width of 10 percent of half-power are given by:

The length of driven active element,

$$L_a = 0.46\lambda \qquad\qquad\qquad ...(6.17)$$

The length of reflector,

$$L_r = 0.475\lambda \qquad\qquad\qquad ...(6.18)$$

The length of director,

$$L_{d1} = 0.44\lambda \qquad\qquad\qquad ..(6.19)$$

$$L_{d2} = 0.44\lambda \qquad \qquad \qquad ...(6.20)$$

$$L_{d3} = 0.43\lambda \qquad \qquad \qquad ...(6.21)$$

$$L_{d4} = 0.40\lambda \qquad \qquad \qquad ...(6.22)$$

Spacing between reflector and driven element,

$$S_L = 0.25\lambda \qquad \qquad \qquad ...(6.23)$$

Spacing between director and driving element

$$= \text{spacing between directors}$$

$$S_d = 0.31\lambda \qquad \qquad \qquad ...(6.24)$$

Diameters of the elements,

$$d = 0.01\lambda \qquad \qquad \qquad ...(6.25)$$

The length of Yagi array

$$= 1.5\lambda$$

The design equations of a typical Yagi-Uda antenna are:

1. Length of driven element or active element,

$$L_a = \frac{478}{f_{\text{MHz}}} \text{ feet} \qquad \qquad \qquad ...(6.26)$$

2. Length of reflector,

$$L_r = \frac{492}{f_{\text{MHz}}} \text{ feet} \qquad \qquad \qquad ...(6.27)$$

3. Length of director,

$$L_d = \frac{461.5}{f_{\text{MHz}}} \text{ feet} \qquad \qquad \qquad ...(6.28)$$

4. Element spacing,

$$S = \frac{142}{f_{\text{MHz}}} \text{ feet} \qquad \qquad \qquad ...(6.29)$$

6.17 LOG-PERIODIC ANTENNA

A typical structure of log-periodic antenna is shown in Fig. 6.30.

It is an array antenna which has structural geometry such that its impedance is periodic with the logarithm of the frequency.

It is a non-linear array in which the spacing of the elements as well as their dimensions are unequal. However, excitation is uniform. It is basically called a frequency-independent antenna. It can be used to receive a good number of TV channels without any deterioration of the received field strength.

Salient features

1. It is a frequency-independent antenna.

2. The input impedance variation of the antenna with the log of frequency is periodic and hence the name. This is shown in Fig. 6.31.

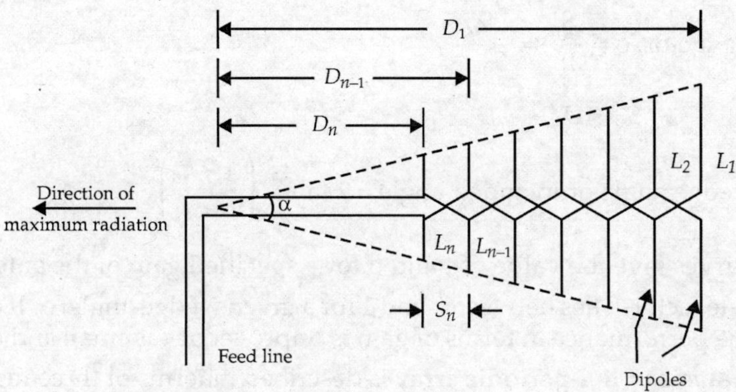

Fig. 6.30 *Log-periodic antenna or array*

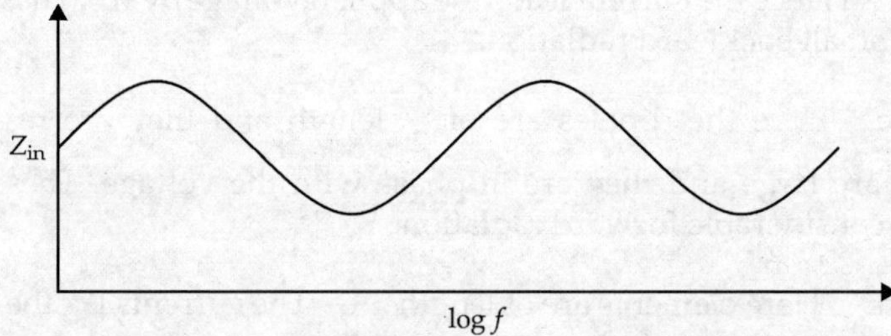

Fig. 6.31 *Impedance characteristics of log-periodic antenna*

3. It is an array of non-identical dipoles which are all excited equally.

4. It is a non-uniform array where the spacing between the elements is unequal.

5. Its impedance, directional patterns and directivity are constant with frequency.

6. The gain of a well-designed antenna lies between 7.5 and 12 dB_i.

7. It is a broad band antenna.

8. It has uni-directional characteristics.

9. There are a variety of log-periodic structures and all of them are not frequency-independent.

10. They are used in VHF and UHF bands.

11. They are used for TV reception and can receive a number of channels.

12. The planar trapezoidal and wire trapezoidal tooth log-periodic antennas can be used.

13. It is more efficient than Rhombic antenna.

Design Equations   The design equations are:

Scale factor, $$\tau = \frac{D_n}{D_{n-1}} = \frac{L_n}{L_{n-1}}, \quad n = 1, 2, 3 \qquad \qquad ...(6.30)$$

$$\text{Spacing factor, } \sigma = \frac{S_n}{2L_n} = \frac{S_n}{S_{n-1}} \qquad \qquad ...(6.31)$$

$$S_n = D_{n-1} - D_n$$

where α = wedge angle or included angle = $2 \tan^{-1}\left(\frac{1-\tau}{4\sigma}\right)$.

Isbel curves give the value of τ and σ for a specified gain of the antenna.

The scale factor, τ lies between 0 and 1 for a given wedge angle, α. If α is large, τ is small. The performance in terms of gain is improved if τ is small and α is large.

The analysis of log-periodic array is described in terms of three regions.

1. **Capacitive**   In this region, the elements are shorter than $\frac{\lambda}{2}$ and they are capacitive. Hence the current leads the applied voltage by $90°$. These elements produce small backward radiation.

2. **Resistive**   Here the dipoles are of $\frac{\lambda}{2}$ length and they are resistive. The currents are large and they are in phase with the voltage. These elements produce considerable forward radiation.

3. **Inductive**   Here elements are of length $> \frac{\lambda}{2}$. The currents lag the voltage by $90°$. The element reflects the incident wave in the backward direction.

The active region band width is given by

$$B_a = 1.1 + 7.7\,(1 - \tau)^2 \cot(\alpha/2)$$

When the designed band width is assumed to be greater than the desired band width, it is given by

Designed bandwidth,

$$B_d = BB_a = B\left[1.1 + 7.7\,(1 - \tau)^2 \cot\frac{\alpha}{2}\right] \qquad \qquad ...(6.32)$$

Here, B = desired bandwidth

Total length of the array is

$$L = \frac{\lambda_{max}}{4}\left(1 - \frac{1}{B_d}\right)\cot\frac{\alpha}{2} \qquad \qquad ...(6.33)$$

Here, $\lambda_{max} = 2l_{max} = \dfrac{v_0}{f_{min}} \qquad \qquad ...(6.34)$

The number of elements in the array is

$$N = 1 + \frac{ln\,(B_d)}{ln\,(1/\tau)} \qquad \qquad ...(6.35)$$

The average characteristic impedance, $(z_0)_{av}$ is

$$(z_0)_{av} = 120 \left[ln \left(\frac{L_n}{d_n} - 2.25 \right) \right]$$

where $\qquad\qquad d_n$ = diameter of the nth dipole.

6.18 LOOP ANTENNA

It is an antenna which is in the form of a loop.

An antenna which consists of one or more turns of wire forming a DC short circuit is called a loop antenna. The loop antenna can be of circular, square or rectangular shape. Typical loop antennas are shown in Fig. 6.32.

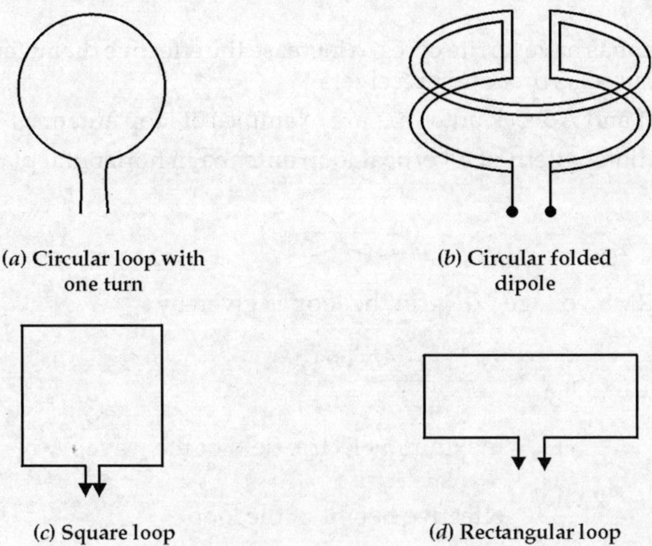

Fig. 6.32 *Loop antennas*

Small vertical loops are used for finding the direction. The loop is oriented until a null or zero field is obtained. This gives the direction of the received signal.

Loop antennas have advantage over the other antennas in direction-finding as they are small in size. These are more suitable for mobile communication applications.

The polarisation of the loop antenna is the same as that of a short dipole.

Horizontal loop antenna produces horizontal polarisation and vertical loop produces vertical polarisation.

Salient features of loop antenna

1. Small loops, whose circumferences are less than 0.1λ at the highest frequencies, are suitable for receiving signals upto about 30 MHz.
2. The loop antennas are characterised by a null along the axis of the loop.

3. Directional characteristics of loop antennas are improved by shielding them electrostatically.

4. A vertical loop antenna is popular and it receives bi-directional signals.

5. A vertical loop antenna, if shielded, receives uni-directional signals.

6. It has excellent directivity.

7. Vertical loop antennas are very useful for direction-finding applications.

8. These are suitable for LF, MF, HF, VHF and UHF ranges.

9. The radiation pattern is in the shape of a doublet.

10. The directional patterns of loop antennas are independent of the exact shape of the loop.

11. In direction-finding applications, a small vertical loop is rotated about the vertical axis. The plane of the loop is perpendicular to the direction of radiation.

12. Loop antennas have ferrite cores to increase the effective diameter of the loop. These are used as broadcast receivers.

13. Cloverleaf and Adcock antennas are examples of loop antennas.

14. The radiation pattern of a vertical loop antenna in horizontal plane is

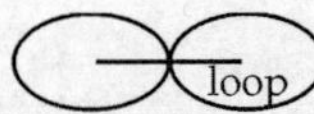

15. Induced RMS voltage, V_{RMS} in the loop is given by

$$V_{\text{RMS}} = \frac{2\pi E_{\text{RMS}}\, AN \cos \phi}{\lambda} \qquad \qquad ...(6.36)$$

Here, E_m = maximum electric field of the wave, V/m

$\dfrac{2\pi AN}{\lambda}$ = effective height of the loop

λ = wavelength, m

A = area of the loop, m^2

N = number of turns

ϕ = angle between plane of the loop and direction of incident wave.

16. The radiation efficiency of a small loop antenna is poor.

17. The dimensions of the antenna should be of the order of λ for using as transmitters.

18. The field expressions of small loop antennas are:

$$E_\phi = \frac{120\pi^2\, IA \sin \theta}{r\lambda^2},\ \text{V/m} \qquad \qquad ...(6.37)$$

$$H_\theta = \frac{\pi I \sin \theta A}{r\lambda^2},\ \text{V/m} \qquad \qquad ...(6.38)$$

Here, $\qquad I = $ retarded current

$$= I_0\, e^{j\,\omega\,(t-r/v_0)} \qquad\qquad ...(6.39)$$

19. The radiation resistance of small loop antenna

$$R_r \approx 31{,}171 \left(\frac{NA}{\lambda^2}\right)^2 \Omega \qquad\qquad ...(6.40)$$

Here $\qquad N = $ number of turns

$\qquad\qquad A = $ area of the loop

$\qquad\qquad \lambda = $ wavelength.

20. Radiation resistance of a loop antenna is

$$R_r = 3{,}720\,\frac{a}{\lambda}, \ \Omega$$

where $\qquad a = $ radius of the loop.

21. The radiated power of loop antenna is

$$P_T = 10K^4\, A^2\, I_m^2, \text{ watts.}$$

22. For small loop $\left(\dfrac{2\pi a}{\lambda} < \dfrac{1}{3}\right)$

The directivity, $\quad D = \dfrac{3}{2} = 1.5$

For large loop, $\quad C = 2\pi a \geq 5\lambda$

$$D = 4.25 \left(\frac{a}{\lambda}\right).$$

23. The maximum effective aperture for small loop antenna is

$$A_{em} = \frac{3\lambda^2}{8\pi}.$$

24. The maximum effective aperture is given by

$$A_{em} = 0.341\lambda a \ \text{m}^2$$

25. Loop antennas are used extensively in radio receivers, aircraft receivers, for direction-finding and also in UHF transmitters.

26. If the current in the loop is uniform and the loop circumference is small compared to the operating wave length, its radiation pattern is almost like that of a magnetic dipole.

6.18.1 Radiation Resistance, R_r of Loop Antenna

Consider a circular loop antenna of Fig. 6.33.

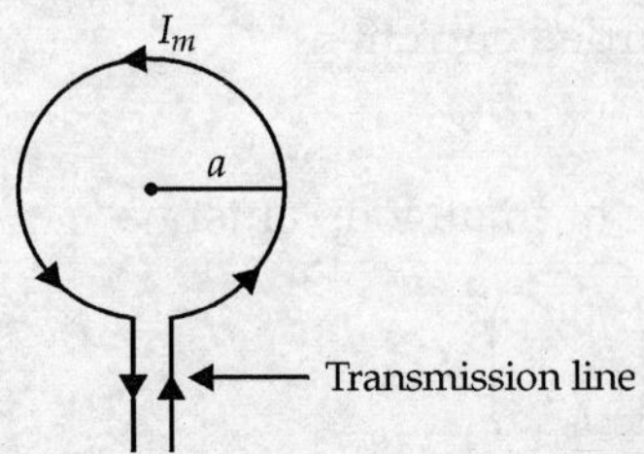

Fig. 6.33 *Loop antenna with feed line*

By definition, radiation resistance is

$$R_r \equiv \frac{\text{radiated power}}{I_{\text{RMS}}^2}$$

$$= \frac{2 \times \text{total radiated power}}{I_m^2} = \frac{2P_T}{I_m^2} \qquad \text{...(6.41)}$$

where $\qquad I_m$ = maximum current in the antenna.

Total radiated power,

$$P_T = \iint_S (P_r)_{av}\, ds \qquad \text{...(6.42)}$$

Here, $\qquad (P_r)_{av}$ = average radiated power density, watts/m^2

$$= \frac{1}{2} H^2 \eta_0 \qquad \text{...(6.43)}$$

where $\qquad H = |\mathbf{H}|, \; \eta_0$ = intrinsic impedance of free space.

For a loop antenna in y-z plane with its centre at the origin, the radiation fields consist of E_ϕ and H_θ components. H_θ is given by

$$H_\theta = \frac{ka I_m}{2r} J_1 (ka \sin \theta) \qquad \text{...(6.44)}$$

where $\qquad J_1 (ka \sin \theta)$ = Bessel function of the first kind

$$\approx \frac{ka \sin \theta}{2} \text{ for small loops} \qquad \text{...(6.45)}$$

$$H_\theta = k \times \frac{a I_m}{2r} \times \frac{ka \sin \theta}{2}$$

$$= \frac{k^2 a^2 I_m \sin \theta}{4r} \qquad \text{...(6.46)}$$

$$(P_r)_{av} = \frac{1}{2} H^2 \eta_0$$

$$= \frac{1}{2} \frac{\eta_0 \, k^4 \, a^4 \, I_m^2 \, \sin^2 \theta}{16 r^2}$$

$$= \frac{120\pi \times k^4 \, a^4 \, I_m^2 \, \sin^2 \theta}{32 r^2}$$

$$= \frac{15\pi}{4} \frac{k^4 \, a^4 \, I_m^2 \, \sin^2 \theta}{r^2} \qquad \qquad \text{...(6.47)}$$

$$P_T = \iint (P_r)_{av} \, ds$$

$$= \int_0^{2\pi} \int_0^{\pi} \frac{15\pi}{4} \frac{k^4 \, a^4 \, I_m^2 \, \sin^2 \theta}{r^2} \, r^2 \sin \theta \, d\theta \, d\phi$$

$$= \int_0^{\pi} \frac{15\pi^2}{2} k^4 \, a^4 \, I_m^2 \, \sin^3 \theta \, d\theta$$

$$= 10\pi^2 \, k^4 \, a^4 \, I_m^2 \qquad \qquad \text{...(6.48)}$$

But the area of the loop is given by

$$A = \pi a^2 \qquad \qquad \text{...(6.49)}$$

So
$$P_T = 10 k^4 \, A^2 \, I_m^2 \qquad \qquad \text{...(6.50)}$$

From Equations (6.41) and (6.50), we have

$$R_r = \frac{2 P_T}{I_m^2}$$

$$= \frac{2}{I_m^2} 10 k^4 \, A^2 \, I_m^2 \qquad \qquad \text{...(6.51)}$$

But
$$k = \frac{2\pi}{\lambda}$$

So
$$R_r = 31{,}171 \left(\frac{A}{\lambda^2} \right)^2, \ \Omega \qquad \qquad \text{...(6.52)}$$

If there are N number of turns in the loop antenna,

$$R_r = 31{,}171 \left(\frac{NA}{\lambda^2} \right)^2, \ \Omega \qquad \qquad \text{...(6.53)}$$

As the circumference of the loop is $2\pi a$,

$$R_r \approx 197 \left(\frac{C}{\lambda} \right)^4 \qquad \qquad \text{...(6.54)}$$

or

$$R_r \approx 3{,}720 \left(\frac{a}{\lambda} \right)$$

...(6.55)

The **directivity** for small loop antenna $(C < 0.33\lambda)$

$$D = 3/2$$

...(6.56)

and for large loop $(C > 5\lambda)$

$$D = 4.25 \left(\frac{a}{\lambda} \right)$$

...(6.57)

The **maximum effective aperture** is given by

$$A_{em} = 0.341 \lambda a, \ \text{m}^2$$

...(6.58)

The **maximum effective aperture** for small loop antennas,

$$A_{em} = \frac{3\lambda^2}{8\pi}$$

For a **square loop of side** a, the **radiation resistance** is given by

$$R_r = 31{,}171 \left(\frac{a}{\lambda} \right)^4, \ \Omega.$$

...(6.59)

6.19 HELICAL ANTENNA

It is an antenna which is in the shape of a helix.

Its polarisation and radiation properties depend on the diameter, pitch, number of turns, wavelength, excitation and spacing between the helical loops.

A typical structure of helical antenna is shown in Fig. 6.34.

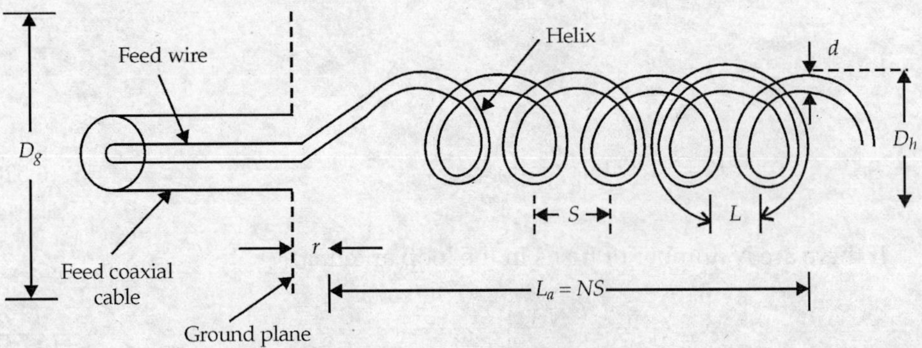

Fig. 6.34 *Helical antenna*

Helical antenna consists of helical loops made of a thick conductor which have the appearance of a screw thread. It is associated with a ground plane made of the conductor. The ground plane is often made of screen or sheet or of radial and concentric conductors. This antenna is fed by a coaxial cable. This can be operated in normal and axial modes. The common antenna parameters are:

C = circumference of helix

α = pitch angle

S = loop separation

L_a = axial length = NR

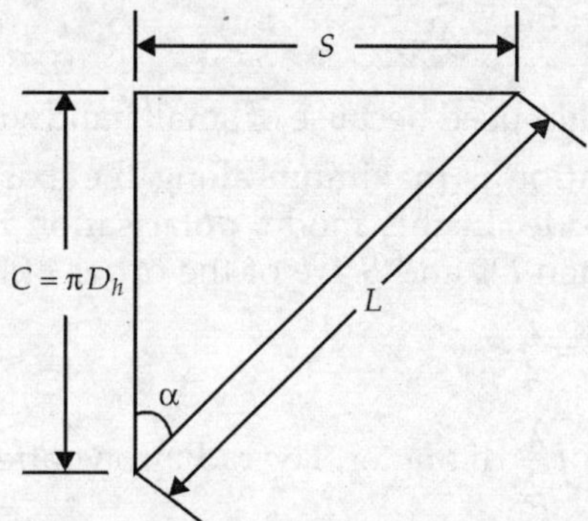

Fig. 6.35   *Pitch angle*

N = number of turns

L = length of one turn

r = distance between ground plane and helix proper

d = diameter of helix conductor

The inter relations between the parameters are

$$L = \sqrt{S^2 + C^2}, \qquad\qquad\qquad ...(6.60)$$

$$\alpha = \tan^{-1}\left(\frac{S}{\pi D_h}\right) \qquad\qquad\qquad ...(6.61)$$

In **normal mode**, the radiation is maximum in the broadside direction. This can be called broadside mode. This mode happens if NL << λ, that is, the dimensions of the helix are small. In this, beam width is small and efficiency is low. These two parameters can be increased if the helix is large.

Axial ratio for elliptical polarisation is given by

$$\text{Axial ratio,} \quad AR = \frac{2S\lambda}{\pi^2 D_h^2} \qquad\qquad ...(6.62)$$

If $AR = 0$, elliptical polarisation becomes linear horizontal polarisation.

If $AR = \infty$, elliptical polarisation becomes vertical linear polarisation.

If $AR = 1$, elliptical polarisation becomes circular polarisation.

Hence, for circular polarisation,

$$AR = 1$$

$$2S\lambda = \pi^2 D_h^2$$

or

$$S = \frac{\pi^2 D_h^2}{2\lambda} = \frac{C^2}{2\lambda} \qquad \qquad ...(6.63)$$

and

$$\alpha = \tan^{-1}\frac{C}{2\lambda} \qquad \qquad ...(6.64)$$

The normal mode is not used because of small bandwidth and low efficiency.

In **axial mode**, radiation is maximum along the axis of the helix. It can be considered as end-fire mode. In this mode, polarisation is almost circular. This mode can be obtained when D_h and S are of the order of λ. Circular polarisation occurs when $\dfrac{C}{\lambda} = 1$ and $S = \dfrac{\lambda}{4}$.

The ground plane is of $\dfrac{\lambda}{2}$ diameter. The radiation patterns in normal and axial modes are shown in Fig. 6.36.

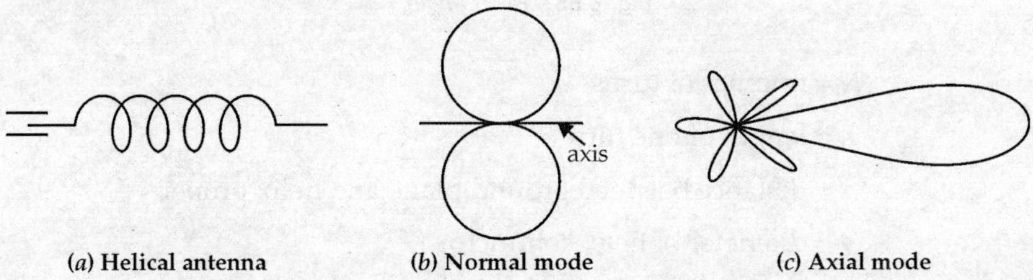

Fig. 6.36 *Radiation patterns in both modes*

In axial mode, the terminal impedance of helix is about 100 to 200 Ω.

The empirical expressions for parameters of helical antenna are:

$$Z_i = \text{input impedance} \approx 140\,\frac{C}{\lambda} \qquad \qquad ...(6.65)$$

$$\text{HPBW} = \text{half-power beam width} = \frac{52\lambda^{3/2}}{C\sqrt{L_a}} \qquad \qquad ...(6.66)$$

$$\text{BWFN} = \text{Null-to-Null beam width} \approx \frac{115\lambda^{3/2}}{C\sqrt{L_a}} \qquad \qquad ...(6.67)$$

$$D = \text{directivity} = 15N\,\frac{C^2 S}{\lambda^3} \qquad \qquad ...(6.68)$$

The axial ratio $AR = \dfrac{2N+1}{N} \qquad \qquad ...(6.69)$

The normalised far-field,

$$E = \sin\left(\frac{\pi}{2N}\right)\cos\theta \; \frac{\sin\left(\frac{N}{2}\psi\right)}{\sin\left(\frac{\psi}{2}\right)} \qquad \ldots(6.70)$$

Here
$$\psi = 2\pi\left[\frac{S}{\lambda}(1-\cos\theta)+\frac{1}{2N}\right] \qquad \ldots(6.71)$$

Salient features of helical antenna

1. Helical antenna is a simple antenna for circular polarisation.
2. It is used in VHF and UHF bands.
3. It is most popularly used in axial mode.
4. In normal mode, beam width and efficiency are small.
5. It is a wide band antenna in axial mode.
6. It is used for extra-terrestrial communications, satellites and space probe communications, radio astronomy and so on.
7. It is not preferred in normal mode.
8. If axial ratio, AR = 0, linear horizontal polarisation results.
9. If AR = ∞, linear vertical polarisation results.
10. If AR = 1, circular polarisation results.
11. It is simple in construction and has high directivity.

Design equations   The design equations of helical antenna in axial mode are:

$$d = 0.02\lambda$$
$$r = 0.12\lambda$$
$$D_h = 0.32\lambda$$
$$D_g \geq 0.8\lambda$$
$$S = 0.22\lambda$$
$$C = \pi D_h = 1.005\lambda$$

$$\alpha = \tan^{-1}\left(\frac{S}{C}\right)$$

$$R = 140\frac{C}{\lambda}\,\Omega = \text{terminal resistance}$$

AR = axial ratio for maximum directivity of the circular polarisation

$$= (2N+1)/N.$$

Applications of helical antenna

1. It is used to transmit and receive VHF waves for ionospheric propagation.

2. It is used for the following communications:

(*a*) Satellite, space communications.

(*b*) Space telemetry at HF and VHF bands

(*c*) Radio astronomy.

6.20 WHIP ANTENNA

It is a short vertical monopole used for mobile communication purposes. A few typical whip antenna structures are shown in Fig. 6.37.

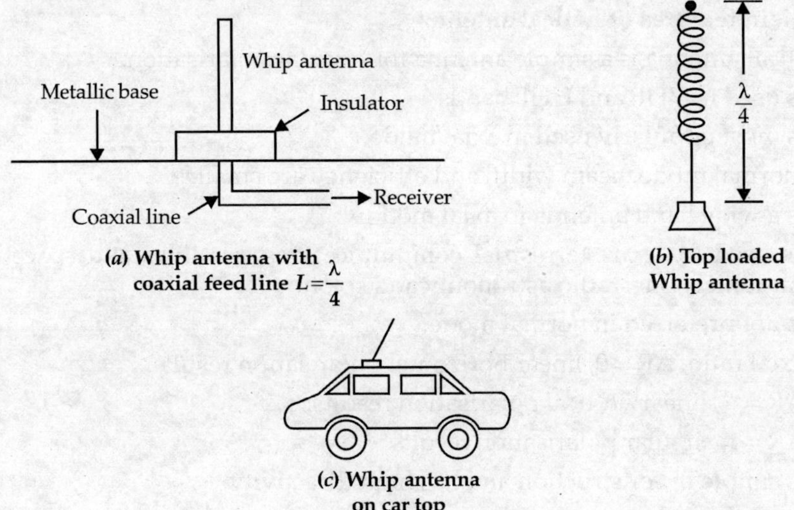

Fig. 6.37   *Whip antennas*

Salient features of whip antenna

1. It gives a gain of three with respect to isotropic radiator in a direction perpendicular to its axis.

2. Its length can be reduced by loading.

3. It can be used in HF and VHF bands.

4. It is used mostly for mobile communications.

5. A continuously wound step tapered helical conductor with a uniform current distribution gives a $50\,\Omega$ match at its resonant frequency. Its standard length is 4 feet for most of the applications.

6.21 FERRITE ROD ANTENNA

It is an antenna which consists of a ferrite rod on which a coil with a number of turns are wound. A typical structure is shown in Fig. 6.38.

This is commonly used in all transistorised radio receivers. It has a ferrite rod with one or more coils. A typical structure of ferrite rod antenna is shown in Fig. 6.38.

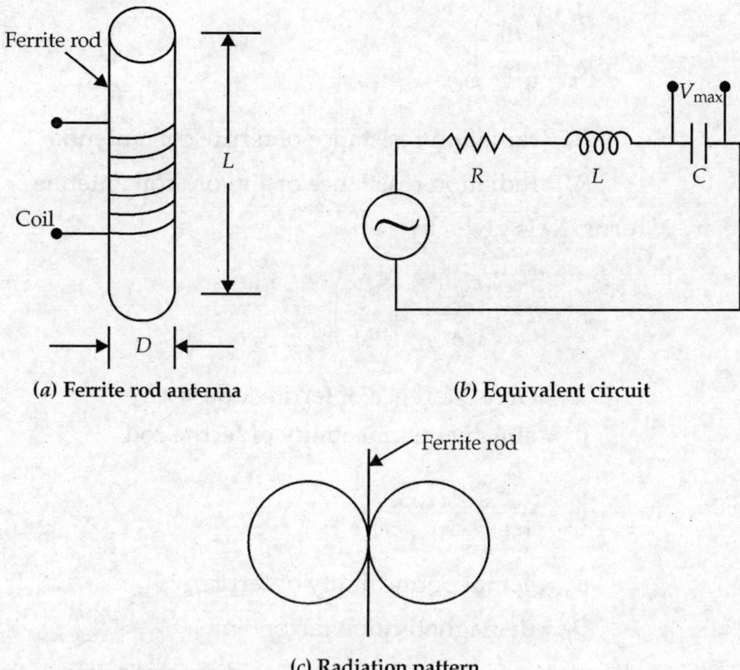

Fig. 6.38 *Ferrite rod antenna with equivalent circuit and radiation pattern*

In the Fig. 6.38,

D = diameter of the ferrite rod

L = length of the rod

The induced voltage is given by

$$V = \frac{2\pi}{\lambda} ESNK\mu_r, \ \text{volt} \qquad \qquad ...(6.72)$$

where E = electric field present at antenna, V/m

μ_r = permeability of the rod

K = modifying factor which takes care of coil length

= 1 for short coils

= 0.7 for coils of full length of rod

S = cross-sectional area of the rod

N = number of turns in the coil.

Effective length of the antenna is defined as

$$l_{eff} = \frac{V}{E} = \frac{2\pi}{\lambda} SK\mu_r, \ \text{m} \qquad \qquad ...(6.73)$$

The relation between radiation resistance of the ferrite coil and that of air core coil is

$$\frac{R_f}{R_r} = \left(\frac{\mu_e}{\mu_o}\right)^2 \qquad \text{...(6.74)}$$

where $\qquad R_f$ = radiation resistance of ferrite coil antenna

$\qquad\qquad R_r$ = radiation resistance of air core coil antenna

If there are N turns, R_f is given by

$$R_f = 2\lambda^2 \left(\frac{C}{\lambda}\right)^4 \left(\frac{\mu_e}{\mu_o}\right)^2 N^2, \ \Omega \qquad \text{...(6.76)}$$

where $\qquad C$ = circumference of ferrite loop = πD

$\qquad\qquad \mu_e$ = effective permeability of ferrite rod

That is, $\qquad\qquad \mu_e = \dfrac{\mu_a}{1 + D_f (\mu_a - 1)} \qquad \text{...(6.76)}$

$\qquad\qquad \mu_a$ = actual permeability of ferrite

$\qquad\qquad D_f$ = demagnetisation factor

The magnetisation factor depends on length-to-diameter ratio.

Salient features of ferrite rod antenna:

1. It is used in all radio receivers.

2. It is compact.

3. Its quality factor, Q is very high. It exhibits high selectivity and more induced voltage.

6.22 TURNSTILE ANTENNA

It is an antenna composed of two dipole antennas perpendicular to each other. They intersect at their mid-points. The currents on the two dipoles are equal and in phase quadrature.

Salient features of turnstile antenna:

1. Turnstile antenna consists of two half-wave dipoles which are perpendicular to each other. The dipoles are excited with a phase difference of 90° (phase quadrature) with equal currents.

2. A typical antenna is shown in Fig. (6.39).

3. The excitation is provided by different non-resonant lines of unequal length.

4. It produces almost an omni-directional pattern.

5. The electric field of turnstile antenna is given by

$$E\left(\theta\right) = \frac{\cos\left(\dfrac{\pi}{2}\sin\theta\right)}{\cos\theta}\sin\omega t + \frac{\cos\left(\dfrac{\lambda}{2}\cos\theta\right)}{\sin\theta}\cos\omega t \qquad \text{...(6.77)}$$

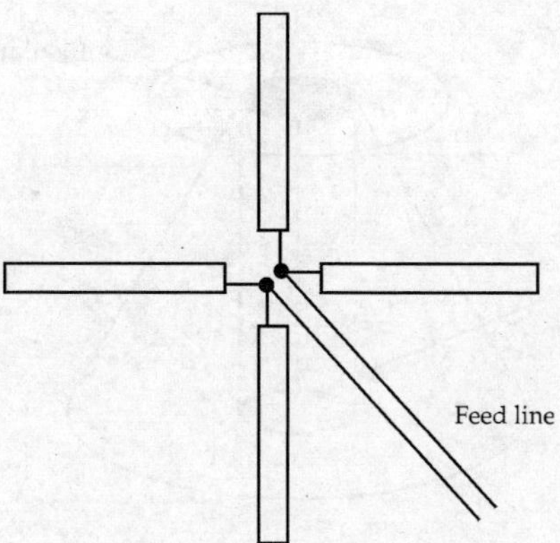

Fig. 6.39   *Turnstile antenna*

6. Directivity is improved by the array of turnstile antennas.

7. This antenna is best suited to match 70Ω dual co-axial line.

8. This is often used for T.V. and FM broadcasting in VHF and UHF bands.

9. This produces horizontal polarisation.

10. The polarisation is disturbed due to loss of power. However, the purity of polarisation is improved by super-turnstile antennas.

11. The super-turnstile antenna can be made of four flat sheets.

12. It is possible to obtain voltage standing wave ratio (VSWR) of about 1.1 over 30% bandwidth.

13. It is used as a mast mounted television transmitting antenna for frequencies about 50 MHz.

14. Bandwidth is improved by an array of super-turnstile antennas with a spacing of λ between the elements.

15. Array of super-turnstile antennas produces more horizontal gain.

6.23 DISCONE ANTENNA

It is an antenna which consists of a disc and a cone. The disc is fixed at the centre conductor of coaxial feed line so that it is perpendicular to its axis. The apex of the cone is connected to the outer shield of the coaxial line. Its variation of impedance and radiation characteristics as a function of frequency is much less when compared with those of a dipole of constant length.

Salient features of discone antenna

1. A disc and a cone together form a discone.

2. It is a ground plane antenna and is evolved from the vertical dipole.

3. It has a radiation pattern similar to that of vertical dipole.

4. A typical discone antenna is shown in Fig. 6.40. It is fed by a coaxial cable.

5. The cone semi-angle is about 30°, disc diameter is about three-fourth of the diameter of the base of the cone.

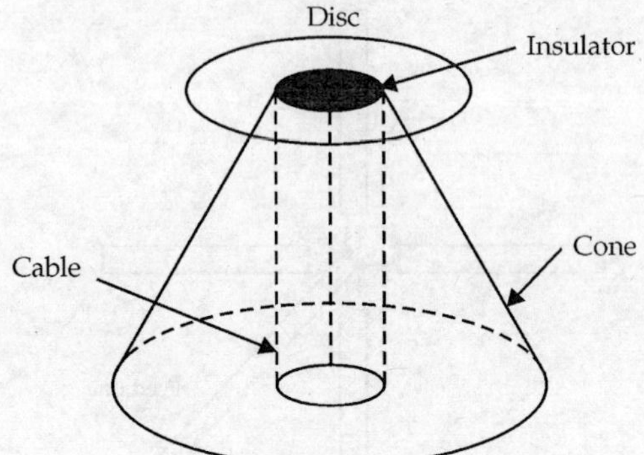

Fig. 6.40 *Discone antenna*

6. It acts as if the disc is a reflector.
7. It is a broad-band antenna.
8. It is simple and easy to fabricate.
9. Voltage standing wave ratio and gain are low.
10. It is basically omni-directional antenna.
11. It can be used at VHF and UHF bands.
12. It is often used in airport communication systems.
13. It is an ideal antenna for mobile communication base stations.
14. It is compact, rugged and economical.
15. Its gain in the horizontal plane is comparable to that of a dipole antenna.
16. Its overall performance as a function of frequency is similar to a high pass filter.
17. It is inefficient below cut-off and voltage standing wave ratio becomes high.
18. The slant height of the cone is about $\lambda/4$ at cut-off.

6.24 NOTCH ANTENNA

It is an antenna which consists of a notch in a metallic sheet.

Salient features of Notch antenna

1. It consists of a notch cut in the edge of a metallic surface.
2. It is an open ended slot antenna (Fig. 6.41).
3. It is a broad-band antenna.

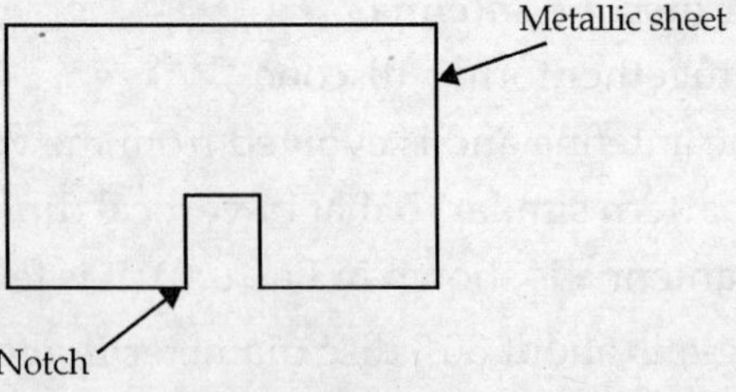

Fig. 6.41 *Notch antenna*

4. A typical radiation pattern of notch antenna is shown in Fig. 6.42.

Fig. 6.42 *Radiation pattern of notch antenna*

5. Notch antennas can be easily made on the body of an aircraft.

6. The notches are filled with dielectric materials to avoid aerodynamic drag.

 **POINTS TO REMEMBER**

1. Isotropic radiator is a hypothetical element which radiates in all directions equally.

2. An omni-directional antenna is an antenna, which has non-directional pattern in azimuth and directional pattern in elevation.

3. An example of omni-directional antenna is dipole.

4. Resonant antennas produce standing waves.

5. Non-resonant antennas produce travelling waves.

6. The size of the antenna at low frequencies is very large.

7. Antenna impedance is increased by folding a dipole.

8. The radiation resistance of folded dipole is $292\,\Omega$.

9. The radiation resistance of a folded dipole is $73K^2$, K being number of arms.

10. Yagi-Uda antenna is popularly used for TV reception.

11. Yagi-Uda antenna is a narrow-band antenna.

12. Log-periodic antenna is a frequency independent antenna.

13. Log-periodic antenna is a wide-band antenna.

14. Rhombic antenna is a travelling wave antenna.

15. Rhombic antenna is a wide-band antenna.

16. Loop antenna is used in mobile communications.

17. Radiation resistance of loop antenna is $R_r = 31{,}171 \left(\dfrac{NA}{\lambda^2} \right)^2$.

18. Helical antenna is used to produce circularly polarised waves.

19. Ferrite rod antenna is used in all transistorised radio receivers.

20. Turnstile antenna is used for T.V., FM broadcasting purposes.

21. Discone antenna is a broad-band antenna.

22. Notch antenna is used in aircrafts.

 ## SOLVED PROBLEMS

Problem 6.1 Design a Rhombic antenna to operate at a frequency of 30 MHz with the angle of elevation, $\Delta = 30°$ with respect to the ground.

Solution
$$H = \frac{\lambda}{4 \sin \Delta} = \frac{\lambda}{4 \sin 30} = \frac{\lambda}{2}$$

At $f = 30$ MHz,

$$\lambda = \frac{v_0}{f} = \frac{3 \times 10^8}{30 \times 10^6} = 10 \text{ m}$$

Height of Rhombic,

$$H = 10 \times 1/2 = 5 \text{ m}$$

Tilt angle, $\phi = 90 - \Delta = 60°$

Length of each wire, l is

$$l = \frac{\lambda}{2 \cos^2 \phi} = \frac{\lambda}{2 \sin^2 \Delta} = 20 \text{ m}$$

So, The design parameters are $\phi = 60°$, $H = 5$ m, $l = 20$ m.

Problem 6.2 Design a Rhombic antenna to operate at 20 MHz when the angle of elevation, $\Delta = 10°$.

Solution Tilt angle, $\phi = 90 - \Delta = 90 - 10 = 80°$

Rhombic height, $H = \dfrac{\lambda}{4 \sin \Delta}$

$$= \frac{\lambda}{0.6945}$$

or $\qquad H = 1.439\lambda$

At $\qquad f = 20$ MHz

$$\lambda = \frac{3 \times 10^8}{20 \times 10^6} = 15 \text{ m}$$

$$H = 21.585 \text{ m}$$

$$l = \frac{\lambda}{2 \sin^2 10}$$

$$l = 16.58\lambda$$

or
$$l = 248.725 \text{ m.}$$

Problem 6.3   Obtain design data of a Rhombic antenna to operate at 30 MHz if the angle of elevation is $10°, 15°, 20°, 25°, 30°, 35°, 40°$.

Solution   (*a*)   Angle of elevation, $\Delta = 10°$

Tilt angle,
$$\phi = 90° - \Delta = 90° - 10°$$

$$\phi = 80°$$

$$f = 30 \text{ MHz}$$

$$\lambda = \frac{v_0}{f} = \frac{3 \times 10^8}{30 \times 10^6} = 10 \text{ m}$$

Rhombic height,
$$H = \frac{\lambda}{4 \sin \Delta} = \frac{\lambda}{4 \sin 10°}$$

$$H = 1.439\lambda$$

or
$$H = 14.396 \text{ m}$$

Wire length,
$$l = \frac{\lambda}{2 \sin^2 \Delta}$$

$$l = 16.58\lambda$$

or
$$l = 165.8 \text{ m}$$

$\therefore$
$$\phi = 80°, \ H = 14.396 \text{ m}, \ l = 165.8 \text{ m}$$

(*b*)   Angle of elevation, $\Delta = 15°$

$$\phi = 90° - 15° = 75°$$

Rhombic height, $H = \dfrac{\lambda}{4 \sin \Delta} = 0.966\lambda$

$$H = 0.966\lambda$$

or
$$H = 9.66 \text{ m}$$

Wire length,
$$l = \frac{\lambda}{2 \sin^2 \Delta}$$

or
$$l = 7.46\lambda$$

$$l = 74.6 \text{ m}$$

$$\phi = 75°, \ H = 9.66 \text{ m}, \ l = 74.6 \text{ m}$$

(*c*)   Angle of elevation $\Delta = 20°$

Tilt angle, $\phi = 90 - 20 = 70°$

Rhombic height, $H = \dfrac{\lambda}{4 \sin \Delta}$

or $H = 0.73\lambda$

$H = 7.3 \text{ m}$

Wire length, $l = \dfrac{\lambda}{2 \sin^2 \Delta}$

or $l = 4.27\lambda$

$l = 42.7 \text{ m}$

$\phi = 70^\circ, \ H = 7.3 \text{ m}, \ l = 42.7 \text{ m}$

(d) Angle of elevation, $\Delta = 25^\circ$

Tilt angle, $\phi = 90^\circ - 25^\circ$

$= 65^\circ$

Rhombic height, $H = \dfrac{\lambda}{4 \sin \Delta}$

or $H = 0.591\lambda$

$H = 5.91 \text{ m}$

Wire length, $l = \dfrac{\lambda}{2 \sin^2 \Delta}$

or $l = 2.79\lambda$

$l = 27.9 \text{ m}$

$\phi = 65^\circ, \ H = 5.91 \text{ m}, \ l = 27.9 \text{ m}$

(e) Angle of elevation, $\Delta = 30^\circ$

Tilt angle, $\phi = 90^\circ - 30^\circ$

$= 60^\circ$

Rhombic height, $H = \dfrac{\lambda}{4 \sin \Delta}$

$H = 0.5\lambda$

$H = 5 \text{ m}$

Wire length, $l = \dfrac{\lambda}{2 \sin^2 \Delta}$

$= 2\lambda$

$$l = 2 \times 10 = 20 \text{ m.}$$

$$\phi = 60^{\circ}, \ H = 5 \text{ m}, \ l = 20 \text{ m}$$

(*f*) Angle of elevation, $\Delta = 35^{\circ}$

Tilt angle, $\qquad\qquad \phi = 90^{\circ} - 35^{\circ} = 55^{\circ}$

Rhombic height, $\quad H = \dfrac{\lambda}{4 \sin \Delta}$

or $\qquad\qquad\qquad H = 0.435\lambda$

$$\qquad\qquad\qquad H = 4.35 \text{ m}$$

Wire length, $\qquad\quad l = \dfrac{\lambda}{2 \sin^2 \Delta}$

$$\qquad\qquad\qquad = 1.52\lambda$$

or $\qquad\qquad\qquad l = 15.2 \text{ m}$

$$\phi = 55^{\circ}, \ H = 4.35 \text{ m}, \ l = 15.2 \text{ m}$$

(*g*) Angle of elevation, $\Delta = 40^{\circ}$

Tilt angle, $\qquad\qquad \phi = 90^{\circ} - 40^{\circ} = 50^{\circ}$

Rhombic height, $\quad H = \dfrac{\lambda}{4 \sin \Delta}$

or $\qquad\qquad\qquad H = 0.39\lambda$

$$\qquad\qquad\qquad H = 3.9 \text{ m}$$

Wire length, $\qquad\quad l = \dfrac{\lambda}{2 \sin^2 \Delta}$

or $\qquad\qquad\qquad l = 1.21\lambda$

$$\qquad\qquad\qquad l = 12.1 \text{ m}$$

$$\phi = 50^{\circ}, \ H = 3.9 \text{ m}, \ l = 12.1 \text{ m.}$$

Problem 6.4   Obtain alignment design parameters of Rhombic antenna to operate at 30 MHz when the required elevation angle is 30°.

Solution $\qquad\qquad\quad f = 30 \text{ MHz}$

$$\lambda = \frac{3 \times 10^8}{30 \times 10^6} = 10 \text{ m}$$

Elevation angle, $\ \Delta = 30^{\circ}$

Tilt angle, $\qquad\quad \phi = 90^{\circ} - \Delta = 90^{\circ} - 30^{\circ}$

$$= 60^\circ$$

$$\phi = 60^\circ$$

Rhombic height, $\qquad H = \dfrac{\lambda}{4 \sin \Delta} = 5 \text{ m}$

Wire length, $\qquad l = \dfrac{\lambda}{2 \sin^2 \Delta} \times k$

$$= \dfrac{\lambda}{2 \sin^2 30^\circ} \times 0.74$$

$$= 1.48\lambda = 14.8 \text{ m}$$

$$\boxed{\phi = 60^\circ, \ H = 5 \text{ m}, \ l = 14.8 \text{ m.}}$$

Problem 6.5 Obtain alignment design parameter of Rhombic antenna to operate at 20 MHz if the elevation angle is 20°.

Solution Frequency, $\qquad f = 20 \text{ MHz}$

$$\lambda = \dfrac{3 \times 10^8}{20 \times 10^6} = 15 \text{ m}$$

Elevation angle, $\qquad \Delta = 20^\circ$

Tilt angle, $\qquad \phi = 90^\circ - 20^\circ = 70^\circ$

$$K = 0.74$$

Rhombic height, $\qquad H = \dfrac{\lambda}{4 \sin \Delta} = 10.96 \text{ m}$

Wire length, $\qquad l = \dfrac{\lambda}{2 \sin^2 \Delta} \times K$

$$l = 47.44 \text{ m}$$

$$\boxed{\phi = 70^\circ, \ H = 10.96 \text{ m}, \ l = 47.44 \text{ m.}}$$

Problem 6.6 Design a three element Yagi-Uda antenna to operate at a frequency of 172 MHz.

Solution Frequency, $\qquad f = 172 \text{ MHz}$

$$\lambda = \dfrac{3 \times 10^8}{172 \times 10^6} = \dfrac{300}{172} = 1.744 \text{ m}$$

The length of driven element,

$$L_a = \dfrac{478}{f_{\text{MHz}}} = \dfrac{478}{172} = 2.78 \text{ feet}$$

Length of reflector, $\quad L_r = \dfrac{492}{172} = 2.86 \text{ feet}$

Length of director,

$$L_d = \frac{461.5}{172} = 2.683 \text{ feet}$$

Element spacing,

$$S = \frac{142}{172} = 0.825 \text{ feet}$$

$$L_a = 2.78', \ \ L_r = 2.86', \ \ L_d = 2.683', \ \ S = 0.825'.$$

Problem 6.7   Design Yagi-Uda antenna of six elements to provide a gain of 12 dB$_i$ if the operating frequency is 200 MHz.

Solution   Required gain = 12 dB$_i$

$$\text{Frequency,} \qquad f = 200 \text{ MHz}$$
$$\lambda = 1.5 \text{ m}$$
$$L_a = 0.416\lambda = 0.69 \text{ m}$$
$$L_r = 0.475\lambda = 0.7125 \text{ m}$$
$$Ld_1 = 0.44\lambda = 0.66 \text{ m}$$
$$Ld_2 = 0.44\lambda = 0.66 \text{ m}$$
$$Ld_3 = 0.43\lambda = 0.645 \text{ m}$$
$$Ld_4 = 0.40\lambda = 0.60 \text{ m}$$
$$S_L = 0.25\lambda = 0.375 \text{ m}$$
$$S_d = 0.31\lambda = 0.465 \text{ m}$$

Diameter of elements,

$$d = 0.01\lambda = 0.015 \text{ m}$$

The length of array $\ = 1.5\lambda = 2.25 \text{ m}$.

Problem 6.8   Design a log-periodic antenna to obtain a gain of 9 dB and to operate over a frequency range of 125 MHz-500 MHz.

Solution   Gain required = 9 dB

Lowest frequency, $f = 125$ MHz

Highest frequency, $f = 500$ MHz

Longest wavelength corresponds to shortest frequency and shortest wavelength corresponds to longest frequency.

$$\lambda_{\text{long}} = \frac{3 \times 10^8}{125 \times 10^6} = 2.4 \text{ m}$$

$$\lambda_{\text{short}} = \frac{3 \times 10^8}{500 \times 10^6} = 0.6 \text{ m}$$

To obtain a gain of 9 dB, the values of scale and spacing factors are taken from Isbel's curves. They are,

$$\tau = 0.861$$
$$\sigma = 0.162$$

Now the wedge angle is,

$$\alpha = 2 \tan^{-1}\left(\frac{1-\tau}{4\sigma}\right) = 24.2^{\circ}$$

We have
$$\tau = \frac{D_n}{D_{n-1}} = \frac{L_n}{L_{n-1}}$$

or
$$L_2 = \tau L_1, \quad L_3 = \tau L_2, \ldots$$

$$L_1 = \frac{\lambda_{\text{long}}}{2} = \frac{2.4}{2} = 1.2 \text{ m}$$

$$L_2 = \tau L_1 = 1.0332 \text{ m}$$
$$L_3 = \tau L_2 = 0.8895 \text{ m}$$
$$L_4 = \tau L_3 = 0.7659 \text{ m}$$
$$L_5 = \tau L_4 = 0.6594 \text{ m}$$
$$L_6 = \tau L_5 = 0.5678 \text{ m}$$
$$L_7 = \tau L_6 = 0.4888 \text{ m}$$
$$L_8 = \tau L_7 = 0.4210 \text{ m}$$
$$L_9 = \tau L_8 = 0.3624 \text{ m}$$
$$L_{10} = \tau L_9 = 0.3120 \text{ m}$$
$$L_{11} = \tau L_{10} = 0.2686 \text{ m}$$

And the element spacing relation is,

$$\sigma = \frac{S_n}{2L_n}$$

or
$$S_n = 2\sigma L_n = (D_n - D_{n-1})$$
$$S_1 = 2\sigma L_1 = 0.3888 \text{ m}$$
$$S_2 = 2\sigma L_2 = 0.3347 \text{ m}$$
$$S_3 = 2\sigma L_3 = 0.2881 \text{ m}$$
$$S_4 = 2\sigma L_4 = 0.2481 \text{ m}$$
$$S_5 = 2\sigma L_5 = 0.2136 \text{ m}$$
$$S_6 = 2\sigma L_6 = 0.1839 \text{ m}$$
$$S_7 = 2\sigma L_7 = 0.1583 \text{ m}$$
$$S_8 = 2\sigma L_8 = 0.1364 \text{ m}$$

$$S_9 = 2\sigma\, L_9 = 0.1174 \text{ m}$$

$$S_{10} = 2\sigma\, L_{10} = 0.1010 \text{ m}$$

$$S_{11} = 2\sigma\, L_{11} = 0.0870 \text{ m}.$$

Problem 6.9 Find the induced voltage in a vertical 10 turn loop antenna due to a field strength of 10 mV/m and frequency 2 MHz. The area of the loop antenna is 1.4 m^2.

Solution Electric field strength,

$$E_{RMS} = 10 \text{ mV/m}$$

$$f = 2 \text{ MHz}$$

$$N = 10 \text{ turns}$$

$$\phi = 0° \text{ when the plane of the loop is in the plane}$$
$$\text{of propagation of electromagnetic wave}$$

$$S = 1.4 \text{ m}^2$$

$$V_{RMS} = \frac{2\pi\, E_{max}\, SN}{\lambda}\cos\phi \text{ volts}$$

$$= \frac{2\pi\,\sqrt{2}\,E_{RMS}\, SN}{\lambda}\cos\phi$$

$$= \frac{2\pi\,\sqrt{2}\times 10\times 1.4\times 10\times 1}{150}$$

$$V_{RMS} = 8.29 \text{ mV}.$$

Problem 6.10 Find the radiation resistance of a loop antenna of diameter 0.5 m operating at 1 MHz.

Solution Diameter of the loop antenna

$$= 0.5 \text{ m}$$

Its radius $\quad = 0.25 \text{ m}$

$$f = 1 \text{ MHz}$$

$$\lambda = 300 \text{ m}$$

$$R_r = 3720\left(\frac{a}{\lambda}\right)$$

$$= 3{,}720\times\frac{0.25}{300} = 3.1\,\Omega.$$

Problem 6.11 Determine the directivity of a loop antenna whose radius is 0.5 m when it is operated at 0.9 MHz.

Solution Radius of loop antenna

$$a = 0.5 \text{ m}$$

$$f = 0.9 \text{ MHz}$$

$$\lambda = 333.33 \text{ m}$$

$$\frac{2\pi a}{\lambda} = 9.42 \times 10^{-3}$$

As $\dfrac{2\pi a}{\lambda} < \dfrac{1}{3}$, $D = 1.5$.

Problem 6.12   If the radius of a small loop is 0.035λ, find its physical area and maximum effective aperture.

Solution   Radius of the loop antenna

$$a = 0.035\lambda$$

Physical area $= \pi a^2$

$$= \pi \times (0.035\lambda)^2$$

$$A = 3.848 \times 10^{-3} \lambda^2$$

Maximum effective aperture,

$$A_{em} = \frac{3\lambda^2}{8\pi}$$

$$A_{em} = 0.119\lambda^2.$$

Problem 6.13   A circular loop antenna has a diameter of 1.5λ. Find its directivity and radiation resistance.

Solution   Radius of the loop antenna,

$$a = \frac{1.5\lambda}{2} = 0.75\lambda$$

$$\frac{C}{\lambda} = \frac{2\pi a}{\lambda} = \frac{2\pi}{\lambda} \cdot 0.75\lambda$$

$$= 1.5\pi$$

The expression for radiation resistance,

$$R_r = 3720 \left(\frac{a}{\lambda} \right)$$

$$= 3720 \times 0.75$$

$$R_r = 2790\,\Omega$$

The directivity of the loop antenna is

$$D = 4.25 \left(\frac{a}{\lambda} \right)$$

$$= 4.25 \times 0.75$$

$$D = 3.1875.$$

Problem 6.14   An array of dipoles of $\frac{\lambda}{2}$ length in end-fire mode is to produce a power gain of 28.

Find the array length, number of elements when spaced at $\frac{\lambda}{2}$ and Null-to-Null beam width.

Solution   For end-fire array, the power gain is given by

$$g_p = 4\left(\frac{L}{\lambda}\right), \qquad L = \text{array length}$$

That is,
$$28 = 4\left(\frac{L}{\lambda}\right)$$

or
$$L = 7.0\lambda$$

Number of elements in the array when spaced at $\frac{\lambda}{2}$

$$= 7.0 \times 2 = 14$$

Null-to-Null beam width

$$= 2\sqrt{\frac{2\lambda}{Nd}}$$

$$= 2\sqrt{\frac{2\lambda}{14 \times \dfrac{\lambda}{2}}}$$

$$= \frac{4}{\sqrt{14}} = \frac{4}{3.7416}$$

$$= 1.07 \text{ rad}$$

$$\text{B.W.} = 61.30°.$$

Problem 6.15   If a helical antenna has a spacing between turns 0.05 m, diameter 0.1 m, number of turns equal to 20 and operates at 1,000 MHz, find the Null-to-Null beam width of the main beam and also half-power beam width and directivity.

Solution
$$S = 0.05 \text{ m}$$
$$D_h = 0.10 \text{ m}$$
$$N = 20$$
$$f = 1{,}000 \text{ MHz}$$
$$\lambda = 0.3 \text{ m}$$

BWFN,
$$\phi_0 = \frac{115\lambda^{3/2}}{C\sqrt{L_a}}$$

where
$$C = \pi D_h$$
$$L_a = NS$$

$$\phi_0 = \frac{115\,(0.3)^{3/2}}{\pi \times (0.1)\,\sqrt{20 \times 0.05}} = 60.14^\circ$$

HPBW, $\qquad \phi = \dfrac{52\lambda^{3/2}}{C\,\sqrt{L_a}} = 27.20^\circ$

Directivity, $\quad D = \dfrac{15NC^2\,S}{\lambda^3} = 54.84$

$$\boxed{\text{BWFN} = 60.2^\circ,\ \text{HPBW} = 27.2^\circ,\ D = 54.84.}$$

OBJECTIVE QUESTIONS

1. Isotropic radiator radiates equally in all directions. (Yes/No)

2. Isotropic radiator and omni-directional radiator are one and the same. (Yes/No)

3. If P_i is the input to isotropic radiator, power density is _______________.

4. Marconi antenna is nothing but _______________.

5. Standing waves are produced in non-resonant antennas. (Yes/No)

6. Travelling waves are produced in resonant antennas. (Yes/No)

7. Rhombic antenna is _______________
   - (*a*) travelling wave antenna
   - (*b*) standing wave antenna
   - (*c*) narrow-band antenna
   - (*d*) used in LF bands

8. Resonant antenna has a length in multiples of $\dfrac{\lambda}{2}$. (Yes/No)

9. When the length of the antenna is λ, the polarity of the current in one-half of the antenna is opposite to that on the other half. (Yes/No)

10. The radiation at right angles from λ antenna is zero because _______________.

11. Antennas transmit efficiently when the length is _______________.

12. The voltage distribution on half-wave dipole is _______________.

13. Resonant antenna is
   - (*a*) Aperiodic
   - (*b*) Periodic
   - (*c*) Travelling wave
   - (*d*) Rhombic

14. If the length of wire antenna is more, beam width is small. (Yes/No)

15. HF band is _______________.

16. UHF band is _______________.

17. Tower antenna is _______________.

18. A typical inductance loaded LF antenna is ________________.

19. One application of VLF is ________________.

20. Null-to-Null beam width of end-fire array is ________________.

21. Null-to-Null beam width of broadside array is ________________.

22. Radiation beam in broadside array is along the axis of the array. (Yes/No)

23. If the number of elements is more in an array, beam width is small. (Yes/No)

24. V antennas are ________________.

25. The excitation to each wire of V antenna is ________________.

26. The radiation pattern of resonant V antenna is ________________.

27. The radiation pattern of non-resonant V antenna is ________________.

28. Arrays of V antennas are not possible. (Yes/No)

29. Inverted V antenna is a travelling wave antenna. (Yes/No)

30. The directivity of Rhombic antenna is greater than that of V antenna. (Yes/No)

31. Rhombic antenna is an HF antenna. (Yes/No)

32. The efficiency of Rhombic antenna is very high. (Yes/No)

33. Rhombic antenna is used for transmission purpose only. (Yes/No)

34. The radiation of Rhombic antenna is ________________.

35. The design parameters of Rhombic antenna are ________________.

36. Radiation resistance of $\dfrac{\lambda}{2}$ folded dipole is ________________.

37. Radiation resistance of three folded $\dfrac{\lambda}{2}$ dipole is ________________.

38. Radiation pattern of folded dipole is the same as that of straight dipole. (Yes/No)

39. The voltage and current in resonant antennas are ________________.

40. The length of non-resonant antenna is in multiples of $\dfrac{\lambda}{2}$. (Yes/No)

41. In end-fire array, all the elements are fed with no additional phase. (Yes/No)

42. The impedance of folded dipole is a function of dipole radius. (Yes/No)

43. The disadvantage of non-resonant V antenna is ________________.

44. Isotropic antenna is used as ________________.

45. The director's reactance in Yagi-Uda antenna is ________________.

46. The reflector's reactance in Yagi-Uda antenna is ________________.

47. The sensitivity of Yagi-Uda is very high. (Yes/No)

48. The band width of Yagi-Uda antenna is limited. (Yes/No)

49. The band width of straight dipole is _______________.

50. In end-fire two-element array, the elements are fed with a phase difference of 180°.

(Yes/No)

51. In broadside array, the elements are in phase. (Yes/No)

52. The impedance and directivity changes with frequency in log- periodic array. (Yes/No)

53. Log-periodic antenna is frequency independent. (Yes/No)

54. Log-periodic antenna is a wide-band antenna. (Yes/No)

55. Rhombic antenna is more efficient than log-periodic antenna. (Yes/No)

56. Log-periodic array is a uniform linear array. (Yes/No)

57. The gain of log-periodic antenna is more when τ is small and α is large. (Yes/No)

58. Log-periodic antenna becomes compact when scale factor is small and wedge angle α is large.

(Yes/No)

59. The designed band width is greater than the desired band width in log-periodic antenna.

(Yes/No)

60. The number of dipoles in log-periodic antenna is a function of

(*a*)   desired gain only   (*b*)   desired band width only

(*c*)   wedge angle only   (*d*)   designed band width and scale factor

61. Length of the log-periodic array is a function of _______________.

62. The length of the antenna at an operating frequency of 0.5 GHz is

(*a*)   570 m   (*b*)   5.70 m

(*c*)   57.0 m   (*d*)   600 m

63. The equivalent of a small loop antenna is _______________.

64. The Marconi antenna is used in _______________.

65. The loop antennas are used for _______________.

66. Loop antennas may have several turns. (Yes/No)

67. Loop antennas may have ferrite cores. (Yes/No)

68. Ferrite cores in loop antennas increase the diameter of the loop. (Yes/No)

69. Loop antennas with ferrite cores are as _______________.

70. The loop antenna is always circular in shape. (Yes/No)

71. The directional properties of loop antennas at medium frequencies are different from those at microwave frequencies. (Yes/No)

72. The direction of the given radiation in loop antenna is indicated by maximum signal.

(Yes/No)

73. The direction of given radiation is indicated by null. (Yes/No)

74. The directional loop antenna is independent of the shape of the loop. (Yes/No)

75. The radiation pattern of loop antenna is the same as that of a half-wave dipole. (Yes/No)

76. Helical antenna is used in ________________.

77. Helical antenna produces circular polarisation. (Yes/No)

78. Helical antenna has wide band width. (Yes/No)

79. Helical antenna is mostly used in normal mode. (Yes/No)

80. Helical antenna is used in axial mode. (Yes/No)

81. Helical antenna can be used in HF, VHF bands. (Yes/No)

82. Helical antenna is only for receiving purposes. (Yes/No)

83. Helical antenna is used for transmission and receiving purposes. (Yes/No)

84. Whip antennas are used in ________________.

85. Whip antennas are used at HF and VHF bands. (Yes/No)

86. Whip antenna is a quarter-wave Marconi antenna. (Yes/No)

87. At 30 MHz, whip antenna has a length of

    (*a*)   2.5 m   (*b*)   25 m

    (*c*)   250 m   (*d*)   10 m

88. Effective height of quarter-wave grounded vertical wire is ________________.

89. An example of Marconi antenna is ________________.

90. Ferrite rod antennas are used in ________________.

91. The selectivity of ferrite rods is very high. (Yes/No)

92. Radiation resistance of ferrite rod depends on ________________.

93. Antenna efficiency is ________________.

94. The approximate practical $\frac{\lambda}{2}$ dipole length after taking end effects into account is ________________.

95. If the length of antenna is more, its directivity is high. (Yes/No)

96. The disadvantage of Rhombic antenna is ________________.

97. Loop antennas are used in ________________.

98. Loop antenna can be of any shape including triangular loop for direction-finding. (Yes/No)

99. For direction-finding loop antenna is rotated. (Yes/No)

100. The purpose of Adcock antenna is ________________.

101. In general, loop antennas are satisfactory for frequencies between 2 and 30 MHz due to polarisation error. (Yes/No)

102. Discone antenna is a broad-band antenna compared to dipole. (Yes/No)

103. Notch antenna is used ______________.

104. Inverted V antenna is balanced fed. (Yes/No)

105. For small square and circular loop antennas, the field patterns are identical. (Yes/No)

106. The radiation patterns depend only on the area and the shape of the small loop has no effect. (Yes/No)

ANSWERS

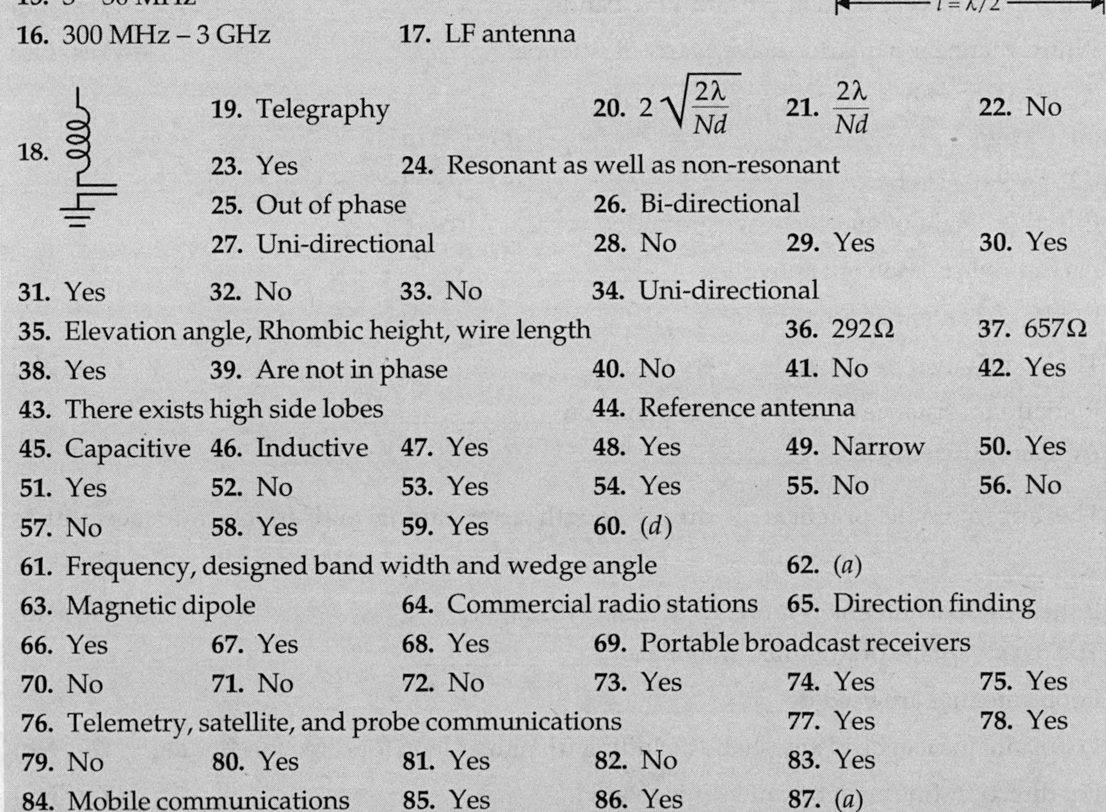

1. Yes **2.** No **3.** $P_i/4\pi r^2$ watts/m^2 **4.** A quarter-wave monopole

5. No **6.** No **7.** (a) **8.** Yes **9.** Yes

10. The currents are out of phase **11.** $\lambda/2$, $\lambda/4$

13. (b) **14.** Yes

12.

15. 3 – 30 MHz

16. 300 MHz – 3 GHz **17.** LF antenna

18. **19.** Telegraphy **20.** $2\sqrt{\dfrac{2\lambda}{Nd}}$ **21.** $\dfrac{2\lambda}{Nd}$ **22.** No

23. Yes **24.** Resonant as well as non-resonant

25. Out of phase **26.** Bi-directional

27. Uni-directional **28.** No **29.** Yes **30.** Yes

31. Yes **32.** No **33.** No **34.** Uni-directional

35. Elevation angle, Rhombic height, wire length **36.** $292\,\Omega$ **37.** $657\,\Omega$

38. Yes **39.** Are not in phase **40.** No **41.** No **42.** Yes

43. There exists high side lobes **44.** Reference antenna

45. Capacitive **46.** Inductive **47.** Yes **48.** Yes **49.** Narrow **50.** Yes

51. Yes **52.** No **53.** Yes **54.** Yes **55.** No **56.** No

57. No **58.** Yes **59.** Yes **60.** (d)

61. Frequency, designed band width and wedge angle **62.** (a)

63. Magnetic dipole **64.** Commercial radio stations **65.** Direction finding

66. Yes **67.** Yes **68.** Yes **69.** Portable broadcast receivers

70. No **71.** No **72.** No **73.** Yes **74.** Yes **75.** Yes

76. Telemetry, satellite, and probe communications **77.** Yes **78.** Yes

79. No **80.** Yes **81.** Yes **82.** No **83.** Yes

84. Mobile communications **85.** Yes **86.** Yes **87.** (a)

88. $\dfrac{2}{\pi} \times$ actual height **89.** Quarter-wave antenna **90.** Radio receivers

91. Yes **92.** Diameter of ferrite rod, its effective permeability and frequency of operation

93. $R_r/(R_r + R_l)$ 94. 0.475λ 95. Yes   96. It requires more space

97. Radio receivers and aircraft receivers for direction finding   98. Yes   99. Yes

100. Direction finding   101. Yes   102. Yes   103. On aircrafts

104. No   105. Yes   106. Yes.

 **EXERCISE PROBLEMS**

1. Design a Rhombic antenna to operate at 25 MHz if the required elevation angle is
   (a) 20°   (b) 25°.

2. Design a Rhombic antenna to operate at 15 MHz if the required elevation angle is
   (a) 30°   (b) 35°.

3. Design a Rhombic antenna to operate at 20 MHz if the tilt angle is
   (a) 60°   (b) 65°.

4. Obtain alignment design parameters of Rhombic antenna to operate at 25 MHz when the required elevation angle is
   (a) 25°   (b) 30°.

5. A multiple circular loop of radius 1.0 cm operates at 100 MHz. If its radiation resistance is 10.0 Ω, find the number of turns.

chapter **7**

Microwave Antennas

"Microwaves are nothing but Electromagnetic waves with wavelengths at micro level and microwave antennas are small in size."

CHAPTER OBJECTIVES

This chapter discusses

✦ The design, construction, applications and analysis of all types of microwave antennas including microstrip antennas

✦ Objective questions and solved problems useful for class tests, final examinations and also for competitive examinations

✦ Exercise problems to develop self problem solving skills

7.1 INTRODUCTION

The size of the antenna depends mainly on the frequency of operation. If the frequency is low, size of the antenna is large and vice-versa.

Microwave antennas are popular for their small size and better radiation characteristics. In the present day miniaturised world, the smaller the antennas in communication, the more is the attraction.

Different important and popular microwave antennas and their characteristics are presented in this chapter.

It is difficult to classify antennas based on frequency. Most of the antennas are used in overlapping frequency bands. However, in the present chapter, the following antennas are considered to be microwave antennas and they are described in detail.

Microwave antennas:
- Reflector antennas
- Horn antennas
- Dielectric and metal lens antennas
- Slot antennas
- Micro-strip antennas

The microwave region extends from 1 GHz to 100 GHz. The transmitting and receiving antennas in microwave frequencies are directive with high gain and narrow beam width in both vertical and horizontal planes.

Types of reflectors:

1. Rod reflector
2. Plane reflector
3. Corner reflector
4. Cylindrical reflector
5. Horn reflector
6. Spherical reflector
7. Parabolic reflector.

7.2 ROD REFLECTOR

It is mainly used in Yagi-Uda antenna. It is placed behind the driven element. Its length is slightly longer than that of the driven element. That is, it is greater than $\frac{\lambda}{2}$. It offers inductive reactance and contributes in increasing the gain. Here rod reflector is not the main antenna but is only a parasitic element. The main disadvantage of the rod element is that it alters the impedance of the driven element.

7.3 PLANE REFLECTOR

It is the simplest reflector to direct electromagnetic energy in a desired direction. But it is difficult to collimate the energy in the forward direction. In fact, polarisation of the primary antenna and its position with respect to the reflecting surface is used to control the pattern characteristics, impedance, power gain and directivity of the complete system. Infinitely large reflector is ideal. It is difficult to collimate the energy by a plane reflector.

A typical plane reflector is shown in Fig. 7.1.

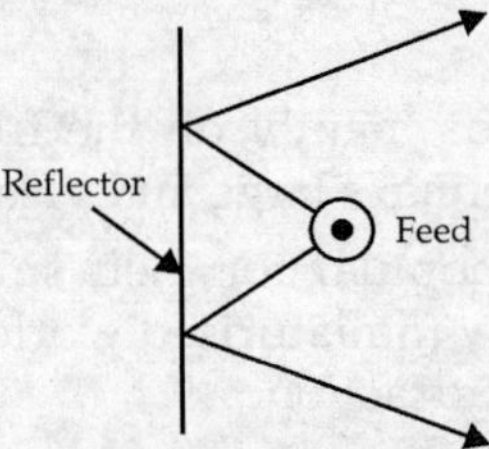

Fig. 7.1 *Plane reflector*

7.4 CORNER REFLECTOR

A corner reflector is a reflecting object which consists of two or three mutually intersecting, conducting flat surfaces.

Dihedral forms of corner reflectors are frequently used in antennas. However, trihedral forms with mutually perpendicular surfaces are used as radar targets.

A typical corner reflector is shown in Fig. 7.2.

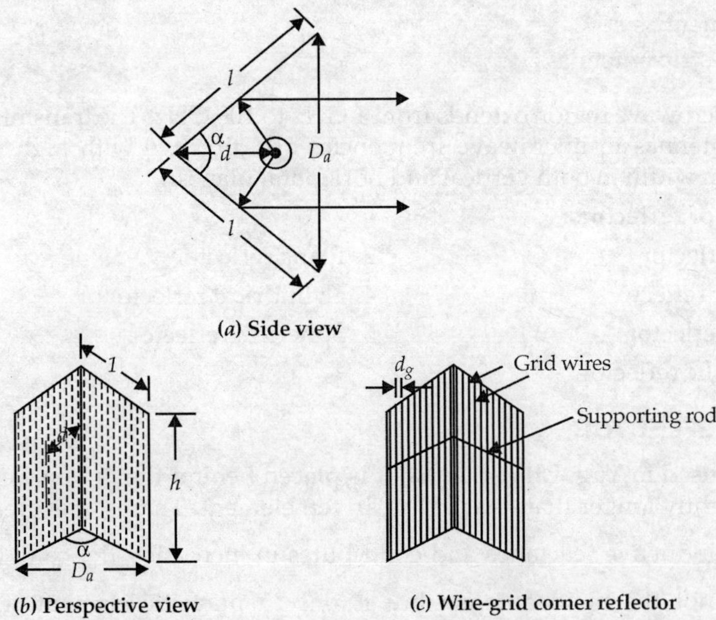

Fig. 7.2 *Corner reflector*

A corner reflector is designed to improve the collimation of electromagnetic energy in the forward direction and to eliminate radiation in the back and side directions.

For the corner reflector shown in Fig. 7.2

$$D_a = \text{aperture size}$$

$$l = \text{length}$$

$$h = \text{height}$$

d = spacing between the vertex and feed point location

d_g = spacing between grid wires

α = included angle

The ranges of the above parameters for a good corner reflector are:

1. $\lambda < D_a < 2\lambda$

2. $l \approx 2d$

3. $\dfrac{\lambda}{3} < d < \dfrac{2\lambda}{3}$ ...(7.1)

4. h is 1.2 to 1.5 times greater than the total length of feed element

5. $d_g \le \dfrac{\lambda}{10}.$

Salient features of corner reflector

1. It is simple to construct.

2. It is used as a passive target for radar and communication applications to return the signal exactly in the same direction by choosing $\alpha = 90°$. Due to this unique feature, most of the defence ships and vehicles are designed with minimum sharp corners to reduce the chances of their detection by enemy radars.

3. It is also used in home television antennas.

4. The most preferred value of α is $90°$.

5. The spacing between the vertex and feed element position is increased if α is decreased and vice-versa, in order to improve efficiency.

6. When α is small, gain is increased by increasing the length of the sides of the reflector.

7. The feed element can be a dipole or an array of collinear dipoles.

8. When the feed elements are cylindrical or biconical dipoles instead of thin wires, band width and radiation resistance are high.

9. When the reflector is large with high λ, the surfaces of corner reflectors are made of grid wires to reduce aerodynamic drag due to wind speeds and overall system weight.

10. If the spacing, d is small, radiation resistance becomes small and hence efficiency is reduced.

11. If the spacing is very large, the system produces undesirable multiple lobes and it loses its directional characteristics.

12. The main lobe is broad for reflectors with finite sides compared to those of infinite dimensions.

13. The array factor of corner reflector antenna is

$$E = 2\left[\cos\left(Kd\sin\theta\,\cos\phi - \cos(Kd\sin\theta\,\sin\phi)\right)\right] \quad ...(7.2)$$

Here, $K = \dfrac{2\pi}{\lambda}.$

14. For small included angle, the side lengths should be longer.

Cylindrical reflector It is a reflector which is part of a cylinder. The cylinder is usually parabolic in shape. However, cylinders of other shapes also can be used.

Horn reflector It is a reflector antenna which consists of a section of a paraboloidal reflector fed with an offset horn which intersects the reflector surface. Horn antenna is a radiating element which is in the shape of a horn. The horn reflector antenna is either pyramidal or conical.

Spherical reflector It is a reflector which is part of a spherical surface.

7.5 PARABOLIC REFLECTOR

It is a reflector antenna which has the shape of paraboloid and employs the properties of parabola.

It can also be defined as a reflector which is part of a paraboloid of revolution.

The parabola, it is a plane curve obtained by the locus of a point which moves so that its distance from another point, called the focus, plus its distance from a straight line, called directrix, is constant.

A paraboloid is a three dimensional surface obtained by revolving the parabola about its axis. The paraboloid is called the parabolic reflector or dish antenna.

The geometry of a parabolic reflector in transmitting mode is shown in Fig. 7.3.

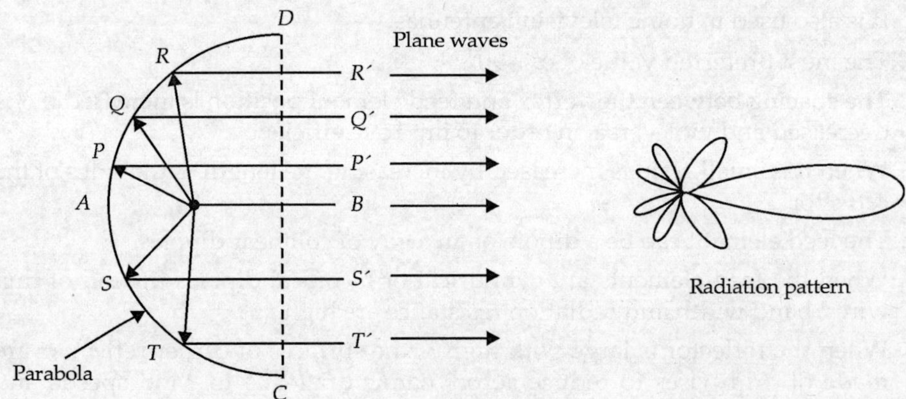

Fig. 7.3 *Geometry of parabolic reflector in transmitting mode and its radiation pattern*

Here AB = axis of the parabola

CD = mouth diameter, D_a

AF = focal length = l_f

A = vertex

F = focus

CAD = parabola

The line CD = directrix

AF/CD = aperture of the parabola

From the definition of a parabola we have,

$$FP + PP^1 = FQ + QQ^1 = FS + SS^1 = \text{constant } (K)$$

K varies with shape.

The equation of the parabola is

$$y^2 = 4l_f\, x \qquad\qquad ...(7.3)$$

and the equation of the paraboloid is

$$y^2 + z^2 = 4l_f\, x \qquad\qquad ...(7.4)$$

A parabolic reflector in receiving mode is shown in Fig. 7.4.

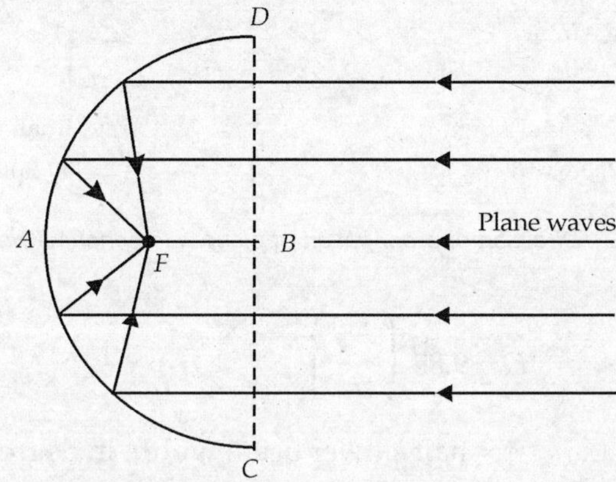

Fig. 7.4   *Geometry of parabolic reflector in receiving mode*

Operation of parabolic reflector   If a feed antenna is placed at the focus, all the waves are incident on the reflector and they are reflected back, forming a plane wave front. By the time the reflected waves reach the directrix, all of them will be in phase, irrespective of the point on the parabola from which they are reflected. Hence the radiation is very high and is concentrated along the axis of the parabola. At the same time, waves will be cancelled in other directions as a result of path and phase differences.

The main purpose of the parabolic reflector is to convert a spherical wave into a plane wave.

The difference between the plane wave and spherical wave are shown in Fig. 7.5.

If the primary or feed antenna is non-directional or isotropic, the beam width of the radiation pattern of the paraboloid is given by:

$$\text{HPBW}, \qquad = \phi = \frac{70\lambda}{D_a}$$

$$\text{BWFN}, \qquad \phi_0 = 2\phi = \frac{140\lambda}{D_a} \qquad\qquad ...(7.5)$$

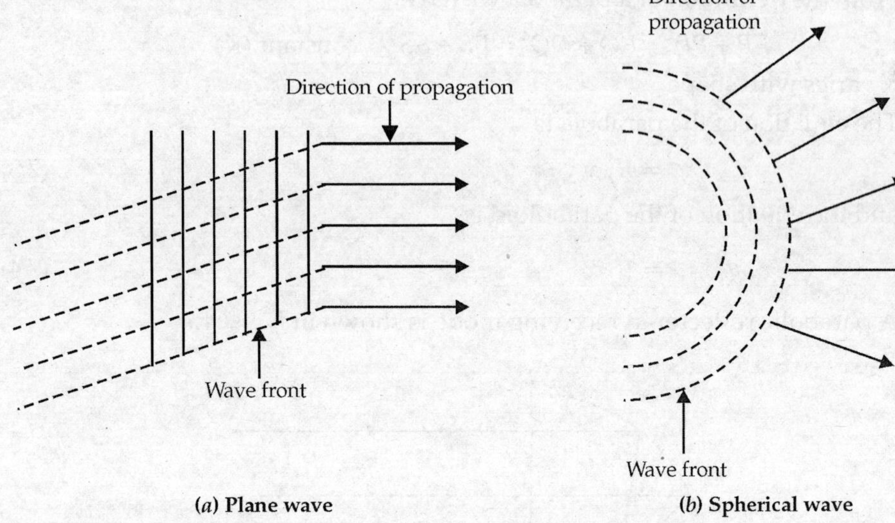

Fig. 7.5 *Direction of propagation of plane and spherical waves*

Directivity, $\qquad D = 9.87 \left(\dfrac{D_a}{\lambda} \right)^2$

Here, $\qquad \phi = $ half-power beam width, in degrees

$\qquad \phi_0 = $ beam width between first nulls, in degrees

$\qquad \lambda = $ wavelength, m

$\qquad D_a = $ mouth diameter, m.

For a large, uniformly illuminated rectangular aperture,

HPBW $\qquad = \phi = \dfrac{57.5\lambda}{L}$ (degrees)

BWFN $\qquad = \phi_0 = \dfrac{115\lambda}{L}$ (degrees) $\qquad\qquad\qquad$ ...(7.6)

Directivity, $D = \dfrac{4\pi A}{\lambda^2}$

Here, $\qquad L = $ length of the aperture, in λ

$\qquad A = $ aperture area, m^2.

The above equations are for ideal illuminations. No primary feed antenna is truly isotropic. Moreover, the illumination in a paraboloid decreases towards the edges and hence it is not illuminated uniformly. This results in the capture area being always smaller than the actual area. That is,

The capture area,

$$A_c = bA \qquad\qquad\qquad\qquad\qquad \text{...(7.7)}$$

Here,　　　　　　A = actual area

b = constant depending on the type of primary antenna

≈ 0.65 for dipole antenna

For a loss-less antenna and tapered illumination,

Power gain,　　$g_p = \dfrac{4\pi}{\lambda^2} A_c$　　　　　　　　　　　　...(7.8)

$$= \dfrac{4\pi}{\lambda^2} bA$$

For circular apertured paraboloid,

$$A = \dfrac{\pi D_a^2}{4} \qquad\qquad ...(7.9)$$

So,　　　　　　$g_p = \dfrac{4\pi K}{\lambda^2} \dfrac{\pi D_a^2}{4}$

$$= \dfrac{\pi^2 K D_a^2}{\lambda^2}$$

$$= 0.65\pi^2 \left(\dfrac{D_a}{\lambda}\right)^2 \qquad [K \approx 0.65 \text{ for dipole feed}]$$

$$= 6.41 \left(\dfrac{D_a}{\lambda}\right)^2$$

So,　　　　　　$g_p = 6.4 \left(\dfrac{D_a}{\lambda}\right)^2$　　　　　　　　...(7.10)

Salient features of paraboloid reflectors

1. The directional beam has a sharp main lobe surrounded by several side lobes.

2. The three dimensional shape of the main lobe resembles a fat cigar in the direction of the axis of the paraboloid.

3. If the primary antenna is non-directional,

$$\text{BWFN} = 140 \left(\dfrac{\lambda}{D_a}\right)$$

and　　　　$\text{HPBW} = 70 \left(\dfrac{\lambda}{D_a}\right).$

4. The gain of the antenna with parabolic reflector is influenced by aperture ratio $\left(\dfrac{D_a}{\lambda}\right)$ and type of illumination.

5. For tapered illumination, the power gain, $g_p = 6.4 \left(\dfrac{D_a}{\lambda} \right)^2$.

6. Directivity with respect to isotropic antenna is $D = 6.4 \left(\dfrac{D_a}{\lambda} \right)^2$.

7. Effective Radiated Power (ERP) = product of input power to the antenna and power gain. It is very high even for small input power.

8. Very large gains and narrow beam width are obtainable with paraboloid reflectors.

9. Paraboloids are not used at low frequencies because of large size.

10. In order to be fully effective and useful, its mouth diameter must be at least 10λ.

11. At the lower end of the television band, say at 63 MHz, the required mouth diameter is 48 m.

12. Performance of paraboloid reflectors depends on the radiation characteristics of primary antenna and its size.

13. Parabolic reflectors have several applications in communications and radars.

14. The reflector is called the **secondary antenna** and feed antenna is called the primary antenna.

15. The radiation pattern of primary antenna placed at the locus of the parabolic reflector is called the **primary pattern**. The radiation pattern of the entire system consisting of primary and secondary antenna is called the **antenna pattern**.

16. A mesh surface is often used to minimise wind effect on the antenna and extra strain on the supports. This also reduces distortion caused by uneven wind force distribution over the surface.

Disadvantages of paraboloid reflectors

1. The radiation beam is a pencil beam and it is surrounded by side lobes. These side lobes create electromagnetic induction and the effect of electromagnetic interference is more prominent in low noise receivers due to the imperfections in the reflector.

2. Deviations from the true shape of a paraboloid should not exceed one sixteenth of λ. Such tolerances may be difficult to achieve in large dishes whose surface is a network of wires instead of a smooth continuous skin.

3. Diffraction is another cause of side lobes and will occur around the edges of the paraboloid. This produces electromagnetic induction.

4. The finite size of the primary antenna also influences the band width.

5. As the feed antenna is not a true point source, it cannot be located exactly at the focus.

6. Defects like aberrations cause the main lobe to be broadened and the side lobes to be reinforced.

7. The primary antenna does not radiate evenly at the reflector and hence distortion is introduced. If a dipole feed is used, the radiation in one plane is different from that in the other and the beam from the reflector becomes broad.

8. Flattened beams are avoided by using horn feed. Here also, paraboloid is not illuminated uniformly and it will taper at the edges. This leads to a small capture area.

7.6 TYPES OF PARABOLIC REFLECTORS

Apart from full paraboloid reflectors, there are other types of reflectors:

1. Cut or truncated paraboloid.
2. Parabolic cylinder.
3. Pill box and cheese antenna
4. Offset paraboloid reflector.
5. Torus antenna.

The advantages of these are low cost and small size. The main disadvantage is that the beam is not directional in both azimuth and elevation.

7.6.1 Cut or Truncated Paraboloid

This is shown in Fig. 7.6. It is not circular in appearance when viewed from a point on the parabolic axis.

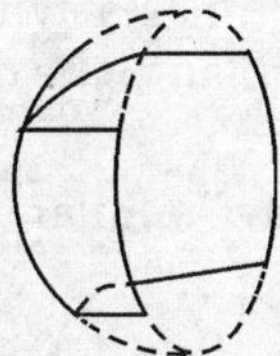

Fig. 7.6   *Cut paraboloid*

7.6.2 Parabolic Cylinder

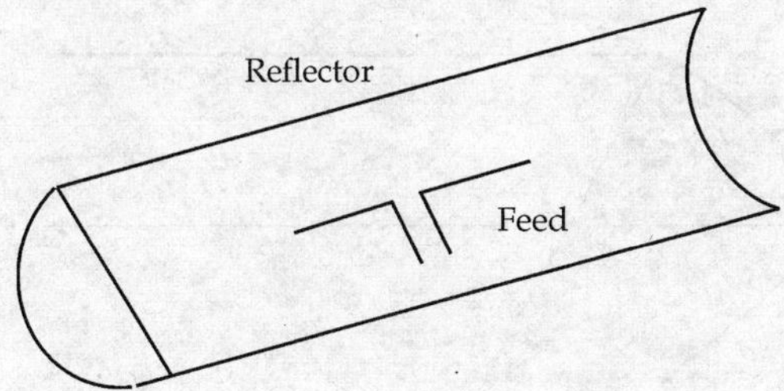

Fig. 7.7   *Parabolic cylinder*

This is formed by moving the parabola sideways. A plane sheet is curved in one dimension to obtain a parabolic cylinder. It is characterised by a focal line instead of a focal point and a vertex line instead of a vertex. When a radiating line source is on the focal line, the parabolic cylinder is illuminated uniformly. This results in a beam in the vertical plane. Its beam width is slightly less than that of a full paraboloid. It gives rise to a wide beam in E-plane and narrow beam in H-plane.

7.6.3 Pillbox Antenna

This is a reflector antenna which has a cylindrical reflector enclosed by two parallel conducting plates perpendicular to the cylinder, spaced less than one wavelength apart. A typical pillbox structure is shown in Fig. 7.8. It looks like a pillbox and

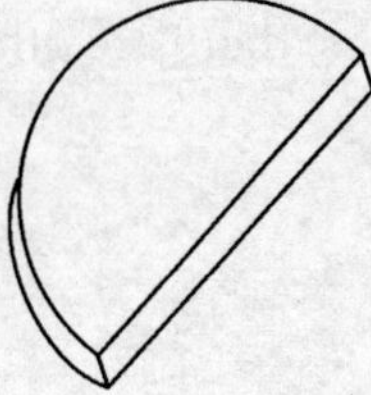

Fig. 7.8 *Pillbox antenna*

hence the name. It is excited by a probe through a coaxial line. It produces a wide beam in *E*-plane and narrow beam in *H*-plane. That is, these reflectors produce shaped beams-wide beam in one plane and narrow in the other. This is used in ship-to-ship radars.

Cheese antenna

This is also a reflector antenna which has a cylindrical reflector enclosed by two parallel conducting plates perpendicular to the cylinder but spaced more than one wavelength apart.

Pillbox and Cheese antennas have similar applications.

7.6.4 Offset Paraboloid

This is one form of cut-paraboloid in cross-section. In this, the focus is located outside the aperture (Fig. 7.9). If the feed antenna is kept at the focus, the reflected

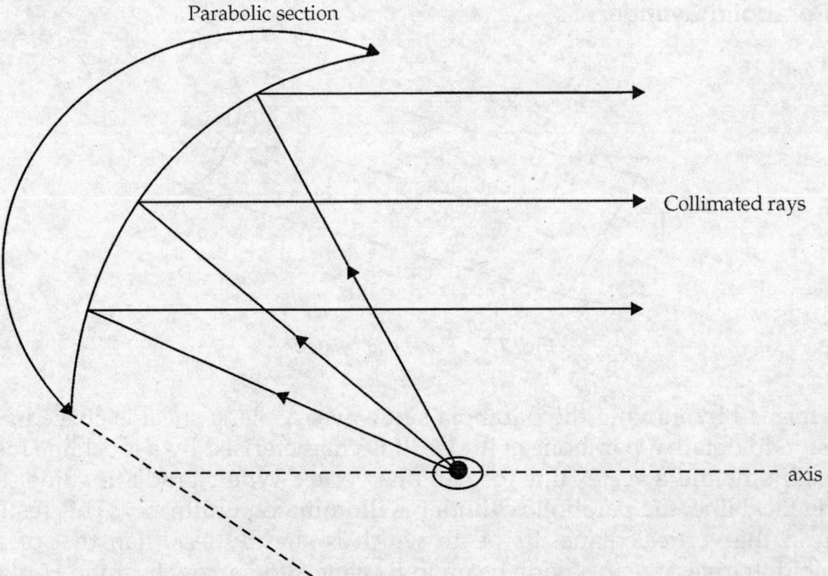

Fig. 7.9 *Offset parabolic reflector*

and collimated rays will be produced without any interference. This reflector is very useful when the size of the feed antenna is comparable with the reflector. In this case feed antenna will not block the reflected rays.

7.6.5 Torus Antenna

It is a better version of an offset reflector. It is similar to cut paraboloid. It is parabolic along one axis and circular along the other. This antenna is useful to transmit or receive a number of beams simultaneously to or from a geostationary satellite orbit. This is possible by placing several feeds at the focus.

7.7 FEED SYSTEMS FOR PARABOLIC REFLECTORS

It is possible to feed the reflector in several ways. Some of them are:

1. Half-wave dipole
2. An array of collinear dipoles
3. Yagi-Uda antenna
4. Centre-fed with spherical reflector
5. Horn
6. Cassegrain feed.

7.7.1 Half-Wave Dipole Feed

This has bi-directional radiation characteristics. Ideally, uni-directional antenna is required as feed antenna. In this case the reflected rays will interfere with the backward radiated rays and disturb the plane wave–front because of phase difference.

7.7.2 Yagi-Uda Antenna Feed

Although this produces a uni-directional pattern, the size of the antenna is a common problem. It blocks the reflected rays.

7.7.3 Array of Collinear Dipoles Feed

This is another possible primary feed antenna. But feeding with a dipole array involves changing from unbalanced system to a balanced system.

7.7.4 Centre-fed with Spherical Reflector

This is shown in Fig. 7.10.

Salient features of centre-fed with spherical reflector

1. The primary antenna is kept at the focus of the paraboloid for better reception or transmission.
2. Direct radiation from the feed spoils the directivity. To prevent direct radiation, a small spherical reflector is used which redirects the direct radiations back to the paraboloid.
3. The spherical shell obstructs the reflected rays. But this is not high. If a spherical shell of diameter 2 cm is placed at the focus of 2 m paraboloid, the obstruction is only one percent.

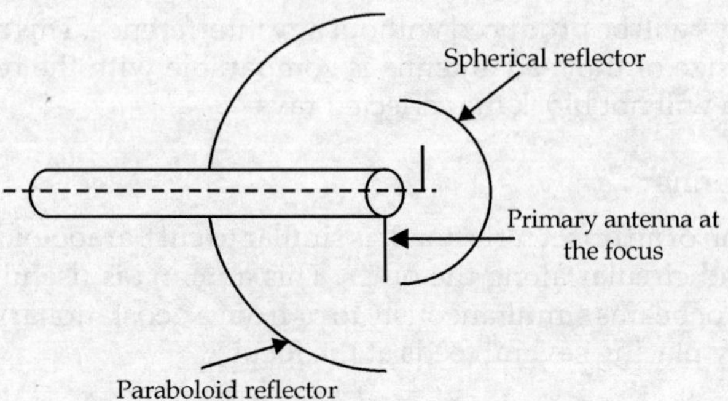

Fig. 7.10 *Centre-fed with spherical reflector*

7.7.5 Horn Feed

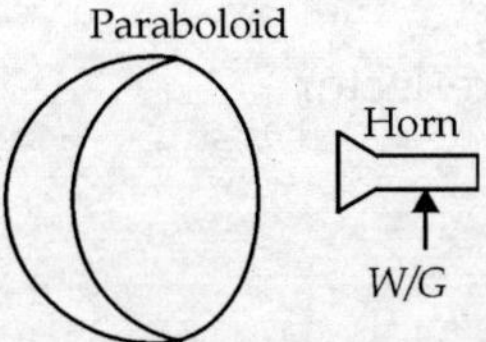

Fig. 7.11 *Horn feed*

Salient features of horn feed

1. Horn has moderate directional characteristics towards the reflector.

2. There is no direct radiation.

3. Horn obstructs the reflected rays when it is placed at the focus. But obstruction is not high. It may be one or two percent of the total reflected energy.

7.7.6 Cassegrain Feed

It is named after the early eighteenth century astronomer. The feed mechanism is shown in Fig. 7.12. It uses:

- a parabolic reflector,

- a hyperbolic reflector and

- a feed antenna, horn with waveguide.

One of the foci of the hyperbolic reflector coincides with the focus of the paraboloid. When electromagnetic rays from a horn antenna are incident on the hyperboloid reflector, they are reflected back and are then incident on the paraboloid. These incident rays are reflected and propagate as a plane wave front.

Hyperbola is a curve traced by a point which moves so that its ratio of the distance from a fixed point, the focus to its distance from a fixed straight line, the directrix is a constant and is greater than unity. The hyperboloid is a three dimensional surface obtained by revolving the hyperbola about its axis. The size of

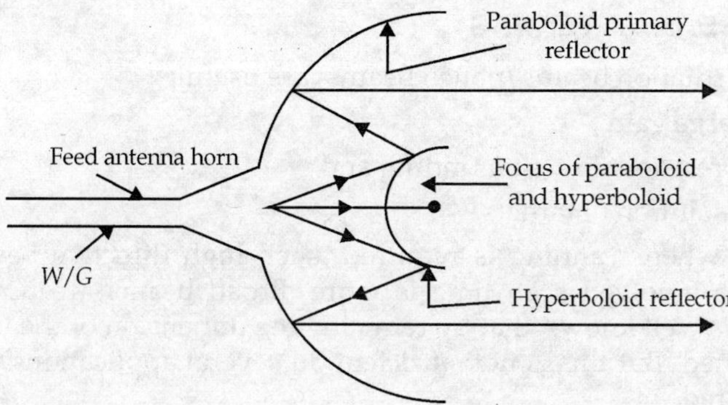

Fig. 7.12 *Cassegrain mechanism in transmission mode*

the hyperboloid depends on its distance from the primary feed antenna, mouth diameter of horn and the frequency of operation.

Applications of Cassegrain feed

1. It is used when it is required to keep the primary antenna in a convenient position.
2. It is used when it is desired to use a short transmission line or waveguide for connecting the receiver or transmitter to the primary antenna.
3. It is used for low-noise receiver applications.

If the active part of the transmitter or receiver is kept at the focus, it is possible to reduce power loss. But the size of the transmitter or receiver prohibits such a placing. This is the reason why Cassegrain feed is best suited for low-noise applications.

Advantages of Cassegrain feed

1. Spill over and minor lobe radiation is less.
2. It is possible to scan the beam or to broaden the beam by moving of the reflecting surfaces.
3. Feed antenna can be kept at a convenient location.

Disadvantages

1. Some of the electromagnetie energy is obstructed by the secondary hyperboloid reflector. This is tolerable when the size of paraboloid is large compared to hyperboloid. The obstruction is reduced by using large paraboloid reflectors and keeping the distance between the horn feed and hyperboloid reflector small. This technique reduces the diameter of the hyperboloid reflector.
2. Large paraboloid is expensive.
3. When vertically polarised waves are radiated by the feed antenna, they are reflected back to the main reflector by the hyperboloid.
4. Polarisation of the waves is twisted by $90°$ when they are reflected by the paraboloid. That is, the reflected waves are horizontally polarised and they propagate through the vertical bars of the hyperboloid.

7.8 SHAPED BEAM ANTENNAS

High directive radiation beams (pencil beams) are useful:

- to obtain large gain
- to obtain precision direction finding and
- for high resolution of targets.

However, when scanning is required, such high directive beams are not preferred as the involved scan time is more. Broad beams are useful in such applications. It is well known that by reducing the dimension of the aperture, the beam is broadened. But this is not sufficient. In several applications well-shaped beams are required.

Antennas that produce shaped beams are called **shaped beam antennas**.

Some of the popular shaped beams are:

1. Fanned beams
2. Sector beams
3. Cosecant beams.

7.8.1 Fanned Beams

The fan beam is a radiation pattern which exhibits broad beam characteristics in one of the principal planes. Such beams are generated from ship antennas. These are used to compensate roll and pitch of the ships.

Fanned beam antennas are of the following forms:

1. An array antenna with optimally designed amplitude and phase distributions.
2. A section of parabolic reflector with the point source at its focus.
3. A parabolic cylinder with a line source producing a rectangular aperture.
4. A parallel plate antenna which consists of a parabolic cylindrical reflector illuminated by a simple feed at the focus and located between the parallel plates which are perpendicular to the generator of the cylinder. This produces a rectangular aperture.

Applications of fanned beams

1. Air search from ground-based or ship-borne antennas.
2. Surface search from air-borne antennas.

Typical fanned antenna beams are shown in Figs. 7.13 and 7.14.

In Fig. 7.13, the beam is from a ground-based antenna for air search. The beam in elevation provides coverage on aircraft. Azimuth coverage is obtained with scanning. The beam shown in Fig. 7.14 is from air-borne antenna for surface search.

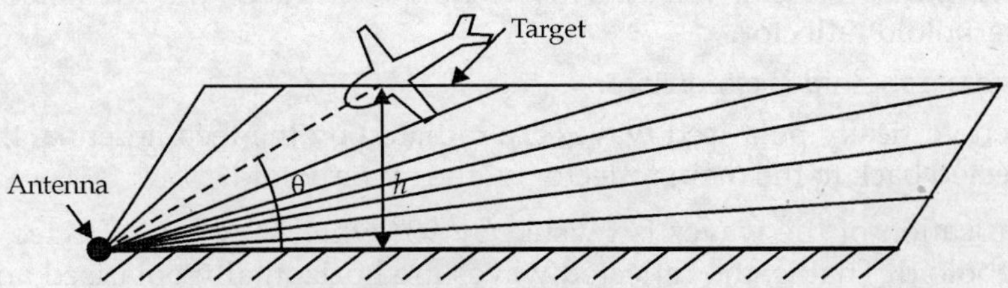

Fig. 7.13 *Fanned beam from ground-based antenna for air search*

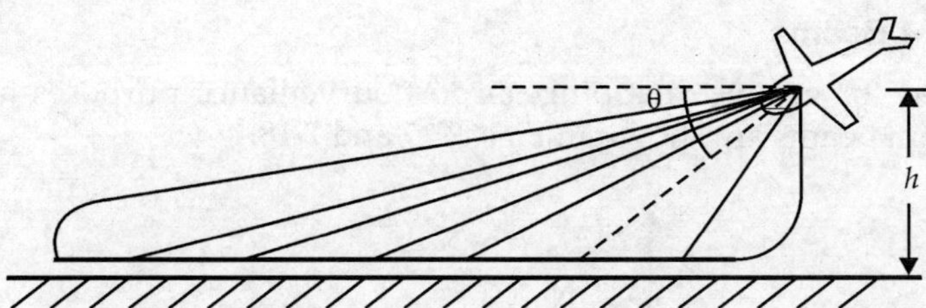

Fig. 7.14 *Fanned beam from air-borne antenna for surface search*

7.8.2 Sector Beam

This beam is basically broad over a desired angular region as shown in Fig. 7.15. This beam is again sharp in azimuth and is broad in elevation to accommodate roll and pitch. This type of beam provides a more constant illumination of the target and is also more conservative.

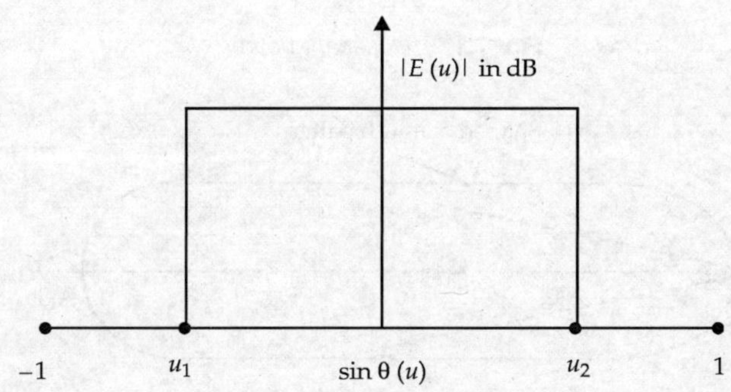

Fig. 7.15 *Sector beam*

Typical sector beams are shown in Figs. 7.15 and 7.16.

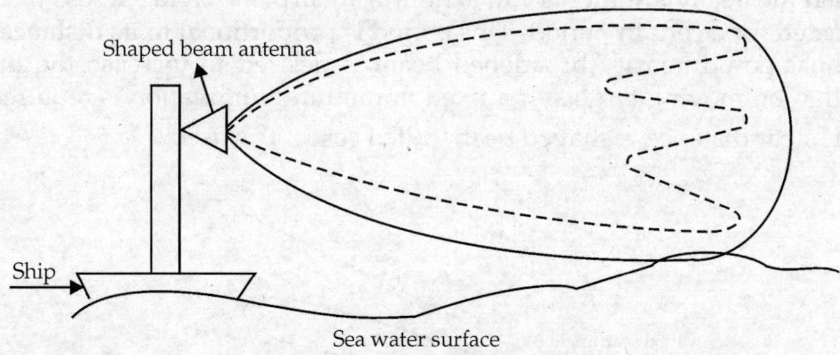

Fig. 7.16 *Sector beam for surface search from ship-borne antenna*

These sector beams are also used for surface search from ship-borne antennas and air search from ground-based antennas.

7.8.3 Cosecant Beams

These are used for ground mapping, airport surveillance purposes and so on. Typical cosecant beams are shown in Figs. 7.17 and 7.18.

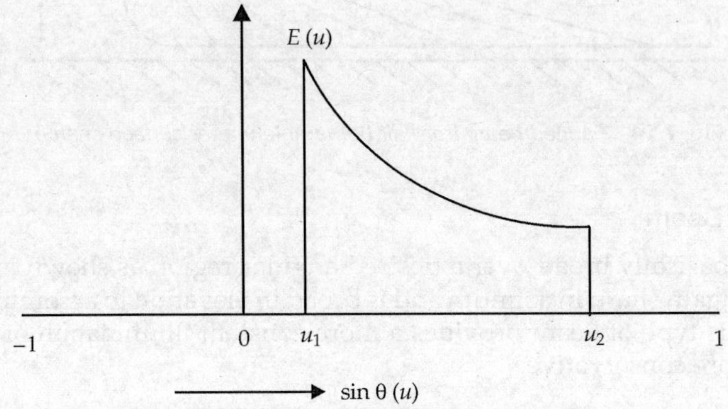

Fig. 7.17   *Cosecant beam*

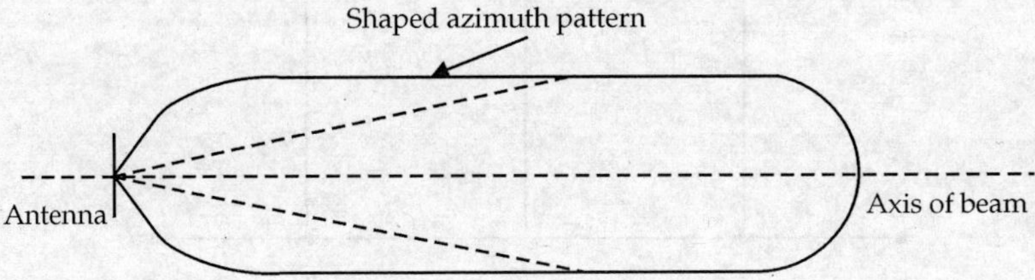

Fig. 7.18   *Double CSC beam*

These shaped beams can be precisely produced from array antennas with appropriate amplitude and phase distributions.

When the beam is stationary in azimuth, an airplane flying across the beam is illuminated for a certain period. This period is proportional to its distance from the antenna. Low intensity broadened beam is desired to increase the time of illumination on moving targets. If a fixed minimum illumination is required at a specified linear distance, a shaped beam called $\csc^2 \theta$ is used.

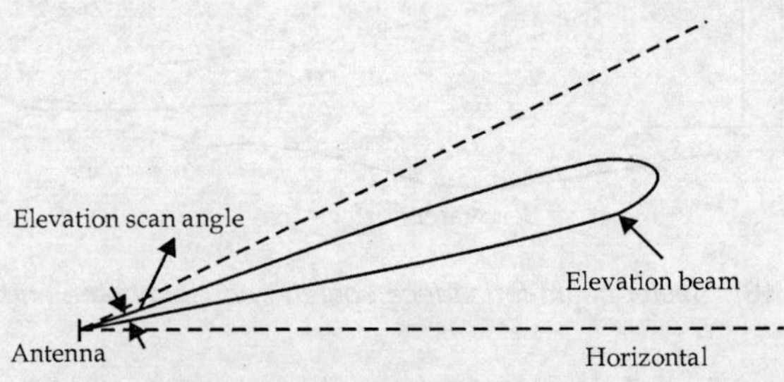

Fig. 7.19   *Elevation sharp beam*

For height finding applications, a ground-based or ship-borne antenna produces a sharp elevation beam for getting precise elevation information and a rapid elevation scan. A typical beam is shown in Fig. 7.19.

7.9 HORN ANTENNA

It is a radiating element which has the shape of a horn. It is a waveguide one end of which is flared out.

A waveguide, when excited at one end and open at the second end, radiates. However, radiation is poor and non-directive pattern results because of the mismatch between the waveguide and free space. The mouth of the waveguide is flared out to improve the radiation efficiency, directive pattern and directivity.

Types of horns:

- Sectoral horn
- Pyramidal horn
- Conical horn.

Sectoral horn is of two types:

(a) Sectoral H-plane horn

(b) Sectoral E-plane horn.

These are shown in Fig. 7.20 and Fig. 7.21.

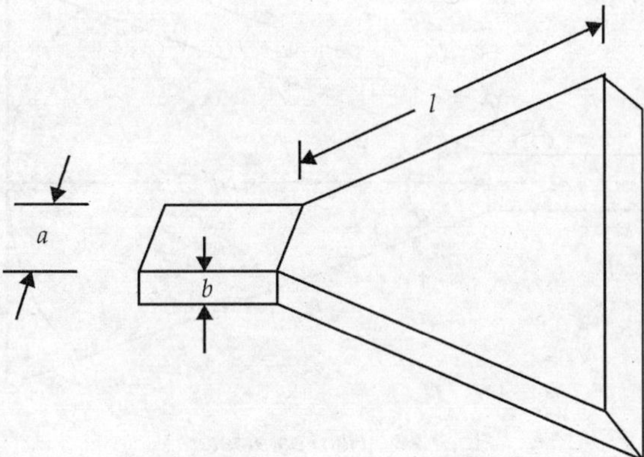

Fig. 7.20   *Sectoral H-plane horn*

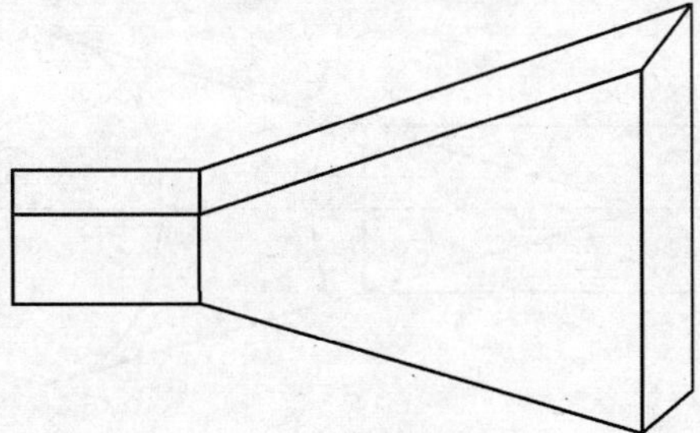

Fig. 7.21   *Sectoral E-plane horn*

A typical pyramidal horn is shown in Fig. 7.22.

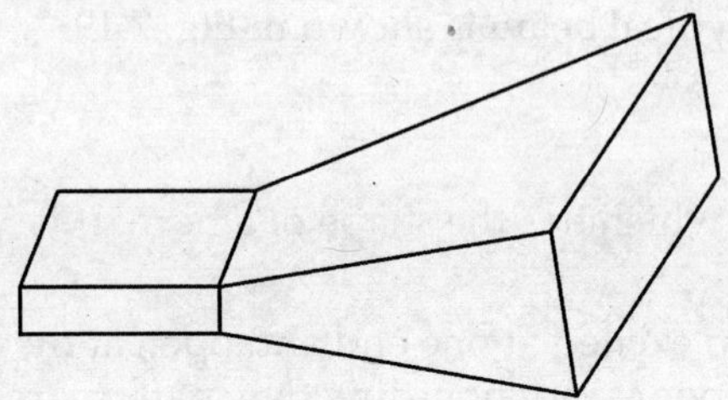

Fig. 7.22 *Pyramidal horn*

The horn parameters are described in Fig. 7.23.

δ = path difference
l = axial length
d = aperture dimension
θ = flare angle

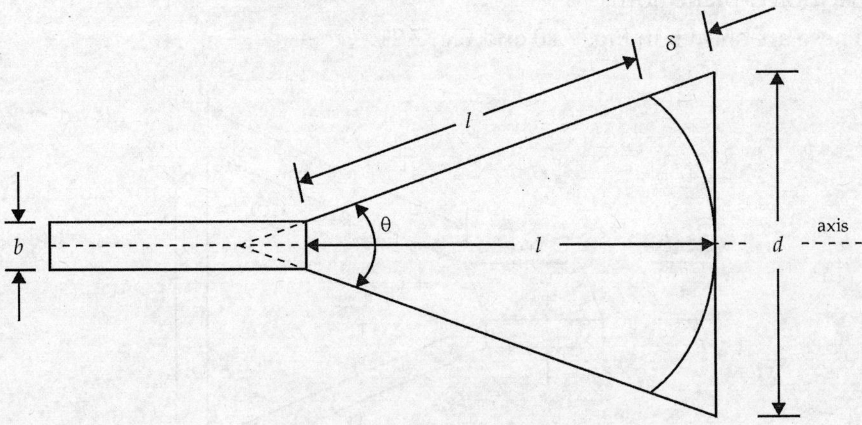

Fig. 7.23 *Horn parameters*

Typical conical horn is shown in Fig. 7.24.

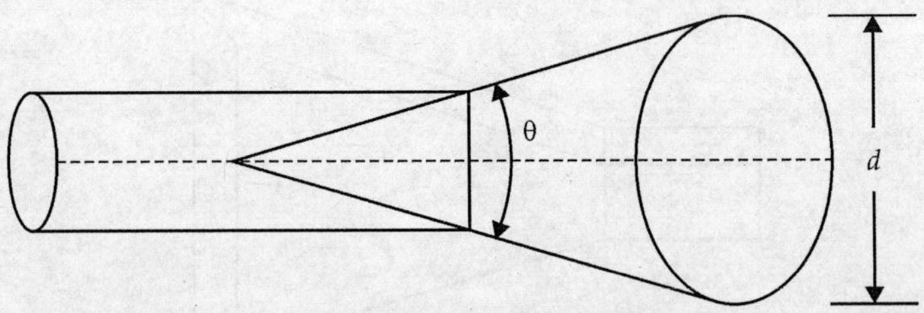

Fig. 7.24 *Conical horn*

Sectoral horn is a horn in which flaring exists only in one direction.

If flaring is along the direction of electric field, it is called **sectoral *E*-plane horn**.

If flaring is along the direction of magnetic field, it is called **sectoral *H*-plane horn**.

If flaring is along *E* and *H*, the horn is called **pyramidal horn**. It has the shape of a truncated pyramid.

If the walls of a circular waveguide are flared out, a **conical horn** is obtained.

Operation   The electromagnetic wave propagation in a waveguide is different from that of free space. In waveguide, propagation is restricted by the conducting walls and waves will not spread. After reaching the mouth, waves spread laterally and wave front becomes spherical. At the mouth of the wave-guide, there exists near-field region where the wave front is a complicated one. This is considered to be the transition region where change of propagation from waveguide to free space takes place. As the waveguide impedance and free space impedance do not match, walls of the waveguide are flared out. This flared out structure is called a horn. The flaring provides impedance matching, more directivity and narrow beam width. Horn produces a uniform plane wave front with a larger aperture in comparison to a waveguide. As the aperture is large, directivity is high.

The design equations of horn antenna are:

$$\theta = 2 \tan^{-1}\left(\frac{d}{2l}\right)$$

$$= 2 \cos^{-1}\left(\frac{l}{l+\delta}\right) \qquad \qquad ...(7.11)$$

and
$$l = \frac{d^2}{8\delta}$$

Half-power beam width of optimum flared horns are

$$\phi_E = \frac{56\lambda}{d_E} \text{ degrees}$$

$$\phi_H = \frac{67\lambda}{d_H} \text{ degrees} \qquad \qquad ...(7.12)$$

Here,
$$\phi_E = \text{HPBW in } E\text{-plane}$$

$$\phi_H = \text{HPBW in } H\text{-plane}$$

$$d_E = \text{aperture in } E\text{-plane in free space wavelength}$$

$$d_H = \text{aperture in } H\text{-plane in free space wavelength}$$

The directivity of horn is $\quad D = \dfrac{7.5 A_a}{\lambda^2}$

$$\qquad \qquad ...(7.13)$$

Power gain, $\qquad\qquad g_p = \dfrac{4.5A}{\lambda^2}$

Here, $A_a = d_E\, d_H$ = physical area of the aperture.

Applications of horns

1. Horns are used at microwave frequencies where moderate gains are sufficient.
2. They are used as feed elements.
3. They are often used in laboratories for the measurement of different antenna parameters.

Salient features of horns

1. Horn becomes small if the flare angle is small. Its radiation pattern is directive, wave front is spherical, mouth area is small and its directivity is small.
2. Flare angle is related to axial length.
3. If $\theta = 15^\circ$, when $l/\lambda = 50$, the beam width is 23° and directivity is 120.
4. Directivity of pyramidal horn is more as the flare is in more than one direction.
5. Its directivity is not as high as that of paraboloid.
6. It is used as radiator.
7. It is easy to use with the waveguide.
8. It is used as primary antenna for paraboloid.
9. The gain of the conical antenna is optimum for a given slant length of flare, l and

$$d \approx \left(\frac{3}{\lambda}\right)^{1/2} \qquad\qquad ...(7.14)$$

Here, d = diameter of the aperture.

10. The directivity of a loss-less horn antenna is its gain and it is given by

$$D = \frac{4\pi A_e}{\lambda^2}$$

$$= \frac{4\pi \eta_a}{\lambda^2} A_a$$

Here, A_e is effective aperture, m^2

A_a is actual area or physical area, m^2

η_a is aperture efficiency $= \dfrac{A_e}{A_a}$

λ is operating wavelength, m.

11. In the case of rectangular horn $A_a = d_E\, d_H$

Here, d_E is aperture size in E-plane

d_H is aperture size in H-plane

The rectangular horn is a pyramidal horn.

12. In the case of conical horn $A_a = \pi d$

    Here, d is aperture diameter.

13. If aperture efficiency is about 0.6,

$$D \approx 7.5\, A_a/\lambda^2.$$

7.10 CORRUGATED HORNS

Corrugation means groove. A typical corrugated horn is shown in Fig. 7.25.

ψ_c = half of the flare angle

t = width of the groove

w = separation between adjacent grooves

b_1 = aperture dimension

d = groove depth.

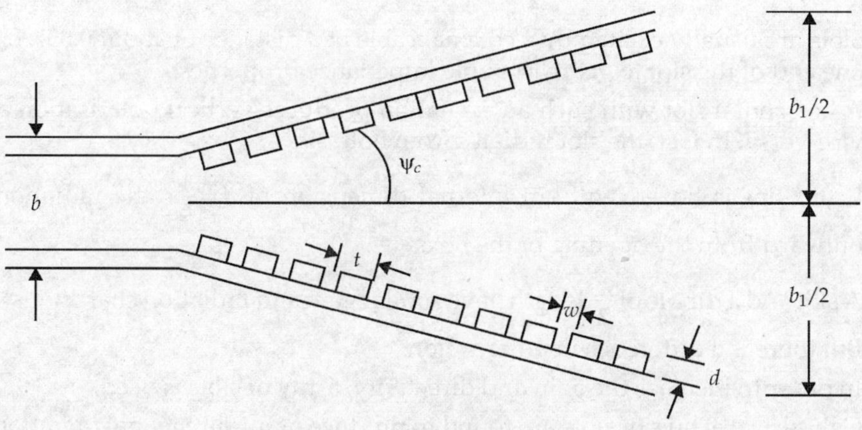

Fig. 7.25   *Corrugated horn*

Salient features of corrugated horn

1. When corrugated horns are used as feed antennas for paraboloids, spill-over efficiency and cross polarisation losses are reduced and aperture efficiency is increased. These systems are used in radio astronomy and satellite communications.

   Spill over is defined as part of the power radiated by a feed which is not intercepted by the secondary radiator.

2. For feed systems using conventional horns, the aperture efficiency is about 50-60%.

3. For feed systems using corrugated horns, the aperture efficiency is 75-80%.

4. The corrugations on the walls of a horn modify the E-field distribution in the E-plane from uniform at the waveguide horn junction to cosine at the aperture.

5. The corrugated surface reactance is given by

$$x = \frac{w}{w+t}\, \eta_0 \tan (kd) \qquad\qquad ...(7.15)$$

Here, $\dfrac{w}{w+t} = 1$, if $t \le w/10$, $w < \lambda/10$ and $w \ll d$

$$\eta_0 = 120\pi\,\Omega$$

$$k = 2\pi/\lambda.$$

7.11 SLOT ANTENNA

It is a radiating element formed by a slot in a metallic surface. An opening cut in a conducting sheet or in one of the walls of the waveguide acts as the antenna. It is excited suitably either by a co-axial cable or through the waveguide.

Salient features of slot antennas

1. A basic slot has a length of $\dfrac{\lambda}{2}$ and its width is much less than $\dfrac{\lambda}{2}$.

2. Slots are usually excited by a co-axial cable at a distance of about 0.05λ from one end of the slot to get reasonable impedance properties.

3. A horizontal slot with such an excitation produces vertical polarisation and vice-versa. In fact, the slot radiates from both sides.

4. If the slot is boxed with an internal dimension of $d = \dfrac{\lambda}{4}$, the radiation is outward from the opening of the box.

5. A slot and a dipole of $\dfrac{\lambda}{2}$ length have similar gain and radiation characteristics. But there is a difference in polarisation.

6. In order to increase the gain and directivity, array of slots is used.

7. Cylindrical arrays of slots are found to produce omni-directional radiation in the horizontal plane with horizontal polarisation.

8. When a high frequency field exists across a thin slot in a conducting plane, it radiates.

9. A slot excited by a two-wire line is shown in Fig. 7.26. The electric field in the $\dfrac{\lambda}{2}$ slot is sinusoidal.

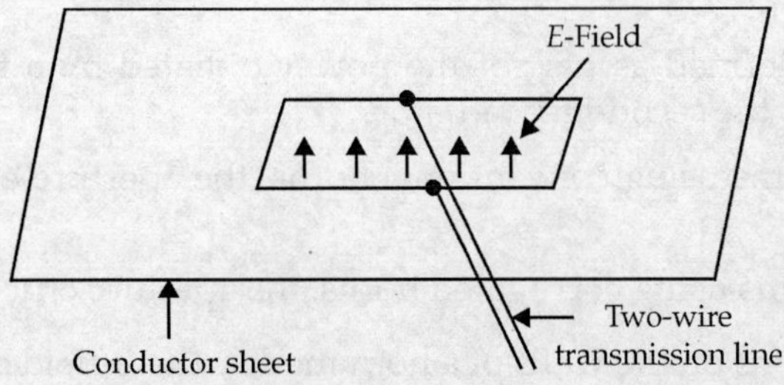

Fig. 7.26 *Slot antenna excited by two-wire line*

10. The impedance of such an excited slot is purely resistive and is 365 ohms.

A complementary dipole antenna made of a conductor is shown in Fig. 7.27.

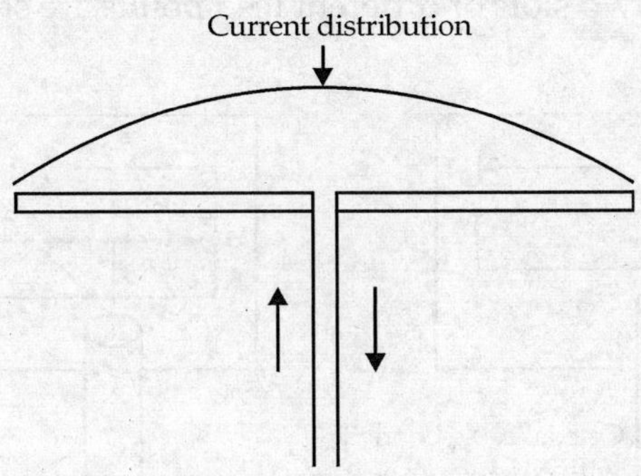

Fig. 7.27 *Dipole antenna*

A dipole and a slot in a conducting sheet have similar shapes except that the metal region and free space are interchanged as shown in Fig. 7.28.

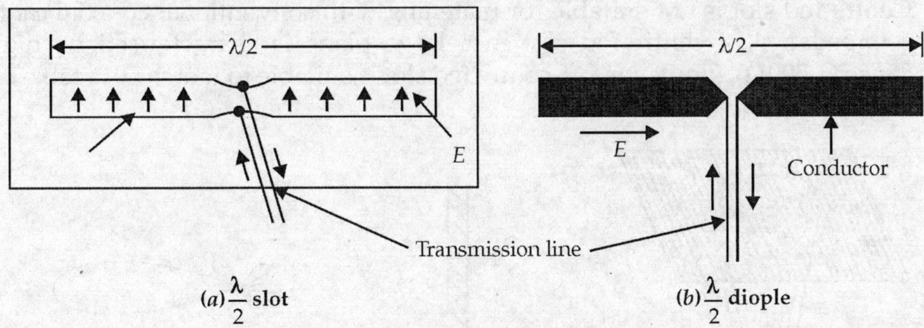

Fig. 7.28 $\frac{\lambda}{2}$ *slot and dipole antenna*

The main difference is that the slot of Fig. 7.28 (*a*) is vertically polarised and dipole of Fig. 7.28 (*b*) is horizontally polarised.

The slot and dipole impedances are related by

$$Z_s \, Z_d = \frac{\eta_0^2}{4}, \quad \eta_0 \text{ is free space} \qquad \qquad ...(7.16)$$

Intrinsic impedance $= 377\,\Omega$

If the width of the dipole is very small, $w \ll \lambda$.

Then $\qquad\qquad Z_d = 73 + j\,42.5\,\Omega$

Slot impedance, Z_s

$$Z_s = \frac{\lambda_0^2}{4Z_d} = \frac{(377)^2}{4\,(73 + j\,42.5)}$$

$$Z_s = 363 - j211 \text{ ohms}$$

By changing the feed point from one end of the slot, its impedance can be changed. For example, two slots of different feed points are shown in Figs. 7.29 (*a*) and (*b*).

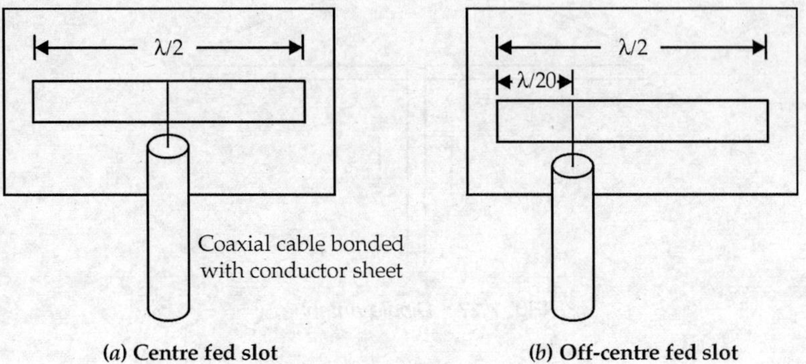

(*a*) Centre fed slot   (*b*) Off-centre fed slot

Fig. 7.29   *Feed points*

Centre-fed slot is not suitable for matching with conventional co-axial cable as the impedance of centre-fed slot in a large plane is characterised by high impedance ($\approx 500\,\Omega$). However, off-centre fed slot is suitable to match $50\,\Omega$ cable by choosing appropriate feed point.

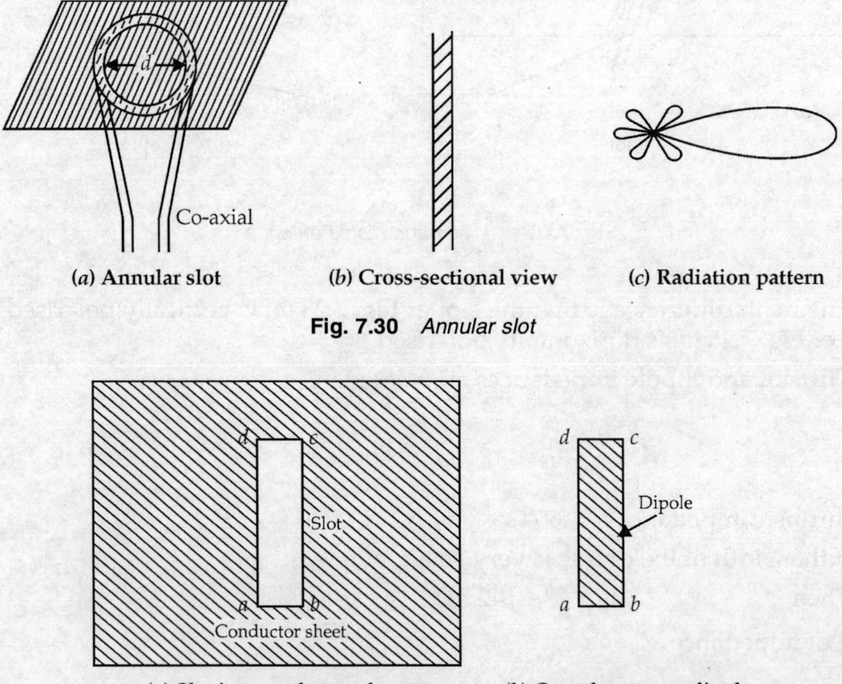

(*a*) Annular slot   (*b*) Cross-sectional view   (*c*) Radiation pattern

Fig. 7.30   *Annular slot*

(*a*) Slot in a conductor sheet   (*b*) Complementary dipole

Fig. 7.31   *Slot and complementary dipole*

In an annular slot, the fields are in the same phase. A typical annular slot is shown in Fig. 7.30. It produces a radiation pattern with narrow beam width.

A rectangular slot in a conducting sheet and its complimentary dipole are shown in Fig. 7.31.

7.12 IMPEDANCE OF SLOT ANTENNA

The slot impedance is

$$Z_s = \frac{\eta_0^2}{4Z_d}$$

$$= \frac{\eta_0^2}{4\,(R_d^2 + X_d^2)}\,(R_d - jX_d),\ \Omega \qquad\qquad ...(7.17)$$

Here, $\eta_0 = 120\pi\,\Omega$

Z_d = impedance of complementary dipole = $R_d + jX_d,\ \Omega$

R_d = real part of $Z_d,\ \Omega$

X_d = reactive part of $Z_d,\ \Omega$.

Proof   Consider a typical slot antenna and a complementary slot dipole of Fig. 7.32.

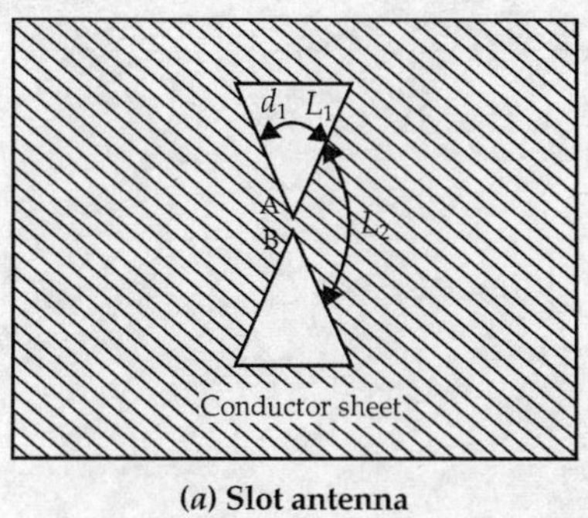

(a) Slot antenna

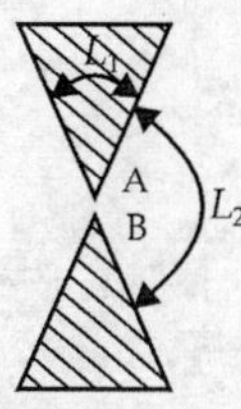

(b) Complementary slot dipole

Fig. 7.32 *Slot and slot dipole*

Let the slot and dipole be excited at AB.

Assume Z_s = slot impedance

Z_d = dipole impedance

V_s = slot terminal voltage

V_d = dipole terminal voltage

I_s = slot terminal current

I_d = dipole terminal current

$\mathbf{E}_s$ = electric field in the slot at any point

$$\mathbf{H}_s = \text{magnetic field in the slot at any point}$$

$$\mathbf{dL} = \text{differential length along } L_1$$

$2L_2 = $ path around a complete circle, V_s at AB is given by

$$V_s = \lim_{L_1 \to 0} \int_{L_1} \mathbf{E}_s \cdot \mathbf{dL} \qquad \qquad \text{...(7.18)}$$

and

$$I_s = \oint \mathbf{H}_s \cdot \mathbf{dL} \qquad \qquad \text{...(7.19)}$$

$$I_s = 2 \lim_{L_2 \to 0} \oint \mathbf{H}_s \cdot \mathbf{dL} \qquad \qquad \text{...(7.20)}$$

The dipole impedance, Z_d is

$$Z_d = \frac{V_d}{I_d} \qquad \qquad \text{...(7.21)}$$

Now

$$V_d = \lim_{L_2 \to 0} \int_{L_2} \mathbf{E} \cdot \mathbf{dL} \qquad \qquad \text{...(7.22)}$$

$$I_d = \oint \mathbf{H}_d \cdot \mathbf{dL} \qquad \qquad \text{...(7.23)}$$

$$= 2 \lim_{L_1 \to 0} \int \mathbf{H}_d \cdot \mathbf{dL}$$

As

$$\lim_{L_2 \to 0} \int_{L_2} \mathbf{E}_d \cdot \mathbf{dL} = \eta_0 \lim_{L_2 \to 0} \int_{L_2} \mathbf{H}_s \cdot \mathbf{dL} \qquad \qquad \text{...(7.24)}$$

and

$$\lim_{L_1 \to 0} \int_{L_1} \mathbf{H}_d \cdot \mathbf{dL} = \frac{1}{\eta_0} \lim_{L_1 \to 0} \int_{L_1} \mathbf{E}_s \cdot \mathbf{dL} \qquad \qquad \text{...(7.25)}$$

We have

$$V_d = \eta_0 \frac{I_s}{2} \qquad \qquad \text{...(7.26)}$$

or

$$\frac{V_d}{I_s} = \frac{\eta_0}{2}$$

Similarly,

$$\frac{V_s}{I_d} = \frac{\eta_0}{2} \qquad \qquad \text{...(7.27)}$$

Now we can write

$$\frac{V_d}{I_s} \cdot \frac{V_s}{I_d} = \frac{\eta_0}{2} \times \frac{\eta_0}{2} = \frac{\eta_0^2}{4}$$

That is,

$$Z_s \cdot Z_d = \frac{\eta_0^2}{4}$$

$$Z_s = \frac{\eta_0^2}{4Z_d} \qquad \text{...(7.28)}$$

If
$$Z_d = R_d + j\,X_d$$

Z_s becomes
$$Z_s = \frac{\eta_0^2}{4}\left[\frac{1}{R_d + j\,X_d}\right] \qquad \text{...(7.29)}$$

or
$$Z_s = \frac{\eta_0^2}{4}\left[\frac{R_d - j\,X_d}{R_d^2 + X_d^2}\right]$$

$$Z_s = \frac{\eta_0^2}{4\,(R_d^2 + X_d^2)}\,[R_d - j\,X_d]. \qquad \text{...(7.30)}$$

7.13 IMPEDANCE OF A FEW TYPICAL DIPOLES

1. If the length of dipole, $l = \dfrac{\lambda}{2}$, diameter of dipole, $d = 0$

$$Z_d = 73 + j\,42.5\,\Omega.$$

2. If $\qquad l \leq \dfrac{\lambda}{2},\ \dfrac{l}{d} = 100$

$$Z_d \approx R_d = 67\,\Omega.$$

3. If $\qquad l \approx \lambda,\ \dfrac{l}{d} = 28$

$$Z_d = 710\,\Omega.$$

The length, diameter and impedance of a cylindrical dipole are shown in Fig. 7.33.

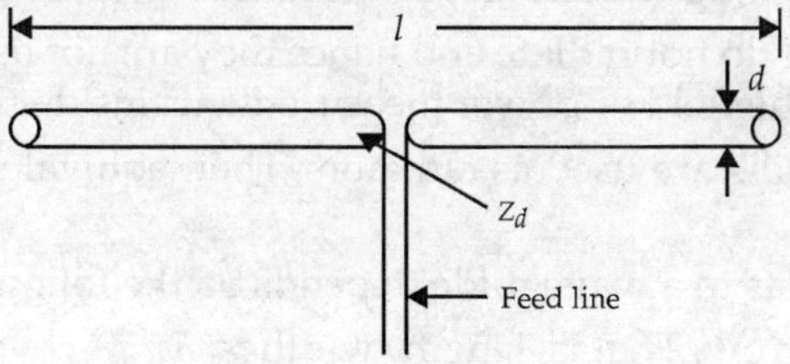

Fig. 7.33 *Cylindrical dipole*

7.14 SLOTS IN THE WALLS OF RECTANGULAR WAVEGUIDE

The slots can be cut in one of the walls of a rectangular waveguide. Different slot configurations are shown in Fig. 7.34.

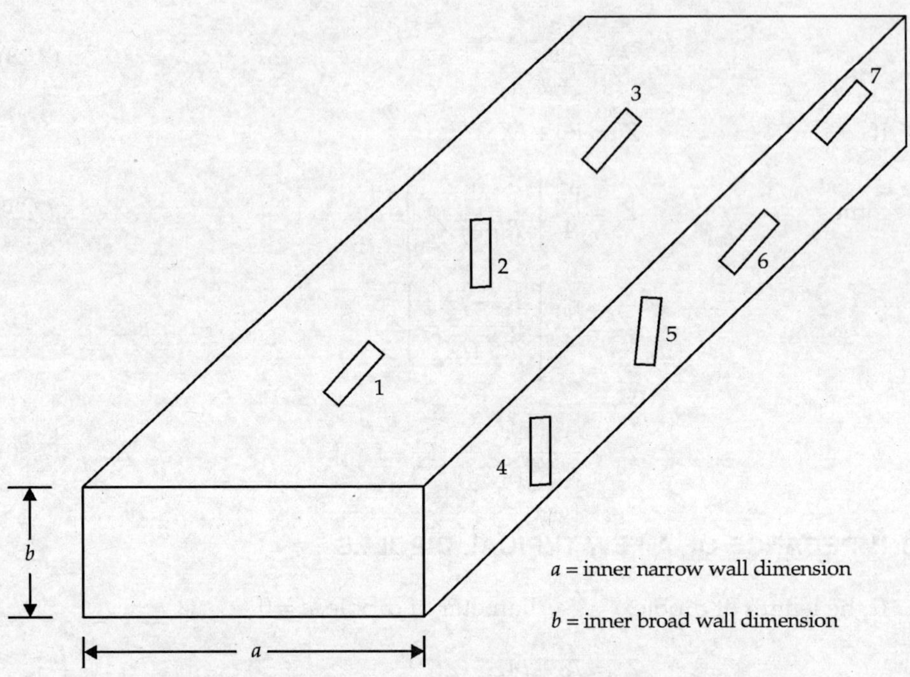

Fig. 7.34   *Slot configurations in a waveguide*

The types of slots are:

Slot 1: Longitudinal slot in the centre of broad wall.

Slot 2: Inclined slot in the broad wall.

Slot 3: Displaced slot in the broad wall.

Slot 4: Vertical slot in the narrow wall of the waveguide.

Slot 5: Inclined slot in the narrow wall of the waveguide.

Slot 6: Longitudinal slot in the centre of the narrow wall of the waveguide.

Slot 7: Displaced longitudinal slot in the narrow wall of the waveguide.

The slots 1 and 4 do not radiate and hence they are not used as radiators. But they can be used for the field or power measurements inside the guide.

Slots 2, 3, 5, 6 and 7 are used as radiators. Their equivalent circuits are given in Fig. 7.35.

The theory of slots in a waveguide depends on the following assumptions:

1. The slots are narrow. That is, length/width $\gg 1$.

2. The slot length $\approx \lambda/2$.

3. The field in the slot is transverse to the long dimension. It varies sinusoidally and is independent of the exciting system.

4. The guide walls are perfectly conducting and infinitely thin.

5. The field in the region behind the face containing the slot is negligible compared to the field outside the guide.

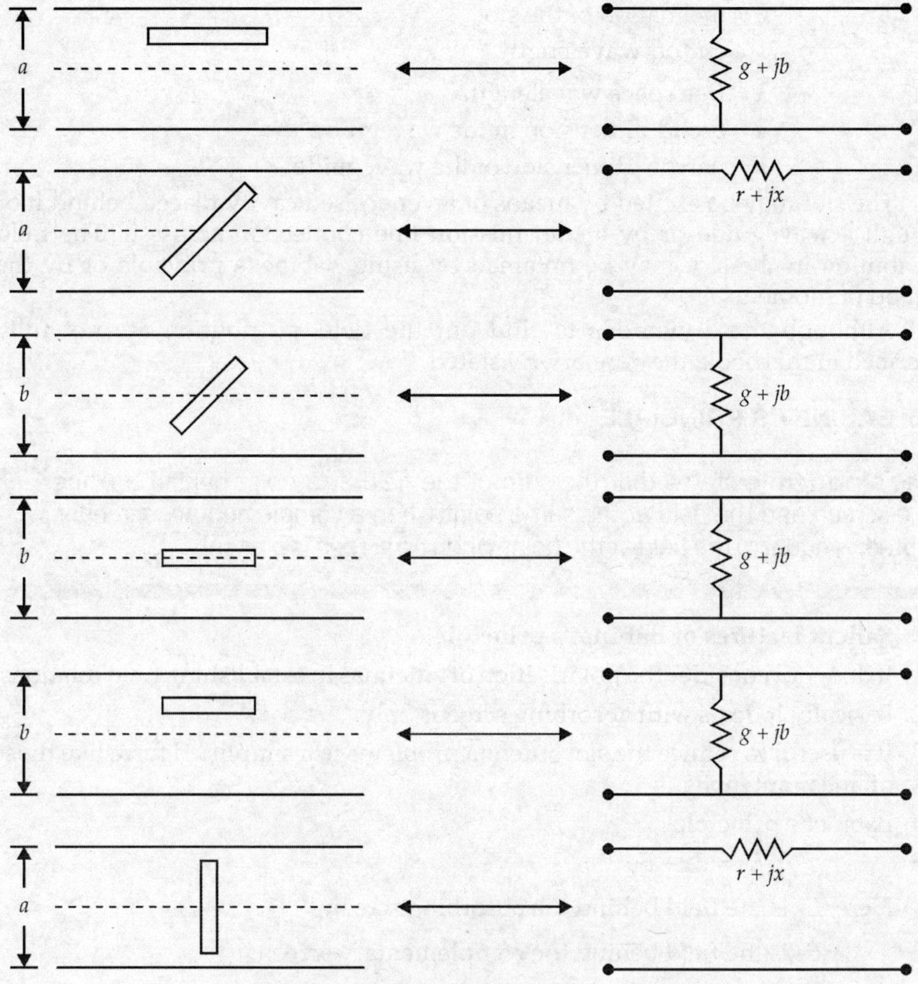

Fig. 7.35   *Slots and their equivalent circuits*

The conductance of an inclined slot in the narrow wall is given by

$$g = \frac{30}{73\pi} \left(\frac{\lambda_g}{\lambda} \right) \frac{\lambda^4}{a^3 b} \left[\frac{\sin\theta \cos\left(\dfrac{\pi\lambda}{2\lambda_g} \sin\theta \right)}{1 - \left(\dfrac{\lambda}{\lambda_g} \right)^2 \sin^2\theta} \right]^2 \qquad \ldots(7.31)$$

Here, λ_g = guide wavelength

$$= \frac{\lambda}{\sqrt{1 - \left(\dfrac{\lambda}{\lambda_c} \right)^2}}$$

$$\theta = \text{inclination of the slot}$$

$$\lambda_c = \text{cut-off wavelength}$$

$$\lambda = \text{free space wavelength}$$

$$a = \text{broad dimension in the waveguide}$$

$$b = \text{narrow dimension of the waveguide}$$

The slot may be excited by means of an energised cavity placed behind it or through a waveguide or by a transmission line connection across it. The field distribution in the slot may be obtained by using Babinet's principle or by the method of moments.

Although these methods to find out the field distribution are not fully described in this book, they are briefly stated.

7.15 BABINET'S PRINCIPLE

Babinet's principle states that the sum of the field at a point behind a plane having a screen and the field at the same point when a complementary screen is substituted, is equal to the field at the point when no screen is present.

Salient features of Babinet's principle

1. It does not consider the polarisation of antennas in establishing field relations.
2. Basically it deals with absorbing screens only.
3. It is useful to reduce the slot antenna problems to a simplified form like those of linear antennas.
4. Babinet's principle,

$$E_3 = E_1 + E_2$$

Here, E_1 is the field behind the absorbing screen

E_2 is the field behind the complementary screen

E_3 is the field when there is no screen.

5. It is not applicable in the presence of the conducting screens.
6. It is valid in optics.
7. It cannot be extended to electromagnetic problems as it is not possible to have absorbing screens at radio frequencies and as polarisation of electromagnetic fields is common.

Meaning of Babinet's principle

If S_1 is a plane of screen,

S_2 is a plane of observation,

S is a point source,

E_1 is the field at a point (x, y, z) behind the screen,

E_2 is the field at (x, y, z) behind the screen when S_1 is replaced by its complementary screen,

E_3 is the field at the same point (x, y, z) when no screen is present.

Babinet's principle states that,
$$E_3 = E_1 + E_2$$

This is illustrated in Fig. 7.36. This principle is valid at the points in the plane S_2 and also at the points behind the screen S_1.

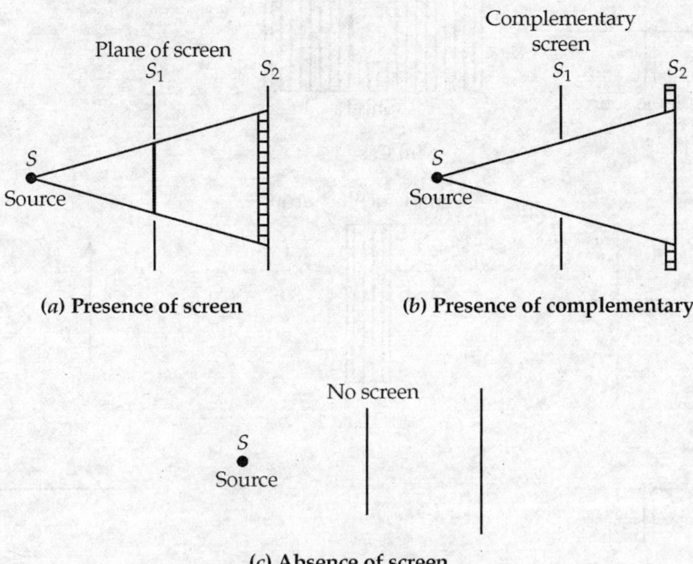

Fig. 7.36   *Illustration of Babinet's principle*

This principle is applicable to the points in the plane S_2 and also to the points behind the screen S_1.

At radio frequencies, it is not possible to have perfectly absorbing screens. Hence, **Babinet's principle** in optics cannot be just extended to the electromagnetic problem as it does not take care of polarisation.

A principle valid for conducting screens and polarised fields is described by **H.G. Booker**.

Illustration of Booker's theory   **Booker's extension** of Babinet's principle is illustrated below:

Consider three cases. The source is a short dipole (Fig. 7.37).

If the screen in an infinite conducting thin sheet with a vertical slot is considered, the field behind the screen at a point, P is E_1.

If the slot is replaced by a complimentary dipole (a thin strip of conducting sheet), the field is E_2 at the same point, P behind the screen.

If there is no screen, the field is E_3 at the same point. Then according to Babinet's principle,
$$E_1 + E_2 = E_3$$

The same principle can be applied to points in front of the screens. When transmitted wave is very high through the slot, $E_1 \approx E_3$. Under these conditions, the complementary dipole acts like a reflector and E_2 is small.

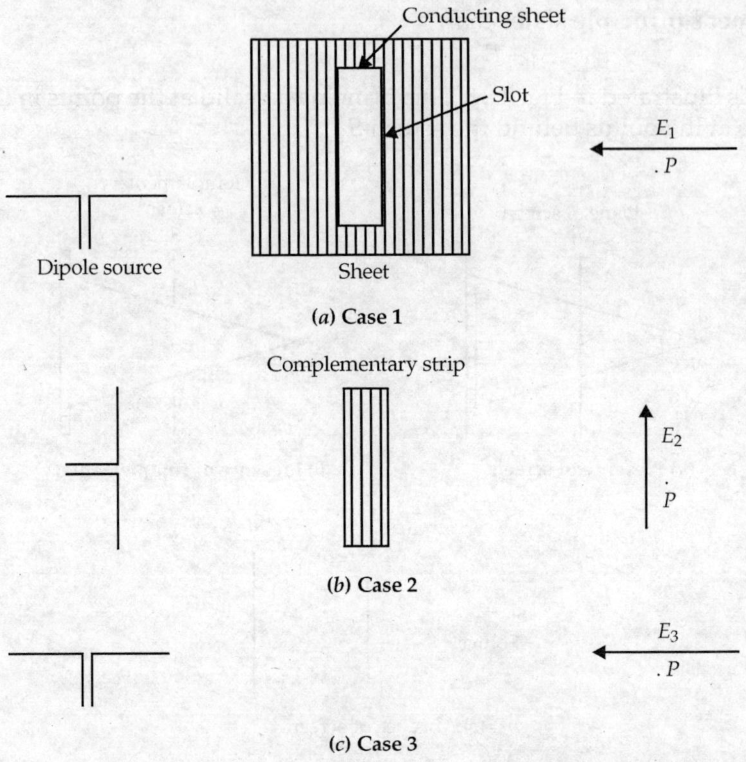

Fig. 7.37 *Illustration of Booker's principle applied to slotted structures*

When the screen and its complement are immersed in a medium whose intrinsic impedance is η, Z_s and Z_c are related by

$$Z_s Z_c = \frac{\eta^2}{4}.$$

...(7.32)

7.16 THE METHOD OF MOMENT (MOM)

This method is used to solve integral equations.

An integral equation is an equation which contains an unknown function under the integral sign. This method is useful for finding field distribution in the slot.

Consider an equation

$$f(g) = a$$

...(7.33)

Here, f is a known linear integral operator, a is a known excitation function, and g is the response function. The aim is to find g once f and a are specified. The linearity of the operator f makes it possible to have a numerical solution.

The method of moments requires that the unknown response function, g is expanded as a linear combination of N terms and it may be written as

$$g(x) = c_1 g_1(x) + c_2 g_2(x) + \dots + c_N g_N(x)$$

$$= \sum_{n=1}^{N} c_n g_n(x) \qquad \qquad \text{...(7.34)}$$

Here, c_n is an unknown constant and $g_n(x)$ is a known function. It is called a basis or expansion function. The domain of $g_n(x)$ is the same as that of $g(x)$. Substituting Equation (7.34) in Equation (7.33)

we get
$$\sum_{n=1}^{N} C_n f(g_n) = a \qquad \qquad \text{...(7.35)}$$

The basis function g_n are chosen in such a way that $f(g_n)$ in Equation (7.35) can be found easily. It can be evaluated numerically. Now the problem is to find the unknown constant c_n.

7.17 LENS ANTENNA

It is an antenna which consists of electromagnetic lens with a feed. In other words, a lens antenna is a three dimensional electromagnetic device whose refractive index is different from unity. It acts like a glass lens in optics.

They are usually made of polystyrene or Lucite. They are used to:

1. Control the aperture illumination.
2. Collimate the electromagnetic rays.
3. Produce directional characteristics.
4. Converge the incoming wavefront at its focus.
5. Produce plane wavefront from spherical wavefront.

Principle   A lens antenna in association with a primary feed antenna in transmitting mode is shown in Fig. 7.38.

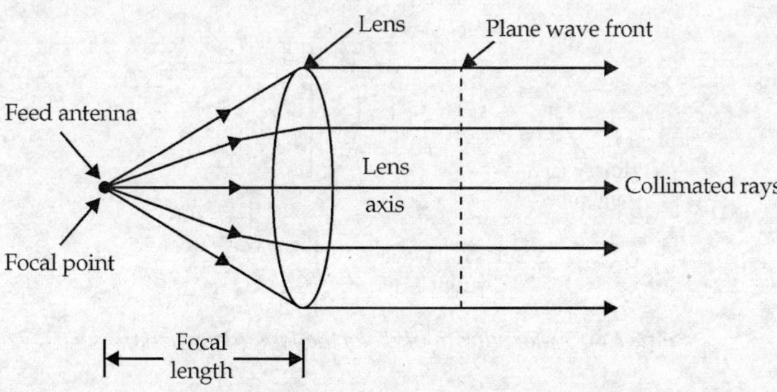

Fig. 7.38 *Lens antenna in transmitting mode*

When the feed antenna is kept at the focal point of the lens antenna, the diverging rays (spherical wave front) are collimated (parallel rays) forming a plane wave front after their incidence on the lens and passing through it. Collimation occurs because of refraction mechanism.

Lens antenna in receiving mode is shown in Fig. 7.39.

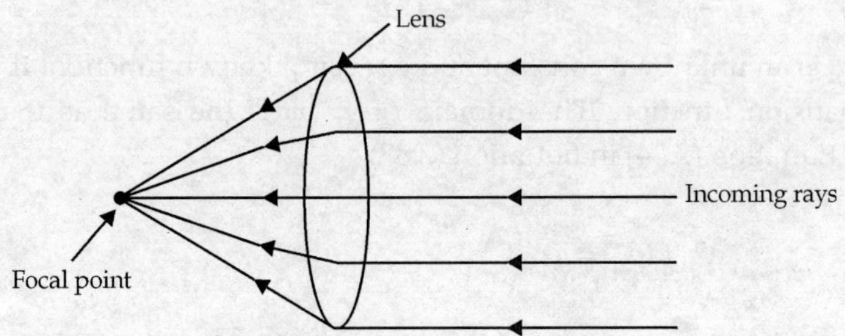

Fig. 7.39 *Lens antenna in receiving mode*

Here again, incoming parallel rays converge at the focal point after passing through the lens due to refraction mechanism. This is an indication of the validity of reciprocity theorem. Collimation is also possible if the lens has refractive index less than unity. Lens antennas are used in association with a point source (ideally). But in practice, it is used with horn-like antennas.

7.17.1 Types of Lens Antennas

These are of two types:

1. Dielectric lens.
2. Metal plate lens.

1. **Dielectric lens** are also known as the **delay lens**. Here, outgoing electromagnetic rays are collimated and are retarded or delayed by the lens material or media. This is illustrated in Fig. 7.40.

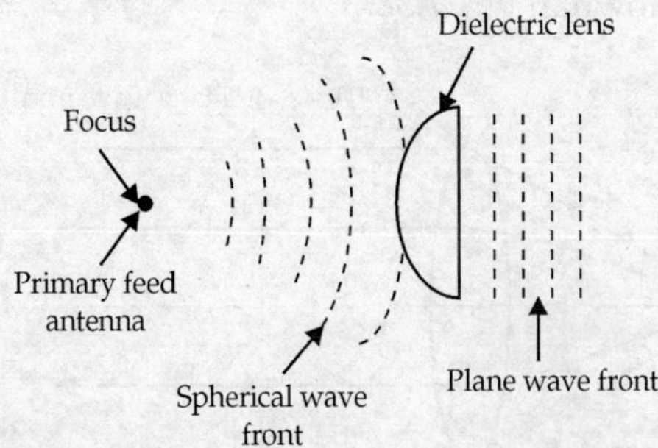

Fig. 7.40 *Mechanism of dielectric lens (delay lens)*

2. *E-plane metal plate lens*   In this, outgoing wavefront is speeded up by the lens material. This is illustrated in Fig. 7.41.

Dielectric lens are again of two types:

(*a*) Non-metallic dielectric type.

(*b*) Metallic type.

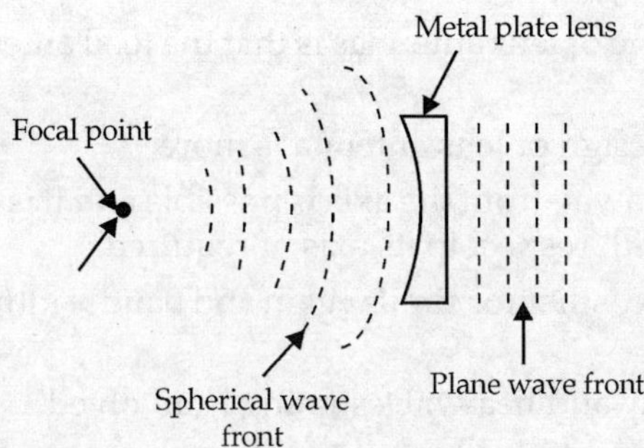

Fig. 7.41   *Mechanism of E-plane metal plate lens*

Salient features of dielectric lens

1. They are usually made of polystyrene or lucite and polyethylene.

   Dielectric constant of polystyrene $\in_r = 2.56$, refractive index, $n = 1.6$

   For polyethylene, $\in_r = 2.25$, $n = 1.5$.

2. They are bulky and heavy for $f < 3$ GHz.

3. Uniform illumination for lens antennas is better if focal length is long.

4. At $f < 10$ GHz the lens become excessively and undesirably thick.

5. This can be avoided by zoned or stepped dielectric lens as in Fig. 7.42.

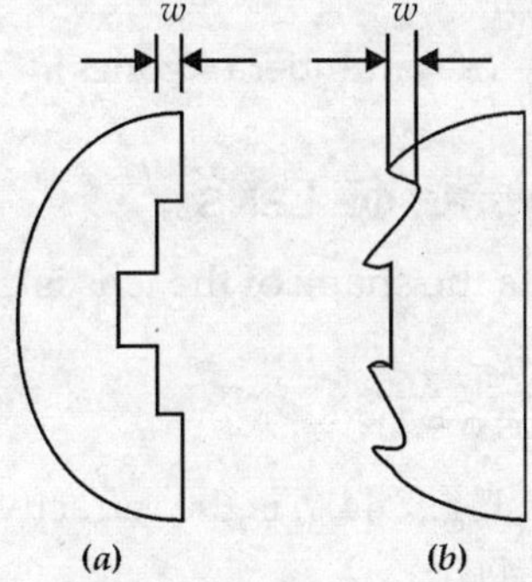

Fig. 7.42   *Zoned or stepped dielectric lens antenna*

6. The width w of a stepped lens is

$$w = \frac{\lambda}{n-1} \qquad \qquad ...(7.36)$$

7. The zoned lens of Fig. 7.42 (*b*) is mechanically stable.

8. Weight of stepped lens is less. Power dissipation is also less.

9. The main disadvantage of dielectric lens antenna is its frequency sensitivity.

10. Un-stepped dielectric lens antenna is wide-band and its shape does not depend on wavelength.

11. The main advantage of lens antennas is that the feed antennas do not obstruct the aperture.

12. Tolerance in the design of lens antenna is more.

13. Feeding at a point away from the axis is possible and it is suitable to move the beam angularly with respect to its axis, if required.

14. Lens antennas are costlier for similar gain and band width in comparison with reflector antennas.

15. They are also bulky and heavy. Designing is involved.

16. A typical horn feed is shown in Fig. 7.43. To avoid undesirable radiation from the sides, horn mouth is extended upto the location of the lens.

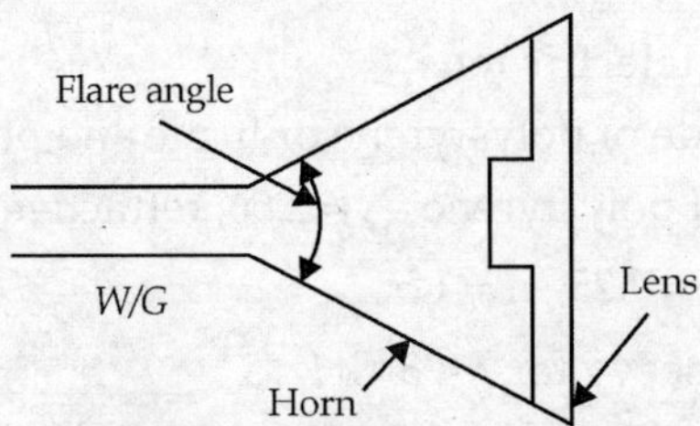

Fig. 7.43   *Horn of lens antenna*

17. The band width of zoned lens is

$$\text{B.W.} = \frac{50n}{1 + Kn} \ (K = \text{number of zones}).$$

7.18 EQUATION OF THE SHAPE OF LENS

The equation which represents the shape of the lens is given by

$$r = \frac{l\,(n-1)}{(n\cos\phi - 1)} \qquad \qquad ...(7.37)$$

Here, r, l, ϕ are shown in Fig. 7.44. n is the refractive index of lens.

Proof   Consider Fig. 7.44.

The electrical paths of all rays from S to AB are equal so that the field is in phase on the entire plane surface. The outgoing rays from the source will have constant phase across the aperture, d for suitably shaped lens.

If the velocity of electromagnetic wave in air is v_0 and in lens medium if it is v, then due to equality of electrical paths

$$SQ + QQ' = SB + BP' = SB + BP + PP'$$

That is, $\qquad SQ = SB + BP$

or $\qquad\qquad\qquad \dfrac{r}{v_0} = \dfrac{l}{v_0} + \dfrac{x}{v}$

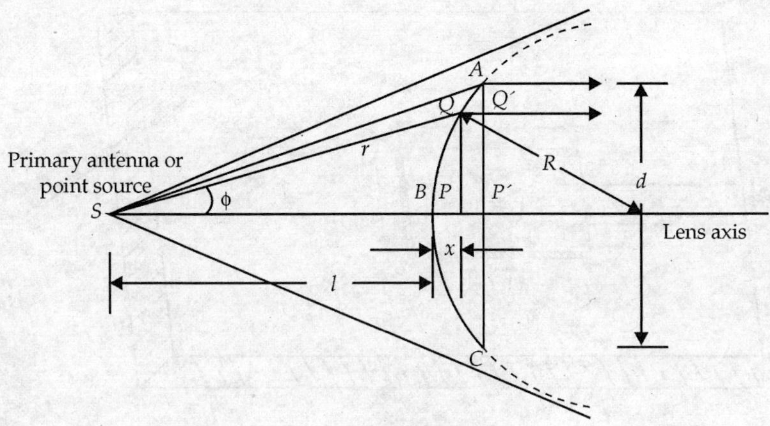

Fig. 7.44   *Ray paths in lens*

or
$$r = l + \left(\frac{v_0}{v}\right) x$$

But the refractive index,

$$n = \left(\frac{v_0}{v}\right)$$

and
$$x = SP - SB = r \cos \phi - l$$

Hence
$$r = l + nx$$
$$= l + n\,(r \cos \phi - l)$$

or
$$r = \frac{l\,(n-1)}{(n \cos \phi - 1)} \qquad\qquad ...(7.38)$$

Hence proved.

This is the equation of hyperbola of focal length, l and radius of curvature, R

$$R = l\,(n-1) \quad \text{if } \phi \text{ is small} \qquad\qquad ...(7.39)$$

7.19 MICROSTRIP OR PATCH ANTENNAS

These are antennas made from patches of conducting material on a dielectric substrate above a ground plane.

Construction   It consists of a very thin ($l \ll \lambda$) metallic strip called a patch placed above a ground plane. The strip and ground plane are separated by a dielectric sheet called substrate as shown in Fig. 7.45. The radiating element and feed lines are usually photoetched on the dielectric substrate. The wide use of printed circuits led to the construction of radiating elements and inter-connecting transmission lines using similar technology.

The side view of a patch antenna is shown in Fig. 7.46.

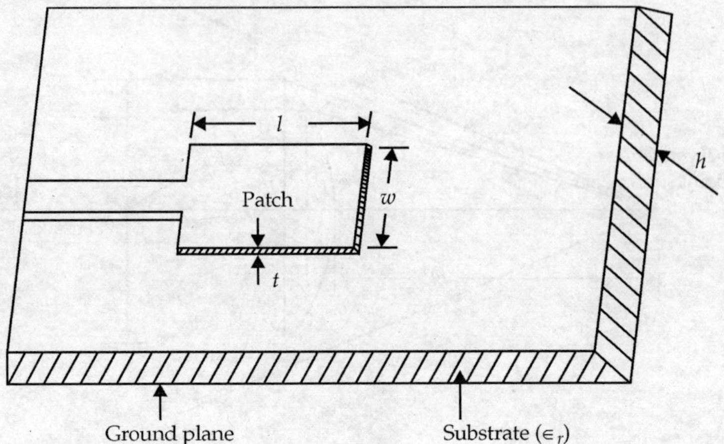

Fig. 7.45 *Microstrip antenna*

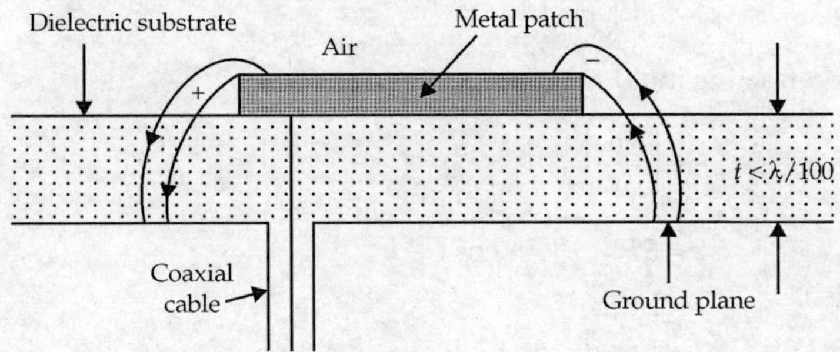

Fig. 7.46 *Side view of a patch antenna*

Typical patch parameters:

1. Dielectric constant of substrate, $\in_r \approx 2$.

2. The thickness of the patch, $t \approx \lambda/100$.

3. The height of the substrate, $h \ll \lambda$.

4. $\lambda < 3$ m.

5. $f \approx 100$ MHz.

6. $l \leq \lambda/2$.

7. $w < \lambda$.

Applications

1. They are used in spacecrafts and aircrafts.

2. They are used in applications where aerodynamic drag due to antennas should be nil.

3. They are used in telemetry, satellite communications and defence radar systems to operate over a frequency range of 1 to 10 GHz.

Advantages

1. Small size.
2. Low weight.
3. Low cost.
4. Ease of installation.
5. It does not give rise to aerodynamic drag when used in aircrafts.
6. These are low profile antennas.
7. These antennas can be flush mounted to a metallic conductor or to other surfaces. They do not require space for feed line. The feed line can be placed behind the ground plane.

Disadvantages

1. Their efficiency is less.
2. Their band width is small and is typically a few percent.

Shapes of patch antennas

1. Square	2. Circular
3. Elliptical	4. Rectangular
5. Triangular	6. Diamond.

These are shown in Fig. 7.47.

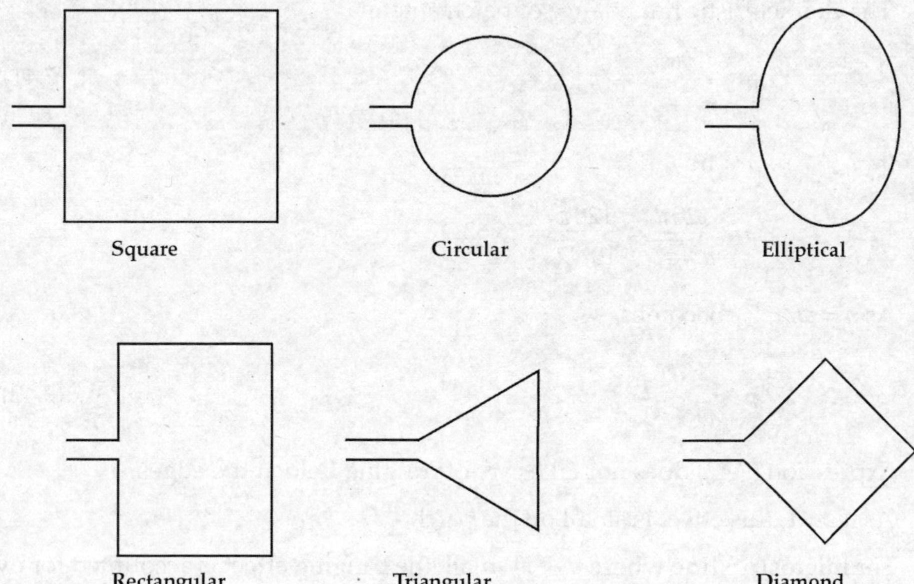

Fig. 7.47 *Different shapes of patch antennas*

The patch antenna acts as a resonant $\lambda/2$ parallel-plate microstrip transmission line. Its characteristic impedance Z_0 is equal to the reciprocal of the number, n of parallel field cell transmission lines.

Each field cell transmission line has a characteristic impedance, Z_i of the medium where

$$Z_i = \sqrt{\frac{\mu}{\epsilon}} = \eta_0 \sqrt{\frac{\mu_r}{\epsilon_r}} \; \Omega \qquad \qquad ...(7.40)$$

An end view of the patch antenna from the left side is shown in Fig. 7.48.

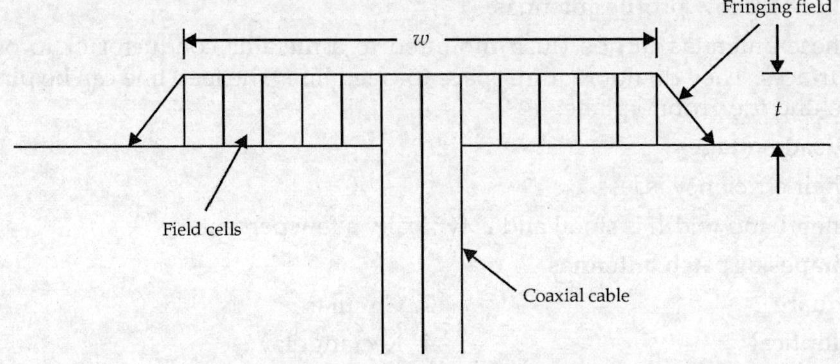

Fig. 7.48 *End view of patch antenna*

The characteristic impedance of patch antenna

$$Z_P = \frac{Z_0}{n \sqrt{\epsilon_r}} \qquad \qquad ...(7.41)$$

If $\qquad n = 10$

$$Z_P = \frac{120\pi}{n \sqrt{\epsilon_r}} = \frac{120\pi}{10 \sqrt{2}} = 26.63\Omega$$

As $n = w/t$, Z_P becomes

$$Z_P = \frac{Z_0 \, t}{w \sqrt{\epsilon_r}} \qquad \qquad ...(7.42)$$

Expression (7.42) does not take care of fringing field at the edges.

As $w \gg t$, this effect is small on the patch.

For **microstrip line** where w/t is small, the fringing effect is accounted for by approximately adding two cells. This gives more accurate formula for the impedance of the microstrip line. That is given by

$$Z_P \text{ (for microstrip line)} = \frac{\eta_0}{\sqrt{\epsilon_r} \left[\dfrac{w}{t} + 2 \right]} \qquad \qquad ...(7.43)$$

Radiation pattern of the patch antenna is broad. Its typical value of beam area Ω_A is 1/2 of half space of or π sterradian.

Therefore, the directivity, D of the patch antenna,

$$D = \frac{4\pi}{\Omega_A} = \frac{4\pi}{\pi} = 4$$

$$D = 10 \log_{10} 4 = 6.021 \text{ dB}$$

The effective height h_e of the antenna is

$$h_e = \sqrt{\frac{2R_r A_e}{\eta_0}} \qquad \qquad \text{...(7.44)}$$

Here, $\qquad R_r$ = radiation resistance, Ω

A_e = effective aperture, λ^2

η_0 = intrinsic impedance of free space, Ω

$R_r = 50\,\Omega$ for a typical patch

$$A_e = \frac{D\lambda^2}{4\pi} = \frac{4\lambda^2}{4\pi} = \frac{\lambda^2}{\pi}$$

$$A_e = \frac{\lambda^2}{\pi} \qquad \qquad \text{...(7.45)}$$

For matching purposes, feed point can be moved from the edge. Arrays of patches can be used to have more directivity.

Monolithic Microwave Integrated Circuit (MMIC) technology is used to manufacture patch antennas and associated circuitry in a compact form. These are used for frequencies ranging from 50 MHz to 100 GHz.

Printing technology is suitable for a variety of antenna elements like dipoles.

The conductance of the patch antennas is given by

$$g = \frac{1}{90} \left(\frac{w}{\lambda} \right)^2 \quad \text{for } w \ll \lambda$$

$$g = \frac{1}{120} \left(\frac{w}{\lambda} \right) \quad \text{for } w \gg \lambda$$

Methods of band width control   It is a narrow band width antenna. However, band width can be increased by:

1. increasing the thickness of parallel plate transmission line

2. cutting holes or slot in the patch. (These holes or slots increase its inductance)

3. using high dielectric constant ($\in_r$) substance

4. adding reactive component to the patch. (This reduces VSWR).

 **POINTS TO REMEMBER**

1. Microwave region extends from 1 GHz to 100 GHz.

2. Parabolic dish antennas are popular to produce narrow beams in the microwave region.

3. Reflectors are used to modify the radiation of the primary antenna.

4. Cassegrain feed mechanism is very popular for low noise applications.

5. Half power beam width of paraboloid is $\phi = \dfrac{70\lambda}{D_a}$.

6. Beam width from null-to-null of paraboloid is $\phi_0 = \dfrac{140\lambda}{D_a}$.

7. Directivity, $D = 9.87 \left(\dfrac{D}{\lambda}\right)^2$.

8. Power gain of paraboloid is $g_p = 6.4 \left(\dfrac{D_a}{\lambda}\right)^2$.

9. Paraboloids are popular in TV reception and radars.

10. Shaped beam antennas are useful to produce desired shapes of the beam.

11. Narrow beams are useful for point-to-point communication and high angular resolution radars.

12. Sector beams are more useful in search radars.

13. Cosecant beams are useful for ground mapping and airport surveillance.

14. Horn is a flared out waveguide.

15. The efficiency of corrugated horn is more than that of conventional horn.

16. The directivity of horn is $D = \dfrac{7.5A}{\lambda^2}$.

17. The power gain of horn is $g_p = \dfrac{4.5A}{\lambda^2}$.

18. Slot antennas are compact and slot arrays are popular in marine radars.

19. The method of moment (MOM) is useful for finding field distribution in slots and other wire antennas.

20. Lens antennas are popular to convert spherical wave front to plane wave front.

21. Polystyrene and lucite are popular lens materials.

22. The shape of the lens is represented by $r = \dfrac{l\,(n-1)}{(n\cos\phi - 1)}$.

23. Microstrip antennas are popular in cellular phones and for installation on body of aircrafts and so on.

24. The desired polarisation can be obtained by different shapes of the microstrip antenna.

25. Characteristic impedance of patch antenna is $Z_p = \dfrac{Z_0}{n\sqrt{\epsilon_r}}$.

 ## SOLVED PROBLEMS

Problem 7.1   Find the null-to-null main beam width of 2 m paraboloid reflector used at 5 GHz. Also find the half power beam width.

Solution   Frequency, $f = 5\ \text{GHz}$

$$\lambda = \frac{3 \times 10^8}{5 \times 10^9} = 0.06\ \text{m}$$

$$\text{BWFN} = 140 \times \left(\frac{\lambda}{D}\right) = 140 \times \frac{0.06}{2} = 4.2^{\circ}$$

$$\boxed{\text{BWFN} = 4.2^{\circ}}$$

$$\text{HPBW} = 70 \times \left(\frac{\lambda}{D}\right) = 70 \times \frac{0.06}{2.0} = 2.1^{\circ}$$

$$\boxed{\text{HPBW} = 2.1^{\circ}.}$$

Problem 7.2   Find the gain of a paraboloid of 2 m diameter operating at 5 GHz when half-wave dipole feed is used.

Solution   Frequency, $f = 5\ \text{GHz}$

$$\lambda = \frac{3 \times 10^8}{5 \times 10^9} = 0.06\ \text{m}$$

The gain of the paraboloid is

$$g_p = 6.4\left(\frac{D}{\lambda}\right)^2 = 6.4 \times \left(\frac{2}{0.06}\right)^2$$

$$\boxed{g_p = 7111.1 = 38.51\ \text{dB.}}$$

Problem 7.3   Find the band width between first nulls and half power points of the radiation pattern of a paraboloid operating at 10 GHz which has a mouth diameter of 0.15 m. Also find the power gain.

Solution   Frequency, $f = 10\ \text{GHz}$

Mouth diameter, $D_a = 0.15\ \text{m}$

$$\lambda = \frac{3 \times 10^8}{10 \times 10^9} = 0.03 \text{ m}$$

$$\frac{\lambda}{D_a} = 0.2$$

$$\text{BWFN} = \phi_0 = 140 \times \left(\frac{\lambda}{D_a}\right) = 28^\circ$$

$$\text{HPBW} = \phi = 70 \times \left(\frac{\lambda}{D_a}\right) = 14^\circ$$

Power gain, $\qquad g_p = 6.4 \left(\frac{D_a}{\lambda}\right)^2 = 160 = 22.04 \text{ dB}$

$$\phi_0 = 28^\circ, \quad \phi = 14^\circ, \quad g_p = 160 = 22.04 \text{ dB}.$$

Problem 7.4 For a paraboloid reflector antenna with 1.8 m diameter operating at 2 GHz, find the power gain in dB.

Solution Frequency, $f = 2$ GHz

Diameter, $\qquad D_a = 1.8$ m

$$\lambda = \frac{3 \times 10^8}{2 \times 10^9} = 0.15 \text{ m}$$

The power gain, $g_p = 6.4 \left(\frac{D_a}{\lambda}\right)^2 = 921.6$

$$g_p \text{ (dB)} = 10 \log_{10} (921.6)$$

$$g_p = 29.64 \text{ dB}.$$

Problem 7.5 A paraboloid operating at 5 GHz has a radiation pattern with Null-to-Null beam width of 10°. Find the mouth diameter of the paraboloid, half power beam width and power gain.

Solution Frequency, $f = 5$ GHz

$$\lambda = \frac{3 \times 10^8}{5 \times 10^9} = 0.06 \text{ m}$$

We have $\qquad \phi_0 = \text{BWFN} = 140 \times \left(\frac{\lambda}{D_a}\right)$

or $\qquad D_a = 140 \times \left(\frac{\lambda}{\text{BWFN}}\right)$

$$= 140 \times \left(\frac{0.06}{10} \right)$$

Mouth diameter, $\quad D_a = 0.84$ m

$$\text{HPBW} = \phi = 70 \times \left(\frac{\lambda}{D_a} \right) = 5°$$

Power gain, $\qquad g_p = 6.4 \left(\frac{D_a}{\lambda} \right)^2$

or $\qquad\qquad\qquad g_p = 1254.4$

$$\boxed{D_a = 0.84 \text{ m}, \ \phi = 5°, \ g_p = 1254.4.}$$

Problem 7.6 For a paraboloid reflector of diameter 6 m, illumination efficiency, $b = 0.65$. The frequency of operation is 10 GHz. Find its beam width, directivity and capture area.

Solution Frequency, $\quad f = 10$ GHz

$$\lambda = \frac{3 \times 10^8}{10 \times 10^9} = 0.03 \text{ m}$$

Mouth diameter, $\quad D_a = 6$ m

Actual area, $\qquad A = \frac{\pi D_a^2}{4} = \pi \times \frac{36}{4} = 28.27 \text{ m}^2$

Capture area, $\qquad A_c = 0.65 \text{ A} = 18.378 \text{ m}^2$

Directivity, $\qquad D = 6.4 \times \left(\frac{D_a}{\lambda} \right)^2 = 2,56,000 = 54.1 \text{ dB}$

$$\text{HPBW} = \phi = 70 \times \left(\frac{\lambda}{D_a} \right) = 0.35°$$

$$\text{BWFN} = \phi_0 = 2\phi = 0.70°$$

$$\phi_0 = 0.7°, \ \phi = 0.35°, \ A_c = 18.378 \text{ m}^2$$

Directivity $\qquad\qquad = 54.1 \text{ dB}.$

Problem 7.7 A paraboloid reflector operates at 4 GHz. Its mouth diameter is 6 m. It is required to measure far-field pattern of the paraboloid. Find the minimum distance required between the two antennas.

Solution The minimum distance required

$$r = \frac{2D_a^2}{\lambda}$$

where
$$D_a = 6.0 \text{ m}$$
$$f = 4 \text{ GHz}$$
$$\lambda = \frac{3 \times 10^8}{4 \times 10^9} = 0.075 \text{ m}$$
$$r = \frac{2D_a^2}{\lambda} = 960.0 \text{ m}.$$

Problem 7.8 A paraboloid reflector is required to have a power gain of 1,000 at a frequency of 3 GHz. Determine the mouth diameter and beam width of the antenna.

Solution Frequency, $f = 3 \text{ GHz}$
$$\lambda = \frac{3 \times 10^8}{3 \times 10^9} = 0.1 \text{ m}$$

Required power gain,
$$g_p = 1,000$$

We have, $$g_p = 6.4 \times \left(\frac{D_a}{\lambda} \right)^2$$

Mouth diameter, $$D_a = \lambda \sqrt{\frac{g_p}{6.4}}$$

$$= 0.1 \sqrt{\frac{1,000}{6.4}} = 1.25 \text{ m}$$

$$\text{HPBW} = 70 \times \left(\frac{\lambda}{D_a} \right) = 5.6^\circ$$

$$\text{BWFN} = 140 \times \left(\frac{\lambda}{D_a} \right) = 11.2^\circ$$

$$\boxed{D_a = 1.25 \text{ m}, \ \ \text{HPBW} = 5.6^\circ, \ \ \text{BWFN} = 11.2^\circ.}$$

Problem 7.9 A paraboloid reflector operates at a frequency of 10 GHz and it provides a power gain of $g_p = 75$ dB. Find the capture area of the paraboloid and beam width.

Solution Frequency, $f = 10 \text{ GHz}$
$$\lambda = \frac{3 \times 10^8}{10 \times 10^9} = 0.03 \text{ m}$$

Power gain, $$g_p = 6.4 \left(\frac{d_a}{\lambda} \right)^2$$

But, $$g_p = 75 \text{ dB}$$

That is,
$$g_p = 10 \log_{10} g = 75$$

or
$$\log_{10} g = 75/10 = 7.5$$

or
$$g = 10^{7.5} = 3.162 \times 10^7$$

So,
$$g_p = 6.4 \left(\frac{D_a}{\lambda}\right)^2 = 3.162 \times 10^7$$

or
$$\left(\frac{D_a}{\lambda}\right)^2 = \frac{3.162 \times 10^7}{6.4}$$

or
$$D_a = 0.03 \times \sqrt{\frac{3.162 \times 10^7}{6.4}}$$

$$D_a = 0.03 \times 10^2 \sqrt{\frac{3162}{6.4}}$$

Mouth diameter, $D_a = 66.68$ m

Actual area, $A = \dfrac{\pi D_a^2}{4} = 3492 \text{ m}^2$

Capture area, $A_c = 0.65 \times A$

$$= 0.65 \times 3492 = 2269.83 \text{ m}^2$$

$$\text{BWFN} = \phi_0 = 140 \times \left(\frac{\lambda}{D_a}\right) = 0.062^\circ$$

$$\text{HPBW} = \phi = 70 \times \left(\frac{\lambda}{D_a}\right) = 0.031^\circ$$

$$\boxed{A_c = 2269.83 \text{ m}^2, \quad \phi_0 = 0.062^\circ, \quad \phi = 0.031^\circ.}$$

Problem 7.10   A parabolic reflector is operated at 2 GHz and it has mouth diameter of 60 m. If it is fed by non-directional antenna, find out HPBW, BWFN and power gain.

Solution   Frequency, $f = 2$ GHz

$$\lambda = \frac{3 \times 10^8}{2 \times 10^9} = 1.5 \times 10^{-1} = 0.15 \text{ m}$$

Mouth diameter, $D_a = 60$ meters

Power gain, $g_p = 6.4 \left(\dfrac{D_a}{\lambda}\right)^2 = 10,24,000$

or
$$g_p = 60.103 \text{ dB}$$

$$\text{HPBW} = \phi = 70 \times \left(\frac{\lambda}{D_a} \right) = 0.175^{\circ}$$

$$\text{BWFN} = \phi_0 = 140 \times \left(\frac{\lambda}{D_a} \right) = 0.35^{\circ}$$

$$g_p = 60.103 \text{ dB}, \quad \phi = 0.175^{\circ}, \quad \phi_0 = 0.35^{\circ}.$$

Problem 7.11   A parabolic reflector with a mouth diameter of 22 meters operates at $f = 5$ GHz. It has illumination efficiency of 0.6. Find the power gain.

Solution   Mouth diameter, $D_a = 22$ m

Frequency, $\qquad\qquad f = 5$ GHz

$$\lambda = \frac{3 \times 10^8}{5 \times 10^9} = 0.06 \text{ m}$$

Illumination efficiency $\quad = 0.6$

Power gain, $\qquad\qquad g_p = $ illumination efficiency $\times \left(\frac{D_a}{\lambda} \right)^2$

$$= 0.6 \times \left(\frac{22}{0.06} \right)^2 = 80{,}666.6 \text{ or } 49.06 \text{ dB}$$

Power gain $\qquad\qquad g_p = 49.06$ dB.

Problem 7.12   For what mouth diameter and capture area of a paraboloid reflector is a BWFN of 12° obtained when it is operated at 2 GHz?

Solution   Frequency, $f = 2$ GHz

$$\lambda = \frac{3 \times 10^8}{2 \times 10^9} = 0.15 \text{ m}$$

$$\text{BWFN} = 140 \times \left(\frac{\lambda}{D_a} \right) = 12^{\circ}$$

$$D_a = \frac{140\lambda}{12} = 1.75 \text{ m}$$

Capture area, $\quad A_c = 0.65 \, A$

Here, $\qquad\qquad A = \frac{\pi D_a^2}{4} = 2.405 \text{ m}^2$

$$A_c = 1.5634 \text{ m}^2$$

$$\text{Mouth diameter,} \qquad D_a = 1.75 \text{ m}$$
$$A_c = 1.5634 \text{ m}^2.$$

Problem 7.13 A paraboloid reflector is required to produce a beam width between the first nulls equal to 3° at an operating frequency of 2.5 GHz. Find the mouth diameter and power gain.

Solution Frequency, $f = 2.5$ GHz

$$\text{BWFN} = 3^o$$

$$\lambda = \frac{3 \times 10^8}{2.5 \times 10^9} = 0.12 \text{ m}$$

But

$$\text{BWFN} = 140 \times \left(\frac{\lambda}{D_a} \right) = 3^o$$

$$D_a = 140 \times \left(\frac{\lambda}{3} \right) = \left(\frac{140 \times 0.12}{3} \right) = 5.6 \text{ m}$$

Power gain,

$$g_p = 6.4 \left(\frac{D_a}{\lambda} \right)^2 = 13{,}937.7 \text{ or } 41.50 \text{ dB}$$

$$D_a = 5.6 \text{ m}, \ g_p = 41.44 \text{ dB}.$$

Problem 7.14 A paraboloid reflector has radiation characteristics whose half power beam width is 5°. Find out its Null-to-Null beam width and power gain.

Solution

$$\text{HPBW} = \phi = 5^o$$

$$\text{BWFN} = \phi_0 = 2\phi = 10^o$$

But

$$\phi = 70 \times \left(\frac{\lambda}{D_a} \right)$$

or

$$\left(\frac{\lambda}{D_a} \right) = \frac{\phi}{70}$$

$$\left(\frac{D_a}{\lambda} \right) = \frac{70}{5} = 14.0$$

Power gain,

$$g_p = 6.4 \left(\frac{D_a}{\lambda} \right)^2 = 1254.4$$

or

$$g_p = 30.98 \text{ dB}$$

$$\text{BWFN} = 10^o, \ g_p = 30.98 \text{ dB}.$$

Problem 7.15 What is the power gain of a paraboloid reflector whose mouth diameter is equal to 8λ?

Solution Power gain, $g_p = 6.4 \left(\dfrac{D_a}{\lambda} \right)^2$

Here $\qquad D_a = 8\lambda$

$$g_p = 6.4 \left(\frac{8\lambda}{\lambda} \right)^2 = 409.6$$

$$g_p = 26.12 \text{ dB.}$$

Problem 7.16 Determine half power and Null-to-Null beam widths of a paraboloid reflector whose aperture diameter is 6λ. Also find its directivity.

Solution Aperture diameter, $= 6\lambda$

$$\text{HPBW} = \phi = 70 \times \left(\frac{\lambda}{D_a} \right) = 70 \times \left(\frac{\lambda}{6\lambda} \right) = 11.66^\circ$$

Null-to-Null beam width,

$$\phi_0 = 2\phi = 23.33^\circ$$

The directivity of the paraboloid,

$$D = 6.4 \times \left(\frac{6\lambda}{\lambda} \right)^2$$

$$= 6.4 \times 36 = 230.4$$

$$\phi = 11.6^\circ, \ \phi_0 = 23.33^\circ, \ D = 230.4.$$

Problem 7.17 The aperture dimensions of a pyramidal horn are 12×6 cm. It is operating at a frequency of 6 GHz. Find the beam width, power gain and directivity.

Solution Frequency, $f = 6$ GHz

$$\lambda = \frac{3 \times 10^8}{6 \times 10^9} = 0.05 \text{ m} = 5 \text{ cm}$$

$$d = 12 \text{ cm}, \ w = 6 \text{ cm}$$

Half power beam width

$$= \text{HPBW}$$

$$\phi_E = 56 \frac{\lambda}{d} = 56 \times \frac{5}{12} = 23.33^\circ$$

$$\phi_E = 67 \frac{\lambda}{w} = 67 \times \frac{5}{6} = 55.83^\circ$$

Power gain, $\quad g_p = \dfrac{4.5wd}{\lambda^2} = 12.96 = 11.12 \text{ dB}$

Directivity, $\quad D = \dfrac{7.5Wd}{\lambda^2} = \dfrac{7.5 \times 12 \times 6}{5^2} = 21.6$

$$\phi_E = 23.33^\circ, \quad \phi_H = 55.83^\circ$$
$$g_p = 11.12 \text{ dB}, \quad D = 21.6.$$

Problem 7.18 Find the power gain of a square horn antenna whose aperture size is 8λ.

Solution The power gain, $g_p = \dfrac{4.5wd}{\lambda^2}$

$$= \dfrac{4.5 \times 8\lambda \times 8\lambda}{\lambda^2} = 288$$

$$g_p = 24.59 \text{ dB}.$$

Problem 7.19 Find the power gain and directivity of a horn whose dimensions are 10×5 cm operating at a frequency of 6 GHz.

Solution The dimensions of horn are

$$d = 10 \text{ cm}, \ w = 5 \text{ cm}, \ f = 6 \text{ GHz}$$

$$\lambda = \dfrac{3 \times 10^8}{6 \times 10^9} = 0.05 \text{ m} = 5 \text{ cm}$$

Power gain, $\quad g_p = \dfrac{4.5wd}{\lambda^2} = 9 = 9.54 \text{ dB}$

Directivity, $\quad D = \dfrac{7.5wd}{\lambda^2} = 15 = 11.76 \text{ dB}$

$$g_p = 9.54 \text{ dB}, \quad D = 11.76 \text{ dB}.$$

Problem 7.20 Find the complementary slot impedance when the dipole impedance is:
(a) $Z_d = 73 + j42.5\,\Omega$
(b) $Z_d = 67\,\Omega$
(c) $Z_d = 710\,\Omega$
(d) $Z_d = 500\,\Omega$
(e) $Z_d = 50 + j20\,\Omega$
(f) $Z_d = 50 - j25\,\Omega$
(g) $Z_d = 300\,\Omega$.

Solution (a) We have

$$Z_s = \text{slot impedance}$$

$$= \frac{\eta_0^2}{4\,(R_d^2 + X_d^2)}\,(R_d - j\,X_d)$$

$$= \frac{35530.6}{(R_d^2 + X_d^2)}\,(R_d - j\,X_d)$$

If $\qquad Z_d = R_d + j\,X_d = 73 + j\,42.5\,\Omega$

$$Z_s = 363.5 - j\,211.6,\ \ \Omega$$

(b) If $\qquad Z_d = 67 + j\,0\,\Omega$

$$Z_s = 530.3\,\Omega$$

(c) If $\qquad Z_d = 710 + j\,0\,\Omega$

$$Z_s = 50\,\Omega$$

(d) If $\qquad Z_d = 500 + j\,0\,\Omega$

$$Z_s = 71\,\Omega$$

(e) If $\qquad Z_d = 50 + j\,20\,\Omega$

$$Z_s = 612.6 - j\,245\,\Omega$$

(f) If $\qquad Z_d = 50 - j\,25\,\Omega$

$$Z_s = 568.5 + j\,284.2\,\Omega$$

(g) If $\qquad Z_d = 300\,\Omega$

$$Z_s = 118.4\,\Omega.$$

OBJECTIVE QUESTIONS

1. Ideally, reflector size is infinitely large. (Yes/No)

2. The polarisation and position of the primary antennas control the radiating properties of the complete system. (Yes/No)

3. Reflector is called primary antenna. (Yes/No)

4. Microwave frequency range is _______________.

5. Corner reflector is better than plane reflectors in collimating electromagnetic energy. (Yes/No)

6. Band width of corner reflector is more when elements are cylindrical dipoles rather than thin wires. (Yes/No)

7. A grid-wired corner reflector reduces the weight of the antenna system. (Yes/No)

8. Efficiency of corner reflector is reduced when spacing of feed element becomes small.

(Yes/No)

9. Multiple lobes are produced when the spacing of feed element from the vertex is large.

(Yes/No)

10. In corner reflectors, the spacing of the feed point should be greater than the length of the sides.

(Yes/No)

11. If the main beam is narrow, the directivity is small.   (Yes/No)

12. Collimation of electromagnetic energy means generation of parallel rays.   (Yes/No)

13. Parabolic reflector is different from paraboloid.   (Yes/No)

14. Dish antenna and paraboloid are one and the same.   (Yes/No)

15. The gain of an antenna with a paraboloid reflector depends on (D_a/λ) and the illumination.

(Yes/No)

16. In Cassegrain feed, the size of the hyperboloid reflector depends on its distance from the horn feed, mouth diameter of horn and frequency.   (Yes/No)

17. The size of hyperboloid reflector is small if its distance from the feed antenna is small.

(Yes/No)

18. Cassegrain feed is best suited for ________________.

19. The disadvantage of Cassegrain feed is the obstruction of electromagnetic energy by hyperbolic reflector.   (Yes/No)

20. If half power band width is 10° in the radiation of pattern of paraboloid beam width from Null-to-Null is ________________.

21. The power gain of paraboloid is given by ________________.

22. Capture area of paraboloid is ________________ where $K = 0.65$ for dipole feed and A is actual area.

23. If the actual area of paraboloid reflector is 10 m^2, its capture area is ________________.

24. Sector beams are used in ________________ antennas.

25. Cosec beams are used for ________________.

26. Narrow beams are used for point-to-point communication purposes.   (Yes/No)

27. For height finding, the antenna beam is ________________.

28. In pyramidal horn, flaring is done in only one plane.   (Yes/No)

29. Power gain of horns is greater than that of paraboloid reflectors.   (Yes/No)

30. Directivity of horns is greater than that of waveguide.   (Yes/No)

31. Power gain of a horn is more than its directivity.   (Yes/No)

32. Feed system with corrugated horn reduces spill over efficiency.   (Yes/No)

33. Feed system with corrugated horn reduces cross-polarisation.   (Yes/No)

34. Horizontal slot produces vertical polarised radiation fields. (Yes/No)

35. Horizontal dipole produces horizontal polarised radiation fields. (Yes/No)

36. If impedance of dipole is inductive, slot impedance is capacitive. (Yes/No)

37. If the impedance of the slot is capacitive, the impedance of complementary dipole is inductive. (Yes/No)

38. From slot antenna, in a conducting plane, its complementary dipole is formed by interchanging air and metallic regions in the slot. (Yes/No)

39. Impedance of the slot antenna can be changed by changing feed point. (Yes/No)

40. Back radiation from a slot in a conductive plane can be avoided by ________________.

41. Slot gain is increased by array of slots. (Yes/No)

42. The radiation pattern of annular slot antenna is ________________.

43. Array of slots is used in ________________.

44. An array of slots when excited with appropriate amplitude and phase is suitable in ________________.

45. Dipole of small length to diameter ratio increases the bandwidth. (Yes/No)

46. Slot of small length to width ratio increases the band width. (Yes/No)

47. Notch antennas are used in aircrafts. (Yes/No)

48. Notch antennas are used in edges of the wing surface of aircraft. (Yes/No)

49. Notch antenna is broad band. (Yes/No)

50. The purpose of dielectric filling of notch is ________________.

51. Microstrip antennas are used because of ________________.

52. Microstrip antennas are used for frequencies above ________________.

53. The band width of microstrip antenna is ________________.

54. In microstrip antennas, Beam width can be increased by ________________ the thickness of the strip.

55. If $\in_r$ of substrate is high in microstrip antenna, Beamwidth increases. (Yes/No)

56. If reactive component is added in microstrip antenna, B.W. is increased. (Yes/No)

57. If reactive component is added in microstrip antennas Voltage standing wave ratio is increased. (Yes/No)

58. The radiation beam of microstrip antenna is ________________.

59. The characteristic impedance Z_0 of microstrip antenna is ________________.

60. Trihedral forms of corner reflectors are used as ________________.

61. Rod reflectors are nothing but parasitic elements. (Yes/No)

62. The length of the rod reflector is greater than $\lambda/2$. (Yes/No)

63. Rod reflector is an active radiating element. (Yes/No)

64. In Cassegrain feed, the dimension of the hyperboloid depends on its distance from the primary feed antenna. (Yes/No)

65. In Cassegrain feed, the dimension of the hyperboloid depends on mouth diameter of the horn. (Yes/No)

66. In Cassegrain feed the dimension of the hyperboloid depends on frequency of operation. (Yes/No)

67. Flare angle of the horn is related to axial length. (Yes/No)

68. The directivity of the paraboloid is greater than that of horn. (Yes/No)

69. The size of the horn becomes large if the flare angle is small. (Yes/No)

70. Horn antenna is called secondary antenna when used with paraboloid. (Yes/No)

71. The disadvantage of lens antenna at low frequencies is ________________.

72. The material of lens antenna is ________________.

73. Lens are preferred over parabolic reflectors at ________________.

74. Lens is used to correct the curved wavefront. (Yes/No)

75. The refractive index of lens material is different from unity. (Yes/No)

76. In fanned beams, the directivity is poor in one of the principal planes. (Yes/No)

77. If the beam width is small, target resolution is high. (Yes/No)

78. Fanned beams are used for ________________.

79. For feed systems using corrugated horns, the aperture efficiency is ________________.

80. Babinet's principle is applicable in electromagnetic problems. (Yes/No)

81. Babinet's principle is valid in optics. (Yes/No)

82. For a slot in conducting sheet, there exists a complementary dipole. (Yes/No)

83. The gain of the horn antenna is ________________.

84. Vertical slot in the narrow wall of rectangular waveguide does not radiate. (Yes/No)

85. Longitudinal centred slot in the broad wall of a rectangular waveguide does not radiate. (Yes/No)

86. The equivalent circuit of an inclined slot in the narrow wall of a rectangular waveguide is a ________________.

87. Resonant length of the slot is ________________.

88. Method of moments is useful to solve ________________.

89. Patch antennas are ________________.

90. Patch is made of dielectric material. (Yes/No)

91. Pyramidal horn is nothing but rectangular horn. (Yes/No)

92. Conical horn is excited conveniently by a circular waveguide. (Yes/No)

93. For lossless antenna, directivity is the same as gain.   (Yes/No)

94. Aperture efficiency is given by _______________.

95. A slot can be excited by a waveguide.   (Yes/No)

96. A slot can be excited by an energised cavity.   (Yes/No)

97. A slot can be excited by a transmission line.   (Yes/No)

98. The efficiency of patch antenna is _______________.

99. The equivalent circuit of symmetrical vertical slot in the broad wall of a rectangular waveguide is _______________.

100. For producing circular polarised waves, the shape of the patch antenna is _______________.

ANSWERS

1. Yes   **2.** Yes   **3.** No   **4.** 1 GHz – 100 GHz   **5.** Yes

6. Yes   **7.** Yes   **8.** Yes   **9.** Yes   **10.** No   **11.** No

12. Yes   **13.** No   **14.** Yes   **15.** Yes   **16.** Yes   **17.** Yes

18. Low noise receiver applications   **19.** Yes   **20.** 20° **21.** $6.4\left(\dfrac{D}{\lambda}\right)^2$

22. KA **23.** $6.5\ \text{m}^2$ **24.** Surface search from ship-borne

25. Airport surveillance   **26.** Yes   **27.** Sharp in elevation   **28.** No

29. No   **30.** Yes   **31.** No   **32.** Yes   **33.** Yes   **34.** Yes

35. Yes   **36.** Yes   **37.** Yes   **38.** Yes   **39.** Yes

40. Boxing the slot suitably   **41.** Yes   **42.** Narrow beam   **43.** Aircrafts

44. Scanning radars without antenna movement   **45.** Yes   **46.** Yes

47. Yes   **48.** Yes   **49.** Yes   **50.** To eliminate aerodynamic drag in aircrafts

51. Small size, less weight, low cost and so on   **52.** 100 MHz   **53.** Small   **54.** Increasing

55. Yes   **56.** Yes   **57.** No   **58.** Broad   **59.** $Z_0 = \eta \sqrt{\dfrac{\mu_r}{\epsilon_r}}\ \Omega$

60. Radar targets   **61.** Yes   **62.** Yes   **63.** No   **64.** Yes

65. Yes   **66.** Yes   **67.** Yes   **68.** Yes   **69.** No   **70.** No

71. Bulkiness   **72.** Lucite   **73.** Millimeter and sub-millimeter frequencies   **74.** Yes

75. Yes   **76.** Yes   **77.** Yes   **78.** Air search from ground   **79.** 75 – 80%

80. No   **81.** Yes   **82.** Yes   **83.** Moderate   **84.** Yes   **85.** Yes

86. Shunt admittance   **87.** $\lambda/2$ **88.** Integral equations

89. Very compact   **90.** No   **91.** Yes   **92.** Yes   **93.** Yes

94. Ratio of effective aperture and physical aperture   **95.** Yes   **96.** Yes

97. Yes   **98.** Low   **99.** Series impedance   **100.** Circular

 **EXERCISE PROBLEMS**

1. The power gain of transmitting antenna $A_p = 20$. The input power is $P_{in} = 200$ W. Find the effective isotropic radiated power EIRP in dB and dB_m.

2. Find the power density at a point at 12 km from the transmitting antenna. Its power gain is 10 and input power is 100 W.

3. The radiation resistance of a transmitting antenna dipole is 80 Ω, loss resistance is 10 Ω, directive gain is 15 and input power is 1 kW. Find antenna efficiency and radiated power.

4. The capture area of a receiving antenna is 10 cm^2 and available power density is 10 μ W/cm^2. Find the capture power.

5. What is the band width in percentage of an antenna operating at a frequency of 100 MHz if 3dB frequencies are 300 MHz and 350 MHz.

6. The diameter of parabolic reflector is 2.0 m. It radiates a power of 100 W at an operating frequency of 3 GHz. Its efficiency is 60% and its aperture efficiency is 60%. Find the antenna power gain and beam width.

7. What is the free space path loss when the transmitting and receiving antennas are separated by 100 km, while operating at a frequency of 10 GHz?

8. A uniformly illuminated parabolic reflector whose aperture size is 2 m is operated at 6 GHz. Find the Null-to-Null beam width and power gain with reference to dipole of half wave length. Assume that the antenna is lossless.

9. A 2 m parabolic reflector operating at $f = 6$ GHz radiates a power of 100 W. It has efficiency of 60% and its aperture efficiency is 60%. Find the antenna power gain in dB.

10. A 3 m parabolic reflector operating at 8 GHz radiates a power of 10 W. It has an efficiency of 55% and its aperture efficiency is 60%. Find the receiver power gain. Also find the EIRP.

11. The power radiated by an antenna is 100 W and dissipated power is 10 W. The antenna has a directional gain of 250. Find antenna efficiency and power gain.

chapter 8

Antenna Measurements

"Antenna parameter measurements are the most reliable and best methods of antenna analysis."

CHAPTER OBJECTIVES

This chapter discusses

- ✦ Merits and demerits of different ranges and methods of measurements

- ✦ The measurement of all antenna parameters including polarization and phase

- ✦ Objective questions and solved problems useful for class tests, final examinations and also for competitive examinations

- ✦ Exercise problems to develop self problem solving skills

8.1 INTRODUCTION

Antenna measurements are a part of the analysis of antenna parameters. **Analysis** is the determination of output knowing the input and system details. On the other hand, **design** is the determination of system details knowing the input and output parameters.

Antenna measurements are required for the following purposes:

1. To calibrate and store data for different types of antennas.
2. To analyse different parameters.
3. To verify design.
4. To adjust critical components and dimensions.
5. To control quality.
6. To find analytical and statistical errors.
7. To indicate the actual performance of antennas.
8. To verify the validity of the assumptions made in the analytical formulations.
9. To make alternate approach for analytical, numerical design and analysis methods. The **Geometrical Theory of Diffraction** (GTD) and **Method of Moments** (MOM) are common analytical methods in antennas.

8.2 DRAWBACKS IN MEASUREMENTS OF ANTENNA PARAMETERS

1. It is sometimes difficult to provide far-field distance or $r > \dfrac{2D_a^2}{\lambda}$, D_a being aperture size and λ being wavelength for far-field pattern measurements.
2. Ground reflections are present while noting the measurements.
3. It is difficult to bring large antennas to the measuring site.
4. Measurements are time consuming.
5. It is difficult to accommodate large antennas in shielded chambers.
6. Open site measurements are not accurate.
7. Measuring equipment is expensive.

8.3 METHODS TO OVERCOME DRAWBACKS IN MEASUREMENTS

1. Determination of far-field patterns from near-field measurements.
2. Making scale model measurements.
3. Using automated measuring equipment.
4. Using computerised techniques.

8.4 SOME METHODS FOR ACCURATE MEASUREMENTS

The following are used for accuracy:

1. Tapered anechoic chambers.
2. Compact ranges.
3. Near-field techniques.
4. Polarisation techniques.

5. Computer-controlled test systems.

6. Sweep-frequency sources.

8.5 MEASUREMENT RANGES

The following ranges are used for antenna parameter measurements:

1. TEM Cell
2. GTEM Cell
3. Outdoor range
4. Indoor range
5. Reflection range
6. Slant range
7. Elevated range
8. Compact range
9. Anechoic chamber
10. Near-field range
11. Ground range
12. Radar cross-section range

All the ranges are required to produce a uniform phase and amplitude across the Antenna Under Test or AUT.

When antennas are large and heavy, source is rotated instead of the antenna. Antenna ranges are never ideal and compromises are always made.

1. **TEM Cell** (Transverse Electromagnetic Cell). It is a rectangular coaxial transmission line. It resembles a stripline. It is tapered at both ends and matched to a 50Ω coaxial line. It allows transmission of electromagnetic energy in TEM mode. The central conductor is held firmly in position by a number of dielectric supports.

The testing unit is placed in the rectangular portion of the transmission line between the bottom plate and the central conductor. It can also be placed between the central conductor and the top plate. A typical TEM cell is shown in Fig. 8.1.

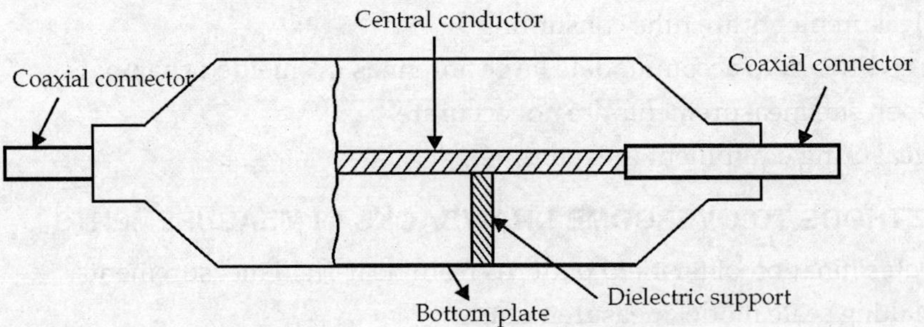

Fig. 8.1 *TEM Cell*

2. **GTEM cell** (Giga Hertz TEM cell) It is a hybrid between an anechoic chamber and a TEM cell. It can be used over a wide range of frequencies. GTEM cell can be made in different dimensions depending on the requirement.

It is a 50Ω tapered rectangular coaxial transmission line in which the central conductor is offset. The rectangular section is connected to a 50Ω coaxial conductor at one end. Cross-section of the central conductor is transformed from a flat strip into a circular shape. The far end of the tapered section is terminated in a distributed matched load which consists of pyramid shaped RF absorbing material. The central conductor is also terminated in a $50\,\Omega$ load which is made up of several carbon resistors.

The resistance values match the current distribution in the central conductor. The flare angle of the tapered section is kept small to make the field formed by TEM wave which has a spherical symmetry with a large radius. A typical GTEM is shown in Fig. 8.2.

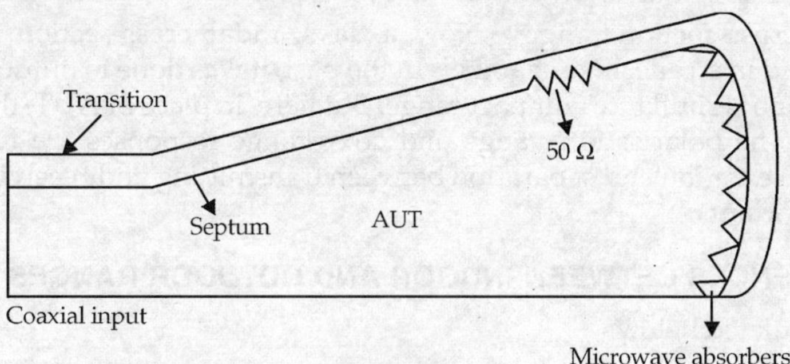

Fig. 8.2   *GTEM Cell*

3. **Outdoor range**   In this, there is no protection from external Electromagnetic Interference (EMI). But antennas of any size can be used for measurement.

4. **Indoor range**   In this, indoor sites are shielded to protect from EMI. But space is a limitation and large antennas cannot be brought in for measurements.

5. **Reflected ranges**   Here the heights of transmitting antenna and AUT are chosen to produce a constructive interference at AUT.

6. **Slant range**   In this, AUT is fixed on a non-conducting tower at a fixed height and the transmitting antenna is kept on the ground. The transmitting antenna is oriented to have maximum radiation towards AUT. Reflected signals are eliminated by directing side lobes towards the ground.

7. **Elevated range**   These are used on smooth terrains. The transmitting antenna and AUT are mounted on two towers. Line of sight facility is provided. Care is taken to reduce the ground reflections. This range depends on tower height.

8. **Compact range**   In this, the transmitting antenna acts as an offset feed for paraboloid. The paraboloid changes spherical waves into plane waves towards AUT. This involves small distances, and hence it is called compact range.

9. **Anechoic chambers**   It is an indoor chamber. The chamber walls, ceiling and floor are filled with RF energy absorbers except at the locations of transmitting antenna and AUT. It is ideal for small antennas. There exists no EMI. It provides all-weather capability and controlled environment. These chambers are either tapered or of rectangular shape.

10. **Near-field range**   It is a small indoor test range in which near-field measurements are made. The data obtained in one of several coordinate systems is converted into far-field data analytically or numerically. This is the fastest growing alternative to outdoor ranges.

11. **Ground range**   Here the transmitting antenna is mounted above the surface which acts like a mirror. The reflection appears as if it comes from an image of the source. The total signal has visible centre near the ground. Near the transmitting antenna, other reflections do not reach the AUT. The advantage of this range is that tall towers are not required.

12. **Radar cross-section range**   Now a days, radar cross-section is usually measured in an anechoic chamber. In the past, it was done in outdoor ranges. This range is similar to compact range. But here, in place of AUT, the target is placed. The polarisation, range and cross-range responses are required as target orientation and separation between transmitting and receiving system are important.

8.6 DIFFERENCES BETWEEN INDOOR AND OUTDOOR RANGES

These are tabulated below:

Table 8.1   Differences between indoor and outdoor ranges

Indoor ranges	Outdoor ranges
These are protected from external EMI	These are susceptible to EMI
Space is limited	Space is unlimited
Only small antennas can be tested	Antennas of any size can be tested
Ground and other reflections are controlled	Ground and other reflections cannot be controlled fully
They have controlled environment	They have uncontrolled environment
They have all-weather capability	They do not have all-weather capability

The selection of range for antenna measurements depends on:

1. Frequency
2. Cost
3. Accuracy required
4. Size of the antenna and so on.

8.7 ANTENNA IMPEDANCE MEASUREMENT

Antenna impedance is of four types.

1. **Intrinsic impedance of an antenna**   It is defined as the critical input impedance of an ideal basic radiating structure.

2. **Antenna impedance or input impedance**   It is defined as the ratio of input voltage of the antenna to its input current. That is,

$$Z_i \equiv \frac{V_i}{I_i}$$

It is the impedance presented by an antenna at its terminals. The point on the antenna where it is fed by a transmission line is known as feed point. The AC load presented by the feed point to the transmission line is known as antenna input impedance. When the antenna input impedance and the output impedance of a transmitter are equal to the characteristic impedance of the transmission line, maximum power is transferred to the antennas and it is radiated without standing waves.

3. **Self impedance**   This is applicable only when an antenna is part of an array. Self impedance of an antenna in an array is defined as its input impedance with all other antenna elements in the open-circuited array.

4. **Mutual impedance**   This is also applicable only when an antenna is in an array. The mutual impedance between any two antennas in an array is defined as the impedance which is equal to the ratio of open-circuit voltage across the first terminal pair to the current supplied to the second when all the terminal pairs are open-circuited.

The impedance of an antenna consists of both real and imaginary parts, that is,

$$Z_a = \text{antenna impedance}$$
$$= R_a + jX_a$$

here R_a consists of two parts, that is,

$$R_a = R_r + R_l$$

here
$$R_r = \text{radiation resistance}$$
$$R_l = \text{loss resistance}$$
$$X_a = \text{antenna reactance}$$

The performance of an antenna is frequency dependent.

Impedance measurement by Wheatstone bridge method

This bridge method is used to measure antenna impedance at frequencies upto 30 MHz. In this method, an unknown impedance is measured by comparison with known impedance.

Measurement procedure

Consider the Fig. 8.3.

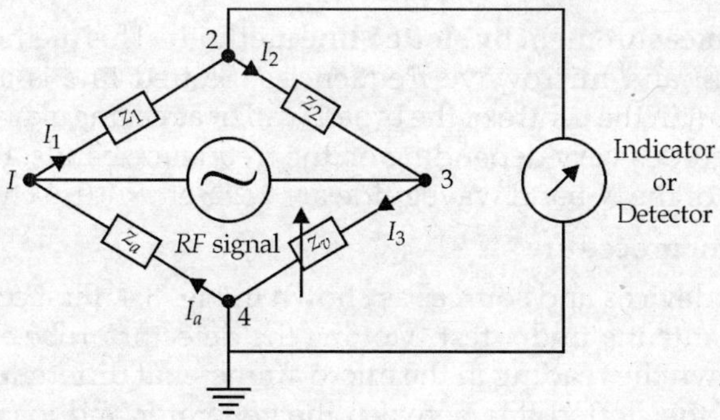

Fig. 8.3   *Wheatstone bridge for antenna impedance measurements*

The bridge consists of four arms. An RF signal of about 1 mV is given between the nodes 1 and 3. A shielded detector of sensitivity of about $5\mu V$ is connected between 2 and 4. Arms connecting Z_1 and Z_2 are known as ratio arms. Z_v is a variable impedance. Here Z_a is unknown. Z_v is varied till a null is obtained in the detector. When the bridge is balanced, the potential difference between 2 and 4, $V_{24} = 0$. That is,

$$\frac{Z_1}{Z_2} = \frac{Z_a}{Z_v}$$

or   Antenna impedance $Z_a = \dfrac{Z_1}{Z_2} \cdot Z_v$ ...(8.1)

Thus unknown impedance, Z_a is found.

As the point 4 is grounded, the method is suitable for the measurement of low frequency grounded vertical antennas. When the measurements are required for balanced antennas, points 1 and 4 are balanced with reference to ground.

As the impedance involves both magnitude and phase, Equation (8.1) is written as

$$\frac{Z_1\,(\phi_1)}{Z_2\,(\phi_2)} = \frac{Z_a\,(\phi_a)}{Z_v\,(\phi_v)}$$

or $Z_1\,Z_v\,(\angle\phi_1 + \angle\phi_v) = Z_2\,Z_a\,(\angle\phi_2 + \angle\phi_a)$ ...(8.3)

$$Z_a = \frac{Z_1\,Z_v}{z_2}$$

and $\angle\phi_a = \angle\phi_1 + \angle\phi_v - \angle\phi_2$ ...(8.4)

Precautions in the measurements

1. Load inductance and distributed capacitance in the bridge should be minimum.

2. The variable resistance should be of inductively compensated type. Here the resistances are adjusted to keep the introduced inductances at constant values.

3. Bridge should be calibrated before the actual measurements.

Impedance measurement by slotted line method   This method is convenient and useful at UHF and microwave frequencies. Slotted line is a coaxial line or longitudinal slot cut in the centre of the broad wall of a rectangular waveguide. The sizes of the waveguides vary depending on the frequency bands. For example, the inner dimensions of the X-band waveguide are $2.286\,\text{cm} \times 1.016\,\text{cm}$.

Measurement procedure

1. Connect the devices and sources as shown in Fig. 8.4. First connect a short in place of the antenna under test. Moving the detector probe along the slotted line, note down the reading in the micro ammeter at different positions of the probe along the slotted line between the generator and load positions. Plot standing wave pattern as in Fig. 8.5.

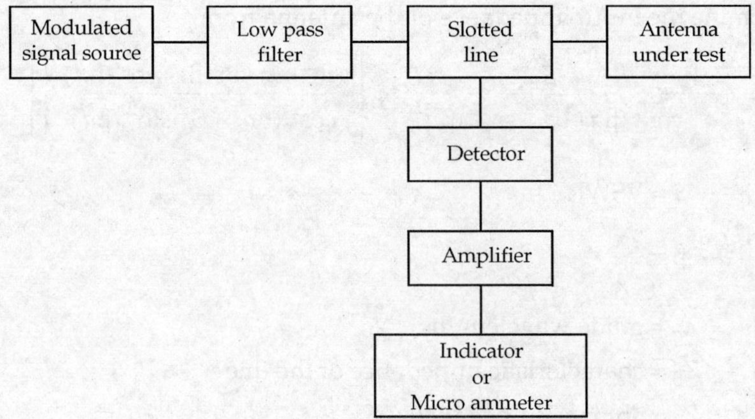

Fig. 8.4 *Slotted line set-up for impedance measurement*

2. Replace the short by the antenna under test and repeat as in step-1. The standing wave pattern for the antenna connected is plotted in Fig. 8.5.

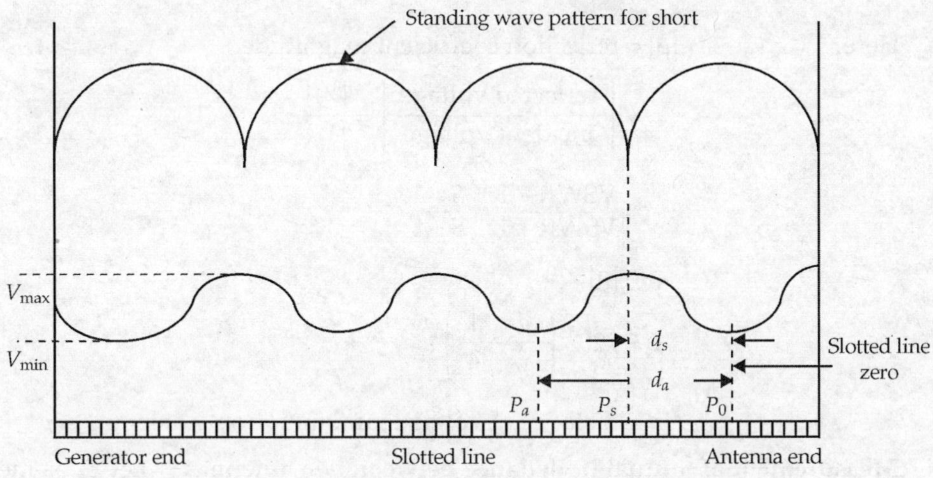

Fig. 8.5 *Standing wave patterns for short and AUT*

3. Find VSWR from

$$\text{VSWR} = S = \frac{I_{max}}{I_{min}} = \frac{V_{max}}{V_{min}}. \qquad \qquad ...(8.5)$$

4. Note down the minimum positions in the standing wave patterns, both for short and AUT. Let them be P_s and P_a respectively. Let the distances of P_s and P_a from the arbitrary slotted line zero be d_S and d_a respectively.

5. Find $d = |P_a - P_s|$ if P_a is closer to generator

$\qquad d = |P_s - P_a|$ if P_s is closer to generator.

6. Determine the input impedance of the antenna from

$$Z_a = Z_0 \left\{ \frac{S}{\cos^2(\beta d) + S^2 \sin^2(\beta d)} + j \left[\frac{(S^2 - 1)\sin(\beta d)\cos(\beta d)}{\cos^2(\beta d) + S^2 \sin^2(\beta d)} \right] \right\} \qquad \text{...(8.6)}$$

Here $\qquad S = \text{VSWR}$

$$\beta = \frac{2\pi}{\lambda_g}$$

$\lambda_g = $ guide wavelength

$Z_0 = $ characteristic impedance of the line

$Z_a = $ antenna impedance.

7. The antenna impedance, Z_a is also found from the knowledge of reflection coefficient. That is,

$$Z_a = Z_0 \left[\frac{1 + |\rho| \angle \theta}{1 - |\rho| \angle \theta} \right] \Omega \qquad \text{...(8.7)}$$

Here, $\qquad |\rho| = $ reflection coefficient magnitude.

$$= \left| \frac{\text{reflected voltage}}{\text{incident voltage}} \right| = \frac{|V_r|}{|V_i|}$$

$$= \frac{\text{VSWR} - 1}{\text{VSWR} + 1} = \frac{S - 1}{S + 1}$$

$$\rho = |\rho| \angle \theta$$

$$\theta = 720° \left(\frac{d}{\lambda_g} - \frac{1}{4} \right)$$

$d = $ distance of voltage minimum from antenna.

Measurement of mutual impedance between two antennas   Let Z_s be the self impedance of antenna 1 or antenna 2 and Z_m be the mutual impedance between the two antennas. Let Z_1 be the measured terminal impedance of antenna 1, when antenna 2 is short circuited. Then,

$$Z_1 = Z_s + \frac{I_2}{I_1} Z_m \qquad \text{...(8.8)}$$

$$0 = Z_s + \frac{I_1}{I_2} Z_m \qquad \text{...(8.9)}$$

$$Z_m^2 = Z_S (Z_s - Z_1)$$

$$Z_m = \sqrt{Z_s (Z_s - Z_1)}. \qquad \text{...(8.11)}$$

8.8 MEASUREMENT OF ANTENNA PATTERN

Antenna pattern is also known as radiation pattern. It is defined as the graphical representation of the radiation properties as a function of space coordinates.

In general, the radiation pattern is determined in the far-field region. The radiation properties include electric field strength, radiation intensity, phase and polarisation.

The antenna patterns consist of radiation lobes. The radiation lobe is only one for an ideal antenna. In fact, no antenna is ideal. Hence, the **radiation lobe** is defined as the portion of the radiation pattern bounded by the regions of relatively weak radiation intensity.

The radiation pattern of any antenna consists of one major lobe and a set of minor or side lobes.

Major lobe or main lobe   It is defined as the radiation lobe which contains the direction of maximum radiation.

Minor lobe   It is defined as any lobe other than the major lobe.

Antenna patterns are of two types:

1. Field pattern
2. Power pattern.

Field pattern is the variation of absolute field strength with θ in free space. That is,

$|E|$ Vs θ is field pattern

Similarly, power (proportional to E^2) pattern is the variation of radiated power with θ in free space. That is,

$$P \text{ Vs } \theta \text{ or } |E|^2 \text{ Vs } \theta \quad \text{ is power pattern.}$$

Measurement procedure   The set-up for measurement is shown in Fig. 8.6. The set-up consists of:

- Modulating source
- Transmitter
- Transmitting antenna
- Antenna under test
- Antenna mount
- Antenna driving unit
- Shaft for antenna rotation
- Antenna position indicating device
- Detector and
- Indicator.

Here, transmitting antenna is fixed and antenna under test is rotated by the driving unit. For each position indicated by the position indicator, the received power is noted from the indicator. The indicator can be a power meter or a microammeter. Then, from the results obtained, field (proportional to current) or

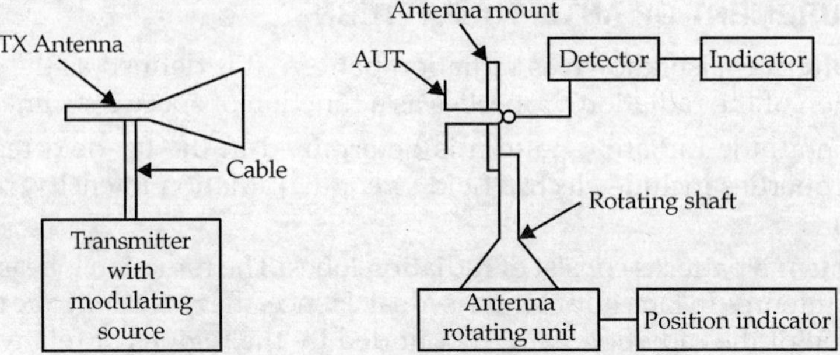

Fig. 8.6 *Set-up for pattern measurements*

power (proportional to I^2) is plotted as a function of θ. This gives the desired patterns of antenna under test. For pattern measurements, the following precautions should be taken.

Precautions in pattern measurements

1. Distance between the transmitting antenna and the receiving antenna (AUT) must be

$$R \geq \frac{2D_a^2}{\lambda}$$

Here D_a = maximum dimension of the aperture of AUT

λ = wavelength

If $R < \dfrac{2D_a^2}{\lambda}$, the measured patterns will be broad and side lobe levels will be high.

2. AUT should be illuminated uniformly.

3. Ground and other reflections should be avoided.

4. Measurements should be taken in shielded chambers like anechoic chambers to eliminate the effect of external EMI.

5. Automatic range equipment should be used to avoid manual errors.

6. The transmitting antenna should be able to produce a uniform wave front to reduce phase error of AUT.

7. The TX antenna should have high gain.

8. The side lobe level of TX antenna should be very small.

9. Horns, paraboloids or arrays of dipoles may be used as TX antennas.

8.9 MEASUREMENT OF RADIATION RESISTANCE OF AN ANTENNA

Radiation resistance is a part of antenna resistance. The antenna resistance is defined as the real part of input impedance of an antenna. This consists of two parts, one is radiation resistance and another one is loss resistance.

Radiation resistance is defined as the ratio of the radiated power by an antenna to the square of effective (root-mean square) current referred to a specified point.

$$\text{Radiation resistance, } R_r = \frac{\text{Radiated power}}{I_{\text{RMS}}^2}.$$

Measurement procedure   The experimental setup for the measurement of radiation resistance of an antenna consists of:

- A source
- A transmitter with antenna
- A receiver with antenna
- An indicator
- The antenna under test (AUT)
- A known standard half-wave antenna.

A typical experimental set-up is shown in Fig. 8.7.

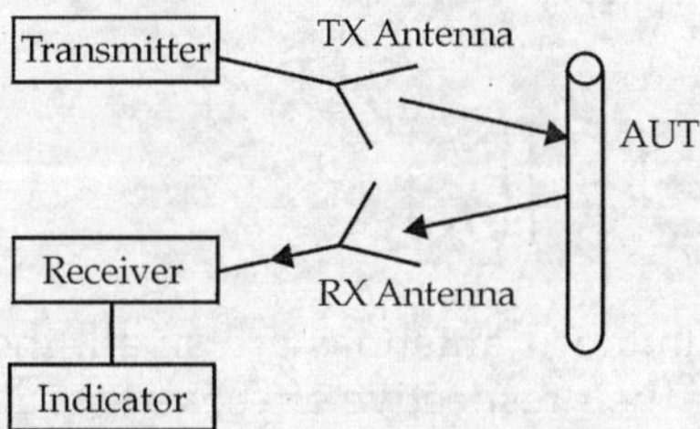

Fig. 8.7   *Measurement set-up for radiation resistance*

1. Antenna under test (AUT) is kept in the field of the transmitting antenna. The reflected power of AUT is received and noted from the indicator. Let it be P_t.

2. AUT is replaced by a standard half-wave antenna which has the same radiation pattern as that of AUT. The reflected power of the standard antenna is received and noted again from the indicator. Let it be P_s. Standard antenna is similarly used as reflector.

3. The power reflected by AUT (test antenna) is given by

$$P_t = k\,\frac{(E_t\,l_t)^2}{R_t} \qquad \qquad ...(8.12)$$

and the power reflected by standard antenna is given by

$$P_s = k\,\frac{(E_s\,l_s)^2}{R_s} \qquad \qquad ...(8.13)$$

here E_t = electric field strength at test antenna

$$E_s = \text{electric field strength at standard antenna}$$

$k =$ a constant which depends on the distance between the transmitting antenna and AUT

$l_t =$ effective length of test antenna

$l_s =$ effective length of standard antenna

$R_s =$ radiation resistance of standard antenna

$R_t =$ radiation resistance of test antenna

From Equations (8.13) and (8.14), we have

$$\frac{P_t}{P_s} = \left(\frac{E_t\, l_t}{E_s\, l_s}\right)^2 \left(\frac{R_s}{R_t}\right) \qquad \text{...(8.14)}$$

As $\qquad E_t = E_s,$

$$\frac{P_t}{P_s} = \left(\frac{l_t}{l_s}\right)^2 \left(\frac{R_s}{R_t}\right) \qquad \text{...(8.15)}$$

or $\qquad R_t = R_{\text{AUT}} = \left(\frac{P_s}{P_t}\right)\left(\frac{l_t}{l_s}\right)^2 R_s \qquad \text{...(8.16)}$

When the test and standard antennas are small, their physical lengths are approximately the same as their respective effective lengths.

8.10 GAIN MEASUREMENT BY TWO ANTENNA METHOD

Gain of an antenna is of five types:

1. Directive gain
2. Directivity
3. Power gain
4. Relative gain
5. Superdirectivity.

1. **Directive gain**   It is defined as the ratio of radiation intensity in a given direction to the average radiated power of the antenna.

The average radiated power over spherical surface is the ratio of total radiated power to 4π. Therefore, directive gain in a given direction is also defined as 4π times the ratio of radiation intensity in that direction to the total radiated power.

That is, $\qquad g_d = \dfrac{4\pi \times \text{radiation of intensity}}{\text{total radiated power}}$

2. **Directivity, D**   It is defined as the maximum directive gain. That is, $D = (g_d)_{\max}.$

3. **Power gain, g_p**   Power gain in a given direction is defined as 4π times the ratio of the radiation intensity in that direction to the total input power to the antenna. That is,

$$g_p = \frac{4\pi \times \text{radiation intensity}}{\text{total input power}}$$

When direction is not specified, power gain is considered to be its maximum value for all practical purposes. Conventionally, power gain is obtained without taking reflection-losses into account. Moreover, power gain is fully realised only on reception when incident and transmission polarisations are the same.

4. **Relative gain of an antenna, g_r**   It is defined as the ratio of the power gain in a given direction to the power gain of a reference antenna in the reference direction. That is,

$$g_r \equiv \frac{\text{power gain of the antenna in a given direction}}{\text{power gain of the reference antenna in the reference direction}}$$

Common reference antennas are electric and magnetic half-wave dipoles, monopoles, isotropic radiators, calibrated horn antennas and so on.

5. **Superdirectivity, D_s**   It is defined as the directivity of an antenna whose value is more than the expected value. The expected value depends on antenna dimensions and excitation.

In this method, two antennas, one for radiation and another for reception are used. When both the antennas are polarisation matched, the expression for the received power is given by **Friis transmission formula**. That is,

$$P_R = P_T \, G_T \, G_R \left(\frac{\lambda}{4\pi R} \right)^2 C_P$$

G_R = gain of AUT

P_T = transmitter power

G_T = gain of transmitting antenna

λ = operating wavelength

R = distance between the transmitter and receiver

C_P = a factor of polarisation losses which is equal to 1
    if T_X and R_X are well matched

$\left(\dfrac{\lambda}{4\pi R} \right)^2$ = basic free space path loss factor

So,
$$G_R = \left(\frac{P_R}{G_T \, P_T} \right) \left(\frac{4\pi R}{\lambda} \right)^2$$

A typical gain measurement set-up is shown in Fig. 8.8.

The gain of AUT in dB is given by

$$G_R = G_{\text{AUT}} = 10 \log_{10} G_R$$

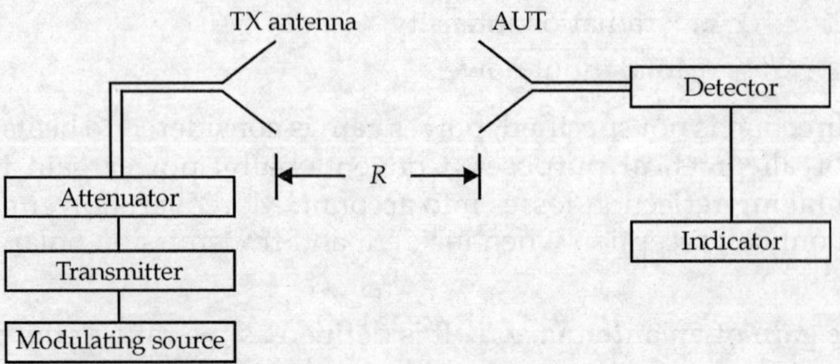

Fig. 8.8 *Set-up for gain measurement by two antenna method*

That is,

$$G_{\text{AUT}} = 10 \log \left(\frac{P_R}{P_T} \right) + 20 \log \left(\frac{4\pi R}{\lambda} \right) - 10 \log G_T$$

By measuring P_R, and knowing P_T, λ, R and G_T, G_{AUT} is obtained. This is a simple and straightforward method.

Precautions

1. R should be greater than $\dfrac{2D_a^2}{\lambda}$.

2. The frequency and hence λ should be stable.

3. The two antennas should have polarisation and impedance matching.

4. The antennas should align for maximum radiation and reception.

5. The reflections and multiple path EMI should be minimised.

6. The sweep frequency generator is preferred to klystron source.

7. Measurements should be made in anechoic chambers.

8. The gain of the transmitting antenna should be accurately known.

8.11 GAIN MEASUREMENT BY THREE ANTENNA METHOD

This method consists of:

1. Three unknown antennas.

2. Using antenna 1 as transmitter and antenna 2 as receiver, the received power W_1 is measured. Let the transmitter power be P_1.

3. Replacing antenna 2 by antenna 3, the received power W_2 is measured for the same transmitted power ($P_2 = P_1$).

4. When antenna 2 is used as transmitter and antenna 3 is used as receiver, receiver power W_3 is measured. Let the transmitter power be P_3. Then we have

$$G_1 G_2 = \left(\frac{W_1}{P_1} \right) \left(\frac{4\pi R}{\lambda} \right)^2$$

$$G_1\,G_3 = \left(\frac{W_2}{P_2}\right)\left(\frac{4\pi R}{\lambda}\right)^2$$

$$G_2\,G_3 = \left(\frac{W_3}{P_3}\right)\left(\frac{4\pi R}{\lambda}\right)^2$$

The above expressions in dB are given

$$G_1 + G_2 = 10\log\left(\frac{W_1}{P_1}\right) + 20\log\left(\frac{4\pi R}{\lambda}\right)$$

$$G_1 + G_3 = 10\log\left(\frac{W_2}{P_2}\right) + 20\log\left(\frac{4\pi R}{\lambda}\right)$$

$$G_2 + G_3 = 10\log\left(\frac{W_3}{P_3}\right) + 20\log\left(\frac{4\pi R}{\lambda}\right)$$

Knowing $\dfrac{W_1}{P_1}$, $\dfrac{W_2}{P_2}$, $\dfrac{W_3}{P_3}$, λ and R, $\mathbf{G_1}$, $\mathbf{G_2}$ and $\mathbf{G_3}$ are obtained.

8.12 GAIN MEASUREMENT BY REFLECTION FROM GROUND

The details of this method are shown in Fig. 8.9.

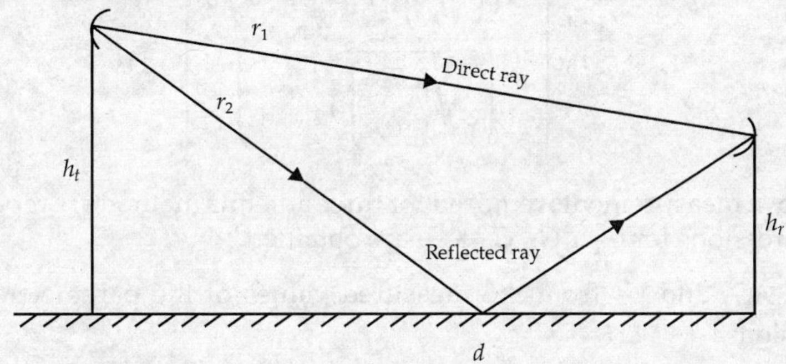

Fig. 8.9 *Direct and reflected rays between $\overline{T}_X$ and R_X antennas*

Procedure

1. The height of the transmitting antenna h_t is adjusted for a fixed receiving antenna height h_r so that the direct and reflected rays are in the same phase at the receiver to obtain maximum field.

2. The directive gains of transmitting and receiving antennas relative to their maximum values along the direct path are measured from pattern measurements. Let them be G_T and G_R respectively.

3. Measure the received power, P_R knowing r_1, r_2 and λ. Here,

$$r_1 = \text{distance between the } T_X \text{ and } R_X \text{ due to direct ray and}$$

$$r_2 = \text{distance through reflected ray.}$$

4. If G_1 is the gain of the transmitting antenna and G_2 is the gain of the receiving antenna, then $G_1 + G_2$ is given by

$$G_1 + G_2 = 10 \log \left(\frac{P_R}{P_T} \right) + 20 \log \left(\frac{4\pi r_1}{\lambda} \right) - 20 \log \left(G_T\, G_R + k\, \frac{r_1}{r_2} \right)$$

where k is a factor which depends on frequency, radiation pattern of antennas, geometrical and electrical properties of the antenna range.

$$G_T = \text{directive gain of transmitting antenna}$$

$$G_R = \text{directive gain of receiving antenna.}$$

5. Now the transmitting antenna height is adjusted to get the minimum field at the receiver. Obtain the direct path and reflected path distances. Let them be r_3 and r_4 respectively.

Also obtain the directive gains of T_X antenna and receiving antennas. Let them be G_T and G_R respectively. Let P_r be the received power.

6. Obtain k from the following Hemming and Heaton's formulation. That is,

$$k = \frac{r_1\, r_4}{r_2\, r_3} \left[\frac{\sqrt{\left(\dfrac{P_R}{P_r} \right)}\, r_1 - \sqrt{G_T\, G_R}\, r_3}{\left(\sqrt{\left(\dfrac{P_R}{P_r} \right)}\, r_2 + r_4 \right)} \right]$$

7. The above measurements required for three antenna method are repeated and the expressions for $G_1 + G_3$, $G_2 + G_3$ are obtained.

8. Find G_1, G_2 and G_3 from the measured values of the parameters and the expressions.

Salient features of the method

1. It is useful to find gain of broad beam antennas.

2. It is useful for frequencies less than 1 GHz.

3. It is limited to linear antennas which couple electric field only.

4. Circular and elliptically polarised antennas are avoided for this method.

5. This method is preferred to horizontally polarised linear antennas. This is because, the reflection coefficient of earth for vertically polarised antennas has quick variation with the angle of incidence.

8.13 DIRECTIVITY MEASUREMENT

The **directivity**, D of an antenna is its maximum directive gain. It is obtained from the field pattern of the antenna. From the measured patterns and their beam width in both the principal planes, D is obtained.

The principal planes are E-plane and H-plane.

E-**plane pattern** For a linearly polarised antenna, *E*-**Plane pattern** is defined as a pattern in the plane which contains the electric field and the direction of maximum radiation.

H-**plane pattern** For a linearly polarised antenna, the *H*-**plane pattern** is defined as the pattern in the plane which contains the magnetic field and the direction of maximum radiation.

Procedure for the measurement of directivity

1. Obtain E and H-plane patterns of AUT as described in Section 8.8.

2. Find the half-power beam widths from the patterns of step 1.

3. Find the directivity of AUT from

$$D = \frac{41,253}{(B.W)_E \times (B.W)_H}$$

here $(B.W)_E$ = half-power beam width in E-plane (degrees)

  $(B.W)_H$ = half-power beam width in H-plane (degrees)

This method is accurate when the patterns consist of only one main lobe.

4. As the field varies with both θ and ϕ, the directivity also varies with θ and ϕ. That is, $D = D(\theta, \phi)$. Hence, the total directivity is obtained from

$$D_t = \frac{\partial D}{\partial \theta} + \frac{\partial D}{\partial \phi} \qquad \qquad ...(8.17)$$

Here $$D_\theta = \frac{4\pi (RI_{max})_\theta}{(P_r)_\theta + (P_r)_\phi} \qquad \qquad ...(8.18)$$

and $$D_\phi = \frac{4\pi (RI_{max})_\phi}{(P_r)_\theta + (P_{rad})_\phi} \qquad \qquad ...(8.19)$$

  $(RI_{max})_\theta$ = maximum radiation intensity of θ component

  $(RI_{max})_\phi$ = maximum radiation intensity of ϕ component

  $(P_r)_\theta$ = radiated power in θ direction

  $(P_r)_\phi$ = radiated power in ϕ direction

The pattern in **elevation plane** is obtained by varying θ over 0 to π for a fixed ϕ.

The pattern in **azimuthal plane** is obtained by varying ϕ over 0 to 2π for a fixed value of θ.

8.14 MEASUREMENT OF ANTENNA BEAM WIDTH

Beam width of an antenna is specified in terms of:

- Half-power beam width and
- Null-to-Null beam width

Half-power beamwidth   in a plane containing the direction of the maximum of the main beam is defined as the angle between the two directions in which the radiation intensity is half of the maximum value of the beam.

HPBW is also defined as the difference between two angles at which the field strength corresponds to $\dfrac{E_{\max}}{\sqrt{2}}$. This is measured using the following procedure.

Measurement procedure

1. Measure the radiation pattern as described in Section 8.8.

   The main lobe of a typical pattern is shown in Fig. 8.10.

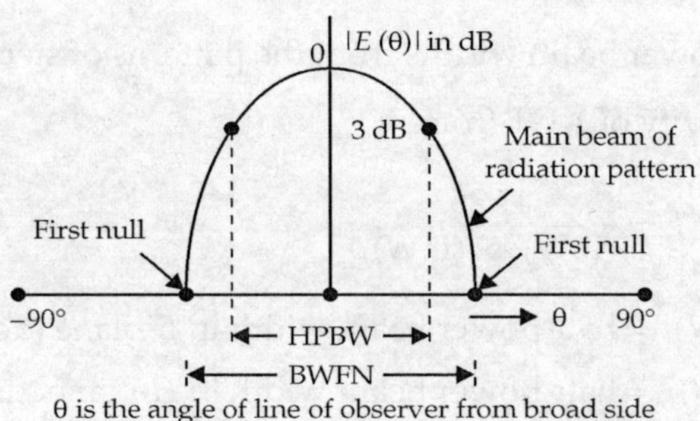

Fig. 8.10 *Measurement of HPBW and BWFN*

2. It is also possible to measure the HPBW without plotting the pattern. Here, AUT is fixed and the transmitting antenna is rotated.

3. Measure the angular position corresponding to $E_{\max}$. Let this be θ_m.

4. The transmitting antenna is rotated in either side of θ_m and the angular position at which the field is equal to $\dfrac{1}{\sqrt{2}} E_{\max}$ is noted down. Let this reading be θ_{HP} in one of the two sides. Then,

> **Half-power beam width,**
>
> $$\text{HPBW} = 2\,(\theta_m - \theta_{HP}) \ \text{if} \ \theta_m > \theta_{LHP}$$
>
> or $\qquad\quad = 2\,(\theta_{HP} - \theta_m) \ \text{if} \ \theta_{HP} > \theta_m$

5. The transmitting antenna is further rotated on either side of θ_m and the angular position at which the field strength is minimum (null) is noted. Let this reading be θ_N on the two sides. Then,

Null-to-Null beam width,
$$BWFN = 2\,(\theta_m - \theta_N) \quad \text{if} \quad \theta_m > \theta_N$$
or
$$= 2\,(\theta_N - \theta_m) \quad \text{if} \quad \theta_N > \theta_m$$

8.15 MEASUREMENT OF SIDE LOBE RATIO (SLR)

The first side lobe level is the maximum value of the first side lobe. **Side Lobe Ratio** (SLR) is defined as the ratio of the first side lobe level to the maximum value of the main beam. That is,

$$SLR = \frac{\text{first side lobe level}}{\text{main beam maximum}}$$

$$SLR \text{ in dB} = 20 \log_{10} SLR$$

A typical pattern is shown in Fig. 8.11. To obtain SLR, it may not be required to obtain complete radiation pattern.

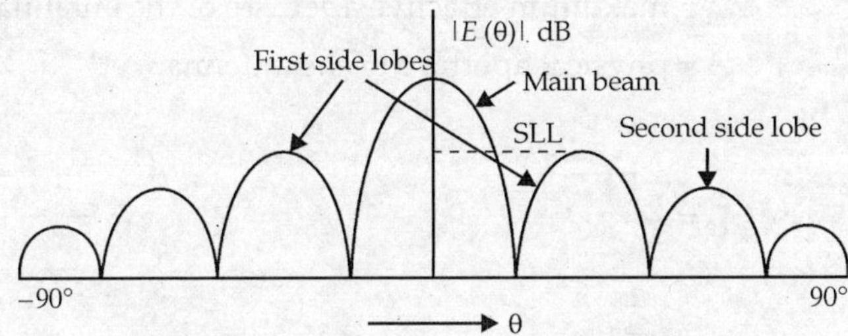

Fig. 8.11 *Measurement of first side lobe level*

Procedure

1. Fix AUT in receiving mode and rotate the transmitting antenna.
2. Note down the $E_{\max}$ and second maximum of the field strength.
3. SLR is obtained from

$$SLR = \frac{\text{second maximum of } E}{E_{\max}}$$

Let this be R.

4. SLR in dB $= 20 \log_{10} R$.

8.16 MEASUREMENT OF RADIATION EFFICIENCY

Antenna efficiency, η is defined as

$$\eta \equiv \frac{\text{radiated power}}{\text{input power}}$$

$$= \frac{W_r}{W_i} = \frac{W_r}{W_r + W_l} = \frac{g_p}{g_d}$$

here $\qquad W_r$ = radiated power

$\qquad\qquad\quad W_i$ = input power

$\qquad\qquad\quad W_l$ = power loss

$\qquad\qquad\quad g_p$ = power gain

$\qquad\qquad\quad g_d$ = directive gain.

Therefore, η is measured by measuring the directive and power gains as described in the previous sections.

8.17 MEASUREMENT OF ANTENNA APERTURE EFFICIENCY,η_a

Antenna aperture efficiency is defined as the ratio of maximum effective area of the antenna to the aperture area. That is,

$$\eta_a = \frac{A_{em}}{A}$$

here $\qquad A_{em}$ = maximum effective aperture of the antenna

$\qquad\qquad A$ = physical aperture of the antenna.

A_e is given by

$$A_e = \frac{\lambda^2}{4\pi} g_d$$

or $\qquad\qquad A_{em} = \frac{\lambda^2}{4\pi} (g_d)_{max}$

But $\qquad (g_d)_{max}$ = directivity, D

So, $\qquad\qquad A_{em} = \frac{\lambda^2}{4\pi} D$

or $\qquad\qquad D = \frac{4\pi}{\lambda^2} A_e$

As $\qquad\qquad \eta_a = \frac{A_{em}}{A}$

$$= \frac{\lambda^2}{4\pi} D \times \frac{1}{A}$$

$$\eta_a = \frac{D}{A} \frac{\lambda^2}{4\pi}$$

Therefore, aperture efficiency is obtained by measuring the directivity D and calculating A and λ. Typical value of η_a lies between 0.5 and 1.0.

Aperture illumination efficiency, η_{ai} Aperture illumination efficiency for a planar aperture is defined as the ratio of its directivity to the directivity obtained with uniform aperture illumination. That is,

$$\eta_{ai} \equiv \frac{\text{Directivity of the aperture}}{\text{Directivity of the aperture with uniform illumination}}.$$

8.18 MEASUREMENT OF POLARISATION OF ANTENNA

Polarisation of antenna is defined as the polarisation of its radiated wave. The polarisation of electromagnetic wave is the direction of its electric field. In general, the direction of electric field with time forms an ellipse.

The ellipse has either clockwise or anti-clockwise sense. When the ellipse becomes a circle, the polarisation is circular. When the ellipse becomes a straight line, the polarisation is linear.

The clockwise rotation of electric field with time is called right-hand polarisation and anti-clockwise rotation of electric field is called left-hand polarisation.

The electric field consists of both E_θ and E_ϕ components.

The direction of rotation along the direction of propagation represents the sense of polarisation.

The axial ratio (AR) and tilt angle, α describe the ellipse. α is measured from the reference direction in the clockwise direction.

The methods of measurement of polarisation are:

1. Polarisation pattern method
2. Linear component method
3. Circular component method
4. Power measurement method

1. **Polarisation method**

   Procedure

(*a*) A rotatable half-wave dipole is connected to a calibrated receiver as in Fig. 8.12.

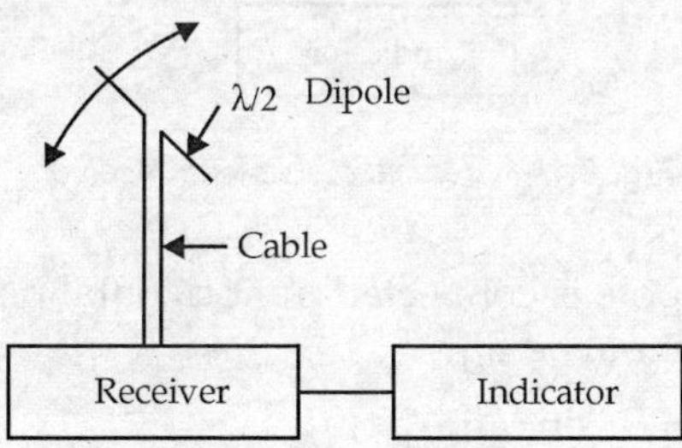

Fig. 8.12 *Polarisation measurement by polarisation pattern method*

(*b*) The dipole is rotated and incident field coming from AUT is measured. AUT is used in transmitting mode.

(*c*) If the variation of received signal forms an ellipse as in Fig. 8.13 the AUT is said to be elliptically polarised.

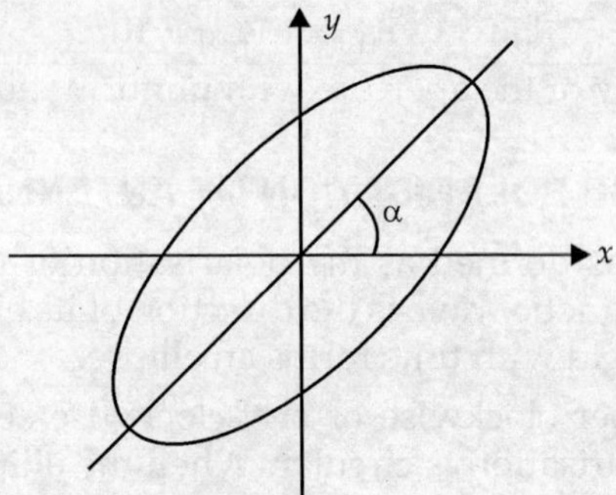

Fig. 8.13 *Tilted ellipse*

(*d*) The sense of polarisation is obtained by using two antennas. Here one is right-hand circular polarised and the other left-hand circular polarised. The antenna which receives a large signal gives the sense of polarisation.

2. **Linear component method**

   Procedure

(*a*) The AUT is used in transmitting mode.

(*b*) The signal coming from AUT is measured by a vertical antenna as in Fig. 8.14. Let the signal be E_v.

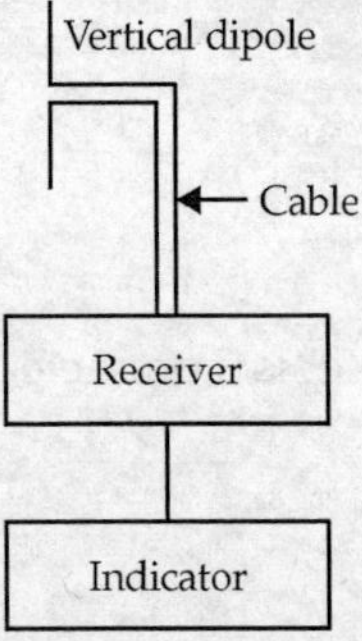

Fig. 8.14 *Vertical dipole with receiver*

(*c*) Now the vertical dipole is connected in horizontal position and the signal is measured. Let the signal be E_H.

Then,
$$E_x = E_v \sin (\omega t - \beta z)$$

$$E_y = E_H \sin (\omega t - \beta z + \alpha)$$

here α = phase difference between the two signals

$$\beta = \frac{2\pi}{\lambda}, \quad \omega = \text{angular frequency.}$$

(*d*) The phase difference, α is measured by a phase comparative method.

(*e*) The signal from the vertical antenna is measured as in step 2. But the signal from the horizontal antenna is connected to a matched terminated slotted line. The probe in the slotted line is connected to the receiver as in Fig. 8.15.

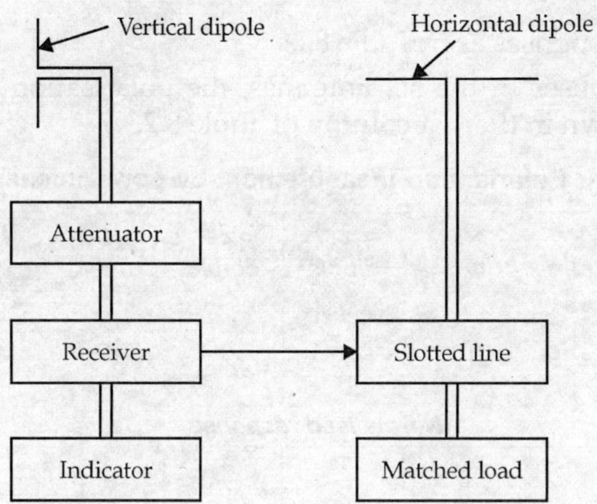

Fig. 8.15   *Phase comparison*

If α lies in $0 < \alpha < 180°$, the direction of rotation is clockwise. If α lies in $0 < \alpha < -180°$, the direction of rotation is anti-clockwise. The angle of tilt ϕ_t is given by

$$\phi_t = \frac{1}{2}\tan^{-1}\left(\frac{2E_1 E_2 \cos\alpha}{E_1^2 - E_2^2}\right) \qquad \ldots(8.20)$$

3. **Circular component method**   In this method, two circularly polarised antennas of opposite sense, for example, left and right-hand helical antennas, are used to receive the signals E_L and E_R from AUT. The set-up for measurement using this method is shown in Fig. 8.16.

The axial ratio is given by

$$AR = \frac{E_R + E_L}{E_R - E_L} \qquad \ldots(8.21)$$

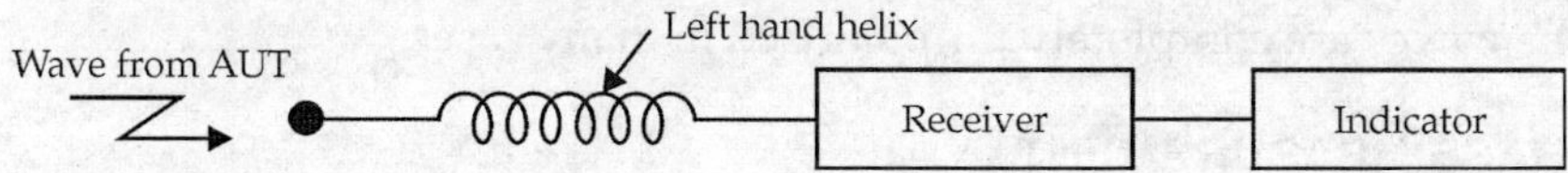

Fig. 8.16   *Polarisation measurement by circular component method*

4. **Power measurement method**

Procedure

(*a*) AUT is used in the transmitting mode.

(*b*) Six types of antennas, namely, vertical dipole, horizontal dipole, a dipole with an inclination of $+45°$, a dipole with an inclination of $-45°$, right circularly polarised helix and left circularly polarised helix are used to measure the incident power.

(*c*) Tabulate the responses as in Table 8.2.

(*d*) From the responses of the six antennas, the polarisation of the antenna is decided as shown in the last column of Table 8.2.

Table 8.2   Polarisation measurement by power measurement

Incident power normalised to unity	VP dipole	HP dipole	$+45°$ dipole	$-45°$ dipole	RCP helix	LCP helix	Polarisation of wave
			Normalised response				
1	1	0	$\frac{1}{2}$	$\frac{1}{2}$	$\frac{1}{2}$	$\frac{1}{2}$	VP
1	0	1	$\frac{1}{2}$	$\frac{1}{2}$	$\frac{1}{2}$	$\frac{1}{2}$	HP
1	$\frac{1}{2}$	$\frac{1}{2}$	1	0	$\frac{1}{2}$	$\frac{1}{2}$	$+45°$ LP
1	$\frac{1}{2}$	$\frac{1}{2}$	0	1	$\frac{1}{2}$	$\frac{1}{2}$	$-45°$ LP
1	$\frac{1}{2}$	$\frac{1}{2}$	$\frac{1}{2}$	$\frac{1}{2}$	1	0	RCP
1	$\frac{1}{2}$	$\frac{1}{2}$	$\frac{1}{2}$	$\frac{1}{2}$	0	1	LCP
1	$\frac{1}{2}$	$\frac{1}{2}$	$\frac{1}{2}$	$\frac{1}{2}$	$\frac{1}{2}$	$\frac{1}{2}$	Unpolarised

Here VP = vertical polarisation

HP = horizontal polarisation

LP = linear polarisation

RCP = right circular polarisation

LCP = left circular polarisation

It is well-known that waves are completely polarised in communication.

The waves from celestial sources are partially polarised in radio astronomy.

The waves are completely un-polarised in many cases.

8.19 PHASE MEASUREMENT

Phase of a wave is defined as the fraction of the period, which has elapsed, measured from some fixed origin.

Phase difference between two sinusoidal waves/signals of the same frequency is defined as the difference of phase.

Phase shift is defined as the change in phase of a periodic wave or in the phase difference between two or more waves.

There are two methods to measure phase patterns:

1. The measurement of phase pattern at short distance
2. The measurement of phase pattern at long distance.

1. **The measurement of phase pattern at short distance**

Procedure   The measurement set-up is shown in Fig. 8.17.

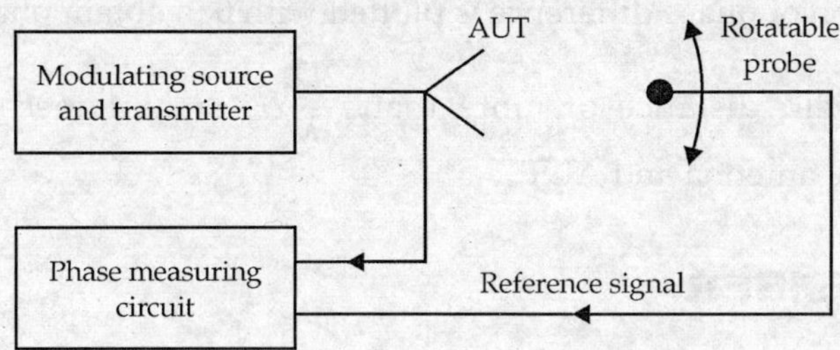

Fig. 8.17   *Phase pattern measurement at short distance*

(*a*) The measurement set-up consists of a modulating source, transmitter, AUT, a rotatable probe and phase measuring circuit.

(*b*) AUT is used as the transmitting antenna.

(*c*) The rotatable probe is used as the receiving antenna.

(*d*) The transmitting signal is used as the reference signal.

(*e*) By varying the probe position, the received signal is fed to the phase measuring circuit.

(*f*) The phase measuring circuit can be a dual-channel heterodyne system.

(*g*) For each angular position of the probe, the phase difference is noted.

(*h*) The variation of phase difference is plotted with θ to obtain phase pattern.

2. **The measurement of phase pattern at long distance**

Procedure   The measurement set-up is shown in Fig. 8.18.

(*a*) The measurement set-up consists of a modulating source and transmitter, transmitting antenna, AUT, a reference antenna and the phase measuring circuit.

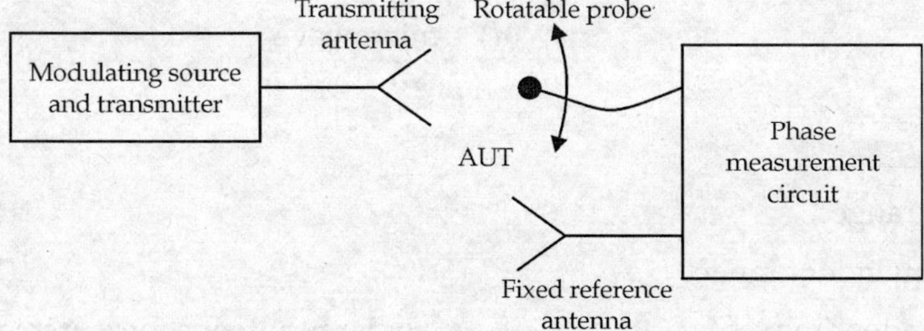

Fig. 8.18   *The phase pattern at long distance*

(*b*) AUT is a rotatable probe.

(*c*) AUT is used as the receiving antenna.

(*d*) By varying the probe position, the received signal is fed to the phase measuring circuit along with the signal from the reference antenna.

(*e*) The position of the reference antenna is fixed.

(*f*) For each position of the probe, the phase difference is noted.

(*g*) The phase measuring circuit is a dual-channel heterodyne system.

(*h*) The variation of phase difference is plotted with θ to obtain phase pattern.

(*i*) Here, far-field distance of more than $\dfrac{2D_a^2}{\lambda}$ is maintained between the transmitting antenna and AUT.

 ## POINTS TO REMEMBER

1. The far-field distance between the transmitting and receiving antennas should be $r > \dfrac{2D_a^2}{\lambda}$.

2. GTD represents geometrical theory of diffraction.

3. The antenna measurements are mainly classified as indoor and outdoor ranges.

4. Anechoic chambers and GTEM cell are popular for measurements.

5. Wheatstone bridge is used to measure antenna impedance upto 30 MHz.

6. The slotted lines are used to measure antenna parameters like VSWR, impedance and reflection coefficient.

7. Antenna gain measurements are made by comparison methods.

8. Antenna polarisation can be easily obtained by received power measurement.

 ## OBJECTIVE QUESTIONS

1. Wheatstone bridge is used to measure antenna impedance in the frequency range of
   (*a*)　upto 30 MHz　　　　　　　　　　(*b*)　between 30 and 50 MHz
   (*c*)　UHF　　　　　　　　　　　　　　(*d*)　microwave

2. GTD is _______________.

3. MOM means _______________.

4. EMI is less in outdoor ranges.　　　　　　　　　　　　　　　　　　　(Yes/No)

5. Anechoic chamber is an indoor range.　　　　　　　　　　　　　　　　(Yes/No)

6. One disadvantage of indoor ranges is _______________ for large antenna measurements.

7. Reflection ranges are used in the frequency range of _______________.

8. In all antenna measurements, the antenna is illuminated by ________________.

9. The disadvantages of compact ranges are ________________.

10. In anechoic chambers, the materials fixed to the walls, ceiling, floor

    (*a*)   absorb EM energy   (*b*)   reflect EM energy

    (*c*)   refract EM energy   (*d*)   diffract EM energy

11. Real part of antenna impedance consists of ________________.

12. Substitution method of antenna measurement is better than the bridge method.   (Yes/No)

13. Impedance measurement by slotted line method is better suited at

    (*a*)   VHF   (*b*)   VLF

    (*c*)   HF   (*d*)   UHF and microwave frequencies

14. Slotted line consists of ________________.

15. VSWR is given by

    (*a*) $\dfrac{V_{min}}{V_{max}}$ (*b*) $\dfrac{V_r}{V_i}$

    (*c*) $\dfrac{V_i}{V_r}$ (*d*) $\dfrac{V_{max}}{V_{min}}$

16. Wavelength in free space and wavelength in a waveguide are the same.   (Yes/No)

17. Reflection coefficient magnitude is given by

    (*a*) $\dfrac{\text{VSWR} - 1}{\text{VSWR} + 1}$ (*b*) $\dfrac{\text{VSWR} + 1}{\text{VSWR} - 1}$

    (*c*) $\dfrac{V_i}{V_r}$ (*d*) $\dfrac{V_{max}}{V_{min}}$

18. Self impedance, Z_{self} of an antenna is the ________________.

19. Field strength pattern is the variation of $|E|$ with θ.   (Yes/No)

20. Power pattern is the variation of $|E|^2$ with θ.   (Yes/No)

21. For far-field antenna measurement, R must be

    (*a*) $\geq \dfrac{2D_a^2}{\lambda}$ (*b*) $\geq \dfrac{2D_a}{\lambda}$

    (*c*) $= \dfrac{\lambda}{2D_a^2}$ (*d*) $2\left(\dfrac{D_a}{\lambda}\right)^2$

22. Effective length of an antenna is always greater than actual length.   (Yes/No)

23. Power gain of an antenna is the ratio of maximum radiation intensity of the antenna and maximum radiation intensity of the reference antenna.   (Yes/No)

24. For a loss-less and matched antenna, gain of the antenna over isotropic source is

    (a)   directivity
    (b)   greater than directivity
    (c)   less than directivity
    (d)   not related to directivity

25. Half-power beam width is _______________ θ_2 and θ_1 being the angles corresponding to 3 dB level.

26. Null-to-Null beam width is _______________ θ_2 and θ_1 being the angle corresponding to two nulls of the main beam.

27. Antenna efficiency is defined as radiation power/input power.   (Yes/No)

28. Antenna efficiency is

    (a) $\dfrac{g_d}{g_p}$
    (b) $\dfrac{g_p}{g_d}$
    (c) g_p
    (d) g_d

29. Antenna aperture efficiency is

    (a) $\dfrac{A_{em}}{A}$
    (b) $\dfrac{A}{D}$
    (c) $\dfrac{D}{A_{eff}}$
    (d) $\dfrac{g_p}{g_d}$

30. General value of aperture efficiency lies between

    (a)   0 to 0.5
    (b)   1 to 2
    (c)   0.1 to 0.3
    (d)   0.5 to 1

31. Linear polarised wave is produced by _______________.

32. Circular polarised wave is produced by _______________.

33. In radio astronomy, waves from celestial sources are _______________ polarised.

34. In communications, the electromagnetic waves are _______________ polarised.

35. The polarisation of horizontal dipole is _______________.

36. The polarisation of vertical dipole is _______________.

37. The polarisation of inclined dipole is _______________.

38. produces _______________ polarised waves.

39. produces _______________ polarised waves.

40. Phase difference is _______________.

41. If the response of a vertical dipole is 1 for a unity normalised input power, the polarisation is

(*a*) horizontal

(*b*) vertical

(*c*) circular

(*d*) elliptical

42. If the response of a vertical dipole is 0 for a unity normalised input power, the polarisation is

(*a*) vertical

(*b*) unpolarised

(*c*) horizontal

(*d*) circular

43. If the response of RCP helix is 0 for a unity normalised incident power, the polarisation of the test antenna is

(*a*) horizontal

(*b*) LCP

(*c*) vertical

(*d*) RCP

44. If the response of RCP helix is 1 for a unity normalised incident power, the polarisation of the test antenna is

(*a*) RCP

(*b*) LCP

(*c*) linear

(*d*) horizontal

45. If the response of any polarised antenna is $\dfrac{1}{2}$ for a unity normalised incident power, the polarisation of test antenna is

(*a*) unpolarised

(*b*) linear

(*c*) horizontal

(*d*) circular

46. Range of VSWR is

(*a*) 0 to 1

(*b*) 1 to ∞

(*c*) 0 to ∞

(*d*) $-\infty$ to ∞

47. VSWR is

(*a*) $\dfrac{1+\rho}{1-\rho}$

(*b*) $\dfrac{1+|\rho|}{1-|\rho|}$

(*c*) $\dfrac{1-|\rho|}{1+|\rho|}$

(*d*) $\dfrac{1-\rho}{1+\rho}$

48. Reflection coefficient is

(*a*) $\dfrac{z_0 - z_L}{z_L + z_0}$

(*b*) $\dfrac{z_L - z_0}{z_L + z_0}$

(*c*) $\dfrac{z_L + z_0}{z_L - z_0}$

(*d*) $\dfrac{z + z_0}{z_0 + z_L}$

49. Transmission coefficient is

(*a*) $1 - \rho$

(*b*) $1 + \rho$

(*c*) $\rho - 1$

(*d*) $1 + |\rho|$

50. If the response of a horizontal dipole is 1 for a unity normalised incident power, the polarisation of the test antenna is (*a*)   horizontal   (*b*)   vertical

    (*c*)   circular   (*d*)   elliptical

51. If the feed point of the antenna is at a current maximum, the input impedance is only real.

    (Yes/No)

52. Antenna band width is often expressed as a percentage of the optimum frequency of operation of the antenna.

    (Yes/No)

53. Antenna gain is inversely proportional to beam width.   (Yes/No)

54. The percent band width of an antenna with an optimum frequency of operation of 500 MHz and − 3 dB frequencies of 475 and 525 MHz is ______________.

55. For an ideal antenna, directive and power gains are equal.   (Yes/No)

56. For an ideal antenna, the radiation resistance and input impedance are equal.   (Yes/No)

57. The impedance at the ends of the antenna is maximum.   (Yes/No)

58. A Hertz antenna is a resonant antenna.   (Yes/No)

59. Standing waves are present along half-wave dipole.   (Yes/No)

ANSWERS

1. (*a*)   **2.** Geometrical theory of refraction   **3.** Method of moment

4. No   **5.** Yes   **6.** Space limitation   **7.** UHF to 16 GHz

8. Uniform plane wave   **9.** Wall reflections   **10.** (*a*)

11. Radiation resistance and loss resistance   **12.** Yes   **13.** (*d*)

14. Longitudinal slot in a waveguide broad wall   **15.** (*d*)   **16.** No

17. (*a*)   **18.** The impedance of isolated antenna   **19.** Yes   **20.** Yes

21. $\geq \dfrac{2D_a^2}{\lambda}$ **22.** No   **23.** Yes   **24.** (*a*)   **25.** $(\theta_2 - \theta_1)^\circ$ **26.** $(\theta_2 - \theta_1)^\circ$

27. Yes   **28.** (*b*)   **29.** (*a*)   **30.** (*d*)   **31.** Dipoles

32. Helical antennas   **33.** Partially   **34.** Completely   **35.** Horizontal

36. Vertical   **37.** Linear   **38.** Right circularly   **39.** Left circularly

40. Product of path difference and wave number   **41.** (*b*)   **42.** (*c*)

43. (*b*)   **44.** (*a*)   **45.** (*a*)   **46.** (*b*)   **47.** (*b*)   **48.** (*b*)

49. (*a*)   **50.** (*a*)   **51.** Yes   **52.** Yes   **53.** Yes   **54.** 10%

55. Yes   **56.** Yes   **57.** Yes   **58.** Yes   **59.** Yes.

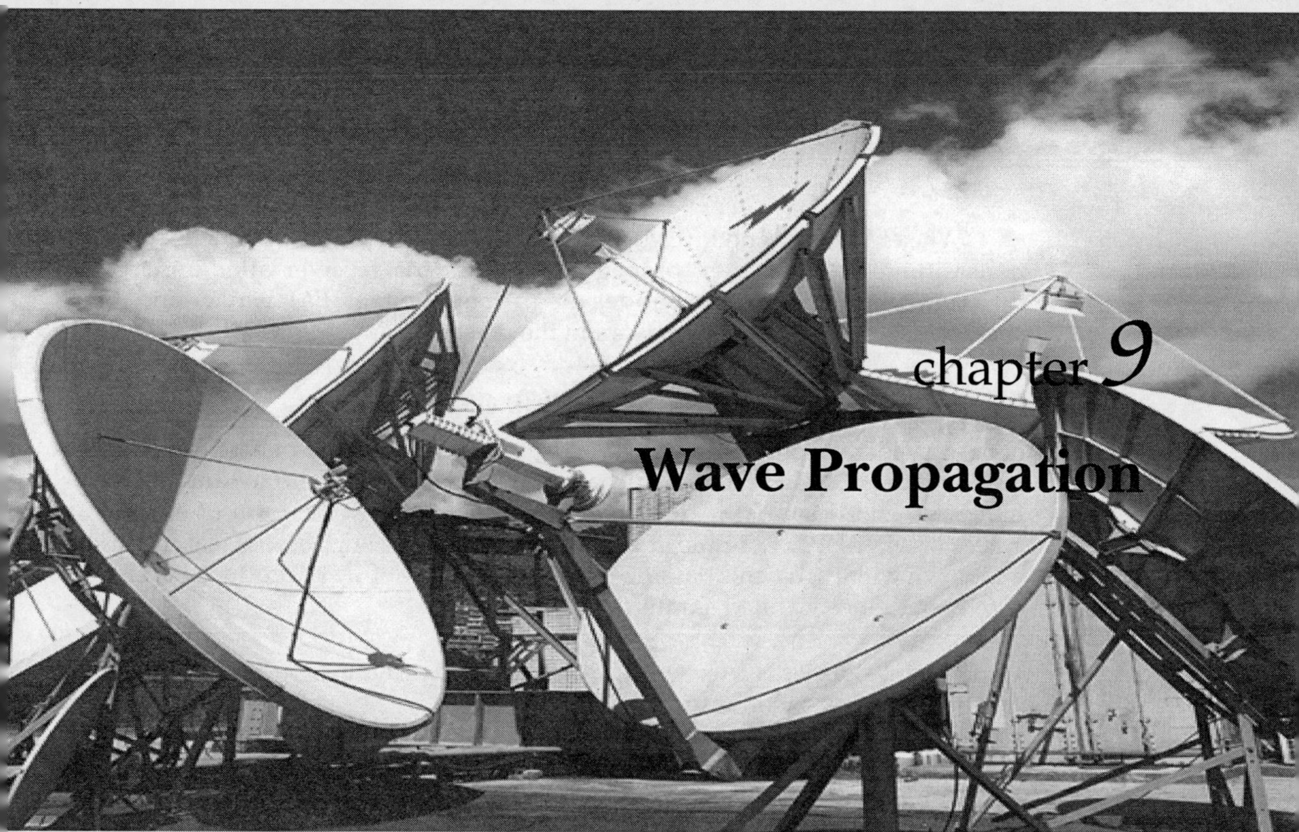

"The wave propagation characteristics between transmitter and receiver are controlled by the transmitting antenna, operating frequencies and media between them."

CHAPTER OBJECTIVES

This chapter discusses

- Different methods of wave propagation between transmitter and receiver
- Merits and demerits of each method
- Characteristics of Ground, Tropospheric and Ionospheric propagation
- Field strength estimation in each method
- The losses and attenuation
- Objective questions and solved problems useful for class tests, final examinations and also for competitive examinations
- Exercise problems to develop self problem solving skills

When Electromagnetic (EM) waves carrying information are generated by the transmitting antenna, they propagate towards the receiver after undergoing different phenomena. The modes of propagation of EM waves and their characteristics are described in detail in the present chapter.

9.1 PROPAGATION CHARACTERISTICS OF EM WAVE

Let us consider the physical example of dropping a stone into a pool of water. When a stone is dropped in water, disturbances take place on the surface and the water moves up and down. This disturbance is transmitted in the form of expanding circles of waves. For example, if a leaf is placed on the surface, the leaf moves up and down with each wave passing under it. This type of wave produced under the above conditions is called a **transverse wave**.

A **transverse wave** is a wave which occurs in directions perpendicular to the direction of the propagation.

An EM wave radiated by a transmitting antenna is a transverse wave. A **transverse wave** is also called **travelling wave**.

When an EM wave is produced by an antenna it moves from the transmitter to the receiver in the following ways:

1. A part of the wave travels along or near the surface of the earth. This wave is called the **ground wave** or **surface wave**.

2. Some waves travel directly from the transmitting to the receiving antenna. That is, these waves do no follow the earth and also do not move towards the sky. These waves are called **space waves**.

3. Some waves travel upwards into space towards the sky and get reflected back to the receiver. These waves are called **sky waves**.

9.2 FACTORS INVOLVED IN THE PROPAGATION OF RADIO WAVES

As explained above, an EM wave/Radio wave travels from the transmitter to the receiver in three different types of waves:

1. **Ground wave or surface wave**

2. **Space wave or tropospheric wave**

3. **Sky wave or ionospheric wave**.

Ground wave which is also called **surface wave** exists when the transmitting and receiving antennas are close to the earth and are **vertically polarised**. This type of wave propagation is useful at broadcast and low frequencies. The broadcast signals received during day-time are due to ground waves. It is useful for communication at VLF, LF and MF.

Space wave is also called **tropospheric wave**. Here, the wave propagates directly from the transmitter to the receiver in the tropospheric region. The portion of the atmosphere above the earth and within 16 km is called **troposphere**. This is useful above the frequency of 30 MHz. **FM reception** is normally by space wave propagation.

Sky wave is called **ionospheric wave**. The signal reception here is by reflection of the waves from the ionosphere. The **ionosphere** is an ionised region which lies approximately between 60 km to 450 km of atmosphere. Long distance communication is possible by this mode of propagation. It is useful for frequencies between 2 to 30 MHz.

When an EM wave travels from the transmitter to the receiver, there are several factors that influence the propagation. The factors are:

1. Earth's characteristics in terms of conductivity, permittivity and permeability.
2. Frequency of operation.
3. Polarisation of transmitting antenna.
4. Height of the transmitting antenna.
5. Transmitter power.
6. Curvature of the earth.
7. Obstacles between the transmitter and receiver.
8. Electrical characteristics of the atmosphere in the tropospheric region.
9. Moisture content in the troposphere.
10. Characteristics of the ionosphere.
11. Earth's magnetic field.
12. Refractive index of troposphere and ionosphere.
13. Permittivity of the tropospheric and ionospheric regions.
14. The distance between the transmitter and receiver.
15. Roughness of the earth.
16. Type of earth like hilly terrain, forest, sea water or river water.

9.3 GROUND WAVE

A wave is said to be **ground wave** or **surface wave** when it propagates from transmitter to receiver by gliding over the surface of the earth. This wave exists when,

- both transmitting and receiving antennas are close to the earth and
- the antennas are vertically polarised.

Ground wave is useful at low frequency broadcast applications. The broadcast signals received in the day-time are by ground waves. Ground wave signal is limited to only a few kilometers. The horizontally polarised antennas are not preferred as the horizontal component of the electric field in contact with the earth is short circuited by the earth. When the wave is in contact with the earth, it induces charges in the earth and constitute a current. The electric and magnetic fields of the wave over the earth are shown in Fig. 9.1.

The earth behaves like a leaky capacitance in carrying the induced current. **Equivalent circuit** of the earth is shown in Fig. 9.2.

The characteristics of the earth are described by its fundamental constants namely permittivity, conductivity and permeability.

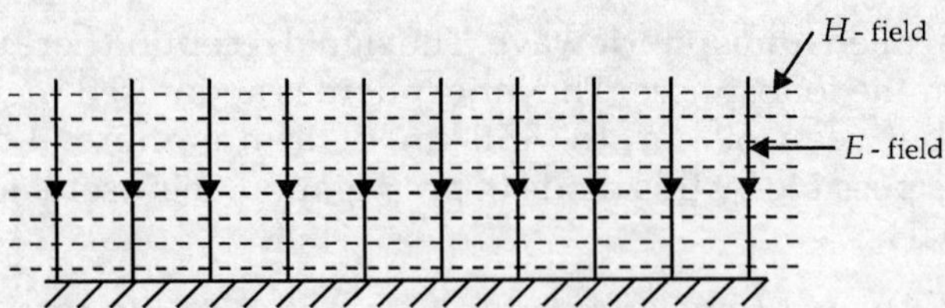

Fig. 9.1   *Vertically polarised wave*

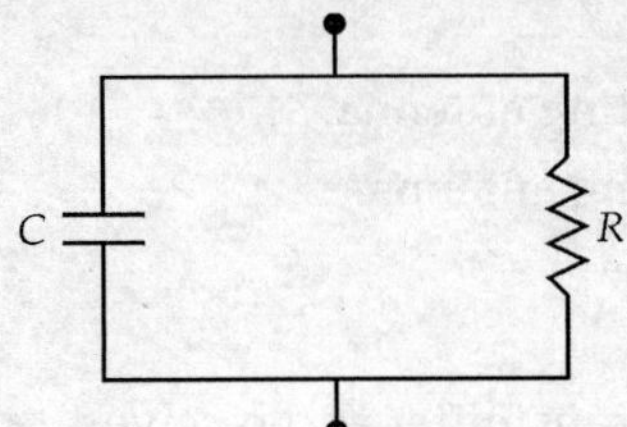

Fig. 9.2   *Equivalent circuit of the earth*

The wave is attenuated as it propagates due to imperfect nature of the earth (Fig. 9.3). The attenuation is mainly as a result of the absorption and reflection of EM energy by the earth. A part of this lost energy is compensated by the diffraction of additional energy coming from the atmospheric region above the earth.

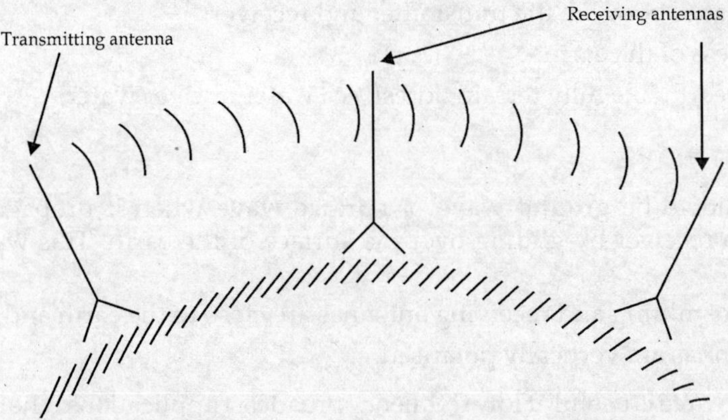

Fig. 9.3   *Ground wave between transmitting and receiving antennas*

9.4 GROUND WAVE FIELD STRENGTH

According to Somerfield analysis, the ground wave field strength for flat earth, is given by

$$E = \frac{AE_0}{d} \qquad \qquad \text{...(9.1)}$$

where $\qquad\qquad E$ = field strength at a point, V/m

E_0 = field strength of the wave at a unit distance from the transmitting antenna, neglecting earth's losses (V/m)

A = factor of the ground losses

d = distance of the point from Transmitting antenna

E_0 depends on

1. Power radiated by the transmitting antenna.
2. Directivity of the antenna in vertical and horizontal planes.

Example   Short vertical antenna is non-directional in horizontal plane. Its far-field is proportional to the cosine of the angle of elevation in vertical plane.

$$E_0 = 300 \text{ mV/m at } d = 1 \text{ km} \text{ and for a radiated power} = 1 \text{ kW}.$$

For other radiated powers, E_0 is proportional to the square root of power or

$$E_0 = 300 \sqrt{P_{kW}}, \text{ mV/m at 1 km}.$$

For antenna with vertical directivity, the field strength along the horizon is 1.41 times the field strength obtained with assumed cosine law.

The factor, A depends on

1. Conductivity, σ mho/m
2. Permittivity of the earth, $\in_r$
3. Frequency of the wave, f
4. Distance from the transmitter, d.

The determination of field strength due to Somerfield analysis mainly consists of determination of A.

This is found from the knowledge of **numerical distance**, p and **phase constant**, b. For vertical polarisation, these are given by

$$p = \frac{\pi}{D_f} \frac{d}{\lambda} \cos b \qquad\qquad ...(9.2)$$

$$b = \tan^{-1}\left(\frac{\in_r + 1}{D_f}\right) \qquad\qquad ...(9.3)$$

where $\qquad\qquad D_f = 1.80 \times 10^{12} \dfrac{\sigma}{f}$

D_f is known as dissipation factor of the dielectric

$$\sigma = \text{conductivity, (mho)/cm of earth}$$
$$\in_r = \text{relative permittivity of earth}$$
$$f = \text{frequency in Hz}$$
$$\frac{d}{\lambda} = \text{normalised distance with respect to } \lambda$$
$$\lambda = \text{wavelength}$$

The salient features of ground wave propagation

1. Ground wave propagates by gliding over the surface of the earth.

2. It exists for vertically polarised antennas.

3. It exists for antennas close to the earth.

4. It is suitable for VLF, LF and MF communications.

5. It can be used even at 15 kHz and upto 2 MHz.

6. Ground wave field strength is $E = \dfrac{AE_0}{d}$.

7. The ground wave field strength varies with characteristics of the earth.

8. Ground waves require relatively high transmitter power.

9. Ground wave propagation losses vary considerably with the type of earth.

10. Ground waves are not affected by the changes in atmospheric conditions.

11. Ground waves can be used to communicate between any two points on the globe if there is sufficient transmitter power.

12. It can be used for radio navigation, for ship-to-ship, ship-to-shore communication and maritime mobile communications.

A typical variation of A with numerical distance, p and phase constant, b is shown in Fig. 9.4.

The factor, **A** can be also calculated approximately from the following expression:

$$p = \frac{0.582 d_{km} f^2 \,(\text{MHz})}{\sigma \,(\text{mho/m})}$$

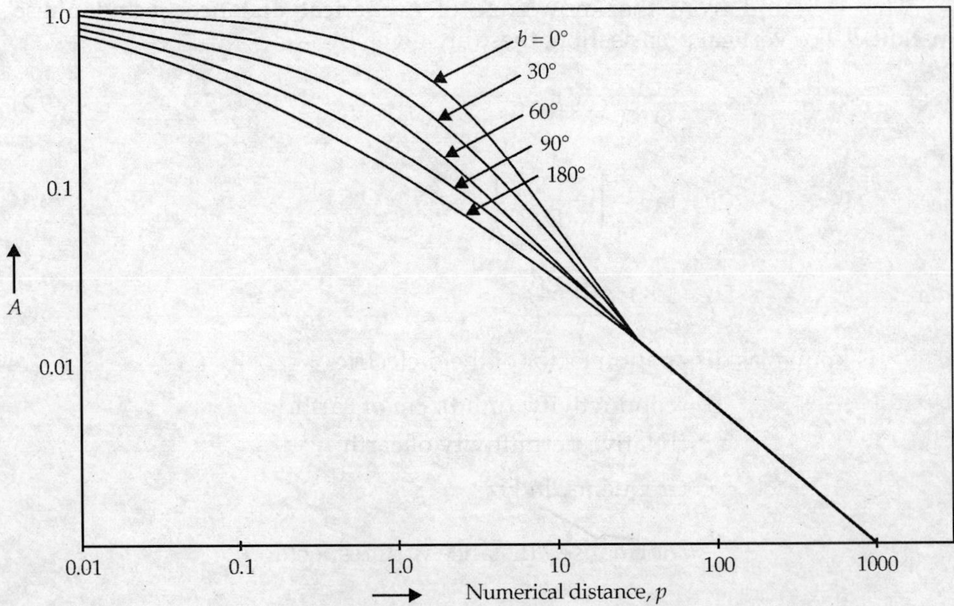

Fig. 9.4 *Variation of A with p and b*

and
$$A \approx \frac{2 + 0.3p}{2 + p + 0.6p^2} \quad \text{for } b < 5^\circ$$

For all values of phase constant, b,

$$A = \frac{2 + 0.3p}{2 + p + 0.6p^2} - \sin b \ \sqrt{\frac{p}{2}} \ e^{-\left(\frac{5}{8}\right)p}$$

where $\qquad p$ = numerical distance

However, for $b < 50^\circ$, $p < 4.5$, A is approximately given by $A = e^{-y}$

where $\qquad y = -0.43p + 0.01p^2$.

9.5 GROUND WAVE FIELD STRENGTH BY MAXWELL'S EQUATIONS

The field strength at a distance, d is given by

$$E = \frac{\eta_0 \, h_t \, I}{\lambda d} \text{ volts/m} \qquad \qquad \text{...(9.4)}$$

where $\qquad h_t$ = effective height of transmitting antenna, m

$\qquad \eta_0$ = characteristics impedance of free space = $120\pi\,\Omega$

$\qquad I$ = antenna current, A

$\qquad d$ = distance from transmitter, m

$\qquad \lambda$ = wavelength, m

When a receiving antenna of height, h_r is placed at a distance of d the **received signal** is given by

$$V \approx \frac{\eta_0 \, h_t \, h_r \, I}{\lambda d}, \text{ volts.} \qquad \qquad \text{...(9.5)}$$

9.6 REFLECTION OF RADIO WAVES BY THE SURFACE OF THE EARTH

If an EM wave is incident on the earth, it is reflected back. The angle of reflection is equal to the angle of incidence. The reflection coefficient is the ratio of the reflected wave to the incident wave. That is,

$$\rho = \frac{\text{Reflected wave}}{\text{Incident wave}}$$

The field strength near the earth is the vectorial sum of the incident and reflected fields.

The **reflection coefficient**, ρ depends on

1. Dielectric constant, $\in_r$

2. Conductivity of the earth, σ

3. Frequency of the wave

4. Polarisation of the wave

5. Angle of incidence of the wave.

For perfect reflecting earth, $|\rho| = 1$ and for practical earth conditions, $|\rho| < 1$ and $\angle \rho \neq 0$.

Typical variations of magnitude of reflection coefficient and phase shift with the angle of incidence on the high conductivity earth at $f = 20$ MHz are shown in Fig. 9.5.

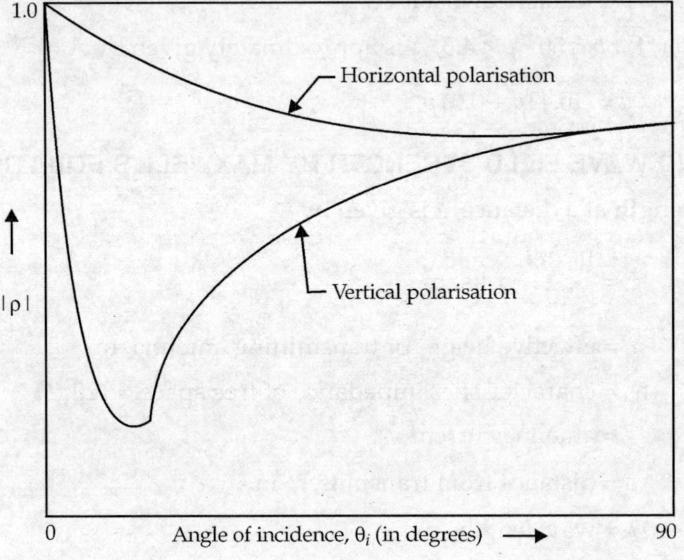

(**a**) *Variation of ρ with θ_i*

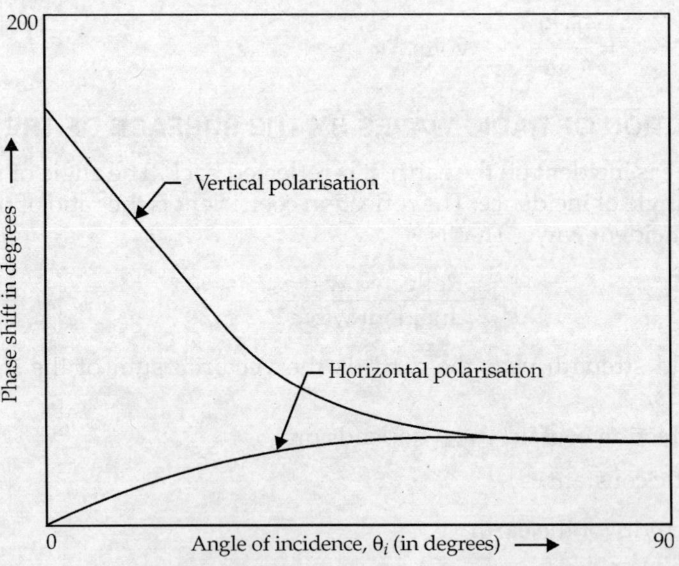

Fig. 9.5 (**b**) *Variation of $\angle \rho$ with θ_i*

It may be noted from Fig. 9.5, that for horizontally polarised waves, the reflection coefficient is the same as that of perfect earth for $\theta_i = 0$. That is, $\rho = 1 \angle 0$. When θ_i increases, $|\rho|$ reduces from 1 and the phase shift become small. The phase is found to be lagging with respect to that of perfect earth.

For vertically polarised wave, at $\theta_i = 0$, the reflection coefficient, ρ is $1 \angle 180^\circ$. At $\theta_i = 90^\circ$, the reflection coefficient for vertical and horizontally polarised waves are identical. The angle of incidence at which there is no reflection is known as **Brewster angle**.

$$\text{Brewester angle, } \theta_b = \theta_i = \tan^{-1} \sqrt{\frac{\in_{r2}}{\in_{r1}}} \, .$$

9.7 ROUGHNESS OF EARTH

Reflections of EM waves vary with the nature of the earth. Earth is classified as rough and smooth.

Roughness of earth According to the **Rayleigh criterion,** it is defined as:

$$R \equiv \frac{4\pi\sigma_s \sin\theta_i}{\lambda} \qquad \qquad ...(9.6)$$

where
σ_s = standard deviation of the surface irregularities relative to the mean surface height
θ_i = angle of incidence measured from earth's surface
λ = wavelength

If $R < 0.1,$ the earth is considered **electrically smooth.**
If $R > 10,$ it is said to be **electrically rough**.

9.8 REFLECTION FACTORS OF EARTH

Earth is neither a good conductor nor a good dielectric. It is a partially conducting dielectric medium and hence its dielectric constant can be considered to be complex. That is,

$$\in' = \left(\in + \frac{\sigma}{j\omega} \right)$$

Keeping this in mind, the reflection coefficient for horizontal polarisation is given by:

$$\rho_h = \frac{E_r}{E_i}$$

$$= \frac{\sin\theta_i - \sqrt{(\in_r - jD_f) - \cos^2\theta_i}}{\sin\theta_i + \sqrt{(\in_r - jD_f) - \cos^2\theta_i}} \qquad \qquad ...(9.7)$$

where
$$D_f = \frac{\sigma}{\omega\in_0}$$

θ_i = angle of incidence measured from earth's surface

For vertical polarisation,

$$\rho_v = \frac{(\epsilon_r - jD_f)\sin\theta_i - \sqrt{(\epsilon_r - jD_f) - \cos^2\theta_i}}{(\epsilon_r - jD_f)\sin\theta_i + \sqrt{(\epsilon_r - jD_f) - \cos^2\theta_i}}. \qquad \dots(9.8)$$

9.9 WAVE TILT OF THE GROUND WAVE

Wave tilt is defined as the change of orientation of the vertically polarised ground wave at the surface of the earth.

Salient features of wave tilt

1. Wave tilt occurs at the surface of the earth.
2. The tilt depends on conductivity and permittivity of the earth.
3. It causes power dissipation.
4. Due to tilt, there exists both horizontal and vertical components of the electric field.
5. These two components are not in phase.
6. The wave tilt changes the original vertically polarised wave into elliptically polarised wave.
7. The resultant typical electric field due to wave tilt is shown in Fig. 9.6.

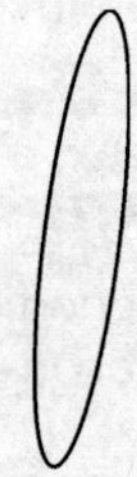

Fig. 9.6 *Electric vector due to wave tilt*

8. The horizontal component of the electrical field, E_h is

$$E_h = J_s\, Z_s.$$

9. Vertical component of electric field, E_v is

$$E_v = H\eta_0$$

where $\qquad Z_s$ = surface impedance of earth

$$= \sqrt{\frac{\omega\mu}{\sqrt{\sigma^2 + \omega^2\,\epsilon^2}}} \; \angle \; \frac{1}{2}\tan^{-1}\frac{\sigma}{\omega\epsilon} \qquad \dots(9.9)$$

$$\sigma = \text{conductivity (mho/m)}$$
$$\epsilon = \text{permittivity (F/m)}$$
$$\mu = \text{permeability of earth (H/m)}$$
$$J_s = \text{surface current density, (A/m)}$$
$$\eta_0 = \text{intrinsic impedance of free space } (\Omega).$$

10. The ratio of x and y components of E is given by

$$\frac{E_h}{E_v} = \frac{1}{377} \sqrt{\frac{\omega\mu}{\sqrt{\sigma^2 + \omega^2 \epsilon^2}}} \angle \frac{1}{2} \tan^{-1} D_f \qquad \ldots(9.10)$$

Example   For a typical earth, with $\sigma = 5 \times 10^{-3}$ mho/m and $\epsilon = 10\epsilon_0, \mu = \mu_0$

$$\frac{E_h}{E_v} = 0.105 \angle 41.83°$$

This means that the horizontal component of electric field is 10.5% of the vertical component and it leads E_y by $41.83°$.

9.10 SPACE WAVE OR TROPOSPHERIC WAVE PROPAGATION

The EM wave that propagates from the transmitter to the receiver in the earth's **troposphere** is called **space wave** or **tropospheric wave**.

Troposphere is the region of the atmosphere within 16 km above the surface of the earth (Fig. 9.7).

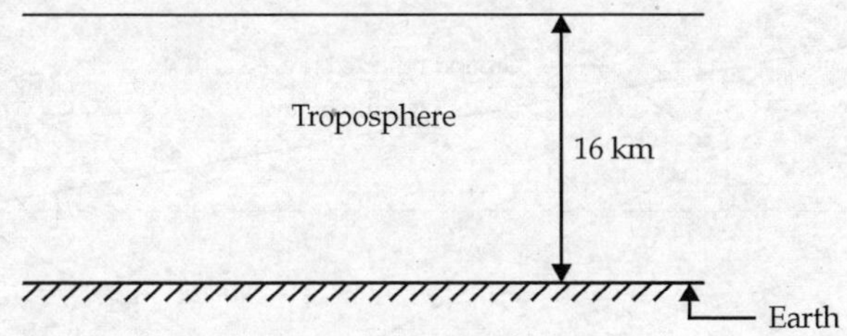

Fig. 9.7   *Troposphere*

In space wave propagation, the field strength at the receiver is contributed by

1. Direct ray from transmitter

2. Ground reflected ray

3. Reflected and refracted rays from the troposphere

4. Diffracted rays around the curvature of the earth, hills and so on.

However, the contribution of the first two rays is predominant.

Space wave propagation is useful at frequencies above 30 MHz. It is useful for FM, TV and radar applications. It is also used in VHF, UHF and higher frequency bands.

9.11 FIELD STRENGTH DUE TO SPACE WAVE

The **field strength** due to space wave,

$$E = \frac{2E_0}{d} \sin \frac{2\pi h_t\, h_r}{\lambda d} \qquad\qquad …(9.11)$$

or

$$E \approx \frac{4\pi h_t\, h_r}{\lambda d^2} E_0 \qquad\qquad …(9.12)$$

where $\quad E_0$ = field strength due to direct ray at unit distance. This depends on directivity of transmitting antenna and transmitter power. $E_0 = 137.6\ \sqrt{P_{kW}}$ mV/m at one mile for half- wave transmitting antenna.

h_t = height of the transmitting antenna

h_r = height of the receiving antenna

d = distance between the two antennas.

Proof   Assume flat earth.

Let h_t be the height of the transmitting antenna,

h_r be the height of the receiving antenna and

d be the distance between the two antennas.

The field strength at the receiver is mostly contributed by direct and ground reflected rays. These two rays are shown in Fig. 9.8.

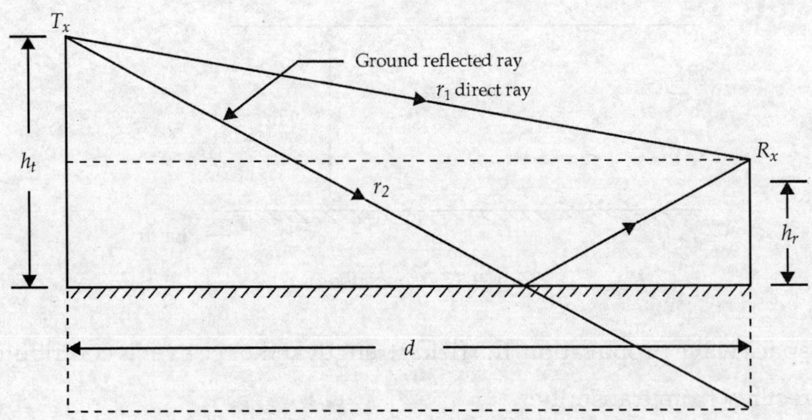

Fig. 9.8   *Direct and ground reflected rays in space wave*

From Fig. 9.8, we have

$$r_1^2 = (h_t - h_r)^2 + d^2$$

or

$$r_1 = d \left[1 + \left(\frac{h_t - h_r}{d} \right)^2 \right]^{1/2} \qquad\qquad …(9.13)$$

From Binomial series, we have

$$(1 \pm x)^{1/2} = 1 \pm \frac{1}{2} x - \frac{1}{2.4} x^2 + \ldots$$

If x is small, the higher order terms can be neglected. Then

$$(1 \pm x)^{1/2} \approx 1 \pm \frac{1}{2} x$$

Therefore Equation (9.13) can be written as

$$r_1 = d \left[1 + \frac{1}{2} \left(\frac{h_t - h_r}{d} \right)^2 \right]$$

$$\approx \left[d + \frac{(h_t - h_r)^2}{2d} \right] \qquad \ldots(9.14)$$

Similarly, $\qquad r_2^2 = d^2 + (h_t + h_r)^2$

or $\qquad r_2^2 = d^2 \left[1 + \frac{(h_t + h_r)^2}{d} \right]$

So, $\qquad r_2 \approx \left[d + \frac{(h_t + h_r)^2}{2d} \right] \qquad \ldots(9.15)$

From Equations (9.14) and (9.15), the path difference between the two ray is given by

$$r_2 - r_1 = d + \frac{(h_t + h_r)^2}{2d} - d - \frac{(h_t - h_r)^2}{2d}$$

$$= \left(\frac{h_t^2 + h_r^2 + 2h_t\,h_r}{2d} \right) - \left(\frac{h_t^2 + h_r^2 - 2h_t\,h_r}{2d} \right)$$

or $\qquad r_2 - r_1 = \frac{4h_t\,h_r}{2d} = \frac{2h_t\,h_r}{d} \qquad \ldots(9.16)$

The phase difference due to path difference, α is

$$\alpha = \text{path difference} \times k$$

where $\qquad k = \text{wave number} = 2\pi/\lambda$

or $\qquad \alpha = \frac{2h_t\,h_r}{d} \times \frac{2\pi}{\lambda}$

Let E_d be the field due to direct ray and E_r be the field due to reflected ray. Then the resultant field E_R at the receiver is given by:

$$E_R = (E_d + E_r\,e^{-j\,\psi}) \qquad \ldots(9.17)$$

When a wave is incident on earth, it is reflected with the same amplitude but with phase reversal. Therefore, the total phase is

$$\psi = 180^\circ + \alpha$$

where α is phase difference due to path difference. Moreover,

$$E_d = E_r = E_s.$$

Now Equation (9.17) becomes

$$E_R = E_s \{1 + e^{-j(180+\alpha)}\}$$

$$= E_s [1 + \cos(180 + \alpha) - j\sin(180 + \alpha)]$$

$$= E_s [1 - \cos\alpha + j\sin\alpha]$$

$$= E_s \lfloor (1 - \cos\alpha)^2 + \sin^2\alpha \rfloor$$

$$= E_s \lfloor \sqrt{2(1 - \cos\alpha)} \rfloor$$

$$= E_s \sqrt{4\sin^2\frac{\alpha}{2}}$$

$$E = 2E_s \sin\frac{\alpha}{2}$$

But

$$E_s = \frac{E_0}{d}$$

So,

$$E = \frac{2E_0}{d}\sin\frac{2\pi h_t h_r}{\lambda d} \qquad \qquad \dots(9.18)$$

As $d \gg h_t$ or h_r, Equation (9.18) becomes

$$E \approx \frac{4\pi h_t h_r}{\lambda d^2} E_0. \qquad \qquad \dots(9.19)$$

9.12 CONSIDERATIONS IN SPACE WAVE PROPAGATION

The **space wave field strength** is affected by the following:

1. Curvature of the earth.
2. Earth's imperfections and roughness.
3. Hills, tall buildings and other obstacles.
4. Height above the earth.
5. Transition between ground and space wave.
6. Polarisation of the waves.

9.12.1 Effect of the Curvature of the Earth

The expression for space wave field strength, Equation (9.19) is derived assuming flat earth. However, when the distance between the transmitting and receiving antennas is considerably large, curvature of the earth has considerable effect on space wave propagation.

The main effects of earth's curvature are:

1. The field strength at the receiver becomes small as the direct ray may not be able to reach the receiving antenna (Fig. 9.9). The earth-reflected rays diverge after their incidence on the earth.

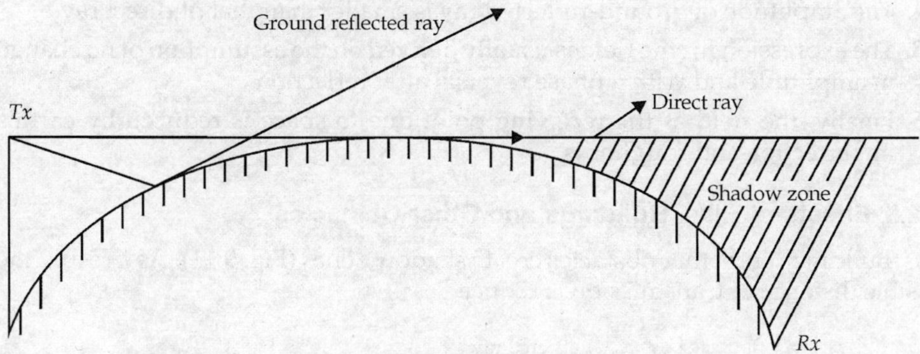

Fig. 9.9   *Effect of earth's curvature*

2. The curvature of the earth creates **shadow zones**. These are the regions where no signal reaches. Shadow zones are also called **diffraction zones**.

3. It reduces the possible distance of transmission.

4. The field strength that is available at the receiver becomes small.

When earth's curvature is considered, **the field strength at the receiver** becomes

$$E = \frac{2E_0}{d} \sin \frac{2\pi h'_t h'_r}{\lambda d} \qquad \qquad ...(9.20)$$

where h'_t and h'_r are the effective heights of transmitting and receiving antennas (Fig. 9.10). Here, effective height is the height of the antenna from the supposed flat earth.

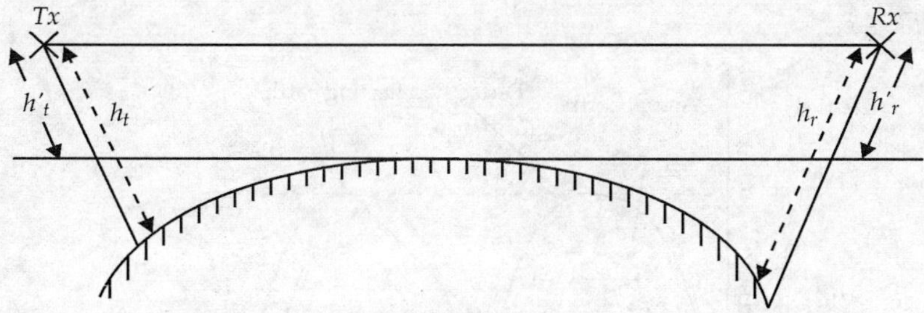

Fig. 9.10   *Effective and actual heights of antennas*

9.12.2 Effect of Earth's Imperfections and Roughness

Salient features

1. Earth is basically imperfect and electrically rough.

2. For perfect earth, reflection coefficient is unity. But actual earth makes it different from unity.

3. When a wave is reflected from perfect earth, its phase change is $180°$. But actual earth makes the phase change different from $180°$.

4. The amplitude of ground-reflected ray is smaller than that of direct ray.

5. The expression for the field is usually derived on the assumption of no change in amplitude and with a phase reversal after reflection.

6. Finally, the field at the receiving point due to space is reduced by earth's imperfection and roughness.

9.12.3 Effects of Hills, Buildings and Other Obstacles

Hills, buildings and other obstacles create shadow zones (Fig. 9.11). As a result, the possible distance of transmission is reduced.

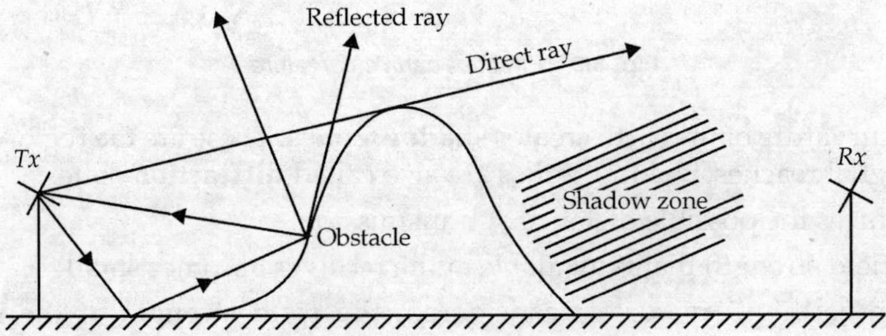

Fig. 9.11 *Effect of obstacles*

9.12.4 Effect of the Height Above the Earth

Salient features

1. The field varies with the height above the earth.

2. The field variation is characterised by the presence of maximas, minimas and nulls.

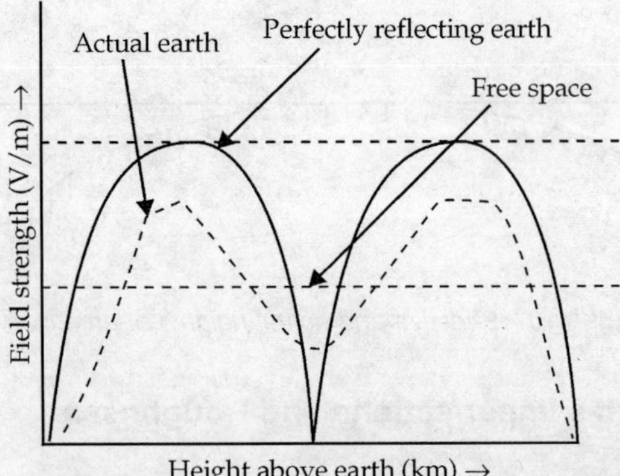

Fig. 9.12 *Typical variation of field strength with height*

3. The maximas and minimas depend on frequency, height of the transmitting antenna, ground characteristics and polarisation of the wave.

4. The variation of the field with height for perfect earth, actual earth and free space conditions is shown in Fig. 9.12.

9.12.5 Effect of Transition between Ground Wave and Space Wave

Salient features

1. When the transmitting antenna is close to earth, ground wave exists and the field strength is independent of the height of the antenna.

2. The antenna height has an effect on the field strength and direct and ground reflected rays predominate over the ground wave. Its effect depends on frequency, polarisation, and constants of the earth.

3. For vertically polarised wave, the ground wave does not dominate at heights of the order of λ or 2λ. That is, at higher heights of antennas, space wave dominates.

4. At heights $< \dfrac{\lambda}{10}$, transition between the ground wave and space wave takes place for horizontally polarised waves.

9.12.6 Effect of Polarisation

Salient features

1. For any angle of incidence other than $\theta_i = 0$ or $90°$, the magnitude of the reflected wave will be less with vertical polarisation than with horizontal polarisation. This reduces the amplitude of the ground-reflected wave.

2. The height below which ground wave action is to be taken into account is much less with horizontal polarisation than with vertical polarisation. It is important at broadcast and lower frequencies.

3. The electromagnetic interference created by ignition systems, domestic and consumer electrical, electronic and communication equipment and so on is, in general, vertically polarised. Horizontal polarisation is useful for discrimination against these disturbances occurring in TV and FM broadcasting.

9.13 ATMOSPHERIC EFFECTS IN SPACE WAVE PROPAGATION

The atmosphere consists of gas molecules and water vapour. This causes the dielectric constant to be slightly greater than unity. The density of air and water vapour vary with height. As a result, the dielectric constant and hence refractive index of air depends on the height. Dielectric constant decreases with height. The variation of refractive index with height gives rise to different phenomena like refraction, reflection, scattering, duct propagation and fading.

By definition, the **refractive index**, n is the square root of the dielectric constant.

$$n \approx \sqrt{\epsilon_r}$$

In order to study some of the characteristics of the troposphere, a new term namely **modified refractive index**, M is introduced.

Modified refractive index (M)   Modified refractive index in the troposphere is defined as the sum of the refractive indices at a given height above the mean geometrical surface and the ratio of the height to the mean geometrical radius. Mathematically, it is defined as

$$M = \left(n - 1 + \frac{h}{a} \right) \times 10^6 \qquad\qquad ...(9.21)$$

where
$$n = \text{refractive index}$$
$$h = \text{height above ground}$$
$$a = \text{radius of earth} = 6.37 \times 10^6 \text{ m}$$

At high altitudes, $\in_r$ or n is independent of height. Here, M increases by 0.048 units/ft. Near the earth, M increases linearly at a constant rate that is less than 0.048 units/ft.

For typical atmospheric conditions, $\dfrac{dM}{dh} = 0.036$ units/ft. This condition is called **standard atmosphere**.

The phenomenon of refraction in the troposphere due to change in refractive index is shown in Fig. 9.13.

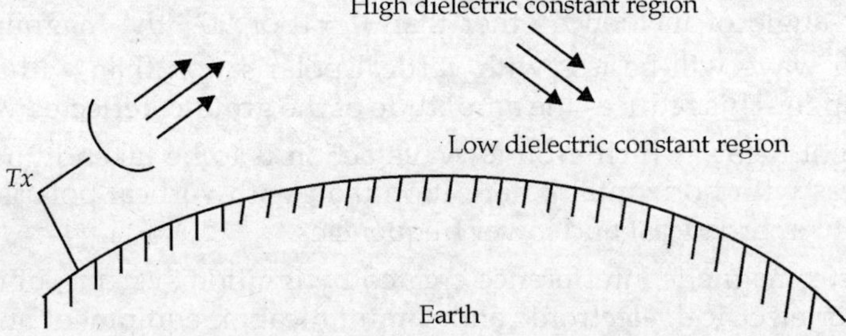

Fig. 9.13   *Refraction in troposphere*

The temperature inversion zone creates super refraction or duct phenomenon as shown in Fig. 9.14.

The EM rays bend away from one region to another region due to change in refractive index with height.

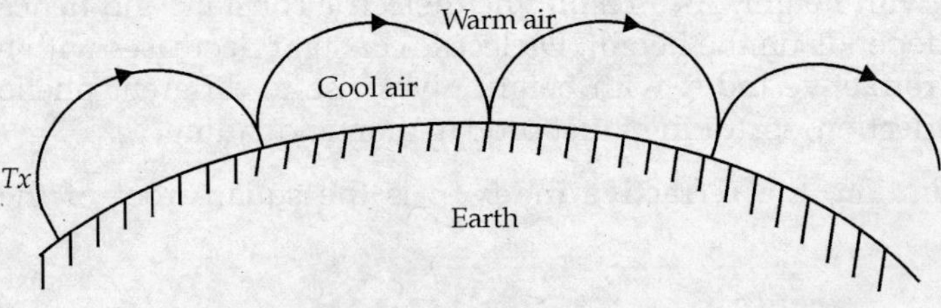

Fig. 9.14   *Super refraction or ducting*

In order to make a particular ray path like a straight line, the coordinates are changed. For this purpose, the **effective radius** of the earth is introduced in the following way.

$$K = \frac{\text{Effective radius of earth}}{\text{actual radius of earth}} = \frac{0.048}{dM/dh} \quad \text{for positive} \ \frac{dM}{dh} \qquad ...(9.22)$$

The **effective radius** of earth is also called **equivalent radius** of the earth.

Effective radius of the earth is defined as equivalent radius of the earth used to correct atmospheric refraction approximately, as refractive index of the atmosphere changes linearly with height.

For standard atmosphere, **effective radius of the earth** is equal to 4/3 times the actual radius of the earth. **Actual radius** of the earth is its **geometrical radius**.

When $\dfrac{dM}{dh} = 0.036$ units/ft for standard atmosphere, k becomes 4/3.

The equivalent radius of the earth becomes infinity for $\dfrac{dM}{dh} = 0$. Under these conditions, the ray path appears to be a straight line over a flat earth.

As radio waves exhibit curved ray paths, radio horizon differs from optical horizon.

9.14 DUCT PROPAGATION

Duct propagation is a phenomenon of propagation making use of the atmospheric duct region. The duct region exists between two levels where the variation of modified refractive index with height is minimum. It is also said to exist between a level, where the variation of modified refractive index with height is minimum, and a surface bounding the atmosphere.

In duct propagation, the ray which is parallel to the earth's surface travels round the earth in a series of hops with successive reflections from the earth. This is shown in Fig. 9.15.

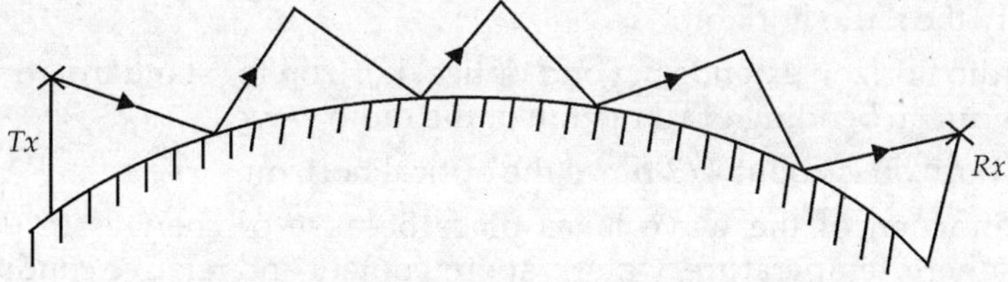

Fig. 9.15 *Duct propagation*

Salient features of duct propagation

1. It happens when $\dfrac{dM}{dh}$ is negative.

2. It happens when dielectric constant changes with height suddenly and rapidly.

3. It is a specific case of refraction of RF energy.

4. It takes place at VHF, UHF and microwave range and in areas contiguous to oceans.

5. It is similar to waveguide propagation of microwaves.

6. It is not a standard propagation.

7. It is a rare phenomenon.

8. It is not a dependable propagation.

9. It happens during monsoons.

10. Long distance communication is possible when duct phenomenon takes place.

11. The transmitting antenna should be within the duct. Otherwise, the signal, even if it is powerful, will not propagate.

12. It happens due to temperature inversion.

13. It occurs due to super refraction.

14. It takes place when low and high moisture regions exist.

9.15 RADIO HORIZON

Horizon means visible. It has another meaning, that is, a line at which earth and sky appear to meet.

Radio horizon of an antenna is defined as the locus of the distant points at which direct rays from the antenna become tangential to a planetary surface. The horizon is a circle on a spherical surface.

The distance to the horizon is affected by the atmospheric refraction.

Salient features of radio horizon

1. Radio horizon is the range by which a direct ray from transmitting antenna reaches receiving antenna.

2. The earth's curvature exhibits a horizon to space-wave propagation. This is actually the radio horizon.

3. The radio horizon extends beyond optical horizon for standard atmosphere. This is due to bending or refraction of the radio wave.

4. Radio horizon is about 4/3 times the optical horizon.

5. The refraction of the wave takes place because of changes in density of troposphere, temperature, water-vapour content and relative conductivity.

6. The radio horizon can be increased by increasing antenna heights.

The radio horizon distance between transmitting and receiving antennas

$$\text{Radio horizon distance, } d_{\text{miles}} = \sqrt{2h_t} \text{ (feet)} + \sqrt{2h_r} \text{ (feet)}$$
$$\text{or radio horizon distance, } \quad d_{\text{km}} = \sqrt{17h_t} \text{ (m)} + \sqrt{17h_r} \text{ (m)}$$

Proof   We have

$$K = \frac{a'}{a} = \frac{\text{equivalent radius of earth}}{\text{actual radius}} = \frac{0.048}{dM/dh}$$

where K is called as **effective earth's radius factor**.

For standard atmosphere,

$$\frac{dM}{dh} = 0.036 \text{ units/ft}$$

Equivalent radius, $a' = a \times \dfrac{0.048}{0.036}$

$$a' = \frac{4}{3}\, a \qquad\qquad\qquad ...(9.23)$$

Here a' is the equivalent radius of earth that accounts for refraction.
Consider Fig. 9.16.

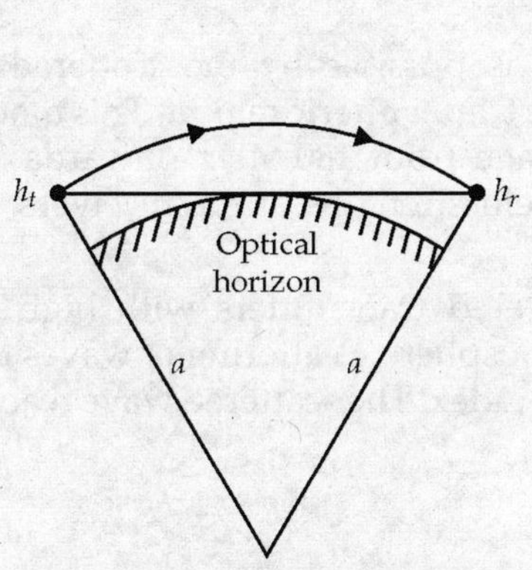

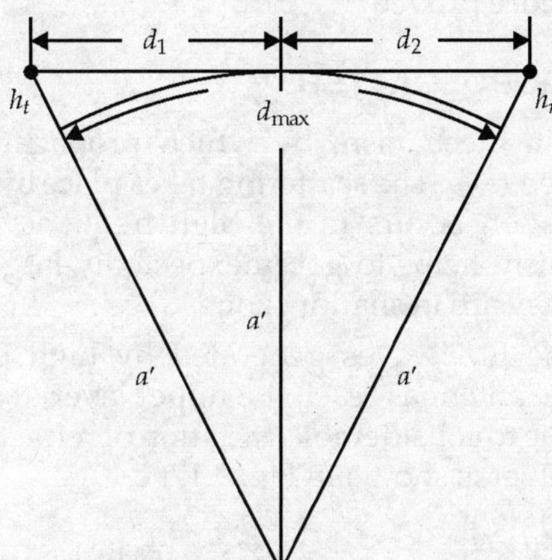

(*a*) Curvature of ray path as a result of refractive index of the air

(*b*) Equivalent straight line ray path for equivalent radius of earth, a'

Fig. 9.16   *Equivalent radius of earth*

From Fig. 9.16 (*b*), we have

$$(a')^2 + d_1^2 = (a' + h_t)^2 \qquad\qquad ...(9.24)$$

or

$$d_1^2 = a'^2 + h_t^2 + 2a'\, h_t - a'^2$$

$$d_1^2 = 2a'\, h_t + h_t^2$$

As

$$a' \gg h_t$$

$$d_1^2 \approx 2a'\, h_t \qquad\qquad\qquad ...(9.25)$$

Similarly, $(a')^2 + d_2^2 = (a' + h_r)^2$

or

$$d_2^2 = 2a'\, h_r + h_r^2$$

As

$$a' \gg h_r,$$

$$d_2^2 = 2a'\, h_r \qquad\qquad\qquad ...(9.26)$$

The maximum distance of tropospheric propagation is

$$d_{max} = d_1 + d_2$$

$$d_{max} = \sqrt{2a'\,h_t} + \sqrt{2a'\,h_r} \qquad \text{...(9.27)}$$

As $\qquad a' = \dfrac{4}{3} \times a, \quad a = 3{,}960 \text{ miles}$

$$= \dfrac{4}{3} \times 3{,}960 \text{ miles}$$

$$d_{max}\ (\text{miles}) = \sqrt{2h_t}\ (\text{feet}) + \sqrt{2h_r}\ (\text{feet}) \qquad \text{...(9.28)}$$

In metric units, $\quad a = 6.37 \times 10^6 \text{ m}$

$$d_{max}\ (\text{km}) = \sqrt{17h_t}\ (\text{m}) + \sqrt{17h_r}\ (\text{m}) \qquad \text{...(9.29)}$$

Hence proved.

9.16 TROPOSCATTER

This is a mechanism by which propagation is possible by the scattered and diffracted rays. The scattering takes place by the **tropospheric region**. This mode of propagation occurs in the high frequency range from 160 MHz onwards. This mechanism helps to get unexpectedly large field strengths at the receivers even when they are in shadow zone.

The EM waves generated by high powered transmitters with high gain directive antennas reach the upper layer of troposphere. Scattering of waves takes place due to considerable variation of refractive index. The scattered wave reaching the receiver is shown in Fig. 9.17.

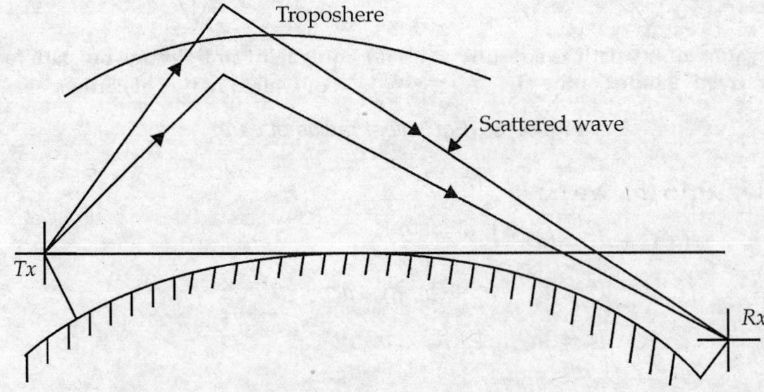

Fig. 9.17 *Troposcatter*

9.17 FADING OF EM WAVES IN TROPOSPHERE

Fading is a loss of signal due to change in electrical characteristics of troposphere. It is mainly due to the following:

1. Variation of dielectric constant.

2. Presence of eddies.

3. Uneven variations of refractive index.

4. Variation of effective earth radius factor, K.

9.18 LINE OF SIGHT (LOS)

It is defined as the distance that is covered by a direct space wave from the transmitting antenna to the receiving antenna.

It depends on:

1. Height of the receiving antenna.

2. Height of the transmitting antenna.

3. Effective earth's radius factor, K.

For standard atmosphere, $K = 4/3$.

The **line of sight distance**, d_{LOS} is given by

$$d_{LOS} = \sqrt{17h_t \text{ (m)}} + \sqrt{17h_r \text{ (m)}} \text{ km.} \qquad \qquad ...(9.30)$$

9.19 IONOSPHERIC WAVE PROPAGATION

Ionospheric wave propagation is also called sky wave propagation. EM waves directed upward at some angle from the earth's surface are called sky waves. Sky wave propagation is useful in the frequency range of 2 to 30 MHz and for long distance communication.

Ionosphere is the upper portion of the atmosphere between approximately 60 km and 400 km above the earth which is ionised by absorbing large quantities of radiation energy from the sun. The major ionisation is from α, β and γ radiations from the sun and cosmic rays and meteors.

Ionisation is a process by which a neutral atom or molecule gains or loses electrons and is left with a net charge.

9.20 CHARACTERISTICS OF IONOSPHERE

The physical properties of the ionosphere vary from time to time as the temperature, ionisation density and composition change regularly. As a result, ionosphere tends to be stratified and it does not have regular and constant distribution.

Ionosphere is divided meteorologically into different regions or layers and each layer exhibits different characteristics.

The layers of the ionosphere are:

1. D-Layer   2. E-Layer

3. E_s-Layer   4. F_1-Layer

5. F_2-Layer.

1. Characteristics of D-Layer

(a) It is the lowest layer of the ionosphere.

(b) It exists at an average height of 70 km.

(c) Its thickness is 10 km.

(d) It exists only in day-time.

(e) Its ionisation properties depend on the altitude of the sun above the horizon.

(f) It is not a useful layer for HF communication.

(g) It reflects some VLF and LF waves.

(h) It absorbs MF and HF waves to some extent.

(i) Its electron density, $N = 400$ electrons/cc.

(j) Its virtual height is 60 to 80 km.

(k) Critical frequency of the layer, $f_c = 180$ kHz.

2. Characteristics of E-Layer

(a) It exists next to D-Layer.

(b) It exists at an average height of 100 km.

(c) Its thickness is about 25 km.

(d) It exists only in day-time.

(e) The ions are recombined into molecules due to the absence of sun at night.

(f) It reflects some HF waves in day-time.

(g) It disappears at nights.

(h) Its electron density, $N = 5 \times 10^5$ electrons/cc.

(i) Its virtual height, $h_v = 110$ km.

(j) Its critical frequency, $f_c \approx 4$ MHz.

(k) Maximum single-hop range ≈ 2350 km.

3. Characteristics of E_s-Layer

(a) It is a sporadic E-Layer.

(b) Its appearance is sporadic in nature.

(c) If at all it appears, it exists in both day and night.

(d) It is a thin layer.

(e) Its ionisation density is high.

(f) It appears close to E-Layer.

(g) If it appears, it provides good reception.

(h) It is not a dependable layer for communication.

4. Characteristics of F_1-Layer

(a) It exists at a height of about 180 km in day-time.

(b) Its thickness is about 20 km.

(c) It combines with F_2-Layer during nights.

(d) HF waves are reflected to some extent.

(e) It absorbs HF to a considerable extent.

(f) It passes on some HF waves towards F_2-layer.

(g) Its virtual height, $h_v \approx 180$ km.

(h) Its critical frequency, $f_c \approx 5$ MHz.

(i) Maximum single-hop range $\approx 3{,}000$ km.

5. Characteristics of F_2-Layer

(a) It is the most import layer for HF communication.

(b) Its average height is about 325 km in day-time.

(c) Its thickness is about 200 km.

(d) It falls to a height of 300 km at nights as it combines with the F_1-Layer.

(e) The height of F_2-Layer varies drastically with the time of the day, the average ambient temperature and sunspot cycle.

(f) It exists at nights also.

(g) It is the topmost layer of the ionosphere.

(h) It is highly ionised.

(i) It offers better HF reflection and hence reception.

(j) Electron density of F-Layer, $N = 2 \times 10^6$ electrons/cc.

(k) Its virtual height, $h_v = 300$ km in day-time and 350 km in night.

(l) Its critical frequency, $f_c \approx 8$ MHz in day-time and $f_c \approx 6$ MHz at nights.

(m) Maximum single-hop range $\approx 3,800$ km during day-time and 4,100 km at night.

The heights of the ionospheric layers and their electron densities are shown in Table 9.1.

Table 9.1   Ionospheric layer heights and their electron densities

Layer	Approximate height above earth	Electron density (electrons/cc)	Day	Night
D	70 km	400	Exists	Absent
E	100 km	5×10^5	Exists	Absent
F_1	180 km	—	Exists	Merges with F_2
F_2	325 km	—	Exists	Exists
F	300 km	2×10^6	Exists	Exists

9.21 REFRACTIVE INDEX OF IONOSPHERE

Refractive index of the ionosphere is defined as the ratio of phase velocity of a wave in vacuum to the velocity in the ionosphere. That is, **refractive index** of ionosphere,

$$n \equiv \frac{v_0}{v_p} = \frac{\dfrac{1}{\sqrt{\mu_0 \in_0}}}{\dfrac{1}{\sqrt{\mu_0 \in}}}$$

$$= \frac{\dfrac{1}{\sqrt{\mu_0 \in_0}}}{\dfrac{1}{\sqrt{\mu_0 \in_0 \in_r}}} \equiv \sqrt{\in_r}$$

When an EM wave enters an ionised region at vertical incidence as in Fig. 9.18, the electric field exerts a force on the charged particles namely electrons and ions. The charges are displaced by this force and hence current flows. Magnitude of the charge of a positive ion is the same as that of an electron but its mass is 100 times that of an electron. Due to its high mass, velocity of the ion is very small and its current contribution is negligible.

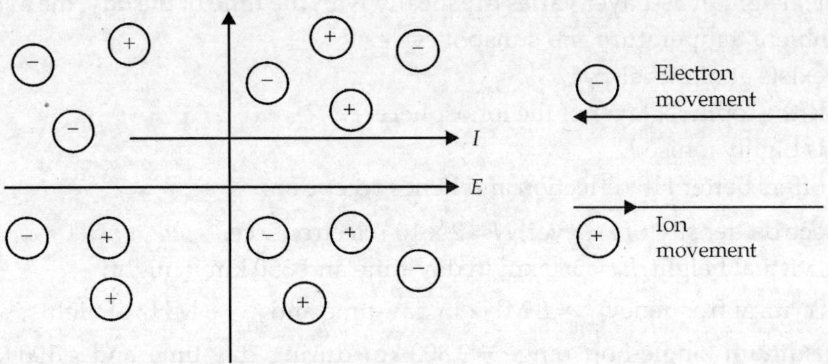

Fig. 9.18 *Effect of electric field of EM wave on the charged particles in the ionosphere*

In the presence of an electric field, electron cloud oscillates but with a phase retardation of 90°. This motion of the electron cloud produces a space current. Moreover, electric field has its own capacitive displacement current which leads the field by 90°. It is, therefore, obvious that the space and displacement currents are out of phase. This is found to reduce the relative permittivity of the ionosphere and it is given by

$$\epsilon_r = 1 - \frac{NQ_e^2}{\epsilon_0 \, m\omega^2} \qquad \text{...(9.31)}$$

Here, ϵ_r = relative permittivity of ionosphere

N = electron density (m^{-3})

m = mass of electron at rest = 9.11×10^{-31} kg

$\omega = 2\pi f$ = angular frequency of the wave

Q_e = magnitude of electron charge = 1.6×10^{-19} C

At $\omega = \omega_p$, $\epsilon_r = 1 - \dfrac{NQ_e^2}{\epsilon_0 \, m\omega_p^2} = 0$

or $\omega_p^2 = \dfrac{NQ_e^2}{\epsilon_0 \, m}$

$$f_p^2 = \frac{NQ_e^2}{(2\pi)^2 \, \epsilon_0 \, m} \qquad \text{...(9.32)}$$

Substituting the following values:

$$Q_e = 1.6 \times 10^{-19} \text{ C}$$

$$m = 9.11 \times 10^{-31} \text{ kg}$$

$$\epsilon_0 = 8.854 \times 10^{-12} \text{ F/m}$$

$$f_p = 9 \sqrt{N} \qquad \qquad \text{...(9.33)}$$

The Equation (9.32) for ϵ_r becomes

$$\epsilon_r = 1 - \frac{f_p^2}{f^2} \qquad \qquad \text{...(9.34)}$$

As the refractive index, n by definition is equal to $\sqrt{\epsilon_r}$,

$$n = \epsilon_r = \sqrt{1 - \frac{f_p^2}{f^2}} \qquad \qquad \text{...(9.35)}$$

Here, f_p is the plasma frequency at which $\epsilon_r = 0$.

Plasma frequency is defined as the natural frequency of oscillation of charged particles in plasma region. It is given by $\quad f_p = \dfrac{1}{2\pi} \sqrt{\dfrac{NQ_e^2}{\varepsilon_0 m}}$ 

Plasma is a completely ionised gas at very high temperature consisting of the charged nuclei and negative electrons.

The highest frequency of the wave that is reflected back from ionospheric layer is determined by the maximum electron density of that layer. This is called critical frequency of the wave and it is given by

$$\text{Critical frequency, } f_c = 9 \sqrt{N_{\max}} \qquad \qquad \text{...(9.36)}$$

where $\qquad\qquad N_{\max}$ = maximum electron density of the layer.

9.22 PHASE AND GROUP VELOCITIES

Phase velocity, v_p   It is defined as the rate at which the EM wave changes phase. It is also defined as the velocity of an equiphase surface along the wave normal of a travelling wave. The wave normal of a travelling wave is the direction normal to an equiphase surface taken in the direction of increasing phase.

$$v_p = \frac{1}{\sqrt{\mu \epsilon}}$$

For the ionosphere, $\mu \approx \mu_0$.

Then
$$v_p = \frac{1}{\sqrt{\mu_0 \in_0 \in_r}}$$

or
$$v_p = \frac{v_0}{\sqrt{\in_r}}$$
...(9.37)

where
$$v_0 = \text{free space velocity}$$

Phase velocity is also defined as

$$v_p \equiv \frac{\text{free space velocity}}{\text{refractive index}}$$

So,
$$v_p = \frac{v_0}{n}$$

When a wave reaches a height where $\in_r = 0$, phase velocity becomes infinite. The charge in a wave travels at the group velocity, v_g and it is given by

$$v_0^2 = v_p\, v_g$$
...(9.38)

It is obvious that when $v_p = \infty$, v_g becomes zero. This means that the energy ceases to propagate upwards.

Group velocity, v_g   The group velocity of a wave is defined as the velocity of propagation of the envelope. The magnitude of the group velocity is the reciprocal of the rate of change of phase constant with angular frequency. This definition is valid if the envelope moves without considerable change of its shape.

If the phase velocity varies with frequency the magnitudes of group and phase velocities are different.

The group and phase velocities have different directions if the phase velocity varies with direction.

9.23 MECHANISM OF IONOSPHERIC PROPAGATION—REFLECTION AND REFRACTION

Ionospheric propagation involves the reflection of the wave by the ionosphere. In actual mechanism, refraction takes place as shown in Fig. 9.19.

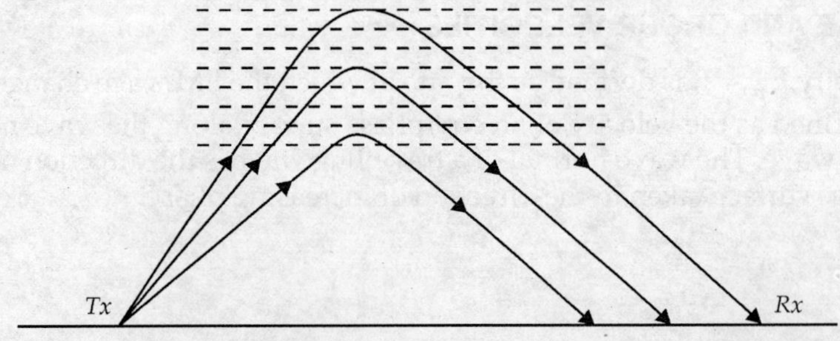

Fig. 9.19   *Ionospheric wave propagation and ray paths*

As ionisation density increases at an angle for the incoming wave, the refractive index of the layer decreases and the dielectric constant also decreases. Hence, the incident wave is gradually bent away from the normal.

If the rate of change of refractive index per unit height in terms of wavelength is sufficient, the refracted ray finally becomes parallel to the layer. Then it bends downwards and returns from the ionised layer at an angle equal to the angle of incidence. Although, some absorption takes place depending on the frequency, the wave is returned by the ionosphere to the receiver on earth. As a result, ionospheric propagation takes place through reflection and refraction of EM waves in the ionospheric layers.

The bending of a wave produced by the ionosphere follows optical laws. The direction of propagating wave at a point in the ionosphere is given by **Snell's law**. That is,

$$n = \frac{\sin \theta_i}{\sin \theta_r}$$

where
θ_i = angle of incidence at lower edge of ionosphere

θ_r = angle of refraction at point P (Fig. 9.20).

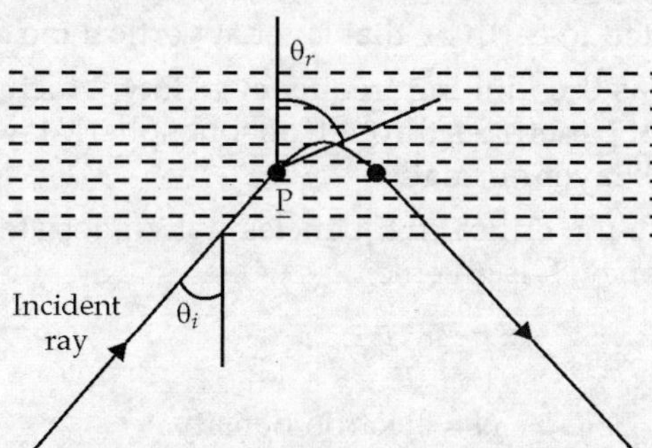

Fig. 9.20   *Refraction of EM wave in ionosphere*

9.24 CHARACTERISTIC PARAMETERS OF IONOSPHERIC PROPAGATION

Generally, propagation characteristics of the layers are described in terms of the following parameters:

1. Virtual height, h_v

2. Critical frequency, f_c

3. Maximum Usable Frequency, MUF

4. Skip distance

5. Lowest Usable Frequency, LUF

6. Critical angle, θ_c

7. Optimum working frequency, OWF or frequency of optimum traffic (FOT).

1. **Virtual height, h_v**   It is defined as the height that is reached by a short pulse of energy which has the same time delay as the original wave. Virtual height of the layer is always greater than the actual height. (Fig. 9.21).

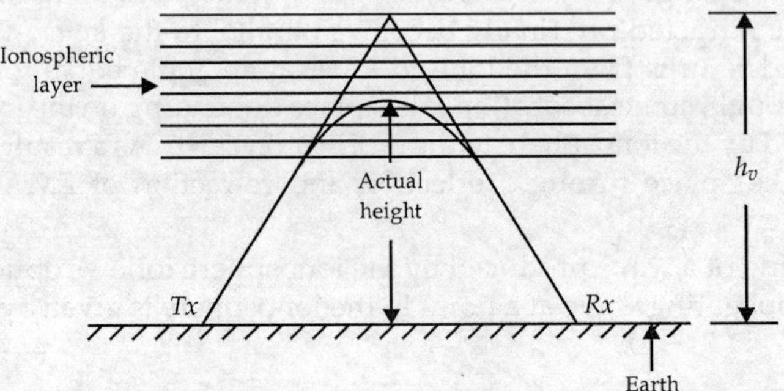

Fig. 9.21   *Actual and virtual heights of ionospheric layer*

Virtual height of the layer is useful to find the angle of incidence required for the wave to return to earth at a specified point.

2. **Critical frequency, f_c** f_c for a given layer is defined as the highest frequency that will be reflected to earth by that layer at vertical incidence.

It is also defined as the limiting frequency below which a wave is reflected and above which it penetrates through an ionospheric layer, when the waves are incident on the layer normally.

Frequencies above the critical frequencies will penetrate through the layer. The critical frequency, f_c is given by,

$$f_c = 9 \sqrt{N}$$

Here N = electron density

It may be noted that f_c is not fixed but depends on electron density. It also differs from time to time during the day.

3. **Maximum usable frequency, MUF**   It is the highest frequency of wave that is reflected by the layer at an angle of incidence other than normal. MUF depends on time of day, distance, direction, season and solar activity.

MUF is the highest frequency that can be used for sky-wave communication between transmitter and receiver. The common values of MUF range between 8 to 30 MHz. However, it may even be 50 MHz at times. The ray paths corresponding to critical and maximum usable frequencies are shown in Figs. 9.22 and 9.23.

From Fig. 9.23, we have,

$$\cos \theta_i = \frac{f_c}{\text{MUF}}$$

or $$\text{MUF} = \frac{f_c}{\cos \theta_i} = f_c \sec \theta_i \qquad \qquad ...(9.39)$$

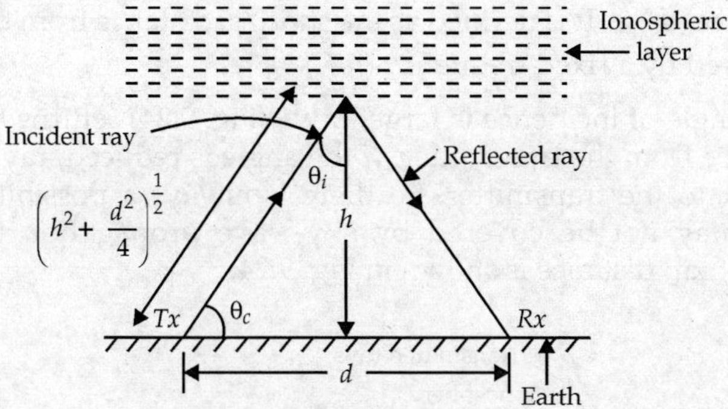

Fig. 9.22 *Ray geometry to find MUF*

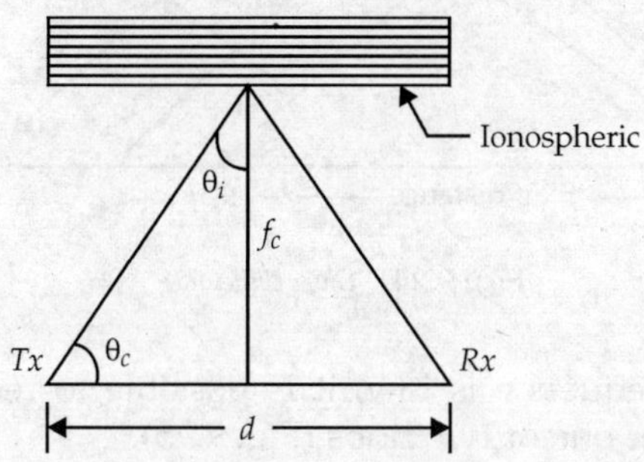

Fig. 9.23 *Ray paths of MUF and f_c*

where θ_i = angle of incidence between the incident ray and normal

$\quad\quad\quad\quad\quad\theta_c$ = critical angle

Equation (9.39) represents the **secant law**. Secant law is useful to find MUF. In fact, it is applicable for flat earth and ionospheric layer. From Fig. 9.22, we have

$$\text{MUF} = \frac{f_c}{\sin \theta_c} \qquad\qquad\text{...(9.40)}$$

But $$\sin \theta_c = \frac{h}{\left(h^2 + \dfrac{d^2}{4}\right)^{1/2}} \qquad\qquad\text{...(9.41)}$$

where h = height of the layer

$\quad\quad\quad\quad\quad d$ = distance between transmitting and receiving antennas.

From Equations (9.40) and (9.41), we have

$$\text{MUF} = f_c \left(\frac{d^2}{4h^2} + 1\right)^{1/2} \qquad\qquad\text{...(9.42)}$$

4. **Skip distance (d_s)**   It is defined as the shortest distance from the transmitter that is covered by a fixed frequency $(> f_c)$.

When the angle of incidence is large, ray 1 (Fig. 9.24) returns to ground at a long distance from the transmitter. If the angle is reduced, ray 2 returns to a point closer to the transmitter. So there is always a possibility that short distances may not be covered by sky-wave propagation under certain conditions. Skip distance is shown in Fig. 9.24.

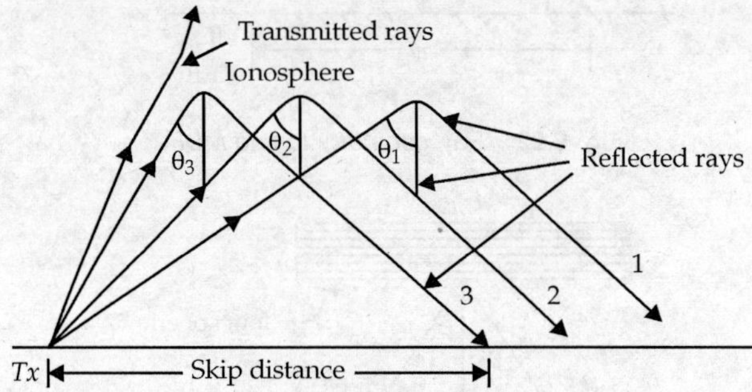

Fig. 9.24   *Skip distance*

If the operating frequency is low, it is possible to receive the ray by two different paths after one or two hops (Fig. 9.25).

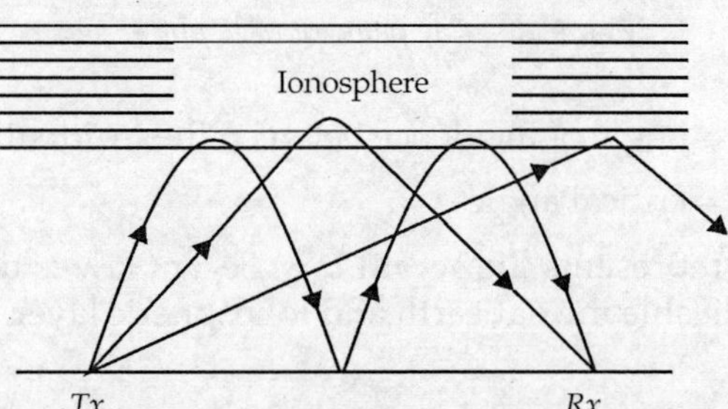

Fig. 9.25   *Multi path sky-wave propagation*

The transmission path is limited by the skip distance and curvature of the earth. The longest single-hop distance is achieved when the transmitted ray is tangential to the surface of the earth.

The skip distance, d_s is found from

$$d_s = \frac{2h}{\tan \theta_c}$$

where h = height of the layer

θ_c = critical angle

When the operating frequency, $f = f_{\text{MUF}}$, the skip distance is expressed in terms of f_{MUF}, f_c and the height of the layer and it is given by

$$d_s = 2h \left[\left(\frac{f_{\text{MUF}}}{f_c} \right)^2 - 1 \right]^{1/2}$$

5. **Lowest usable frequency, LUF** At certain low frequencies, the combination of ionospheric absorption, atmospheric noise, miscellaneous static and receiver $\frac{S}{N}$ requirements conspire to reduce radio communications. The lowest frequency that can be used for communication is called LUF.

6. **Critical angle, θ_c** Critical angle, θ_c is defined as the angle of incidence of a wave at which the wave will not be reflected when $\theta > \theta_c$ and it will be reflected when $\theta < \theta_c$. It depends on the thickness of layer, height and frequency of the wave.

The concept of critical angle is illustrated in Fig. 9.26. When $\theta > \theta_c$, the wave is not reflected and when $\theta < \theta_c$, it is reflected.

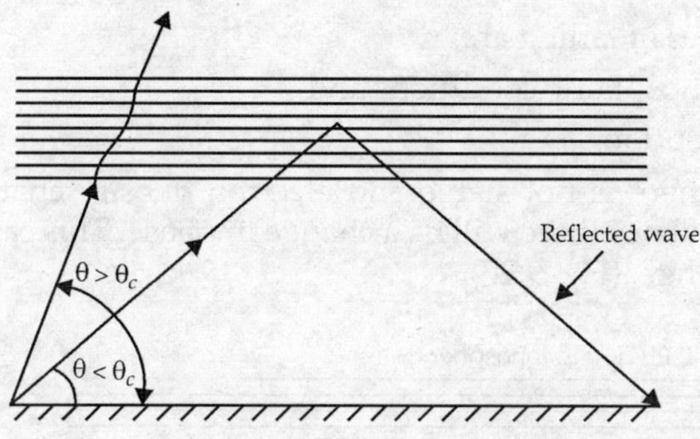

Fig. 9.26 *Effect of critical angle*

7. **Optimum working frequency (OWF) or frequency of optimum traffic (FOT)** The frequency of wave which is normally used for ionospheric communication is known as **optimum working frequency**.

It is generally chosen to be about 15% less than the MUF. It is always desirable to use as high a frequency as possible since the attenuation is inversely proportional to the square of the frequency.

9.25 SKY WAVE FIELD STRENGTH

The sky wave field strength is reduced with distance because of the following facts:

1. Due to spreading of rays during propagation.
2. Due to collisions of vibrating electrons in ionised regions.
3. The signal strength is inversely proportional to distance.
4. Sky wave absorption increases with increase in distance between the transmitter and receiver.
5. At high frequencies, there is loss of energy due to collisions at the top of the D-layer. Here the product of collisional frequency and electron density is a

maximum. This type of loss is called non-deviative absorption when the ray is moving through the absorbing regions.

6. The attenuation constant due to absorption is given by

$$\alpha = K \left(\frac{f_E}{f} \right)^2 \text{ dB/m} \qquad \qquad ...(9.44)$$

f_E = critical frequency of E-Layer

f = frequency of EM wave

K = constant which is a function of collisional frequency.

9.26 FADING AND DIVERSITY TECHNIQUES

Fading is the change in signal strength at the receiver. Most of the receivers are designed with an automatic volume control (AVC) circuit which reduces the effect of fading if the change in signal strength is small. Fading up to 20 dB is common.

The main causes of fading are:

1. Variation in ionospheric conditions and

2. Multi path reception.

As the ionosphere is not stable and electron density changes, signal path length changes and hence there will be a change in phase. This causes the received field strength to change. (Fig. 9.27).

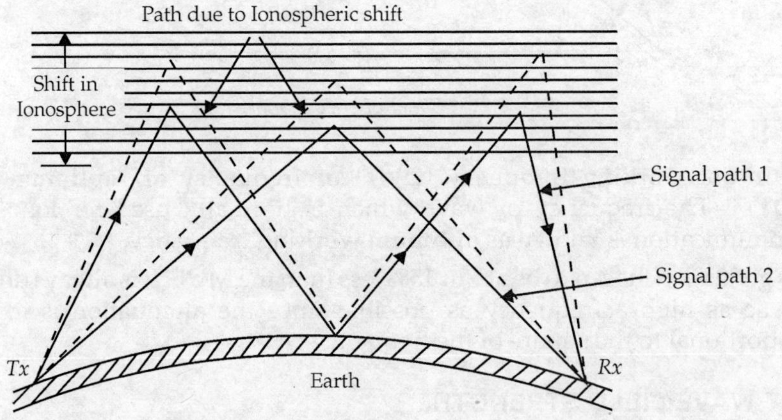

Fig. 9.27 *Fading*

Fading is classified in terms of the duration of the variation in signal strength. They are:

1. **Rapid fluctuations**　　These are due to multi-path interference and they occur for a few seconds.

2. **Short-term fluctuations**　　These are due to variation in the characteristics of the propagating medium and they occur for a few hours.

3. **Long-term fluctuations** These are due to seasonal variations in the propagation medium and they occur for a few days.

Fade out (total fading) occurs during sudden ionospheric disturbances, ionospheric storms, sun spot cycle and so on.

Types of fading are:

1. Selective fading
2. Interference fading
3. Absorption fading
4. Polarisation fading
5. Skip fading

The features of these fadings are given below :

1. **Selective fading:**

(*a*) It produces serious distortion of modulated signal.

(*b*) It is more prominent at high frequencies at which sky-wave propagation is used.

(*c*) It is large with AM signals at high percentage of modulation.

(*d*) AM signals are more distorted by selective fading than SSB signals.

(*e*) Selective fading can be reduced by the use of SSB system.

2. **Interference fading:**

(*a*) It is produced by the interference between rays.

(*b*) It is also produced by the interference between waves reaching the receiver by different paths.

(*c*) It is also produced by the interference between a ground wave and sky wave.

(*d*) It occurs due to fluctuations of layer height at a fixed frequency.

(*e*) As the path length of the wave varies, the relative phase of waves reaching the receiver also varies.

(*f*) Interference fading can be minimised by different diversity techniques.

3. **Absorption fading** This takes place due to absorption of waves by the ionosphere.

4. **Polarisation fading:**

(*a*) This takes place due to change of polarisation of EM wave.

(*b*) This is caused by cross-polarised waves.

(*c*) When polarisation changes, the signal amplitude changes in the receiver.

(*d*) This type of fading is reduced by polarisation diversity.

5. **Skip fading:**

(*a*) This occurs near the skip distance.

(*b*) The variation of height of density of the layer causes skip fading.

(*c*) This is minimised by AVC and AGC in the receiver.

It is difficult to control short and long term fluctuations. But the fading due to rapid fluctuations can be reduced by different diversity techniques.

The diversity techniques are:

1. Frequency diversity
2. Space diversity
3. Polarity diversity
4. Time diversity

1. **Frequency diversity**   In this, the transmitter will send two or more frequencies simultaneously with the same modulating information. As the different frequencies will fade differently, one will always be strong. This scheme is shown in Fig. 9.28.

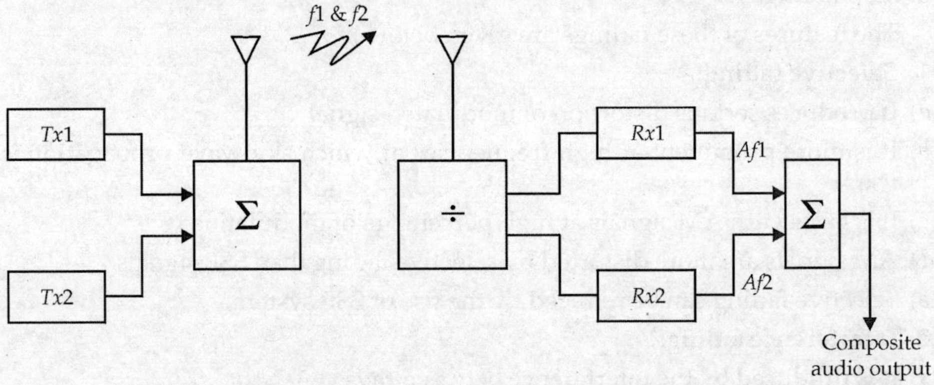

Fig. 9.28   *Frequency diversity to reduce fading*

2. **Space diversity technique**   A single transmitter frequency is used. At the receiving site, two or more receiving antennas spaced at one-half wavelength apart are used. The signal will fade at one antenna while it increases at the

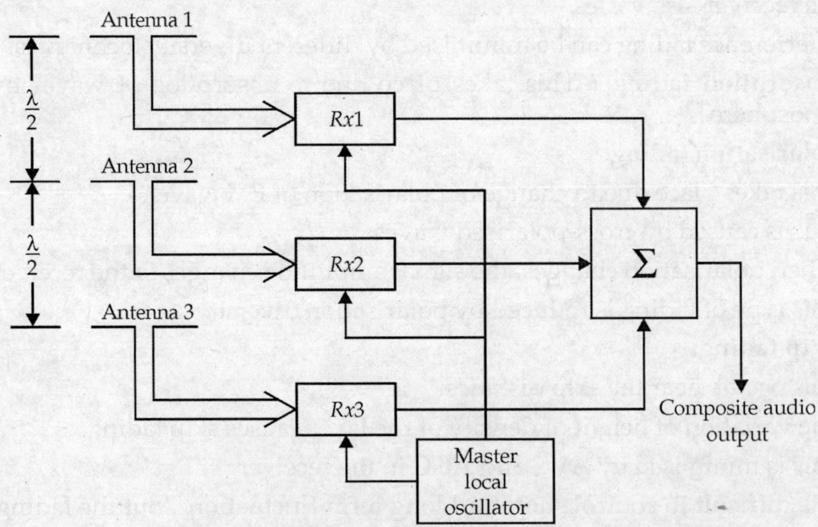

Fig. 9.29   *Space diversity*

other antenna. A three antenna system may be used. Three separate but identical receivers tuned by the same master local oscillator are connected to three antennas. Audio mixing on the basis of the strongest signal keeps the audio output constant while RF signal fades. This scheme is shown in Fig. 9.29.

3. **Polarity diversity system**   In this, vertical and horizontal polarisation antennas are used to receive the signal. As in the case of space diversity system, the two receivers are combined to produce constant output. This scheme is shown in Fig. 9.30.

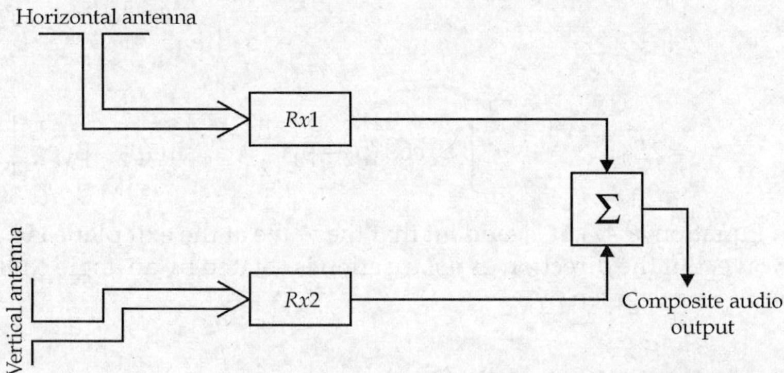

Fig. 9.30   *Polarity diversity*

4. **Time diversity**   In this, the same signals are transmitted at different times. As fading is time-dependent, some signals may be strong and fading is less.

9.27 FARADAY ROTATION

Rotation of the plane of polarisation is defined as **Faraday rotation**. It is also defined as the process of rotation of polarisation ellipse of EM wave in a magneto-ionic medium. This process occurs in the ionospheric regions when a plane wave enters the ionosphere.

It is a variable effect and leads to loss of signal power at the receiving antenna due to polarisation mismatch.

A linearly polarised EM wave can be regarded as the vector sum of two counter-rotating circularly polarised waves. If such a wave propagates in the direction of the magnetic field in a loss-less plasma region, the two circularly polarised components travel with different phase velocities and hence the plane of polarisation rotates with distance.

On an ionospheric layer of L meters, let the incident plane wave along z-direction be represented by

$$E = 2E_0\, e^{-j\beta z} \qquad\qquad ...(9.45)$$

Decomposing this into left and right circularly polarised waves, Equation (9.45) can be written as

$$E = E_0\,(\mathbf{a}_x + j\,\mathbf{a}_y)\, e^{-j\beta z} + E_0\,(\mathbf{a}_x - j\,\mathbf{a}_y)\, e^{-j\beta z} \qquad\qquad ...(9.46)$$

This wave enters the ionospheric layer at $z = 0$ and travels as two circularly polarised waves with different propagation constants. Neglecting reflections at each interface, the electric field at the exit plane is given by

$$E = E_0 \left(\mathbf{a}_x - j\,\mathbf{a}_y\right) e^{-j\beta_1 L} + E_0 \left(\mathbf{a}_x + j\,\mathbf{a}_y\right) e^{-j\beta_2 L}$$

$$= E_0\, e^{-j\,(\beta_1 + \beta_2)\frac{L}{2}} \left[\mathbf{a}_x \left(e^{j\,(\beta_2 - \beta_1)\frac{L}{2}} + e^{-j\,(\beta_2 - \beta_1)\frac{L}{2}} \right) \right.$$

$$\left. -j\,\mathbf{a}_y \left(e^{j\,(\beta_2 - \beta_1)\frac{L}{2}} - e^{-j\,(\beta_2 - \beta_1)\frac{L}{2}} \right) \right]$$

$$= 2E_0\, e^{-j\,(\beta_1 + \beta_2)\frac{L}{2}} \left[\mathbf{a}_x \cos (\beta_2 - \beta_1)\frac{L}{2} + \mathbf{a}_y \sin (\beta_2 - \beta_1)\frac{L}{2} \right] \quad \text{...(9.47)}$$

From Equation (9.47), it is evident that the wave at the exit plane is a linearly polarised wave. But the direction of polarisation is rotated by an angle ϕ relative to x-axis. The angle ϕ is given by

$$\tan \phi = \frac{E_y}{E_x} = \tan (\beta_2 - \beta_1)\frac{L}{2}$$

or
$$\phi = (\beta_2 - \beta_1)\frac{L}{2} \qquad\qquad \text{...(9.48)}$$

Faraday rotation is most pronounced when f is close to f_c. Here, f_c is called cyclotron frequency and it is given by

$$f_c = -\frac{Q_e B}{2\pi m} \qquad\qquad \text{...(9.49)}$$

where Q_e = electron charge

m = mass of the electron

B = earth's magnetic flux density = 5×10^{-5} Wb/m^2

f_c = 1.4 MHz.

Here, $(\beta_2 - \beta_1)$ is the largest. At high frequencies, β_1 and β_2 have almost the same values and hence the magnitude of rotation is much smaller. The rotation angle depends on many unpredictable variables. This leads to a loss in received signal power at the receiving antenna due to mismatched polarisation. At frequencies of hundreds of megahertz, most of Faraday rotation occurs in 90-1,000 km altitude range.

9.28 IONOSPHERIC ABNORMALITIES

The electrical characteristics of the ionosphere depend on solar radiation and hence they vary continuously. The variations of the ionosphere are classified as follows.

9.28.1 Normal

The normal variation in the characteristics of the ionosphere occur due to the following:

1. Diurnal
2. Seasonal
3. Thickness and
4. Height variations of the ionospheric layers.

9.28.2 Abnormal

The abnormal variations in the characteristics of the ionosphere occur mainly due to changes in solar activity. The common abnormal variations are:

1. Ionospheric storms
2. Sudden ionospheric disturbances
3. Sunspot cycle
4. Fading
5. Whistlers
6. Tides and winds.

9.29 IONOSPHERIC STORMS

These are due to the high absorption of sky waves and abnormal changes at the critical frequencies of E and F_2 layers. These storms usually persist for a few days.

9.30 SUDDEN IONOSPHERIC DISTURBANCES (SID)

The sudden appearance of solar flares causes **SIDs**. The solar flares occur suddenly and sporadically. These occur during solar peak activity. SIDs block out the signals completely. They persist for a few minutes to an hour. SID causes complete fading and it is called Dellinger fade-out. Ultraviolet radiation is intensive due to solar flares in D-Layer. SID does not occur in the layer of low air density and hence it is not found in E, F_1 and F_2 layers.

9.31 SUN SPOT CYCLE

Sun spot cycle is a eleven years cycle during which radiation varies drastically. The variation due to ultraviolet rays, flares, particle radiation and sun spots is very high and it is low due to light. During sun spot maxima, the critical frequencies are the highest and they are lowest during sun spot minima. To minimise the effect of sun spot cycle, the operating frequency is carefully chosen.

9.32 WHISTLERS

Whistlers are the transient electromagnetic disturbances which occur naturally. Whistlers consist of EM pulses of audio frequency radiation along the direction of the magnetic field of the earth between conjugate points in the northern and southern hemispheres.

Long whistlers, short whistlers and noise whistlers are a few types of whistlers. The composition of the upper atmosphere can be provided by whistlers.

9.33 TIDES AND WINDS IN THE IONOSPHERE

Tides and winds are common in the atmosphere. Solar tide effects are more pronounced. The winds in the ionosphere are caused by the tides. The presence of ionospheric winds are due to the motion of turbulence in F_2-layer. Tidal effect introduces a small peak of maximum ionisation density in the layer at mid-night.

9.34 EFFECT OF EARTH'S MAGNETIC FIELD

The average magnetic field of earth is about 40 A/m. This makes the ionosphere to behave like an anisotropic medium, that is, permittivity varies in different directions.

The earth's magnetic field causes the electrons to trace complicated trajectories with cyclotron or gyro frequency $f_c = 1.4$ MHz at $H = 40$ A/m.

Gyro frequency is defined as the lowest natural frequency at which charged particles spiral in a fixed magnetic field.

It is a vector quantity and it is defined mathematically as

$$f_g = \frac{1}{2\pi} \frac{Q_e \mathbf{B}}{m}$$

where

Q = charge of the particles, coloumbs

$\mathbf{B}$ = magnetic flux density, Wb/m^2

m = mass of the particle, kg.

Cyclotron frequency is defined as the lowest natural frequency of a wave at which charged particles spiral in a fixed magnetic field.

It is obvious that the cyclotron frequency is same as gyro frequency.

Specific effects of earth's magnetic field

1. The earth's magnetic field exerts a deflecting force on the moving electrons. This force is given by

$$\mathbf{F} = Q_e \, (\mathbf{V} \times \mathbf{B})$$

where

$\mathbf{F}$ = force on the electron in the ionosphere whose charge is Q_e (Newton)

$\mathbf{V}$ = velocity of the electron, m/s

$\mathbf{B}$ = magnetic flux density, Wb/m^2

$= \mu_0 \mathbf{H}$, $\mathbf{H}$ is the magnetic field intensity of earth (A/m).

It is clear from this equation, that the direction of the force is perpendicular to the velocity of the electron and also to the magnetic field.

2. The magnetic field component of the earth which is perpendicular to the electric field of the incident wave makes vibrating electrons to follow elliptical paths as in Fig. 9.31.

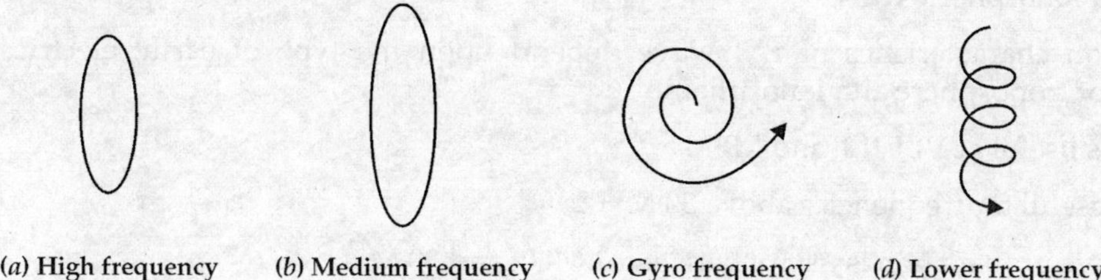

(a) High frequency   (b) Medium frequency   (c) Gyro frequency   (d) Lower frequency

Fig. 9.31 *Effect of earth's magnetic field*

3. The electrons of the ionosphere absorb some energy from the EM wave. This absorbed energy is re-radiated with a polarisation that is rotated by $90°$ in space with respect to that of the incident EM wave. This re-radiated cross polarised component also differs in time phase from the field of the incident wave. As a result, the plane polarised wave is changed to elliptically polarised wave.

4. As the average velocity of the electrons is inversely proportional to the frequency, the effect of earth's magnetic field is more pronounced at low frequencies.

5. At high frequency, electrons vibrate in narrow elliptical paths under the influence of earth's magnetic field.

6. As the frequency is decreased, the minor axis of the ellipse increases and electron vibration amplitude also increases.

7. When the frequency is reduced to cyclotron resonant frequency 1.4 MHz, the electrons follow a spiral path of steadily increasing radius as in Fig. 9.31 (c), along with increase in velocity.

8. This frequency is given by

$$f_g = \frac{Q_e B}{2\pi m} = \frac{\mu_0 Q_e H}{2\pi m}$$

Here μ_0 permeability of ionosphere, H/m

Q_e = electron charge, (C)

m = mass of electron (gm)

H = earth's magnetic field, (A/m)

B = magnetic flux density of the earth, (Wb/m^2).

The average value of

$$B = 0.5 \times 10^{-4} \, Wb/m^2 \text{ or Tesla.}$$

 ## POINTS TO REMEMBER

1. Electromagnetic wave in free space is a transverse wave.

2. The basic modes of wave propagation between transmitter and receiver are ground wave, space wave and ionospheric wave.

3. The propagation characteristics of EM wave depend upon the type of earth, electrical characteristics of troposphere and ionosphere.

4. Ground wave is useful at VLF, LF and MF.

5. Space wave is useful for frequencies above 30 MHz.

6. Ionospheric wave propagation is useful between 2 MHz and 30 MHz.

7. Equivalent circuit of earth is a capacitance in parallel with a resistance.

8. The field strength due to ground wave is $E = \dfrac{AE_0}{d}$ volt/m.

9. The value of the factor of earth's losses lies between 0 and 1.

10. Ground wave field strength is $E = \dfrac{\eta_0 \, h_t \, I}{\lambda d}$ volt/m.

11. The received voltage due to ground wave is $V = \dfrac{\eta_0 \, h_t \, h_r \, I}{\lambda d}$ volts.

12. The roughness of earth is $R = \dfrac{4\pi\sigma_s \sin \theta_i}{\lambda}$.

13. Wave tilt is the change of orientation of the vertically polarised ground wave at the surface of the earth.

14. Troposphere is the region of atmosphere less than 16 km above the earth.

15. The field strength due to space wave is $E = \dfrac{4\pi h_t \, h_r}{\lambda d^2} E_0$.

16. The curvature of the earth creates shadow zones.

17. The refractive index of a medium is $\sqrt{\epsilon_r}$.

18. Modified refractive index of troposphere is $M = \left(n - 1 + \dfrac{h}{a} \right) \times 10^6$.

19. For standard atmosphere, $\dfrac{dM}{dh} = 0.036$ units/ft.

20. Radio horizon is always greater than optical horizon.

21. Radio horizon, $d_{\text{miles}} = \sqrt{2h_{f \text{(feet)}}} + \sqrt{2h_{r \text{(feet)}}}$.

22. Radio horizon, $d_{km} = \sqrt{17h_{t\,(m)}} + \sqrt{17h_{r\,(m)}}$.

23. The effective earth radius factor, K is 4/3 for standard atmosphere.

24. The line of sight distance $d_{LOS} = \sqrt{17h_{t\,(m)}} + \sqrt{17h_{r\,(m)}}$.

25. D is the lowest layer and F_2 is the highest layer in the ionosphere.

26. F-layer has the highest electron density.

27. D-layer has the lowest electron density.

28. The critical frequency of the layer is $f_c = 9\sqrt{N_{max}}$.

29. The relation between group velocity, phase velocity and free space velocity is $v_0^2 = v_p\, v_g$.

30. The phase velocity, $v_p = \dfrac{v_0}{n}$.

31. $MUF = f_c \sec \theta_i$.

32. The skip distance, $d_s = \dfrac{2h}{\tan \theta_c}$.

33. Faraday rotation is the rotation of plane of polarisation.

34. Gyro frequency is the lowest natural frequency at which charged particles spiral in a fixed magnetic field.

35. Gyro frequency is given by $f_g = \dfrac{\mu_0\, Q_e\, H}{2\pi m}$.

 ## SOLVED PROBLEMS

Problem 9.1 A transmitter operating at a frequency of 1.7 MHz is required to provide a ground wave field strength of 0.5 mV/m at a distance of 10 km. A short vertical transmitting antenna has an efficiency of 50%. The conductivity of the ground is 5×10^{-5} (mho)/cm and its relative permittivity is 10. Find the transmitter power required.

Solution The phase constant,

$$b = \tan^{-1}\left(\frac{\epsilon_r + 1}{D_f}\right)$$

$$f = 1.7 \times 10^6\ \text{Hz},\quad \lambda = 1.764 \times 10^2\ \text{m}$$

$$D_f = 1.80 \times 10^{12}\,\frac{\sigma}{f} = \text{dissipation factor}$$

$$= \frac{1.80 \times 10^{12} \times 5 \times 10^{-5}}{1.7 \times 10^6}$$

$$= 52.9$$

$$b = \tan^{-1}\left(\frac{10+1}{52.9}\right) = \tan^{-1}(0.2079) = 11.74^{\circ}$$

The numerical distance, p is

$$p = \frac{\pi}{D_f}\frac{d}{\lambda}\cos b$$

$$= \frac{\pi}{52.9}\frac{10 \times 10^3}{1.764 \times 10^2}\cos(11.74^{\circ}) = 3.296$$

Formula method for A

$$A = \frac{2 + 0.3p}{2 + p + 0.6p^2} - \sin b \sqrt{\frac{p}{2}}\, e^{-\left(\frac{5}{8}\right)p}$$

$$= 0.21075$$

For $b = 11.74^{\circ}$, $p = 3.296$, the value of A is taken from standard plot of Fig. 9.4. That is,

$$A \approx 0.21$$

But the field strength at 10 km

$$E = \frac{AE_0}{d}$$

Here $\qquad E_0 = 300\,\sqrt{P_{kW}},\ d = 10\text{ km}$

or $\qquad 0.5 \times 10^{-3} = \dfrac{0.21 \times 300 \times 10^{-3}\,\sqrt{P_{kW}}}{10}$

or $\qquad \sqrt{P_{kW}} = \dfrac{0.5 \times 10}{300 \times 0.21} = \dfrac{5}{63} = 0.07936$

$$P_{kW} = 0.0063$$

The efficiency of the antenna is only 50%.

The transmitter power required

$$= 2 \times P_{kW} = 2 \times 0.0063$$

$$= 0.0126\text{ kW}$$

$$\boxed{P_{TX} = 12.6\text{ W.}}$$

Problem 9.2 A broadcast transmitter supplies 100 kW to an antenna that radiates 50% of this power. The antenna has directional characteristic such that the field strength without ground losses is given by $E_0 = 300 \times 1.28\,\sqrt{P_{kW}}$ mV/m at 1 km. Find the field strength of the ground wave at 100 km for the following types of earth conditions (*i*) for $f = 500$ kHz, (*ii*) for $f = 1500$ kHz.

(*a*) Sea water earth: $\epsilon_r = 81$, $\sigma = 45 \times 10^{-3}$ (mho)/cm

(*b*) Good soil: $\epsilon_r = 20$, $\sigma = 10^{-4}$ (mho)/cm

(*c*) Poor soil: $\epsilon_r = 10$, $\sigma = 0.2 \times 10^{-4}$ (mho)/cm

(d) Cities, industrial areas: $\epsilon_r = 5,\ \sigma = 10^{-5}$ (mho)/cm

(e) Rockey soil, flat sandy: $\epsilon_r = 10,\ \sigma = 2 \times 10^{-5}$ (mho)/cm

(f) Medium hills, forestation: $\epsilon_r = 13,\ \sigma = 50 \times 10^{-5}$ (mho)/cm

Solution (i) Transmitter power, $P_T = 100$ kW

Efficiency of antenna $\qquad\qquad = 50\%$

So, radiated power $\qquad\qquad\qquad = 50$ kW

The field strength without ground losses is

$$E_0 = 300 \times 1.28\ \sqrt{P_{kW}}\ \text{mV/m at 1 km}$$

$$= 2{,}715.29\ \text{mV/m}$$

Frequency, $\qquad\qquad\qquad f = 500$ kHz.

(a) **Sea water earth**

$$\sigma = 45 \times 10^{-3}\ \text{(mho)/cm}$$

$$\epsilon_r = 81$$

$$D_f = 1.8 \times 10^{12}\ \frac{\sigma}{f}$$

$$= \frac{1.8 \times 10^{12} \times 45 \times 10^{-3}}{500 \times 10^3}$$

$$= 162 \times 10^3$$

Phase constant, $\ b = \tan^{-1}\left(\dfrac{\epsilon_r + 1}{D_f}\right) = 0.029^\circ$

Numerical distance

$$p = \frac{\pi}{D_f}\ \frac{d}{\lambda}\ \cos b$$

$$= \frac{\pi}{1{,}62{,}000} \times \frac{100 \times 10^3 \times 500 \times 10^3}{3 \times 10^8}\ \cos (0.029)$$

$$= 0.0032$$

$$A = \frac{2 + 0.3p}{2 + p + 0.6p^2} - \sin b\ \sqrt{\frac{p}{2}}\ e^{-\left(\frac{5}{8}\right)p}$$

Subtracting the values of p and b, we get

$$A = 0.9988$$

The field strength,

$$E = \frac{AE_0}{d}$$

$$= \frac{0.9988 \times 2,715.29}{100}$$

$$E = 27.1 \text{ mV/m.}$$

(b) **For good soil**

$$\sigma = 10^{-4} \text{ (mho)/cm}, \quad \epsilon_r = 20$$

$$D_f = 1.8 \times 10^{12} \frac{\sigma}{f} = 360$$

Phase constant, $\qquad b = \tan^{-1}\left(\frac{\epsilon_r + 1}{D_f}\right) = 3.33^\circ$

Numerical distance, $\quad p = \frac{\pi}{D_f} \frac{d}{\lambda} \cos b = 1.452$

$$A = 0.4963$$

The field strength at 100 km

$$= \frac{AE_0}{d}$$

$$= \frac{0.4963 \times 2,715.29}{100}$$

$$E = 13.5 \text{ mV/m.}$$

(c) **For poor soil**

$$\sigma = 0.2 \times 10^{-4} \text{ (mho)/cm}, \quad \epsilon_r = 10$$

$$D_f = 1.8 \times 10^{12} \frac{\sigma}{f} = 72.0, \quad b = 8.68^\circ, \quad p = 7.18$$

$$A = 0.01002$$

$$E = \frac{AE_0}{d} = \frac{0.1002 \times 2,715.29}{100}$$

$$E = 2.7 \text{ mV/m.}$$

(d) **For cities, industrial areas**

$$\epsilon_r = 5, \quad \sigma = 10^{-5} \text{ (mho)/cm}$$

$$D_f = 1.8 \times 10^{12} \frac{\sigma}{f} = 36$$

$$b = 9.46^\circ$$

$$p = 14.34$$

$$A = 0.045$$

$$E = \frac{AE_0}{d} = \frac{0.045 \times 2,715.29}{100}$$

$$\boxed{E = 1.222 \text{ mV/m.}}$$

(e) **For rocky soil, flat sand earth**

$$\epsilon_r = 10, \quad \sigma = 2 \times 10^{-5} \text{ (mho)/cm}$$

$$D_f = 1.8 \times 10^{12} \frac{\sigma}{f} = 7,200$$

Phase constant, $\qquad b = \tan^{-1}\left(\frac{\epsilon_r + 1}{D_f}\right) = 0.0875°$

$$p = \frac{\pi}{D_f} \frac{d}{\lambda} \cos b = 0.0727$$

$$A = \frac{2 + 0.3p}{2 + p + 0.6p^2} - \sin b \ \sqrt{\frac{p}{2}} \ e^{-\frac{5}{8}p} = 0.9737$$

$$E = \frac{AE_0}{d} = \frac{0.9737 \times 2,715.29}{100}$$

$$\boxed{E = 26.4 \text{ mV/m.}}$$

(f) **For medium hills and forestation earth**

$$\sigma = 5 \times 10^{-5} \text{ (mho)/cm}$$

$$\epsilon_r = 13$$

$$D_f = 1.8 \times 10^{12} \frac{\sigma}{f} = \frac{1.8 \times 10^{12} \times 5 \times 10^{-5}}{500 \times 10^3} = 180$$

$$b = \tan^{-1}\left(\frac{\epsilon_r + 1}{D_f}\right) = 4.44°$$

$$p = \frac{\pi}{D_f} \frac{d}{\lambda} \cos b = 2.90$$

$$A = 0.2733$$

Now $\qquad E = \frac{AE_0}{d} = \frac{0.2733 \times 2,715.29}{100}$

$$\boxed{E = 7.4 \text{ mV/m.}}$$

(ii) Frequency, $f = 1,500$ kHz

$$E_0 = 300 \times 1.28 \ \sqrt{P_{kW}}$$

$$= 2,715.29 \text{ mV/m.}$$

(*a*) **Sea water earth**

$$\epsilon_r = 81$$

$$\sigma = 45 \times 10^{-3} \text{ (mho)/cm}$$

$$D_f = 1.8 \times 10^{12} \frac{\sigma}{f} = 54{,}000.00$$

$$b = \tan^{-1}\left(\frac{\epsilon_r + 1}{D_f}\right) = 0.087^\circ$$

$$p = \frac{\pi}{D_f} \frac{d}{\lambda} \cos b = 0.0291$$

$$A = \frac{2 + 0.3p}{2 + p + 0.6p^2} - \sin b \ \sqrt{\frac{p}{2}} \ e^{-\frac{5}{8}p}$$

$$= 0.9895$$

$$E = \frac{AE_0}{d} = 26.9 \text{ mV/m.}$$

(*b*) **Good soil**

$$\epsilon_r = 20.0$$

$$\sigma = 1.0 \times 10^{-4} \text{ (mho)/cm}$$

$$D_f = 1.8 \times 10^{12} \frac{\sigma}{f} = 120.0$$

$$b = \tan^{-1}\left(\frac{\epsilon_r + 1}{D_f}\right) = 9.92^\circ$$

$$p = \frac{\pi}{D_f} \frac{d}{\lambda} \cos b = 12.89$$

$$A = \frac{2 + 0.3p}{2 + p + 0.6p^2} - \sin b \ \sqrt{\frac{p}{2}} \ e^{-\frac{5}{8}p}$$

$$= 0.0510$$

$$E = \frac{AE_0}{d} = 1.4 \text{ mV/m.}$$

(*c*) **Poor soil**

$$\epsilon_r = 10$$

$$\sigma = 0.2 \times 10^{-4} \text{ (mho)/cm}$$

$$D_f = 1.8 \times 10^{12} \frac{\sigma}{f} = 24.0$$

$$b = \tan^{-1}\left(\frac{\in_r + 1}{D_f}\right) = 24.62°$$

$$p = \frac{\pi}{D_f}\frac{d}{\lambda}\cos b = 59.49$$

$$A = \frac{2 + 0.3p}{2 + p + 0.6p^2} - \sin b \ \sqrt{\frac{p}{2}}\ e^{-\frac{5}{8}p}$$

$$= 0.0091$$

$$E = \frac{AE_0}{d} = 0.2\ \text{mV/m.}$$

(*d*) **Cities, industrial area**

$$\in_r = 5$$

$$\sigma = 1.0 \times 10^{-4}\ \text{(mho)/cm}$$

$$D_f = 1.8 \times 10^{12}\frac{\sigma}{f} = 12.0$$

$$b = \tan^{-1}\left(\frac{\in_r + 1}{D_f}\right) = 26.56°$$

$$p = \frac{\pi}{D_f}\frac{d}{\lambda}\cos b = 117.08$$

$$A = \frac{2 + 0.3p}{2 + p + 0.6p^2} - \sin b \ \sqrt{\frac{p}{2}}\ e^{-\frac{5}{8}p}$$

$$= 0.0044$$

$$E = \frac{AE_0}{d} = 0.1\ \text{mV/m.}$$

(*e*) **Rocky soil, flat sandy**

$$\in_r = 10$$

$$\sigma = 2 \times 10^{-3}\ \text{(mho)/cm}$$

$$D_f = 1.8 \times 10^{12}\frac{\sigma}{f} = 2,400.0$$

$$b = \tan^{-1}\left(\frac{\in_r + 1}{D_f}\right) = 26.26°$$

$$p = \frac{\pi}{D_f}\frac{d}{\lambda}\cos b = 0.6545$$

$$A = \frac{2 + 0.3p}{2 + p + 0.6p^2} - \sin b \ \sqrt{\frac{p}{2}} \ e^{-\frac{5}{8}p}$$

$$= 0.7526$$

$$E = \frac{AE_0}{d} = 20.4 \ \text{mV/m}.$$

(f) Medium hills, forestation

$$\epsilon_r = 13$$

$$\sigma = 5 \times 10^{-5} \ \text{(mho)/cm}$$

$$D_f = 1.8 \times 10^{12} \ \frac{\sigma}{f} = 60.00$$

$$b = \tan^{-1}\left(\frac{\epsilon_r + 1}{D_f}\right) = 13.13^\circ$$

$$p = \frac{\pi}{D_f} \frac{d}{\lambda} \cos b = 25.49$$

$$A = \frac{2 + 0.3p}{2 + p + 0.6p^2} - \sin b \ \sqrt{\frac{p}{2}} \ e^{-\frac{5}{8}p}$$

$$= 0.2331$$

$$E = \frac{AE_0}{d} = 0.6 \ \text{mV/m}.$$

Problem 9.3 Find the maximum range of tropospheric transmission for which the height of the transmitting antenna is 100 ft and that of the receiving antenna is 50 ft.

Solution   We have

$$d_{\max} \ (\text{miles}) = \sqrt{2h_t \ (\text{feet})} + \sqrt{2h_r \ (\text{feet})}$$

$$= \sqrt{2 \times 100} + \sqrt{2 \times 50}$$

$$d_{\max} = 14.142 + 10$$

$$d_{\max} = 24.142 \ \text{miles}.$$

Problem 9.4   Find the radio horizon distance of a transmitting antenna whose height is 80 m.

Solution $d \ (\text{km}) = \sqrt{17h_t \ (\text{m})}$

$$= \sqrt{17 \times 80}$$

$$= 36.88$$

Radio horizon distance $d = 36.88 \ \text{km}.$

Problem 9.5   Find the maximum distance that can be covered by a space wave when the antenna heights are 80 m and 50 m.

Solution
$$d_{max} = \sqrt{17h_t} + \sqrt{17h_r}$$
$$= \sqrt{17 \times 80} + \sqrt{17 \times 50}$$

or
$$d_{max} = 36.88 + 29.15$$

$$d_{max} = 66.03 \text{ km.}$$

Problem 9.6   A receiving antenna is located at 80 km from the transmitting antenna. The height of the transmitting antenna is 100 m. What is the required height of the receiving antenna?

Solution   We have
$$d_{max} = \sqrt{17h_t} + \sqrt{17h_r}$$

or
$$\sqrt{17h_r} = d_{max} - \sqrt{17h_t}$$

or
$$17h_r = (d_{max} - \sqrt{17h_t})^2$$

$$h_r = \frac{(d_{max} - \sqrt{17h_t})^2}{17}$$

$$h_r = \frac{(80 - \sqrt{17 \times 100})^2}{17}$$

The height of the receiving antenna, h_r is 88.41 m.

Problem 9.7   Determine (*a*) the radio horizon distance for a transmitting antenna height of 300 feet (*b*) the radio horizon distance of a receiving antenna with a height of 100 feet (*c*) the maximum range of space wave communication for the above antenna heights.

Solution   (*a*) Radio horizon distance for the transmitting antenna in miles,
$$d_t \text{ (miles)} = \sqrt{2h_t} = \sqrt{2 \times 300}$$

or
$$d_t = 24.49 \text{ miles.}$$

(*b*) Radio horizon distance for receiving antenna in miles
$$d_r \text{ (miles)} = \sqrt{2h_r} = \sqrt{2 \times 100}$$

$$d_r = 14.142 \text{ miles.}$$

(*c*) The maximum range
$$= d_t + d_r = 24.49 + 14.142$$

$$d_{max} = 38.63 \text{ miles.}$$

Problem 9.8   A communication system is to be established at a frequency of 60 MHz with a transmitter power of 1 kW watts. The field strength of the directive antenna is 3 times that of a half-wave antenna. $h_t = 50$ m, $h_r = 5$ m. A field strength of 80μ V/m is required to give satisfactory reception. Find the range of the system.

Solution   Frequency, $f = 60$ MHz

Transmitter power $P_{TX} = 1$ kW

$$h_t = 50 \text{ m, } h_r = 5 \text{ m}$$

Required field strength,
$$E = 80\,\mu\,V/m$$
$$E_0 = 3 \times 137.6 \sqrt{P_{kW}} \;\; mV/m \text{ at one mile}$$

$$E_0 = 3 \times 137.6 \sqrt{P_{kW}} \times \frac{8}{5} \times 10^3 \;\; mV/m \text{ at 1 m}$$

$$= 660.48 \times 10^3 \;\; mV/m$$

The field strength due to a space wave is

$$E = \frac{4\pi h_t\, h_r}{\lambda d^2}\, E_0$$

or
$$d^2 = \frac{4\pi h_t\, h_r}{\lambda E}\, E_0$$

$$= \frac{4\pi \times 50 \times 5 \times 660.48 \times 10^3 \times 10^{-3}}{5 \times 80 \times 10^{-6}}$$

$$= \frac{200\pi \times 660.48 \times 10^6}{80}$$

$$= 5187.39 \times 10^6$$

$$d = 72.023 \times 10^3 \; m$$

or
$$d = 72.023 \; km$$

The range of the space wave is $d = 72.023$ km.

Problem 9.9 Find the maximum wavelength at which propagation is possible by means of a ground-based duct 30 m high when $\Delta M = 30$.

Solution Height of the duct,

$$h_d = 30 \text{ m}, \;\; \Delta M = 30$$

The maximum wavelength at which duct propagation is possible is given by

$$\lambda_{max} = 2.5 h_d \sqrt{\Delta M \times 10^{-6}}$$

$$= 2.5 \times 30 \; \sqrt{30 \times 10^{-6}}$$

$$= 410.79 \times 10^{-3}$$

$$\lambda_{max} = 0.410 \text{ m.}$$

Problem 9.10 If the critical frequency of an ionised layer is 1.5 MHz, find the electron density of the layer.

Solution We have

$$f_c = 9 \sqrt{N}$$

where
$$f_c = 1.5 \text{ MHz}$$

$$N_{max} = \frac{f_c^2}{81}$$

$$= \frac{1.5^2 \times 10^{12}}{81}$$

$$N_{\max} = 2.777 \times 10^{10} \text{ electrons/m}^3.$$

Problem 9.11 When the maximum electron density of the ionospheric layer corresponds to refractive index of 0.92 at the frequency of 10 MHz, find the range if the frequency is MUF itself. The height of the ray reflection point on the ionospheric layer is 400 km. Assume flat earth and negligible effect of earth's magnetic field.

Solution We have the relation of refractive index, n

$$n = \sqrt{1 - \frac{81 N_{\max}}{f^2}}$$

Here $n = 0.92$, $\qquad f = 10$ MHz

The above equation is rewritten as

$$N_{\max} = \frac{(1 - n^2) f^2}{81}$$

$$= \frac{0.1536}{81} \times 10^{14}$$

$$N_{\max} = 18.96 \times 10^{10} \text{ electrons/m}^3$$

The critical frequency,

$$f_c = 9 \sqrt{N_{\max}}$$

$$= 9 \times \sqrt{18.96 \times 10^{10}}$$

$$= 3.92 \times 10^6 \text{ Hz}$$

$$\text{MUF} = f_c \sec \theta_i = 10 \times 10^6 \text{ Hz}$$

or
$$\sec \theta_i = \frac{\text{MUF}}{f_c}$$

$$= 2.55$$

$$= \frac{\left(h^2 + \dfrac{d^2}{4} \right)^{1/2}}{h}$$

$$d = 1{,}876.59 \text{ km}.$$

Problem 9.12 Find out the relative permittivity of D, E and F-layers of the ionosphere for an EM wave of frequency 50 MHz.

The electron density of D layer $= 400$ electrons/cm^3

The electron density of E layer $= 5 \times 10^5$ electrons/cm^3

The electron density of F layer $= 2 \times 10^6$ electrons/cm^3.

Solution We have

$$n = \sqrt{\epsilon_r} = \left(1 - \frac{81N}{f^2}\right)^{1/2}$$

or

$$\epsilon_r = \left(1 - \frac{81N}{f_{kHz}^2}\right)$$

$$N = 400 \text{ electrons/cm}^3$$

$$\epsilon_r = 1 - \frac{81 \times 400}{(5 \times 10^4)^2}$$

$$= 1 - 1.29 \times 10^{-5}$$

$$\boxed{\epsilon_r \approx 1.0}$$

For *E* layer, $$N = 5 \times 10^5 \text{ electrons/cm}^3$$

$\therefore$

$$\epsilon_r = 1 - \frac{81N}{f_{kHz}^2}$$

$$= 1 - \frac{81 \times 5 \times 10^5}{(5 \times 10^4)^2}$$

$$= 1 - \frac{405 \times 10^5}{25 \times 10^8}$$

$$= 1 - 0.0162$$

$$\boxed{\epsilon_r = 0.9838} \text{ for } E\text{-layer}$$

For *F* layer, $$N = 2 \times 10^6 \text{ electrons/cm}^3$$

$$\epsilon_r = 1 - \frac{81N}{f_{kHz}^2}$$

$$= 1 - \frac{81 \times 2 \times 10^6}{25 \times 10^8}$$

$$\boxed{\epsilon_r = 0.9345} \text{ for } F\text{-layer}$$

Problem 9.13 A sky-wave is incident on *D*-layer at an angle of $30°$. Find the angle of refraction if the frequency of the transmitted wave is 50 MHz.

Solution For *D*-layer, $N = 400$ electrons/cm^3 and we have

$$\sqrt{\epsilon_r} = \left(1 - \frac{81N}{f_{kHz}^2}\right)^{1/2}$$

$$\sqrt{\epsilon_r} = \left(1 - \frac{81 \times 400}{(5 \times 10^4)}\right)^{1/2}$$

$$\epsilon_r \approx 1.0$$

As refractive index, $n = \sqrt{\epsilon_r}$

$$n = 1.0$$

According to Snell's law

$$n \sin \theta_r = \sin \theta_i$$

$$\sin \theta_r = \sin \theta_i$$

$$= \sin 30°$$

The angle of refraction $\theta_r = 30°.$

Problem 9.14   Determine the critical frequency of EM wave for D ($N = 400$ electrons/cm^3), E (5×10^5 electrons/cm^3) and F (2×10^6 electrons/cm^3) layers.

Solution   The expression for critical frequency is given by

$$f_c = 9 \sqrt{N}$$

For D-layer,

$$f_c = 9 \times \sqrt{400} = 180 \text{ kHz}$$

For E-layer

$$f_c = 9 \times \sqrt{5 \times 10^5}$$

$$= 63.64 \times 10^2 \text{ kHz}$$

$$= 6.364 \text{ MHz}$$

For F-layer

$$f_c = 9 \times \sqrt{2 \times 10^6}$$

$$= 12.8 \text{ MHz}.$$

Problem 9.15   Find the critical frequency if the maximum electron density is

$$1.3 \times 10^6 \text{ electrons/cm}^3.$$

Solution $(f_c)_{\text{kHz}} = 9 \sqrt{N_{\text{max}}}$

$$= 9 \times \sqrt{1.3 \times 10^6}$$

$$= 10.26 \times 10^3 \text{ kHz}$$

$$f_c = 10.26 \text{ MHz}.$$

Problem 9.16   An HF radio communication is to be established between two points on the earth's surface. The points are at a distance of 2,600 km. The height of the ionospheric layer is 200 km and critical frequency is 4 MHz. Find MUF.

Solution
$$\text{MUF} = f_c \left[\left(1 + \frac{d}{2h} \right)^2 \right]^{1/2}$$

$$d = 2{,}600 \text{ km}$$

$$f_c = 4 \text{ MHz} \quad \text{and} \quad h = 200 \text{ km}$$

$$\boxed{\text{MUF} = 10.95 \text{ MHz.}}$$

Problem 9.17 Find the frequency of the propagating wave for D-Layer to have refractive index of 0.5.

Solution For D-Layer, $N = 400$ electrons/cm^3

$$\text{We have refractive index, } n = \left(1 - \frac{81N}{f_{\text{kHz}}^2} \right)^{1/2}$$

$$(0.5)^2 = 1 - \frac{81N}{f_{\text{kHz}}^2}$$

or
$$f^2 = \frac{81 \times 400}{0.75}$$

$$\boxed{f = 207.84 \text{ kHz.}}$$

Problem 9.18 Determine the range of line of sight if the height of the transmitting antenna is 60 m and the height of the receiving antenna is 6 m. Assume standard atmosphere.

Solution The range of LOS, in km is given by

$$d = \sqrt{17h_t} + \sqrt{17h_r}$$

Here $h_t = 60$ m, $h_r = 6$ m

$\therefore$ $d = 31.94 + 10.1$

$$\boxed{d = 42.04 \text{ km.}}$$

Problem 9.19 What is the critical angle of propagation for D-Layer if the transmitter and receiver are separated by 500 km?

Solution Height of the D-Layer

$$= 70 \text{ km}$$

The distance between transmitter and receiver

$$= 500 \text{ km}$$

The critical angle,

$$\theta_c = \sin^{-1} \left\{ \frac{h}{\left[h^2 + \left(\frac{d^2}{4} \right) \right]^{1/2}} \right\}$$

$$= \sin^{-1}\left\{\frac{70}{\left[4,900 + \left(\dfrac{500^2}{4}\right)\right]^{1/2}}\right\}$$

$$\theta_c = 15.64°$$

It can also be calculated from

$$\theta_c = \tan^{-1}\left(\frac{2h}{d}\right)$$

$$\theta_c = \tan^{-1}\left(\frac{2 \times 70}{500}\right)$$

$$\theta_c = 15.64°.$$

OBJECTIVE QUESTIONS

1. Ground wave is effective when the transmitting and receiving antennas are:
   (a)   vertically polarised
   (b)   horizontally polarised
   (c)   elliptically polarised
   (d)   circularly polarised

2. Ground wave propagation is used when:
   (a) f is in UHF range
   (b) f is in microwave range
   (c) f is in LF range
   (d) f is in VHF range

3. The equivalent circuit of earth is:
   (a)   a capacitance in shunt with a conductance
   (b)   series R-C circuit
   (c)   series R-L circuit
   (d)   series R-L-C circuit

4. The factor, A that takes care of earth's losses in ground wave propagation depends on:
   (a) σ, $\in$, f, d and polarisation
   (b) σ, $\in$
   (c) f, d
   (d) d

5. The factor, A depends on d as
   (a)   inversely proportional to d^2
   (b)   inversely proportional to d
   (c)   inversely proportional to d^3
   (d)   proportional to d

6. Ground wave field strength depends on:
   (a)   the height of transmitting antenna
   (b)   the height of receiving antenna
   (c)   the heights of transmitting and receiving antennas
   (d)   none of these

7. The reflection coefficient of incident EM wave on the earth depends on the angle of incidence.
   (Yes/No)

8. The reflection coefficient of incident wave on the earth depends on polarisation of the wave.
   (Yes/No)

9. The reflection coefficient of the incident wave on the earth depends on $\in_r$, σ, f.   (Yes/No)

10. Roughness of earth depends on
    (a)   angle of incidence, θ_i only
    (b)   frequency, f only
    (c)   standard deviation of earth's irregularities, σ_s only
    (d) θ_i, f and σ_s

11. Earth is considered to be smooth if the roughness is _______________.

12. Earth is considered to be rough if the roughness is _______________.

13. Space wave field strength depends on _______________.

14. Space wave field strength depends on _______________.
    (a) h_t only   (b) h_r only
    (c) h_t and h_r only   (d) h_t, h_r, d, f, P

15. Curvature of earth creates _______________.

16. Hills and tall buildings affect space wave field strength.   (Yes/No)

17. Atmosphere is said to be standard atmosphere when $\dfrac{dM}{dh} =$ _______________.

18. For standard atmosphere, the ratio of equivalent radius and actual radius of the earth is
    _______________.

19. Refractive index in terms of relative permittivity is _______________.

20. Duct propagation takes place when _______________.

21. Radio horizon distance for standard atmosphere when $h_t = 50$ m and h_r is 5 m is
    (a)   38.37 km   (b)   38.37 miles
    (c)   383.7 km   (d)   383.7 miles

22. Maximum wavelength at which duct propagation is possible, is _______________.

23. E-Layer of ionosphere is the lowest layer.   (Yes/No)

24. D-Layer of the ionosphere is best suited for HF communication.   (Yes/No)

25. The thickness of D-Layer is the smallest.   (Yes/No)

26. The thickness of F-Layer is the highest.   (Yes/No)

27. Actual height of the ionospheric layers is greater than virtual height.   (Yes/No)

28. D-Layer exists at all times.   (Yes/No)

29. *F*-Layer exists at nights.   (Yes/No)

30. The electron density of *D*-Layer is the highest.   (Yes/No)

31. Phase velocity of EM wave in ionosphere is _______________.

32. Phase velocity, group velocity and free space velocity are related by _______________.

33. The relation between MUF and f_c is _______________.

34. The skip distance, d_s in terms of height of the layer and critical angle is _______________.

35. If the critical frequency of the ionospheric layer is 1.5 MHz, the maximum electron density of the layer in electrons/m^3 is

   (*a*) 2.777×10^{10} (*b*) 27.77×10^{10}

   (*c*) 0.2777×10^{10} (*d*) 277.7×10^{10}

36. Fading is nothing but

   (*a*)   amplification of field   (*b*)   multiplication of field

   (*c*)   subtraction of two fields   (*d*)   change of field strength

37. Fading is usually compensated by _______________.

38. Troposcatter propagation is related to _______________.

   (*a*)   SIDS   (*b*)   Faraday rotation

   (*c*)   Fading   (*d*)   Atmospheric storms

39. EM waves of UHF range normally propagate by means of

   (*a*)   ground waves   (*b*)   space waves

   (*c*)   sky waves   (*d*)   surface waves

40. The absorption of EM waves by the atmosphere depends on

   (*a*)   the distance of the transmitter   (*b*)   their frequency

   (*c*)   the polarisation of the waves   (*d*)   strength of EM wave

41. Short-waves for long distance communication depend on

   (*a*)   ionospheric waves   (*b*)   ground waves

   (*c*)   direct waves   (*d*)   space waves

42. Troposcatter propagation is used at frequency range of

   (*a*)   VHF only   (*b*)   VLF only

   (*c*)   MF only   (*d*)   UHF and VHF

43. *D*-Layer lies approximately between

   (*a*)   65 to 75 km   (*b*)   100 to 110 km

   (*c*)   40 to 50 km   (*d*)   150 to 160 km

44. Attenuation in atmosphere is inversely proportional to the square of the frequency.

                                                                             (Yes/No)

45. The relative permittivity of ionospheric layer is

(a) $\epsilon_r = \left(1 - \dfrac{81N}{f^2}\right)$

(b) $\epsilon_r = \sqrt{1 - \dfrac{81N}{f^2}}$

(c) $\epsilon_r = \left(1 - \dfrac{f^2}{81N}\right)$

(d) $\epsilon_r = 9\sqrt{N}$

46. Wave unaffected by night or day is

(a) ground wave

(b) sky wave

(c) space wave

(d) tropospheric wave

47. When an EM wave propagates from air into ionosphere, its velocity

(a) decreases

(b) increases

(c) remains the same

(d) reduces to zero

48. During day-time the ionospheric layer that does not exist is

(a) F

(b) F_1

(c) F_2

(d) D

49. Critical frequency of the ionospheric layer is

(a) $f_c = 81\sqrt{N_{max}}$

(b) $f_c = 81N_{max}$

(c) $f_c = 9\sqrt{N_{max}}$

(d) $f_c = 9N_{max}$

50. If an EM wave whose frequency is 30 MHz is incident with an angle of $60°$. MUF is

(a) 60 MHz

(b) 20 MHz

(c) 30 MHz

(d) 10 MHz

51. When an EM wave of frequency 20 MHz is incident at angle of $30°$, MUF is

(a) 10 MHz

(b) 15 MHz

(c) 23 MHz

(d) 43 MHz

52. When an EM wave whose MUF is 25 MHz is incident at $40°$, the critical frequency is

(a) 19.15 MHz

(b) 1.915 MHz

(c) 2 MHz

(d) 30 MHz

53. If the critical frequency of an ionized layer is 15 MHz, the electron density of the layer is

(a) 2.78×10^6 electrons/m^3

(b) 0.278×10^6 electrons/m^3

(c) 27.8×10^6 electrons/m^3

(d) 3.2×10^6 electrons/m^3

54. The critical angle of an EM wave for an ionospheric layer of height h when the distance between the transmitter and receiver is d, is given by _______________.

55. Critical frequency, f_c is determined from _______________.

56. Frequency of optimum traffic, FOT is the frequency at which _______________.

57. Frequency of optimum traffic, FOT is _______________.

58. OWF is the same as ________________.

59. The Horizon distance is affected by atmospheric refraction.   (Yes/No)

60. On a spherical surface Horizon is ________________.

61. Radio Horizon of an antenna is the locus of the farthest points at which direct rays become ________________ to the planetary surface.

62. Faraday rotation takes place in all media.   (Yes/No)

63. The effective radius of the earth for a standard atmosphere is ________________ times the actual radius of the earth.

64. Gyro-frequency is a vector.   (Yes/No)

65. Gyro-frequency depends on the magnetic field.   (Yes/No)

66. Gyro frequency depends on the mass of the charged particles.   (Yes/No)

67. Gyro-frequency depends on the charges of the particles.   (Yes/No)

68. LOS distance is affected by atmospheric refraction.   (Yes/No)

69. Group and phase velocities have equal magnitudes even when phase velocity varies with frequency.   (Yes/No)

70. Group and phase velocities have the same directions even when the phase velocity varies with direction.   (Yes/No)

ANSWERS

1. (a) **2.** (c) **3.** (a) **4.** (a) **5.** (a) **6.** (a)

7. Yes   **8.** Yes   **9.** Yes   **10.** (d) **11.** Less then 0.1

12. Greater then 10   **13.** Direct and ground reflected rays   **14.** (d)

15. Shadow zones   **16.** Yes   **17.** $\dfrac{dM}{dh} = 0.036$ units/ft   **18.** $\dfrac{4}{3}$

19. $\sqrt{\epsilon_r}$ **20.** $\dfrac{dM}{dh} =$ negative   **21.** (a) **22.** $2.5h_d \sqrt{\Delta M \times 10^{-6}}$

23. Yes   **24.** No   **25.** Yes   **26.** Yes   **27.** No   **28.** No

29. No   **30.** No   **31.** v_0/n **32.** $v_0^2 = v_p\, v_g$ **33.** MUF $= f_c \sec \theta_i$

34. $d_s = 2h/\tan \theta_c$ **35.** (a) **36.** (d) **37.** Diversity schemes

38. (c) **39.** (b) **40.** (b) **41.** (a) **42.** (d) **43.** (a)

44. Yes   **45.** (a) **46.** (a) **47.** (b) **48.** (a) **49.** (c)

50. (a) **51.** (c) **52.** (a) **53.** (a) **54.** $\theta_c = \tan^{-1} \dfrac{2h}{d}$

55. Ionogram   **56.** The signal strength is strong   **57.** About 85% of MUF

58. FOT   **59.** Yes   **60.** Circle   **61.** Tangential   **62.** No   **63.** 4/3

64. Yes   **65.** Yes   **66.** Yes   **67.** Yes   **68.** Yes   **69.** No

70. No

 **EXERCISE PROBLEMS**

1. A transmitter operating at $f = 1.0$ MHz is required to provide a ground wave field strength of 1.0 mV/m at a distance of 20 km. A short vertical transmitting antenna has an efficiency of 60%. $\sigma = 4 \times 10^5$ mho/cm, $\epsilon_r = 15$. Determine the transmitter power required.

2. Find the field strength due to space wave at a distance of 100 km when the transmitting and receiving antennas are 100 m and 20 m respectively. $E_0 = 137.6 \sqrt{P_{kW}}$ mV/m at one mile. The radiated power is 100 kW. It operates at a frequency of 50 MHz.

3. Determine the Radio horizon distance of a transmitting antenna if its height is 100 m.

4. Find the maximum wavelength at which duct propagation is possible by means of a ground based duct of height $h = 2.5$ km when the modified refractive index is 20.

5. Determine the electron density of a layer if the critical frequency of the ionised layer is 2.0 MHz.

6. Communication by ionospheric propagation is required for a distance of 2,000 km. Height of the layer is 220 km and critical frequency is 5 MHz. Find MUF.

7. What is the maximum distance that can be covered by space wave communication if transmitting and receiving antenna heights are 300 feet and 100 feet respectively?

8. Find the Radio horizon distances if receiving antenna height is (a) 80 m (b) 120 m.

9. Determine the transmitting antenna height if (a) the receiving antenna is at a distance of 50 km and its height is 50 m, (b) the receiving antenna is at a distance of 60 km and the its height is 30 m.

1. $\nabla \times \nabla \times \mathbf{E}$ is

   (a) $\nabla \nabla . \mathbf{E} - \nabla^2 \mathbf{E}$

   (b) $\nabla^2 \mathbf{E} - \nabla \nabla . \mathbf{E}$

   (c) $\nabla^2 . \mathbf{E} + \nabla \mathbf{E}$

   (d) $\nabla . \mathbf{E} - \mathbf{E} . \nabla$

2. Unit vector of $\mathbf{E}$ is

   (a) $\dfrac{\mathbf{E}}{|\mathbf{E}|}$

   (b) $|\mathbf{E}| (\mathbf{a}_x + \mathbf{a}_y + \mathbf{a}_z)$

   (c) $\mathbf{E} . \mathbf{E}$

   (d) $\dfrac{|\mathbf{E}|}{\mathbf{E}}$

3. $\mathbf{E} \times \mathbf{H}$ is

   (a) $EH \cos \theta$

   (b) $EH \sin \theta$

   (c) $EH \sin \theta \; \mathbf{a}_n$

   (d) $EH \cos \theta \; \mathbf{a}_n$

4. $\mathbf{E} \times (\mathbf{A} + \mathbf{C})$ is

   (a) $\mathbf{E} \times \mathbf{C} + \mathbf{E} \times \mathbf{A}$

   (b) $\mathbf{E} . \mathbf{A} + \mathbf{E} \times \mathbf{C}$

   (c) $\mathbf{A} . \mathbf{E} + \mathbf{C} . \mathbf{E}$

   (d) $\mathbf{A} \times \mathbf{E} - \mathbf{E} \times \mathbf{C}$

5. Gradient of a scalar is

   (a) not defined

   (b) a vector

   (c) a scalar

   (d) not periodic

6. Divergence of a vector is

   (a) not defined

   (b) a scalar

   (c) a vector

   (d) the same as gradient of a vector

7. The unit of del
(*a*) does not exist (*b*) meter (*c*) 1/meter (*d*) dB

8. ∇^2 operates
(*a*) only on a scalar (*b*) only on a vector
(*c*) on a scalar and also on a vector (*d*) only on a constant

9. $\mathbf{a}_x \cdot \mathbf{a}_x$ is
(*a*) $\mathbf{a}_x$ (*b*) 1 (*c*) 0 (*d*) $\mathbf{a}_y$

10. $\mathbf{a}_x \cdot \mathbf{a}_y$ is
(*a*) 0 (*b*) 1 (*c*) $\mathbf{a}_z$ (*d*) $-\mathbf{a}_z$

11. $\mathbf{a}_x \times \mathbf{a}_y$
(*a*) 1 (*b*) 0 (*c*) $\mathbf{a}_z$ (*d*) $-\mathbf{a}_z$

12. $\mathbf{a}_x \times \mathbf{a}_x$
(*a*) 0 (*b*) 1 (*c*) $\mathbf{a}_z$ (*d*) $\mathbf{a}_y$

13. For static fields
(*a*) $\nabla \times \mathbf{H} = \dot{\mathbf{D}}$ (*b*) $\nabla \times \mathbf{H} = \mathbf{J}$
(*c*) $\nabla \times \mathbf{H} = 0$ (*d*) $\nabla \times \mathbf{H} = \mathbf{E}$

14. In free space
(*a*) $\nabla \times \mathbf{E} = 0$ (*b*) $\nabla \times \mathbf{E} = \rho_\upsilon$
(*c*) $\nabla \cdot \mathbf{D} = \rho_\upsilon$ (*d*) $\nabla \times \mathbf{E} = -\dot{\mathbf{B}}$

15. Unit of $\mathbf{E}$ is
(*a*) volt (*b*) Amp/m
(*c*) volt/m (*d*) volt/coulomb

16. Unit of $\mathbf{H}$ is
(*a*) weber (*b*) Ampere
(*c*) volt/m (*d*) Amp/m

17. The unit of $\mathbf{D}$ is
(*a*) Wb/m (*b*) Amp/m (*c*) C/m^2 (*d*) C/m

18. $\mathbf{D}$ is
(*a*) $\in \mathbf{E}$ (*b*) $\in \mathbf{H}$ (*c*) $\mu \mathbf{H}$ (*d*) $\in \dot{\mathbf{E}}$

19. $\nabla \cdot \dot{\mathbf{D}}$ is
(*a*) ρ_s (*b*) ρ_υ (*c*) ρ_1 (*d*) 0

20. $\nabla \times \mathbf{E}$ is
(*a*) $\dot{\mathbf{B}}$ (*b*) $-\dot{\mathbf{B}}$ (*c*) $\dot{\mathbf{D}} + J$ (*d*) J

21. The electric flux density, $\mathbf{D}$ is

   (*a*) $\in \mathbf{E}$ (*b*) $\in E$ (*c*) $\dfrac{\mathbf{E}}{\in_0}$ (*d*) $\mu \mathbf{E}$

22. The electric flux is

   (*a*) Q (*b*) $\in \mathbf{D}$ (*c*) $\in \mathbf{E}$ (*d*) $\in Q$

23. The unit of electric flux is

   (*a*) Weber   (*b*) Gauss   (*c*) Tesla   (*d*) Coulomb

24. In free space

   (*a*) $\nabla \cdot \mathbf{E} = 0$ (*b*) $\nabla \times \mathbf{E} = 0$

   (*c*) $\nabla \cdot \mathbf{E} = \rho_\upsilon$ (*d*) $\nabla \cdot \mathbf{E} = \rho_\upsilon / \in_0$

25. In free space

   (*a*) $\nabla \cdot \mathbf{B} = 0$ (*b*) $\nabla \cdot \mathbf{B} = \mu_0 H$

   (*c*) $\nabla \times \mathbf{B} = \mu_0 \mathbf{H}$ (*d*) $\nabla \cdot \mathbf{B} = \dfrac{\mathbf{H}}{\mu_0}$

26. For free space

   (*a*) $\sigma = 0$ (*b*) $J = 1 \; \text{Amp/m}^2$

   (*c*) $\mu_r = \mu_0$ (*d*) $\in_r = \in_0$

27. The unit of conduction current density is

   (*a*) Amp/m   (*b*) Amp/m^2 (*c*) Amp/m^3 (*d*) Amp

28. The unit of displacement current density is

   (*a*) Amp/m^2 (*b*) Amp/m   (*c*) Amp   (*d*) Amp-m

29. The unit of conduction current is

   (*a*) Amp   (*b*) Amp/m   (*c*) Amp/m^2 (*d*) Amp-m

30. The unit of permittivity is

   (*a*) Farad   (*b*) Henry   (*c*) Farad/m   (*d*) Henry/m

31. The unit of permeability is

   (*a*) Henry/m   (*b*) Farad/m   (*c*) Henry   (*d*) Weber

32. The conduction current density is

   (*a*) $\sigma \mathbf{E}$ (*b*) $\sigma \mathbf{D}$ (*c*) $\dfrac{\mathbf{E}}{\sigma}$ (*d*) $\in \mathbf{E}$

33. The displacement current density is

   (*a*) $\in \dfrac{\partial \mathbf{E}}{\partial t}$ (*b*) $\mathbf{D}$ (*c*) $\in \mathbf{D}$ (*d*) $\in_0 \mathbf{E}$

34. For uniform plane wave propagating in z-direction
    (a) $E_z = 0$
    (b) $E_z \neq f(y)$
    (c) $E_z \neq f(x)$
    (d) $E_z = 0$, $H_z = 0$

35. The line integral of **E** around a closed loop is
    (a) zero
    (b) Q
    (c) equal to current
    (d) ρ_v

36. The unit of attenuation is
    (a) dB/m
    (b) V/m
    (c) Amp/m
    (d) Coulomb/m

37. Velocity of a plane wave in a medium whose $\epsilon_r = 4$, $\mu_r = 1$ is
    (a) 3×10^8 m/s
    (b) 1.5×10^8 m/s
    (c) 6×10^8 m/s
    (d) 2×10^8 m/s

38. Velocity of uniform plane wave in free space is
    (a) 3×10^8 m/s
    (b) 3×10^8 cm/s
    (c) 3×10^6 cm/s
    (d) 3×10^{10} m/s

39. The unit of depth of penetration is
    (a) dB
    (b) meter
    (c) Neper
    (d) radian

40. **E . H** for a plane wave is
    (a) zero
    (b) 1
    (c) does not exist
    (d) EH

41. Equation of continuity is
    (a) $\nabla \cdot \mathbf{J} = -\dot{\rho}_v$
    (b) $\nabla \cdot \mathbf{J} = -\rho_v$
    (c) $\nabla \cdot \mathbf{J} = \dot{\rho}_v$
    (d) $\nabla \cdot \mathbf{J} = \dot{\rho}_s$

42. Magnetic current density is given by
    (a) $\mathbf{B}$
    (b) $\dfrac{\partial \mathbf{D}}{\partial t}$
    (c) $\dfrac{\partial \mathbf{B}}{\partial t}$
    (d) $-\dfrac{\partial \mathbf{B}}{\partial t}$

43. The unit of magnetic current density is
    (a) Amp/m^2
    (b) Volt/m^2
    (c) Amp
    (d) Amp/m

44. Boundary on **E** is
    (a) $\mathbf{a}_n \times (\mathbf{E}_1 - \mathbf{E}_2) = 0$
    (b) $\mathbf{E}_{t1} = \mathbf{E}_{t2}$
    (c) $\mathbf{a}_n \cdot (\mathbf{E}_1 - \mathbf{E}_2) = 0$
    (d) $\mathbf{a}_n \cdot \mathbf{E}_1 = 0$

45. The complete boundary condition on **H** is
    (a) $\mathbf{H}_{t1} = \mathbf{H}_{t2}$
    (b) $\mathbf{H}_{t1} - \mathbf{H}_{t2} = \mathbf{J}_s$
    (c) $\mathbf{a}_n \times (\mathbf{H}_1 - \mathbf{H}_2) = \mathbf{J}_s$
    (d) $\mathbf{a}_n \times (\mathbf{H}_1 - \mathbf{H}_2) = 0$

46. The complete boundary condition on **B** is

(a) $\mathbf{a}_n \times (\mathbf{B}_1 - \mathbf{B}_2) = 0$

(b) $\mathbf{a}_n \cdot (\mathbf{B}_1 - \mathbf{B}_2) = 0$

(c) $\mathbf{a}_n \cdot (\mathbf{B}_1 - \mathbf{B}_2) = J_s$

(d) $\mathbf{B}_{n1} = \mathbf{B}_{n2}$

47. The complete boundary condition on **D** is

(a) $\mathbf{a}_n \cdot (\mathbf{D}_1 - \mathbf{D}_2) = 0$

(b) $\mathbf{a}_n \cdot (\mathbf{D}_1 - \mathbf{D}_2) = \rho_s$

(c) $\mathbf{a}_n \times (\mathbf{D}_1 - \mathbf{D}_2) = \rho_s$

(d) $\mathbf{a}_n \times (\mathbf{D}_1 - \mathbf{D}_2) = 0$

48. The complete boundary condition on **J** is

(a) $\mathbf{a}_n \times (\mathbf{J}_1 - \mathbf{J}_2) = 0$

(b) $\mathbf{a}_n \cdot (\mathbf{J}_1 - \mathbf{J}_2) = 0$

(c) $\mathbf{a}_n \times (\mathbf{J}_1 - \mathbf{J}_2) = J_s$

(d) $\mathbf{a}_n \cdot (\mathbf{J}_1 - \mathbf{J}_2) = J_s$

49. Lorentz Gauge condition is

(a) $\nabla \cdot \mathbf{A} = -\mu \in \dot{V}$

(b) $\nabla \times \mathbf{A} = \mathbf{H}$

(c) $\nabla \times \mathbf{A} = \mathbf{B}$

(d) $\nabla \cdot \mathbf{A} = -\nabla \dot{V}$

50. Equation of continuity is

(a) $\oint_s \mathbf{J} \cdot \mathbf{ds} = I$

(b) $\oint_s \mathbf{J} \cdot \mathbf{ds} = Q$

(c) $\oint_s \mathbf{J} \cdot \mathbf{ds} = J$

(d) $\oint_s \mathbf{J} \cdot \mathbf{ds} = \rho_\upsilon$

51. Intrinsic impedance of a medium is given by

(a) $\sqrt{\dfrac{\mu}{\in}}$

(b) $\sqrt{\dfrac{j\omega\mu}{\sigma + j\omega\in}}$

(c) $\sqrt{\dfrac{\in}{\mu}}$

(d) $\sqrt{\mu\in}$

52. The characteristic impedance of free space is

(a) $277\,\Omega$

(b) $120\,\Omega$

(c) $377\,\Omega$

(d) $120\pi^2\,\Omega$

53. The wave equation in free space is

(a) $\nabla^2 \mathbf{E} = \mu_0 \in_0 \dot{\mathbf{E}}$

(b) $\nabla^2 \mathbf{E} = \mu_0 \in_0 \ddot{\mathbf{E}}$

(c) $\nabla^2 \mathbf{V} = -\rho_\upsilon / \in$

(d) $\nabla^2 \mathbf{E} = \mu\sigma\ddot{\mathbf{E}}$

54. **E** and **H** are always perpendicular to each other

(a) yes

(b) some times

(c) only for uniform plane wave

(d) none of these

55. Velocity of propagation of a plane wave is

(a) $\dfrac{\beta}{\omega}$

(b) $\dfrac{\omega}{\beta}$

(c) $\dfrac{\lambda}{f}$

(d) $\dfrac{f}{\lambda}$

56. Attenuation of plane wave in free space is

(a) zero (b) infinite

(c) propagation constant (d) β itself

57. A medium is a good conductor if

(a) $\dfrac{\sigma}{\omega\epsilon} \gg 1$ (b) $\dfrac{\sigma}{\omega\epsilon} \ll 1$ (c) $\dfrac{\sigma}{\omega\epsilon} = 1$ (d) $\dfrac{\sigma}{\omega\epsilon} = 0$

58. A medium is a good dielectric if

(a) $\dfrac{\sigma}{\omega\epsilon} \gg 1$ (b) $\dfrac{\sigma}{\omega\epsilon} \ll 1$ (c) $\dfrac{\sigma}{\omega\epsilon} = 0$ (d) $\dfrac{\sigma}{\omega\epsilon} = 1$

59. The ratio of conduction current to displacement current is

(a) $\dfrac{\sigma}{\omega\epsilon}$ (b) $\dfrac{\omega\epsilon}{\sigma}$ (c) 0 (d) 1

60. Dissipation factor of a dielectric is

(a) $\dfrac{\sigma}{\omega\epsilon}$ (b) $\dfrac{\omega\epsilon}{\sigma}$ (c) 1 (d) 0

61. Depth of penetration in good conductors is

(a) inversely proportional to conductivity

(b) inversely proportional to square root of conductivity

(c) not a function of conductivity

(d) directly proportional to conductivity

62. Attenuation constant in good dielectrics is

(a) directly proportional to conductivity

(b) inversely proportional to conductivity

(c) inversely proportional to square root of conductivity

(d) not a function of conductivity

63. Phase velocity of uniform plane wave is

(a) directly proportional to ϵ

(b) inversely proportional to ϵ

(c) is not a function of ϵ

(d) is not inversely proportional to square root of ϵ

64. Phase constant of a uniform plane wave in a medium is

(a) directly proportional to frequency

(b) not a function of frequency

(c) is inversely proportional to frequency

(d) is inversely proportional to square root of frequency

65. The characteristic impedance of a medium is

(a) $\sqrt{\dfrac{\mu}{\epsilon}}$

(b)   is a function of frequency

(c)   is not a function of frequency

(d)   is independent of μ, ϵ and f

66. Poynting vector gives

(a)   instantaneous power density

(b)   average power density

(c)   total power

(d)   total power density

67. Standing waves are produced when

(a)   there are no reflections

(b)   there are full reflections

(c)   there is only transmission

(d)   the waves are incident on good dielectrics

68. The minimum value of voltage standing wave ratio is

(a)   0   (b) -1 (c)   1   (d) $-\infty$

69. When the load impedance $Z_L = Z_0$, the VSWR is

(a)   10   (b)   1   (c)   0   (d) ∞

70. Brewster angle is

(a)   angle of incidence for which there is no reflection

(b)   angle of reflection for which there is no reflection

(c)   equal to reflected angle

(d)   equal to refraction angle

71. The range of HF is

(a)   3-30 kHz

(b)   30-300 kHz

(c)   3-30 MHz

(d)   30-300 MHz

72. The range of VHF is

(a)   3-30 MHz

(b)   30-300 MHz

(c)   300 MHz-3 GHz

(d)   300 kHz-300 MHz

73. The standard antenna for reference is

(a)   isotropic antenna

(b)   half-wave dipole

(c)   dish antenna

(d)   Yagi-Uda antenna

74. The most common popular prevalent TV antenna is

(a)   dipole   (b)   monopole   (c)   Yagi-Uda   (d)   horn

75. Broadband antenna is

(a)   log-periodic   (b)   dipole   (c)   Yagi-Uda   (d)   horn

76. The non-resonant antenna is

(a)   dipole   (b)   Yagi-Uda   (c)   monopole   (d)   rhombic

77. The common microwave link antenna is
   (*a*)   dipole   (*b*)   log-periodic   (*c*)   rhombic   (*d*)   parabolic dish

78. Antenna for direction finding is
   (*a*)   Yagi-Uda   (*b*)   rhombic   (*c*)   dish   (*d*)   loop antenna

79. The directivity of a half-wave dipole is
   (*a*)   10   (*b*)   1   (*c*)   1.5   (*d*)   1.64

80. The directivity of a current element is
   (*a*)   1.64   (*b*)   1.5   (*c*)   2.0   (*d*)   5.0

81. Reflectors in Yagi-Uda antenna are
   (*a*)   1   (*b*)   more than 1
   (*c*)   3   (*d*)   10

82. Directors in Yagi-Uda antenna are
   (*a*)   more than 1   (*b*)   1
   (*c*)   5   (*d*)   3

83. Radar antenna is
   (*a*)   parabolic dish   (*b*)   dipole
   (*c*)   horn   (*d*)   waveguide

84. The length of the folded dipole in Yagi-Uda antenna is
   (*a*) $l = 468/f$ (MHz) feet   (*b*) $l = 492/f$ (MHz) feet
   (*c*) $l = 342/f$ (MHz) feet   (*d*) $l = 192/f$ (MHz) feet

85. The impedance of folded dipole in Yagi-Uda antenna is
   (*a*)   inductive reactance   (*b*)   capacitive reactance
   (*c*)   purely resistive   (*d*) $73\,\Omega$

86. The radiation resistance of a current element is
   (*a*) $80\pi^2 \left(\dfrac{dl}{\lambda}\right)^2 \Omega$ (*b*) $80\pi \left(\dfrac{dl}{\lambda}\right)^2 \Omega$

   (*c*) $80\pi^2 \left(\dfrac{dl}{\lambda}\right) \Omega$ (*d*) $80\pi \left(\dfrac{dl}{\lambda}\right) \Omega$

87. Circularly polarised antenna is
   (*a*)   dipole   (*b*)   parabolic dish
   (*c*)   Yagi-Uda   (*d*)   helical

88. If the current element is y-directed, the resultant vector magnetic potential is
   (*a*) y-directed   (*b*) x-directed
   (*c*) z-directed   (*d*) θ-directed

89. For time-varying fields

(*a*) $\mathbf{E} = -\nabla V$ (*b*) $\mathbf{E} = -\nabla V - \dot{\mathbf{A}}$

(*c*) $\mathbf{E} = -\nabla V - \dot{\mathbf{D}}$ (*d*) $\mathbf{E} = -\nabla V - \dot{\mathbf{H}}$

90. The radiation pattern of a loop antenna is

(*a*)   circular   (*b*)   multi-directional

(*c*)   like that of monopole   (*d*)   cardiod

91. The directivity of isotropic radiator is

(*a*)   1   (*b*)   zero   (*c*)   more than 1   (*d*) ∞

92. Effective length of a half-wave dipole is

(*a*) $\lambda/2$ (*b*) $> \lambda/2$ (*c*) $< \lambda/2$ (*d*) 0.55λ

93. A Balun is

(*a*)   a resistor   (*b*)   an impedance transformer

(*c*)   an antenna   (*d*)   frequency converter

94. The polarisation of horizontal dipole is

(*a*)   horizontal   (*b*)   vertical   (*c*) θ-polarisation   (*d*)   circular

95. Crossed dipoles produce _______________ polarisation.

(*a*)   Linear   (*b*)   Circular

(*c*)   Horizontal   (*d*)   Vertical

96. For broadside linear array, excitation phase is

(*a*) $\alpha = -\beta d$ (*b*) $\alpha = \beta d$ (*c*)   zero   (*d*) 90°

97. For end-fire array, the excitation phase is

(*a*)   zero   (*b*) $\alpha = -\beta d$ (*c*) $\alpha = \beta d$ (*d*) 180°

98. The size of the antenna becomes _______________ when the frequency is increased.

(*a*)   Small   (*b*)   Large

(*c*)   Impracticably large   (*d*)   Constant

99. The size of the antenna is

(*a*)   inversely proportional to frequency

(*b*)   directly proportional to frequency

(*c*)   independent of frequency

(*d*)   inversely proportional to the square of the frequency

100. Poynting vector gives

(*a*)   direction of electric field   (*b*)   polarisation of electromagnetic wave

(*c*)   power flow   (*d*)   rate of energy flow

101. The length of folded dipole in Yagi-Uda antenna is

(*a*)   frequency dependent   (*b*)   frequency independent

(*c*)   equal to λ (*d*)   equal to 0.2λ

102. The spacing between folded dipole and reflector is

(a) λ (b) $\lambda/2$ (c) $>\lambda/2$ (d) $<\lambda/2$

103. Voltage standing wave ratio is

(a) V_{min}/V_{max} (b) V_{max}/V_{min}

(c) $V_{reflected}/V_{incident}$ (d) V_{max}

104. Antenna is a

(a) transducer   (b) filter   (c) regulator   (d) amplifier

105. The radiation resistance of a half-wave dipole close to earth is

(a) $73\,\Omega$ (b) $<73\,\Omega$ (c) $>73\,\Omega$ (d) infinity

106. If the directivity is high, the beam width is

(a) high   (b) low   (c) constant   (d) very high

107. Director in Yagi-Uda antenna is

(a) active element   (b) driven element

(c) parasitic element   (d) identical to dipole

108. Reflector in Yagi-Uda antenna is

(a) active element   (b) driven element

(c) identical to dipole   (d) parasitic element

109. Log-periodic antenna is

(a) narrow band   (b) wide band

(c) frequency independent   (d) frequency dependent

110. In vertical dipole, the electric field is

(a) parallel to the dipole   (b) perpendicular to the dipole

(c) θ-directed   (d) circular

111. The effective length of a vertical radiator is

(a) increased by capacitive hat

(b) increased by loading with lumped inductance

(c) increased by supplying more power

(d) increased by resistance loading

112. The Null-to-Null beam width in end-fire array is

(a) $\dfrac{2\lambda}{Nd}$ (b) $\sqrt{\dfrac{2\lambda}{Nd}}$ (c) $2\sqrt{\dfrac{\lambda}{Nd}}$ (d) $2\sqrt{\dfrac{2\lambda}{Nd}}$

113. The Null-to-Null beam width in broadside array is

(a) $\dfrac{2\lambda}{Nd}$ (b) $2\sqrt{\dfrac{2\lambda}{Nd}}$ (c) $\dfrac{2\lambda^2}{Nd}$ (d) $\sqrt{\dfrac{Nd}{2\lambda}}$

114. The length of resonant dipole is

(a) $\lambda/2$　　　(b) λ　　　(c) $\lambda/4$　　　(d) 2λ

115. The first side lobe level in uniform linear array is

(a) 0.212　　　(b) 0.121　　　(c) 0.312　　　(d) 0.51

116. The side lobe level in binomial array is

(a) zero　　　(b) -13.5 dB　　　(c) -20 dB　　　(d) zero dB

117. In binomial array, the central elements are excited

(a) strongly　　　(b) weakly　　　(c) uniformly　　　(d) easily

118. In horizontal polarised wave, the electric field is

(a) parallel to the ground　　　(b) perpendicular to the ground

(c) in θ-direction to the ground　　　(d) elliptical

119. The maximum directive gain of current element is

(a) 1.76 dB　　　(b) 2.15 dB　　　(c) 3 dB　　　(d) 0 dB

120. Band width of an antenna is

(a) $\dfrac{f_0}{Q}$　　　(b) $\dfrac{Q}{f_0}$　　　(c) $f_0 Q$　　　(d) $\dfrac{f_0^2}{Q}$

121. Antenna can be used as

(a) sound sensor　　　(b) light sensor

(c) temperature sensor　　　(d) colour sensor

122. For far-field of z-directed current element

(a) $H_\phi = -\sin\theta\, \dfrac{\partial A_z}{\partial r}$　　　(b) $H_\phi = 0$

(c) $H_\phi = H_\theta$　　　(d) $H_\phi = H_r$

123. Induction and far-field have equal magnitudes at

(a) $r = \dfrac{\lambda}{2\pi}$　　　(b) $r = \dfrac{\lambda}{6\pi}$　　　(c) $r = \dfrac{\lambda}{\pi}$　　　(d) $r = \infty$

124. Induction and radiation fields have equal magnitudes at

(a) $r = \dfrac{\upsilon_0}{\omega}$　　　(b) $r = \dfrac{\upsilon_0}{2\pi}$　　　(c) $r = \dfrac{\upsilon_0}{\beta}$　　　(d) $r = \dfrac{\upsilon_0}{\lambda}$

125. If the output signal level is 1 W, power gain is

(a) 0 dB　　　(b) 1 dB　　　(c) 10 dB　　　(d) ∞ dB

126. LF antennas are usually used for

(a) vertical polarisation　　　(b) horizontal polarisation

(c) circular polarisation　　　(d) elliptical polarisation

127. The real part of antenna impedance consists of

(a) R_r only

(b) R_r and R_l

(c) R_l only

(d) zero ohms of resistance

128. Power and field patterns are related as

(a) $P \propto E^2$

(b) $P \propto E$

(c) $P \propto \sqrt{E}$

(d) $P \propto \dfrac{1}{E}$

129. For radiation pattern measurements, the distance of the far-field region is

(a) $r > \dfrac{2D^2}{\lambda}$

(b) $r < \dfrac{D^2}{\lambda}$

(c) $r = \dfrac{\lambda}{\pi}$

(d) $r = \dfrac{D}{\lambda}$

130. GTEM cell means

(a) Geometric transverse electromagnetic cell

(b) Giga Hertz TEM cell

(c) Grounded TEM cell

(d) Geo TEM cell

131. Wheatstone bridge is used to measure antenna impedance at a frequency of

(a) Giga hertz

(b) upto microwave range

(c) upto millimeter range

(d) upto 30 MHz

132. If the field measurements are made at $r < \dfrac{2D^2}{\lambda}$

(a) side lobe levels will be high

(b) band width will be small

(c) band width and side lobe level are small

(d) no side lobes appear

133. Null-to-Null band width is

(a) equal to 3 dB band width

(b) greater than 3 dB band width

(c) less than 3 dB band width

(d) not related to 3 dB band width

134. Antenna efficiency is

(a) $\dfrac{g_p}{g_d}$

(b) $\dfrac{g_d}{g_p}$

(c) g_p

(d) g_d

135. Phase difference is

(a) β

(b) $\dfrac{\omega}{\beta}$

(c) path difference $\times \beta$

(d) $\dfrac{\beta}{\omega}$

136. If the response of a vertical dipole is 1 for a unity normalised input power, the polarisation is

(a) vertical

(b) horizontal

(c) circular

(d) elliptical

137. If the response of RCP helix is zero, the polarisation of test antenna is

    (*a*)   LCP   (*b*)   RCP   (*c*)   horizontal   (*d*)   vertical

138. If the response of RCP helix is maximum, the polarisation of test antenna is

    (*a*)   LCP   (*b*)   RCP   (*c*)   horizontal   (*d*)   vertical

139. If the response of LCP helix is maximum, the polarisation of test antenna is

    (*a*)   LCP   (*b*)   RCP   (*c*)   horizontal   (*d*)   vertical

140. If the response of a horizontal dipole is maximum, the polarisation of the test antenna is

    (*a*)   horizontal   (*b*)   vertical

    (*c*)   circular   (*d*)   elliptical

141. If the response of any type of antenna is 0.5 for unity normalised power, the polarisation of the test antenna is

    (*a*)   linear   (*b*)   horizontal

    (*c*)   vertical   (*d*)   unpolarised

142. The excitation levels of a three element binomial array are

    (*a*)   1, 2, 1   (*b*)   1, 3, 1   (*c*)   1, 4, 1   (*d*)   2, 3, 2

143. The excitation levels of a four-element array are

    (*a*)   1, 3, 1, 1   (*b*)   1, 2, 2, 1

    (*c*)   1, 3, 3, 1   (*d*)   1, 4, 4, 1

144. The basic transmission loss between transmitter and receiver is

    (*a*) $10 \log \left(\dfrac{4\pi d}{\lambda} \right)^2$ (*b*) $10 \log \left(\dfrac{\lambda}{4\pi d} \right)^2$

    (*c*) $10 \log (G_{TX}\, G_{RX})$ (*d*)   zero

145. Actual transmission loss between transmitter and receiver is

    (*a*) $10 \log \left\{ \left(\dfrac{\lambda}{4\pi d} \right)^2 \dfrac{1}{G_T\, G_R} \right\}$ (*b*) $10 \log \left\{ \left(\dfrac{4\pi d}{\lambda} \right)^2 \dfrac{1}{G_T\, G_R} \right\}$

    (*c*) $10 \log \left\{ \left(\dfrac{4\pi d}{\lambda} \right) G_T\, G_R \right\}$ (*d*) $10 \log \left\{ \left(\dfrac{\lambda}{4\pi d} \right) G_T\, G_R \right\}$

146. Friis formula is

    (*a*) $L_a = \left(\dfrac{\lambda}{4\pi d} \right)^2$ (*b*) $L_a = \left(\dfrac{4\pi d}{\lambda} \right)^2$

    (*c*) $L_a = 10 \log_{10} \left\{ \left(\dfrac{4\pi d}{\lambda} \right)^2 \dfrac{1}{G_T\, G_R} \right\}$ (*d*) $L_a = 10 \log_{10} \left(\dfrac{4\pi d}{\lambda} \right)^2$

147. Noise figure of antenna is

(a) $\dfrac{T_e}{T_o}$ (b) $1 + \dfrac{T_e}{T_o}$ (c) $1 - \dfrac{T_e}{T_o}$ (d) $1 + \left(\dfrac{T_e}{T_o}\right)^2$

148. Actual transmission loss in dB between transmitter and receiver is
   (a) greater than basic transmission loss   (b) less than basic transmission loss
   (c) equal to basic transmission loss   (d) infinite

149. Schelkunoff polynomial method gives
   (a) nulls in the desired directions   (b) nulls in the undesired directions
   (c) desired side lobes   (d) desired beam width

150. Visible region is
   (a) realisable part of unit circle   (b) realisable part of array
   (c) realisable part of excitation   (d) realisable part of spacing

151. Tschebyscheff polynomial method gives
   (a) desired side lobe ratio   (b) desired beam width
   (c) desired overall pattern   (d) desired phase function

152. The advantage of uniform linear array is
   (a) the required number of sources is one   (b) SLR is small
   (c) number of side lobes are less   (d) grating lobes are present

153. The current distribution in half-wave dipole is
   (a) sinusoidal   (b) constant
   (c) triangular   (d) parabolic

154. When the array length is high, the Null-to-Null beam width is
   (a) small   (b) high   (c) constant   (d) infinity

155. Conducting properties of earth are
   (a) constant with frequency   (b) change with frequency
   (c) change with the type of antenna   (d) change with excitation of the antenna

156. X-band frequency range is
   (a) 8-12 GHz   (b) 2-4 GHz
   (c) 4-8 GHz   (d) 1-2 GHz

157. Ku band frequency range is
   (a) 12-18.5 GHz   (b) 18.5-26.0 GHz
   (c) 8-12 GHz   (d) 4-8 GHz

158. UHF frequency range is
   (a) 300 MHz-3 GHz   (b) 30-300 MHz
   (c) 3-30 MHz   (d) 1-20 MHz

159. Ground wave propagation is used at

(a) MW frequencies
(b) VHF
(c) UHF
(d) microwave frequency

160. Ground wave propagation is used for

(a) long distance communication
(b) short distance communication
(c) global communication
(d) country to country communication

161. The best polarisation used for ground wave propagation is

(a) vertical (b) horizontal (c) circular (d) elliptical

162. The ground wave field strength is

(a) inversely proportional to distance
(b) inversely proportional to the square of distance
(c) directly proportional to distance
(d) directly proportional to the square of distance

163. To obtain ground wave propagation, the transmitting and receiving antennas should be

(a) far-away from earth
(b) close to the earth
(c) at isolated heights
(d) at long distance from each other

164. The equivalent circuit of the earth is

(a) a resistance
(b) an inductance
(c) a capacitance
(d) a capacitance in shunt with resistance

165. In ground wave propagation, earth introduces

(a) no losses
(b) losses
(c) no reflections
(d) refracted rays

166. Earth's losses in ground wave propagation are estimated from

(a) numerical distance
(b) phase constant
(c) numerical distance and phase constant
(d) distance between Tx and Rx only

167. Earth's losses depend on

(a) conductivity only
(b) dielectric constant only
(c) frequency only
(d) conductivity, permittivity, frequency and polarisation of the antennas

168. Roughness of earth depends on

(a) frequency only
(b) surface irregularities
(c) angle of incidence only
(d) frequency, angle of incidence and surface irregularities

169. Earth is electrically smooth if $R = \dfrac{4\pi\sigma_s \sin\theta_i}{\lambda}$ is

    (a) > 100 (b) > 1000 (c) > 50 (d) < 0.1

170. Earth is electrically rough if $R = \dfrac{4\pi\sigma_s \sin\theta_i}{\lambda}$ is

    (a) < 0.1 (b) 0 (c) > 10 (d) > 0.1

171. The dielectric constant is

    (a)   only real   (b)   only imaginary

    (c)   a good conductor   (d)   complex quantity

172. The ground wave due to effect of earth is

    (a)   elliptically polarised   (b)   vertically polarised

    (c)   horizontally polarised   (d)   circularly polarised

173. Space wave propagation is used at

    (a) $f < 10$ MHz   (b) $f < 20$ MHz

    (c) $f > 30$ MHz   (d) $f < 20$ kHz

174. Space wave propagation is useful at

    (a)   LF   (b)   HF   (c)   VLF   (d)   VHF and UHF

175. The troposphere is

    (a)   part of earth

    (b)   part of ionosphere

    (c)   part of atmosphere at about 50 km above the earth

    (d)   part of atmosphere less than 16 km above the earth

176. Field strength due to space wave is

    (a)   proportional to distance

    (b)   inversely proportional to distance

    (c)   inversely proportional to the square of distance

    (d)   not a function of distance.

177. Space wave field strength is a function of

    (a)   curvature of earth only

    (b)   frequency only

    (c)   heights of antennas only

    (d)   curvature of earth, frequency, heights of antennas and so on

178. The refractive index of a medium is

    (a) $\in_r^2$ (b) $\sqrt{\in_r}$ (c) $(\in_r)^{3/2}$ (d) $\sqrt{\in_0}$

179. Radius of earth is

    (a) 6.37×10^6 km   (b) 6.37×10^6 m   (c) 6.37 km   (d) 6.37×10^8 m

180. For standard atmosphere, $\dfrac{dM}{dh}$ is

    (a)   0.036 ft   (b)   0.036 units/ft   (c)   0.36 units/ft   (d)   0.036 units/m

181. At high altitudes, $\dfrac{dM}{dh}$ is

    (a)   high   (b)   low

    (c)   constant with height   (d)   zero

182. Duct propagation is useful

    (a)   to reduce the effect of curvature of earth

    (b)   to create shadow zones

    (c)   to lower the frequency

    (d)   lower the distance of transmission

183. The ratio of equivalent radius of earth for standard atmosphere is

    (a) $\dfrac{4}{3}$ (b) $\dfrac{3}{4}$ (c) $\sqrt{2}$ (d)   2

184. The radio horizon distance for standard atmosphere in miles is

    (a) $\sqrt{2h_{\text{feet}}}$ (b) $\sqrt{2h_{\text{inch}}}$ (c) $\sqrt{2h_{m}}$ (d) $\sqrt{2h_{\text{miles}}}$

185. The radio horizon distance for standard atmosphere in km is

    (a) $\sqrt{2h_{\text{km}}}$ (b) $\sqrt{17h_{m}}$ (c) $\sqrt{2h_{m}}$

186. Duct propagation takes place if $\dfrac{dM}{dh}$ is

    (a)   negative   (b)   positive   (c)   zero   (d) ∞

187. The most common season for duct propagation to take place is

    (a)   monsoon   (b)   summer

    (c)   winter   (d)   mid-summer

188. Duct propagation is similar to

    (a)   free space propagation   (b)   propagation in waveguides

    (c)   propagation in water   (d)   uniform plane wave

189. LOS means

    (a)   length of the season   (b)   line of sight

    (c)   length of second time interval   (d)   line of seismic activity

190. D-layer is

    (a)   highest layer   (b)   medium layer

    (c)   lowest layer   (d)   a part of troposphere

191. The thickness of D-layer is

    (a)   about 10 km   (b)   about 100 km

    (c)   zero   (d)   infinite

192. The electron density of D-layer is

    (*a*)   4000 electrons/cc   (*b*)   400 electrons/cc

    (*c*)   400 electrons/m^3 (*d*)   4000 electrons/m^3

193. D-layer exists in

    (*a*)   only day-time   (*b*)   only night

    (*c*)   all times   (*d*)   summer

194. D-layer reflects

    (*a*)   HF   (*b*)   UHF

    (*c*)   microwave   (*d*)   VLF and LF

195. E-layer exists at

    (*a*)   about 20 km   (*b*)   about 100 km

    (*c*)   about 300 km   (*d*)   about 400 km

196. The cross product of **E** and **H** gives

    (*a*)   complex Poynting vector   (*b*)   Poynting vector

    (*c*)   wave equation   (*d*)   polarisation

197. The magnetic field inside a perfect conductor is

    (*a*)   zero   (*b*)   uniform

    (*c*)   non-uniform   (*d*)   reduced exponentially

198. The electric field in a perfect conductor is

    (*a*)   uniform   (*b*)   non-uniform

    (*c*)   reduced exponentially   (*d*)   zero

199. The electric field is irrotational if

    (*a*) $\nabla \cdot \mathbf{E} = 0$ (*b*) $\nabla \mathbf{E} = 0$ (*c*) $\nabla \times \mathbf{E} = 0$ (*d*) $\nabla^2 \mathbf{E} = 0$

200. One-dimensional wave equation propagating in x-direction is

    (*a*) $\nabla^2 \mathbf{E} = \gamma^2 \mathbf{E}$ (*b*) $\nabla^2 \mathbf{E} = \ddot{\mathbf{E}}$

    (*c*) $\dfrac{\partial^2 \mathbf{E}}{\partial x^2} = \mu_0 \epsilon_0 \dfrac{\partial^2 \mathbf{E}}{\partial t^2}$ (*d*) $\dfrac{\partial^2 \mathbf{E}}{\partial y^2} = 0$

201. In a conductor, if the charge is not moving, the radiation is

    (*a*)   very high   (*b*)   zero

    (*c*)   the same as when the charge moves   (*d*)   moderate

202. If the charge is moving with a uniform velocity in an infinite straight wire, the radiator is

    (*a*)   infinite   (*b*)   moderate   (*c*)   zero   (*d*)   high

203. If the charge is moving in a curved wire, radiation

    (*a*)   exists   (*b*)   does not exist

    (*c*)   is infinite   (*d*)   same as when the wire is straight.

204. If the charge oscillates with time in a straight wire, it

(*a*)   radiates

(*b*)   does not radiate

(*c*)   stores energy

(*d*)   oscillates

205. If the charge accelerates, there exists

(*a*)   no radiation

(*b*)   radiation

(*c*)   stored energy

(*d*)   acceleration of antenna

206. If the charge decelerates, radiation

(*a*)   is zero

(*b*)   exists

(*c*)   does not exist in any antenna

(*d*)   exists only in some wire antennas

207. The primary equation for electromagnetic radiation in a very thin *z*-directed wire of length L is

(*a*) $L \dfrac{dI_z}{dt} = L\rho_L \, \mathbf{a}_z$

(*b*) $L \dfrac{dI_z}{dt} = L\rho_L \, \mathbf{a}_y$

(*c*) $L \dfrac{dI_z}{dt} = \rho_L \, \mathbf{a}_y$

(*d*) $L \dfrac{dI_z}{dt} = L\rho_L$

208. Radiation with broad frequency spectrum is very strong if

(*a*)   the pulses are of shorter duration

(*b*)   the pulses are of longer duration

(*c*)   the pulses have more amplitude

(*d*)   the pulses have small amplitude

209. For frequency independent antennas, the band width is

(*a*)   zero   (*b*) ∞ (*c*)   finite   (*d*)   moderate

210. The radiation intensity of an isotropic radiator is

(*a*) $\dfrac{P_r}{4\pi r^2}$ (*b*) $\dfrac{P_r}{4\pi r}$ (*c*) $\dfrac{P_r}{4\pi}$ (*d*) P_r

211. An omni-directional antenna is a

(*a*)   parabolic dish

(*b*)   dipole

(*c*)   horn

(*d*)   Yagi-Uda antenna

212. Loop antenna is

(*a*)   isotropic radiator

(*b*)   directional radiator

(*c*)   omni-directional radiator

(*d*)   point source

213. Broadside arrays are

(*a*)   omni-directional

(*b*)   point sources

(*c*)   directional antennas

(*d*)   isotropic antennas

214. In linear polarisation, there exists

(*a*)   three components

(*b*)   only one component

(*c*)   two components differing by 90° phase

(*d*)   two components differing by 270° phase

215. If there exists two orthogonal linear components which are in time phase, polarisation is

    (*a*)   linear   (*b*)   circular   (*c*)   elliptical   (*d*)   not present

216. Effective area of an antenna is

    (*a*)   ratio of power delivered to load to power density of incident wave

    (*b*)   ratio of radiation intensity to the power density of incident wave

    (*c*) g_p/g_d

    (*d*) g_d/g_p

217. Aperture efficiency, η_a of an antenna is

    (*a*)   the ratio of g_p and g_d

    (*b*)   maximum effective area to physical area

    (*c*)   effective area to physical area

    (*d*)   physical area to effective area

218. In far-field region, the angular field distribution is independent of

    (*a*)   transmitter power   (*b*)   distance from the antenna

    (*c*)   angular region   (*d*)   antenna type

219. Fresnel region is

    (*a*)   far-field region   (*b*)   near-field region

    (*c*)   the region of constant field   (*d*)   the region of no field

220. Fraunhofer region is

    (*a*)   far-field region   (*b*)   near-field region

    (*c*)   the region of constant field   (*d*)   the region of no field

221. Reactive near-field region exists when

    (*a*) $R > 0.62 \sqrt{\dfrac{D^3}{\lambda}}$ (*b*) $R < 0.62 \sqrt{\dfrac{D^2}{\lambda}}$

    (*c*) $R < 0.62 \sqrt{\dfrac{D^3}{\lambda}}$ (*d*) $R > 0.62 \sqrt{\dfrac{D^2}{\lambda}}$

222. Fresnel region exists when

    (*a*) $R \leq 0.62 \sqrt{\dfrac{D^3}{\lambda}}$ (*b*) $R \geq 0.62 \sqrt{\dfrac{D^3}{\lambda}}$ and $R < \dfrac{2D^2}{\lambda}$

    (*c*) $R \geq \dfrac{2D^2}{\lambda}$ (*d*) $R \geq 0.62 \sqrt{\dfrac{D^3}{\lambda}}$

223. Fraunhofer region exists when

    (*a*) $R > \dfrac{2D^2}{\lambda}$ (*b*) $R < \dfrac{2D^2}{\lambda}$

    (*c*) $R \geq 0.62 \sqrt{\dfrac{D^3}{\lambda}}$ (*d*) $R \leq 0.62 \sqrt{\dfrac{D^3}{\lambda}}$

224. Unit of directivity is

(a) watts (b) watts/m^2 (c) watts/m^3 (d) nil

225. Unit of solid angle is

(a) degrees (b) radian (c) sterradian (d) nil

226. If h is the height of an antenna, the number of lobes are

(a) $\approx \left(\dfrac{h}{\lambda} \right)$ (b) $\approx 2 \left(\dfrac{h}{\lambda} \right)$ (c) $= 4$ (d) $= 2 \left(\dfrac{\lambda}{h} \right)$

227. Divergence factor of a field from earth is

(a) $\dfrac{\text{reflected field from flat earth}}{\text{reflected field from curved surface}}$ (b) $\dfrac{\text{reflected field from curved surface}}{\text{reflected field from flat earth}}$

(c) $\dfrac{\text{reflected field from curved surface}}{\text{incident field on curved surface}}$ (d) $\dfrac{\text{incident field on curved surface}}{\text{reflected field from curved surface}}$

228. If R_r is radiation resistance, μ_e is effective permeability of ferrite core, the radiation resistance of ferrite loop is

(a) $R_r \left(\dfrac{\mu_0}{\mu_e} \right)^2$ (b) $R_r \left(\dfrac{\mu_e}{\mu_0} \right)^2$ (c) R_r (d) $R_r \mu_0$

229. The resultant field of an array antenna is

(a) the product of element pattern and array factor
(b) array factor
(c) sum of element patterns
(d) element pattern

230. The excitation required to orient a beam in θ_0 direction is

(a) $kd \cos \theta_0$ (b) $-kd \cos \theta_0$ (c) $-kd$ (d) kd

231. Super directivity of an array can be obtained by

(a) reducing the spacing (b) increasing the spacing
(c) reducing the number of elements (d) decreasing array length

232. Super directivity obtained by reducing the spacing and increasing the number of elements results in

(a) high reactive power and Q
(b) low reactive power and Q
(c) small Q
(d) high reactive power and lower Q

233. Circular antennas are most sensitive to

(a) linearly polarised waves (b) elliptically polarised waves
(c) circularly polarised waves (d) unpolarised waves

234. Circular antenna has usually a length of

(a) $\dfrac{\lambda}{2}$ (b) λ (c) 2λ (d) $\dfrac{\lambda}{4}$

235. The horizontal pattern of circular antenna is
- (a) circle
- (b) four equal lobe pattern
- (c) figure-eight pattern
- (d) six equal-lobe pattern

236. Two end-fire circular antenna elements with $90°$ phasing produce
- (a) uni-directional pattern
- (b) figure eight pattern
- (c) multi-directional pattern
- (d) no radiation pattern

237. Circular antennas are widely used at
- (a) VLF
- (b) HF
- (c) microwave frequency
- (d) UHF

238. Directors in Yagi-Uda antenna
- (a) reduces the characteristic impedance of driven antenna element
- (b) increases the characteristic impedance of driven antenna element
- (c) has no effect on the characteristic impedance of driven element
- (d) act as open circuit

239. Directors and reflectors are used to
- (a) reduce the impedance
- (b) increase the impedance
- (c) increase the gain
- (d) form an array

240. Due to the use of parasitic elements the band width of Yagi-Uda antenna is
- (a) increased
- (b) not affected
- (c) made ideal
- (d) limited

241. Yagi-Uda antenna has
- (a) poor front-to-back ratio
- (b) good front-to-back ratio
- (c) infinite front-to-back ratio
- (d) zero front-to-back ratio

242. A good front-to-back ratio
- (a) increases co-channel interference
- (b) reduces co-channel interference
- (c) has no effect on co-channel interference
- (d) none of these

243. *V*-antenna yields
- (a) bi-directional pattern
- (b) uni-directional pattern
- (c) good signal strength compared to dipole
- (d) less band width compared to dipole

244. *V*-antenna is popular for
- (a) satellite reception
- (b) FM reception
- (c) mobile reception
- (d) radar signal reception

245. If the power gain of an antenna is 0.5 dB, the power ratio is

    (*a*)   0.216   (*b*)   12.6   (*c*)   1.26   (*d*)   1.06

246. If the voltage gain of an antenna is 1.0 dB, the voltage ratio is

    (*a*)   1.26   (*b*)   0.126   (*c*)   1.06   (*d*)   1.0

247. If the power gain of an antenna is 30 dB, the power ratio is

    (*a*)   1.477   (*b*)   1000   (*c*)   100   (*d*)   10

248. If the power gain of an antenna is 20, the power gain in dB is

    (*a*)   13   (*b*)   130   (*c*)   20   (*d*)   200

249. If a dipole is tilted forward, the band width becomes

    (*a*)   zero   (*b*)   infinite

    (*c*)   more   (*d*)   reduced

250. Voltage distribution on a 1.5λ dipole is

(*a*) 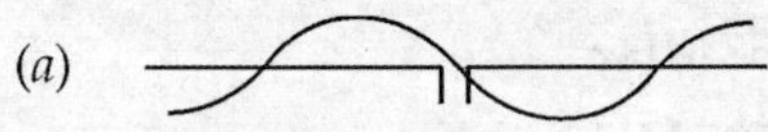   (*b*) 

(*c*) 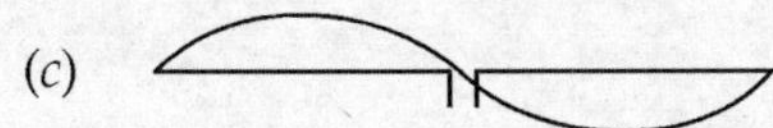   (*d*) 

251. Current distribution in 1.5λ dipole is

(*a*) 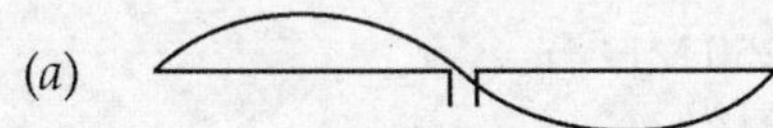   (*b*) 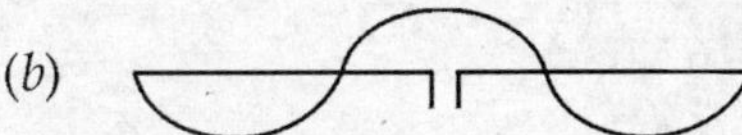

(*c*) 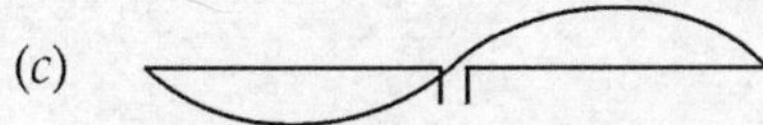   (*d*) 

252. Radiation pattern of a full-wave dipole is

(*a*) 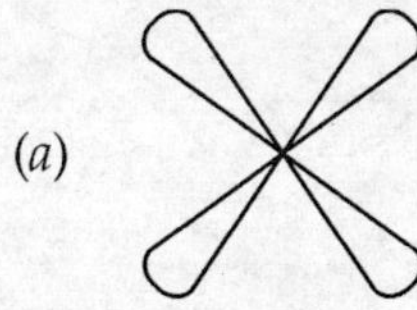   (*b*) 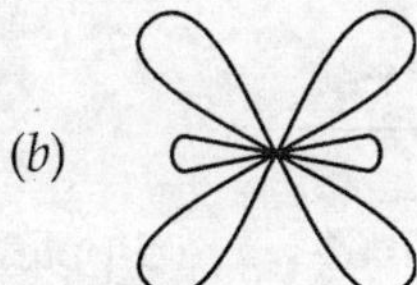

(*c*) 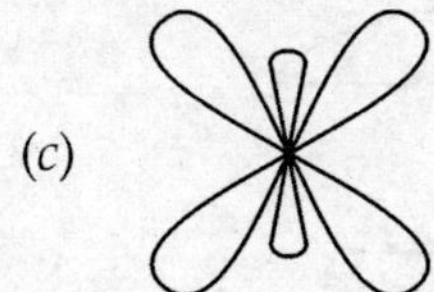   (*d*) 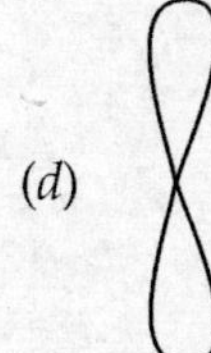

253. Voltage distribution on a full-wave dipole is

(a) 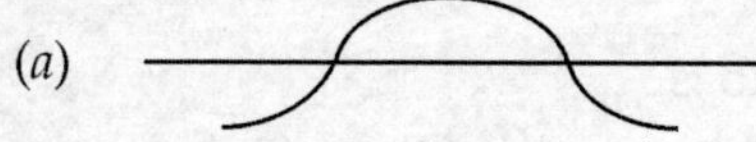

(b) 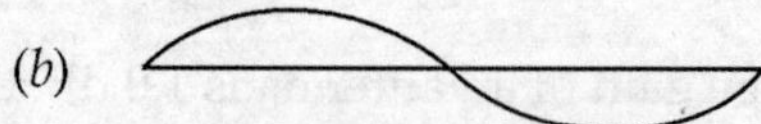

(c) 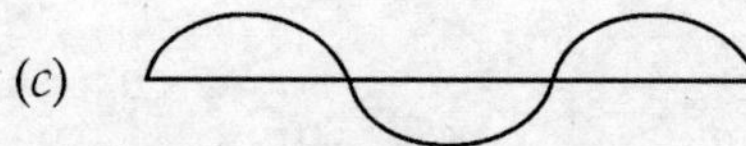

(d) 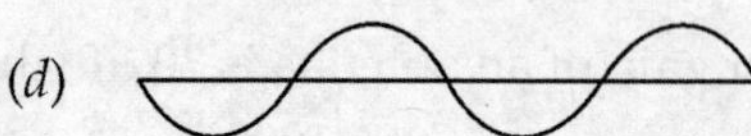

254. Current distribution on a full-wave dipole is

(a) 

(b) 

(c) 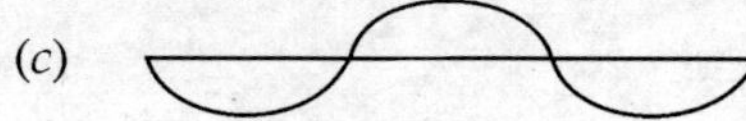

(d) 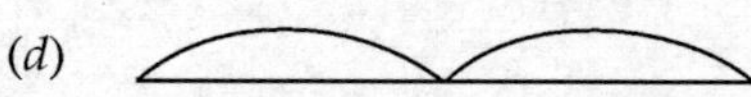

255. FM band is

(a)   78-98 MHz

(b)   88-108 MHz

(c)   108-128 MHz

(d)   100-200 MHz

256. The frequency band of lower five TV channels (VHF)

(a)   50-78 MHz

(b)   54-98 MHz

(c)   54-88 MHz

(d)   60-84 MHz

257. The frequency band of upper seven TV channels (VHF) is

(a)   150-216 MHz

(b)   174-250 MHz

(c)   174-216 MHz

(d)   100-216 MHz

258. The voltage distribution of a half-wave dipole is

(a) 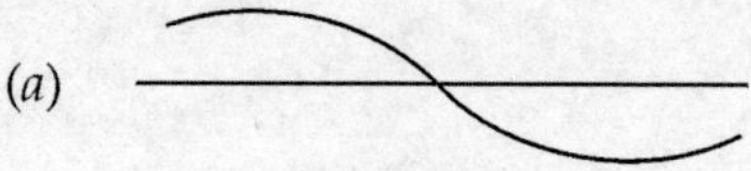

(b) 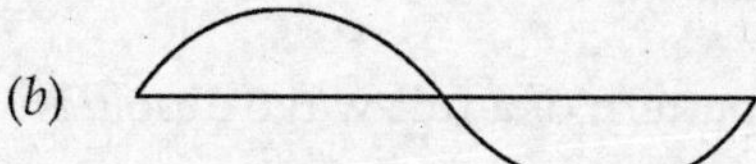

(c) 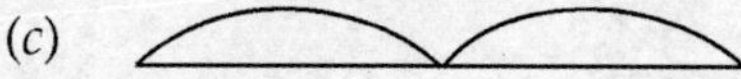

(d) 

259. Current distribution in a half-wave dipole is

(a) 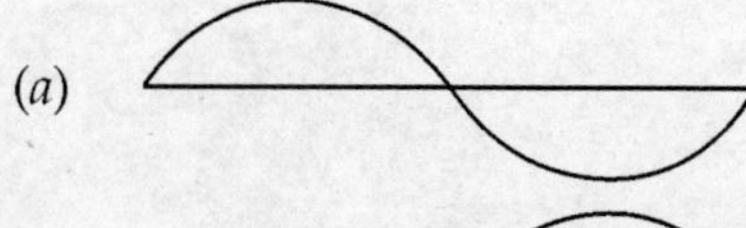

(b) 

(c) 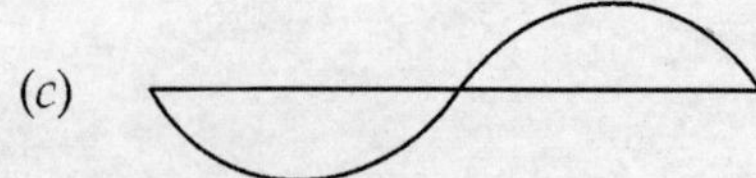

(d) 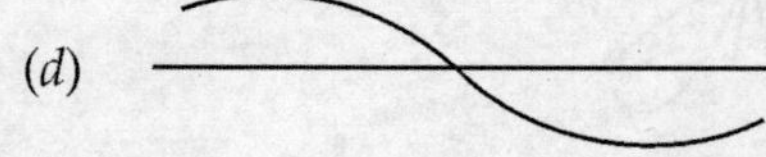

260. The line of sight distance for 500 ft and 30 ft transmitting antenna and receiving antenna respectively is

    (*a*)   31 miles   (*b*)   41 miles   (*c*)   51 miles   (*d*)   21 miles

261. If a 300Ω line is terminated in a 75Ω dipole, SWR is

    (*a*)   4   (*b*)   0.25   (*c*)   8   (*d*)   2

262. The power density at a distance of 1 km from 1 kW isotropic radiator is

    (*a*) 795 mW/m^2 (*b*) 79.5 mW/m^2

    (*c*) $795\mu \text{ W/m}^2$ (*d*) $79.5\mu \text{ W/m}^2$

263. For radio wave propagation, fresh water is considered to be

    (*a*)   very poor   (*b*)   poor   (*c*)   very good   (*d*)   average

264. For radio wave propagation, cities are considered to be

    (*a*)   poor   (*b*)   very poor   (*c*)   very good   (*d*)   good

265. The phase velocity of a wave in a medium whose $\epsilon_r = 0$ is

    (*a*) ∞ (*b*)   0   (*c*)   finite   (*d*) v_0

266. Directivity of a loop antenna whose radius is 0.5 m at $f = 0.9$ MHz is

    (*a*)   1.0   (*b*)   1.5   (*c*)   2.5   (*d*)   3.5

267. The number of log-periodic antenna elements depends on

    (*a*)   gain only   (*b*)   wedge angle only

    (*c*)   band width only   (*d*)   band width, scale and space factors

268. Rhombic antenna is

    (*a*)   standing wave antenna   (*b*)   narrow band antenna

    (*c*) $\dfrac{\lambda}{2}$ antenna   (*d*)   travelling wave antenna

269. Antenna radiation efficiency is high when its length is

    (*a*) $\dfrac{\lambda}{2}$ (*b*) λ (*c*) $3\dfrac{\lambda}{2}$ (*d*) ∞

270. Antenna resonates when its length is integer multiples of

    (*a*) λ (*b*) $\dfrac{\lambda}{2}$ (*c*) $\dfrac{\lambda}{4}$ (*d*) $\dfrac{\lambda}{3}$

271. Tower antenna is used for

    (*a*)   broadcast communication   (*b*)   satellite communication

    (*c*)   microwave communication   (*d*)   millimeterwave communication

272. For a 100Ω antenna with 2 A of current, radiated power is

    (*a*)   400 watts   (*b*)   200 watts

    (*c*)   50 watts   (*d*)   25 watts

273. For a perfect conductor, the power transmission coefficient is

    (*a*)   1   (*b*)   zero

    (*c*) ∞ (*d*)   reflection coefficient

274. For a perfect conductor, the power reflection coefficient $|\rho|^2$ is

    (*a*)   equal to transmission coefficient   (*b*) $1 - T$

    (*c*)   1   (*d*)   0

275. The Rayleigh criterion in the case of reflection of electromagnetic wave from semi-rough surface is

    (*a*) $\cos \theta_i > \dfrac{\lambda}{8d}$ (*b*) $\cos \theta_i = \dfrac{\lambda}{8d}$

    (*c*) $\cos \theta_i < \dfrac{\lambda}{8d}$ (*d*) $\cos \theta_i = 0$

276. Ground wave propagation requires

    (*a*)   high transmitter power   (*b*)   high frequency

    (*c*)   better seasonal conditions   (*d*)   rainy conditions

277. Maximum usable frequency is

    (*a*) $f_c \cos \theta_i$ (*b*) $f_c \sec \theta_i$

    (*c*) $f_c \sec \theta_r$ (*d*) $f_c \operatorname{cosec} \theta_i$

278. For an operating frequency of 6 GHz, the basic transmission loss at a distance of 50 km from the transmitter is

    (*a*)   132 dB   (*b*)   152 dB   (*c*)   142 dB   (*d*)   42 dB

279. The percent band width of an antenna with an optimum frequency of operation of 500 MHz and $- 3$ dB of frequencies of 300 and 350 MHz is

    (*a*)   20%   (*b*)   100%   (*c*)   500%   (*d*)   10%

280. The received power of a receiving antenna whose effective area is 0.2m^2 for an available power density of $100 \mu \text{ W/m}^2$ is

    (*a*) $200 \mu \text{ W/m}^2$ (*b*) $20 \mu \text{ W/m}^2$

    (*c*) $50 \mu \text{ W/m}^2$ (*d*) $500 \mu \text{ W/m}^2$

281. For an ideal antenna, the directivity is

    (*a*)   power gain   (*b*)   1

    (*c*)   1.64   (*d*)   1.5

282. For an ideal antenna, the radiation resistance is

    (*a*) 73Ω (*b*) 36.5Ω

    (*c*) 293Ω (*d*)   input impedance

283. The power gain in dB of isotropic radiator is

    (*a*)   0   (*b*)   1   (*c*)   1.5   (*d*)   1.64

284. The radiation resistance of a small loop antenna is

(a) $31,200\dfrac{A^2}{\lambda^4}$

(b) $73\,\Omega$

(c) $36.5\,\Omega$

(d) $292\,\Omega$

285. Half-power beam width of optimum flare horn in E-plane, is

(a) $\dfrac{56\lambda}{d_E}$

(b) $\dfrac{28\lambda}{d_E}$

(c) $\dfrac{122\lambda}{d_E}$

(d) 112°

286. Half-power beam width of optimum flare horn in H-plane, is

(a) $\dfrac{28\lambda}{d_H}$

(b) $\dfrac{56\lambda}{d_H}$

(c) 56λ

(d) 28λ

287. The normalised radiated power of a dipole is

(a)   1

(b)   1.5

(c) $\sin^2\theta$

(d)   1.64

288. The directive gain of electric dipole is

(a)   1.5

(b) $1.5\sin^2\theta$

(c)   1.64

(d)   1.0

289. A magnetic dipole is

(a)   a small circular loop

(b)   a piece of wire

(c)   a piece of conducting rod

(d)   the same as electric dipole

290. Virtual height of ionospheric layer is measured by an instrument namely

(a)   speedometer

(b)   radar

(c)   altimeter

(d)   ionosonde

291. If T = time required for the round trip, v_0 = velocity of the wave, the virtual height of layer is

(a) $h_\upsilon = \dfrac{v_0\,T}{2}$

(b) $h_\upsilon = v_0\,T$

(c) $h_\upsilon = \dfrac{T}{v_0}$

(d) $h_\upsilon = v_0^2\,T$

292. Gyro frequency is

(a) $\dfrac{q_e}{2\pi m}\,\mathbf{B}$

(b) $\dfrac{2\pi m\,\mathbf{B}}{q_e}$

(c) $2\pi\mathbf{B}\,q_e\,m$

(d) $\dfrac{2\pi}{mq_e}\,\mathbf{B}$

293. If the resistance part of antenna is $100\,\Omega$, radiation resistance is $80\,\Omega$, the antenna efficiency is

(a)   0.8

(b)   10/8

(c)   0.4

(d)   8/18

294. If ϕ is the angle between the axis of a receiving dipole and the direction of electric field, the polarisation loss factor is

(a) $\sin\phi$

(b) $\cos\phi$

(c) $\tan\phi$

(d) $\sec\phi$

295. The effective length of a half-wave dipole is

    (*a*) 0.4λ (*b*) 0.45λ (*c*) $\dfrac{\lambda}{\pi}$ (*d*) 0.55λ

296. Effective area of a Hertzian dipole is

    (*a*) $0.2\lambda^2$ (*b*) $0.25\lambda^2$ (*c*) $0.119\lambda^2$ (*d*) $0.3\lambda^2$

297. Directive gain is equal to power gain if

    (*a*) $\eta = \infty$ (*b*) $\eta = 1$ (*c*) $\eta = g_p$ (*d*) $\eta = g_d$

298. Directive gain and directivity are equal for

    (*a*)   directional antenna   (*b*)   dipole

    (*c*)   parabolic dish   (*d*)   isotropic antenna

299. For an isotropic antenna operating at $\lambda = \sqrt{4\pi}$, the effective area is

    (*a*) 4π (*b*) 1 (*c*) $(4\pi)^2$ (*d*) 2

300. Equivalent circuit of a half-wave dipole is

    (*a*) 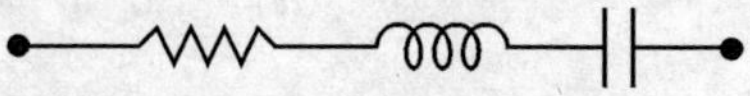   (*b*)

    (*c*) 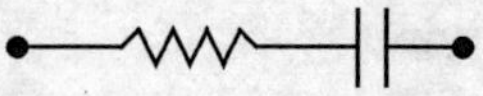   (*d*)

301. For direction finding applications, the required radiation beam should be

    (*a*)   narrow   (*b*)   broad   (*c*)   cosecant   (*d*)   ramp

302. Directivity is

    (*a*)   inversely proportional to beam width

    (*b*)   inversely proportional to square of beam width

    (*c*)   directly proportional to beam width

    (*d*)   directly proportional to square of beam width

303. If the direction of propagation of an electromagnetic wave is in z-direction, the polarisation is in

    (*a*) z-direction   (*b*) y-direction

    (*c*) x-direction   (*d*)   circular polarisation

304. If the quality factor of an antenna is 1000, resonant frequency is 10 MHz, its band width is

    (*a*)   100 kHz   (*b*)   10 kHz   (*c*)   10 Hz   (*d*)   10 MHz

305. The maximum effective area of an antenna operating at $\lambda = 10$ cm with directivity of 100 is

    (*a*) 1000 cm^2 (*b*) $\left(\dfrac{1}{4\pi}\right)$ m^2 (*c*) 4π m^2 (*d*) 10π m^2

306. The radiation resistance of an antenna which radiates 10 kW when a current of 10 ampere flows in it, is

    (*a*)   100 Ω   (*b*)   1,000 Ω

    (*c*)   10 Ω   (*d*)   100 KΩ

307. When an antenna radiates 10 kW in forward and 1 kW in backward directions, the front-to-back ratio of the antenna is

    (*a*)   1 dB   (*b*)   10 dB   (*c*)   100 dB   (*d*)   0 dB

308. The maximum gain of 100 element uniform linear array is

    (*a*)   10   (*b*)   100   (*c*)   1,000   (*d*)   1

309. The radiation pattern of a travelling antenna is

(*a*)
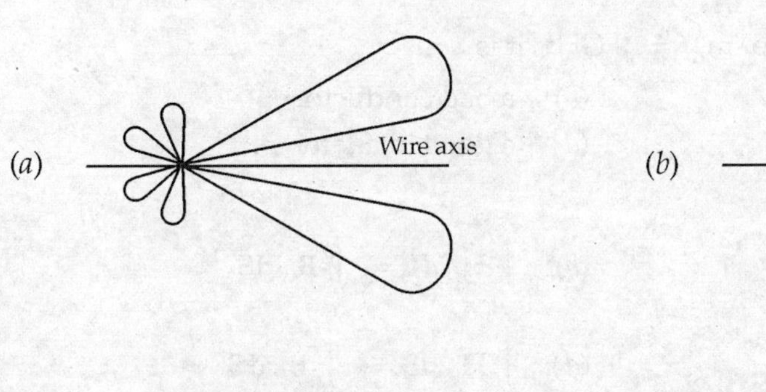

(*b*)
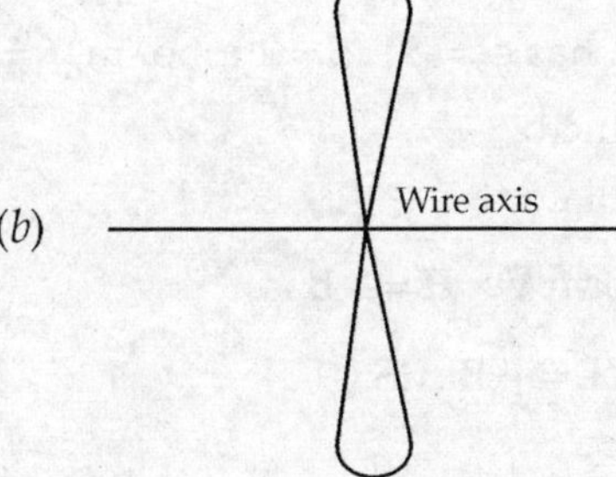

(*c*)
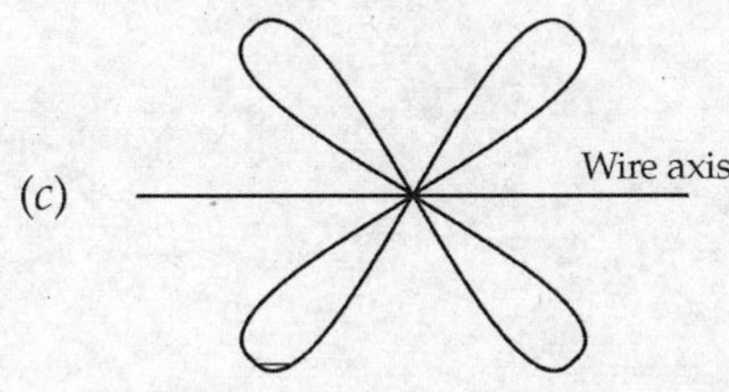

(*d*)
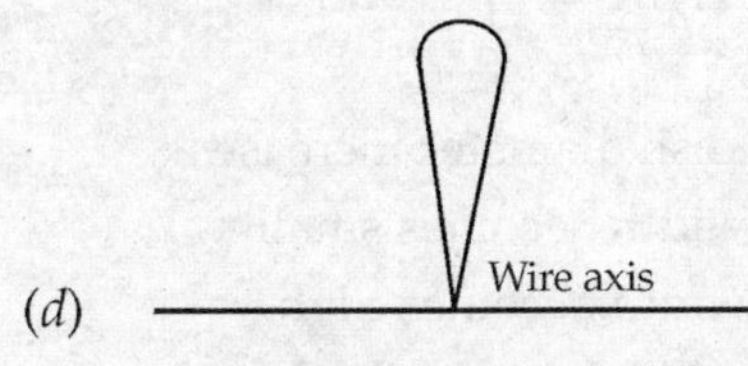

310. If half-power beam width of parabolic antenna is 12°, its Null-to-Null beam width is

    (*a*)   12°   (*b*)   6°   (*c*)   24°   (*d*)   48°

311. If Null-to-Null beam width of a parabolic antenna is 6.5°, its half-power beam width is

    (*a*)   3.25°   (*b*)   6.5°   (*c*)   13°   (*d*)   26°

312. If the mouth diameter of a parabolic antenna is 2.5 m and if it is operating at a frequency of 10 GHz, the power gain in dB is

    (*a*)   46.19   (*b*)   25   (*c*)   250   (*d*)   100

313. If the mouth diameter of a parabolic antenna is 2.5 m, and if it is operating at $\lambda = 0.25$ m, half-power beam width is

    (*a*)   7.0°   (*b*)   14.0°   (*c*)   3.5°   (*d*)   21°

314. If the mouth diameter of a parabolic antenna is 2.5 m, and if it is operating at $\lambda = 0.25$ m, Null-to-Null Beam width is

   (a) 7° (b) 14° (c) 21° (d) 3.5°

315. If the critical frequency of an ionospheric layer is 10 MHz and the angle of incidence is 30°, MUF is

   (a)   11.54 MHz   (b)   115.4 MHz   (c)   11.54 kHz   (d)   115.4 kHz

316. The integral form of $\nabla \times \mathbf{H} = \dot{\mathbf{D}} + \mathbf{J}$ is

   (a) $\oint \mathbf{H} \cdot \mathbf{dL} = \oint_S (\dot{\mathbf{D}} + \mathbf{J}) \cdot \mathbf{dS}$ (b) $\oint (\nabla \times \mathbf{H}) \times \mathbf{dL} = \oint_S (\dot{\mathbf{D}} + \mathbf{J}) \cdot \mathbf{dS}$

   (c) $\oint (\nabla \cdot \mathbf{H}) \cdot \mathbf{dL} = \int_S (\dot{\mathbf{D}} + \mathbf{J}) \cdot \mathbf{dS}$ (d) $\oint_S \mathbf{H} \cdot \mathbf{dS} = \int_L (\dot{\mathbf{D}} + \mathbf{J}) \cdot \mathbf{dS}$

317. If a medium has $\in_r = 81$, $\sigma = 2$ mho/m, $f = 10$ GHz, it is

   (a)   a conductor   (b)   a bad conductor

   (c)   an insulator   (d)   a good insulator

318. If integral form $\nabla \times \mathbf{H} = -\dot{\mathbf{B}}$ is

   (a) $\int_S \mathbf{H} \cdot d\mathbf{L} = \int \mathbf{B} \cdot \mathbf{dS}$ (b) $\oint_L \mathbf{H} \cdot d\mathbf{L} = -\int_S \dot{\mathbf{B}} \cdot \mathbf{dS}$

   (c) $\int_L (\nabla \times \mathbf{H}) \, d\mathbf{L} = -\int_S \dot{\mathbf{B}} \cdot \mathbf{dS}$ (d) $\int_S \mathbf{H} \cdot \mathbf{dS} = -\int_S \dot{\mathbf{B}} \cdot \mathbf{dS}$

319. If parabolic dish diameter increases

   (a)   beam width becomes small

   (b)   beam width becomes high

   (c)   beam width becomes high and sometimes small

   (d)   beam width remains constant

320. Troposcatter communication is used at

   (a)   UHF and VHF   (b)   LF

   (c)   HF   (d)   VLF

321. The relative permittivity of the ionosphere is

   (a)   more than 1   (b)   equal to 1

   (c)   equal to ∞ (d)   less than 1

322. The radiated electric field is

   (a) $\propto \sqrt{P_r}$ (b) $\propto P_r$ (c) $\propto \dfrac{1}{P_r}$ (d) $\propto \dfrac{1}{\sqrt{P_r}}$

323. The radiation resistance of a current element is

   (a) $\propto dl$ (b) $\propto (dl)^2$ (c) $\propto \dfrac{1}{dl}$ (d) $\propto \dfrac{1}{(dl)^2}$

324. Ground wave propagation is affected by
- (a) weather conditions
- (b) rain
- (c) temperature changes
- (d) ground constant

325. Poynting vector has the unit of
- (a) watts/m^3
- (b) watts
- (c) watts/m^2
- (d) volt-ampere

326. The current element has a directive gain of
- (a) 3/2
- (b) 2/3
- (c) 1
- (d) 1.64

327. The polarisation of horizontal dipole is
- (a) vertical
- (b) horizontal
- (c) θ-polarisation
- (d) elliptical

328. The ionospheric layer that exists during day and night is
- (a) D
- (b) E
- (c) F_1
- (d) F_2

329. To receive horizontally polarised wave, the receiving antenna should be polarised
- (a) vertically
- (b) horizontally
- (c) circularly
- (d) elliptically

329. The unit of $\iint (\mathbf{E} \times \mathbf{H}) \cdot \mathbf{dS}$
- (a) watts/m^2
- (b) watts/m^3
- (c) watts
- (d) volt-ampere

331. The skin depth of an ideal conductor is
- (a) ∞
- (b) zero
- (c) finite
- (d) $\sqrt{\omega\mu\sigma}$

332. The unit of propagation is
- (a) nil
- (b) 1/m
- (c) 1/m^2
- (d) rad

333. The conducting properties of a medium depends on
- (a) σ only
- (b) $\in$ only
- (c) f only
- (d) σ, $\in$ and f

334. The characteristics of a good dielectric depend on
- (a) σ only
- (b) σ, $\in$, f
- (c) $\in$ only
- (d) f only

335. The surface current density of a good dielectric medium is
- (a) zero
- (b) infinity
- (c) finite
- (d) -1

336. The surface volume charge density of a good dielectric medium is
- (a) -1
- (b) infinity
- (c) finite
- (d) zero

337. In the boundary condition, $D_{n1} - D_{n2} = \rho_s$, the unit of ρ_s is
- (a) C/m^3
- (b) C/m^2
- (c) C/m
- (d) C

338. In the boundary condition, $H_{t1} - H_{t2} = J_s$, the unit of J_s is

    (*a*)   ampere   (*b*)   amp/m   (*c*)   amp/m^2 (*d*)   H/m

339. In the boundary condition, $B_{n1} - B_{n2} = 0$, the unit of B_{n1} or B_{n2} is

    (*a*)   wb   (*b*)   wb/m   (*c*)   tesla/m   (*d*)   wb/m^2

340. In the boundary condition, $E_{t1} - E_{t2} = 0$, the unit of E_{t1} or E_{t2} is

    (*a*)   volt/m^2 (*b*)   volt/m   (*c*)   volt   (*d*)   volt-m

341. When a wave propagating in free space enters the ionosphere, its velocity
    (*a*)   becomes zero   (*b*)   increases
    (*c*)   decreases   (*d*)   is constant

342. When an electromagnetic wave of critical frequency 20 MHz has an incident angle of $30°$, its maximum usuable frequency is
    (*a*) $20 \sec \theta_i$ (*b*) $20 \cos \theta_i$ (*c*) $20 \sin \theta_i$ (*d*)   20 MHz

343. The critical frequency of D-layer ($N_m = 400$) is
    (*a*)   180 kHz   (*b*)   36.00 kHz   (*c*)   360 kHz   (*d*)   180 Hz

344. H is

    (*a*) $\nabla \times \mathbf{E}$ (*b*) $\dfrac{\mathbf{B}}{\mu}$ (*c*) $\nabla . \mathbf{B}$ (*d*) $\dfrac{\mathbf{P}}{\mathbf{E}}$

345. The permittivity of a space is
    (*a*)   1   (*b*) ϵ_0 (*c*) > 1 (*d*) ∞

346. The electric field of a circularly polarised wave is represented by

    (*a*) $(\mathbf{a}_x + j\mathbf{a}_y)\, e^{j\,(\omega t - \beta z)}$ (*b*) $(\mathbf{a}_x + \mathbf{a}_y)\, e^{j\,(\omega t - \beta z)}$

    (*c*) $\mathbf{a}_x\, e^{j\,(\omega t - \beta z)}$ (*d*) $\mathbf{a}_y\, e^{j\,(\omega t - \beta z)}$

347. A quarter-wave line yields
    (*a*)   zero impedance   (*b*)   infinite impedance
    (*c*)   impedance inversion   (*d*)   real and reactive impedance

348. Turnstile antenna consists of
    (*a*)   four dipoles   (*b*)   crossed dipoles
    (*c*)   three dipoles   (*d*)   long dipole

349. The critical frequency of the ionospheric layer of electron density N, is
    (*a*) $\propto N_{max}$ (*b*) $\propto \sqrt{N_{max}}$

    (*c*) $\propto \dfrac{1}{N_{max}}$ (*d*) $\propto N$

350. The tangential electric field at a perfect conductor is

(a) 1 (b) ∞

(c) zero (d) $-\infty$

351. An electromagnetic wave, when incident on a perfect conductor is

(a) reflected completely (b) transmitted completely

(c) reflected and transmitted (d) refracted completely

352. Dielectric lens act as

(a) directive antennas (b) non-directive antennas

(c) dipoles (d) monopoles

353. The electric field of elliptically polarised electromagnetic wave is represented by

(a) $(\mathbf{a}_x + j\mathbf{a}_y)\, e^{j(\omega t - \beta z)}$

(b) $(E_x\, \mathbf{a}_x + j E_y\, \mathbf{a}_y)\, e^{j(\omega t - \beta z)}$

(c) $E_x\, \mathbf{a}_x\, e^{j(\omega t - \beta z)}$

(d) $E_y\, \mathbf{a}_y\, e^{j(\omega t - \beta z)}$

354. The polarisation of radio broadcast antennas is

(a) horizontal (b) elliptical

(c) vertical (d) nil

355. A transmission line can be converted into

(a) a dipole antenna (b) a dish antenna

(c) a horn (d) lens

356. The antenna as shown in Fig. 1 is

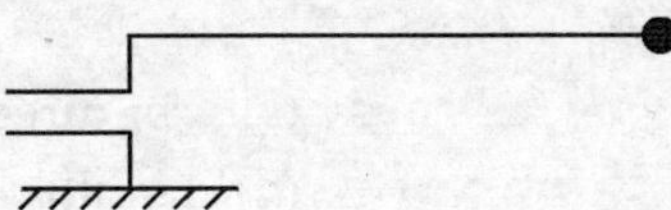

Fig. 1

(a) bi-directional (b) uni-directional

(c) non-resonant (d) not a radiator

357. The antenna as shown in Fig. 2 is

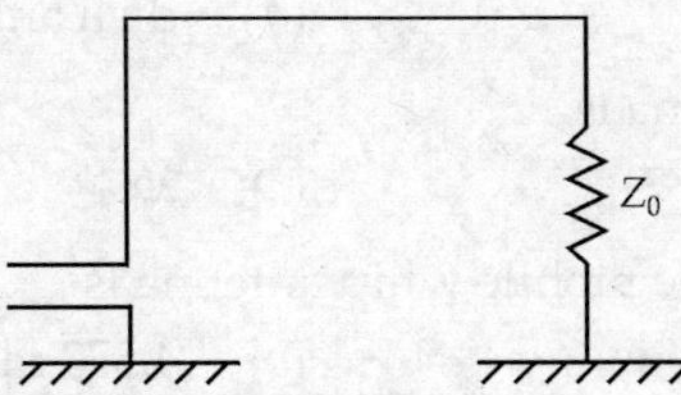

Fig. 2

    (*a*)   bi-directional   (*b*)   uni-directional
    (*c*)   resonant   (*d*)   not a radiator

358. The pattern as shown in Fig. 3 is

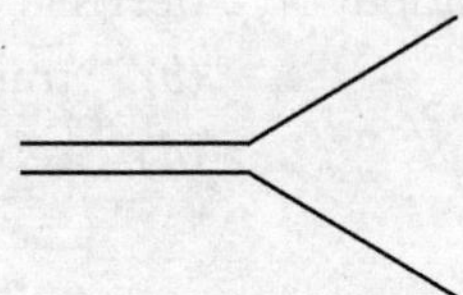

Fig. 3

    (*a*)   uni-directional   (*b*)   bi-directional
    (*c*)   multi-directional   (*d*)   non-directional

359. The pattern as shown in Fig. 4 is

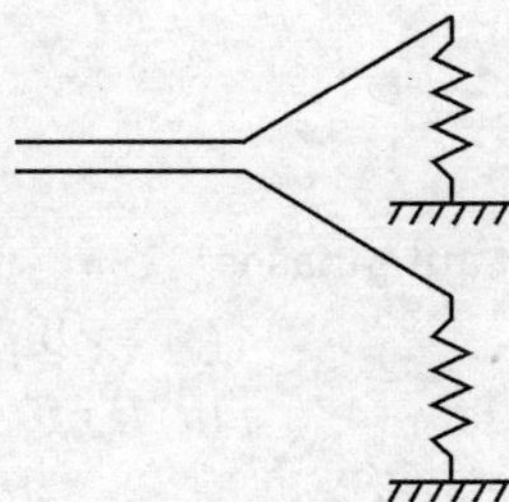

Fig. 4

    (*a*)   uni-directional   (*b*)   bi-directional
    (*c*)   multi-directional   (*d*)   non-directional

360. The pattern of rhombic antenna is
    (*a*)   uni-directional   (*b*)   bi-directional
    (*c*)   non-directional   (*d*)   multi-directional

361. In a travelling wave antenna, if the length of wire increases, the major lobes
    (*a*)   become closer to the wire axis   (*b*)   become vertical to the wire axis
    (*c*)   move away from the wire axis   (*d*)   do not appear

362. The common mobile antenna is a
    (*a*)   dipole   (*b*)   *V* antenna
    (*c*)   whip antenna   (*d*)   dish antenna

363. The length of the mobile antenna is
    (*a*) λ (*b*) $\lambda/2$ (*c*) $\lambda/4$ (*d*) $>\lambda$

364. At $f = 30$ MHz, the length of the mobile whip antenna is
    (*a*)   0.4572 m   (*b*)   4.572 m
    (*c*)   45.72 m   (*d*)   0.4572 m

ANSWERS

1. (*a*)	**2.** (*a*)	**3.** (*c*)	**4.** (*a*)	**5.** (*b*)	**6.** (*b*)
7. (*c*)	**8.** (*c*)	**9.** (*b*)	**10.** (*a*)	**11.** (*c*)	**12.** (*a*)
13. (*b*)	**14.** (*d*)	**15.** (*c*)	**16.** (*d*)	**17.** (*c*)	**18.** (*a*)
19. (*b*)	**20.** (*b*)	**21.** (*a*)	**22.** (*a*)	**23.** (*d*)	**24.** (*a*)
25. (*a*)	**26.** (*a*)	**27.** (*b*)	**28.** (*a*)	**29.** (*a*)	**30.** (*c*)
31. (*a*)	**32.** (*a*)	**33.** (*a*)	**34.** (*d*)	**35.** (*a*)	**36.** (*a*)
37. (*b*)	**38.** (*a*)	**39.** (*b*)	**40.** (*a*)	**41.** (*a*)	**42.** (*c*)
43. (*b*)	**44.** (*a*)	**45.** (*c*)	**46.** (*b*)	**47.** (*b*)	**48.** (*b*)
49. (*a*)	**50.** (*a*)	**51.** (*b*)	**52.** (*c*)	**53.** (*b*)	**54.** (*c*)
55. (*b*)	**56.** (*a*)	**57.** (*a*)	**58.** (*b*)	**59.** (*a*)	**60.** (*a*)
61. (*b*)	**62.** (*a*)	**63.** (*d*)	**64.** (*a*)	**65.** (*b*)	**66.** (*a*)
67. (*b*)	**68.** (*c*)	**69.** (*b*)	**70.** (*a*)	**71.** (*c*)	**72.** (*b*)
73. (*a*)	**74.** (*c*)	**75.** (*a*)	**76.** (*d*)	**77.** (*d*)	**78.** (*d*)
79. (*d*)	**80.** (*b*)	**81.** (*a*)	**82.** (*a*)	**83.** (*a*)	**84.** (*a*)
85. (*c*)	**86.** (*a*)	**87.** (*d*)	**88.** (*a*)	**89.** (*b*)	**90.** (*d*)
91. (*a*)	**92.** (*c*)	**93.** (*b*)	**94.** (*a*)	**95.** (*b*)	**96.** (*c*)
97. (*b*)	**98.** (*a*)	**99.** (*a*)	**100.** (*d*)	**101.** (*b*)	**102.** (*d*)
103. (*b*)	**104.** (*a*)	**105.** (*b*)	**106.** (*b*)	**107.** (*c*)	**108.** (*d*)
109. (*c*)	**110.** (*a*)	**111.** (*a*)	**112.** (*d*)	**113.** (*a*)	**114.** (*a*)
115. (*a*)	**116.** (*a*)	**117.** (*a*)	**118.** (*a*)	**119.** (*a*)	**120.** (*a*)
121. (*c*)	**122.** (*a*)	**123.** (*a*)	**124.** (*a*)	**125.** (*a*)	**126.** (*a*)
127. (*b*)	**128.** (*a*)	**129.** (*a*)	**130.** (*b*)	**131.** (*d*)	**132.** (*a*)
133. (*b*)	**134.** (*a*)	**135.** (*c*)	**136.** (*a*)	**137.** (*a*)	**138.** (*b*)
139. (*a*)	**140.** (*a*)	**141.** (*d*)	**142.** (*a*)	**143.** (*c*)	**144.** (*a*)
145. (*b*)	**146.** (*c*)	**147.** (*b*)	**148.** (*b*)	**149.** (*a*)	**150.** (*a*)
151. (*a*)	**152.** (*a*)	**153.** (*a*)	**154.** (*a*)	**155.** (*b*)	**156.** (*a*)
157. (*a*)	**158.** (*a*)	**159.** (*a*)	**160.** (*b*)	**161.** (*a*)	**162.** (*b*)
163. (*b*)	**164.** (*d*)	**165.** (*b*)	**166.** (*c*)	**167.** (*d*)	**168.** (*d*)
169. (*d*)	**170.** (*c*)	**171.** (*d*)	**172.** (*a*)	**173.** (*c*)	**174.** (*d*)
175. (*d*)	**176.** (*c*)	**177.** (*d*)	**178.** (*b*)	**179.** (*b*)	**180.** (*b*)
181. (*a*)	**182.** (*a*)	**183.** (*a*)	**184.** (*a*)	**185.** (*b*)	**186.** (*a*)
187. (*a*)	**188.** (*b*)	**189.** (*b*)	**190.** (*c*)	**191.** (*a*)	**192.** (*b*)
193. (*a*)	**194.** (*d*)	**195.** (*b*)	**196.** (*b*)	**197.** (*a*)	**198.** (*d*)
199. (*c*)	**200.** (*c*)	**201.** (*b*)	**202.** (*c*)	**203.** (*a*)	**204.** (*a*)
205. (*b*)	**206.** (*b*)	**207.** (*a*)	**208.** (*a*)	**209.** (*b*)	**210.** (*c*)
211. (*b*)	**212.** (*c*)	**213.** (*a*)	**214.** (*b*)	**215.** (*a*)	**216.** (*a*)
217. (*b*)	**218.** (*b*)	**219.** (*b*)	**220.** (*a*)	**221.** (*c*)	**222.** (*b*)

223. (*a*)	**224.** (*d*)	**225.** (*c*)	**226.** (*b*)	**227.** (*b*)	**228.** (*b*)
229. (*a*)	**230.** (*b*)	**231.** (*a*)	**232.** (*a*)	**233.** (*c*)	**234.** (*b*)
235. (*c*)	**236.** (*a*)	**237.** (*d*)	**238.** (*a*)	**239.** (*c*)	**240.** (*d*)
241. (*b*)	**242.** (*b*)	**243.** (*a*)	**244.** (*b*)	**245.** (*c*)	**246.** (*c*)
247. (*b*)	**248.** (*a*)	**249.** (*c*)	**250.** (*a*)	**251.** (*b*)	**252.** (*a*)
253. (*a*)	**254.** (*b*)	**255.** (*b*)	**256.** (*c*)	**257.** (*c*)	**258.** (*a*)
259. (*b*)	**260.** (*b*)	**261.** (*a*)	**262.** (*d*)	**263.** (*a*)	**264.** (*b*)
265. (*a*)	**266.** (*b*)	**267.** (*d*)	**268.** (*d*)	**269.** (*a*)	**270.** (*b*)
271. (*a*)	**272.** (*a*)	**273.** (*b*)	**274.** (*b*)	**275.** (*a*)	**276.** (*a*)
277. (*b*)	**278.** (*c*)	**279.** (*d*)	**280.** (*b*)	**281.** (*a*)	**282.** (*d*)
283. (*a*)	**284.** (*a*)	**285.** (*a*)	**286.** (*b*)	**287.** (*c*)	**288.** (*b*)
289. (*a*)	**290.** (*d*)	**291.** (*a*)	**292.** (*a*)	**293.** (*a*)	**294.** (*b*)
295. (*c*)	**296.** (*c*)	**297.** (*b*)	**298.** (*d*)	**299.** (*b*)	**300.** (*d*)
301. (*a*)	**302.** (*a*)	**303.** (*c*)	**304.** (*b*)	**305.** (*b*)	**306.** (*a*)
307. (*b*)	**308.** (*b*)	**309.** (*a*)	**310.** (*c*)	**311.** (*a*)	**312.** (*a*)
313. (*a*)	**314.** (*b*)	**315.** (*a*)	**316.** (*a*)	**317.** (*a*)	**318.** (*b*)
319. (*a*)	**320.** (*a*)	**321.** (*d*)	**322.** (*a*)	**323.** (*b*)	**324.** (*d*)
325. (*c*)	**326.** (*a*)	**327.** (*b*)	**328.** (*d*)	**329.** (*b*)	**330.** (*c*)
331. (*b*)	**332.** (*b*)	**333.** (*d*)	**334.** (*b*)	**335.** (*a*)	**336.** (*d*)
337. (*b*)	**338.** (*b*)	**339.** (*d*)	**340.** (*b*)	**341.** (*b*)	**342.** (*a*)
343. (*a*)	**344.** (*b*)	**345.** (*b*)	**346.** (*a*)	**347.** (*c*)	**348.** (*b*)
349. (*b*)	**350.** (*c*)	**351.** (*a*)	**352.** (*a*)	**353.** (*b*)	**354.** (*c*)
355. (*a*)	**356.** (*a*)	**357.** (*b*)	**358.** (*b*)	**359.** (*a*)	**360.** (*a*)
361. (*a*)	**362.** (*c*)	**363.** (*c*)	**364.** (*a*)		

- Balanis, C.A. (1997), *Antenna Theory*, Singapore: John Wiley & Sons.
- Blake, L.J. (1966), *Antennas*, New York: John Wiley & Sons.
- Collin, R.E. (1988), *Antennas and Wave Propagation*, New Delhi: McGraw Hill
- Collin, R.E. and F.J. Zucker (eds) (1969), *Antenna Theory*, Vol. I and II, New York: John Wiley.
- Elliot, R.S. (1981), *Antenna Theory and Design*, PHI, New Delhi.
- Harrington, R.F. (1961), *Time Harmonic Electromagnetic Fields*, New York: McGraw Hill.
- Harper, A.E. (1941), *Rhombic Antenna Design*, New York: D. Van Nostrand Company.
- Jasik, H. (1961), *Antenna Engineering Handbook*, McGraw Hill, New York.
- Johnson, R.C. and H. Jasik (1984), *Antenna Applications Reference Guide*, New York: McGraw Hill.
- Edward Jordan C. and Balmain G. Keith (1997), *Electromagnetic Waves and Radiating Systems*, New Delhi: PHI.
- Kennedy, G. and B. Davis (2000), *Electronic Communication Systems*, New Delhi: Tata McGraw Hill.
- King, R.W.P. (1956), *Theory of Linear Antennas*, Cambridge, Mass: Harvard University Press.
- Kraus J.D. (1988), *Antennas*, New Delhi: McGraw Hill.
- Kreyszig, Erwin (1999), *Advanced Engineering Mathematics*, Singapore: John Wiley & Sons, INC.
- Ma M.T. (1974), *Theory and Applications of Antenna Arrays*, New York: John Wiley & Sons.
- Markov G. (1965), *Antennas*, Moscow: Progress Publishers.
- Sadiku, N.O. and Mathew (2003), *Elements of Electromagnetics*, New York: Oxford University Press.

- Noll, E.M. and Mathew Mandel, *Television and FM Antenna Guide*, New York: McMillan Company.
- Prasad, K.D. (1983), *Antenna and Wave Propagation*, New Delhi: Satya Prakashan Publications.
- Rumsey, V.H. (1966), *Frequency Independent Antennas*, New York: Academic Press.
- Schelkunoff, S.A. (1952), *Advanced Antenna Theory*, New York: John Wiley & Sons.
- Silver, S. (1949), *Microwave Antenna Theory and Design*, New York: McGraw Hill.
- Steinberg, B.D. (1976), *Principles of Aperture and Array System Design*, New York: John Wiley & Sons.
- Thourel, L. (translated by Laistre Bannting H. de) (1960), *The Antenna*, London: Chapman & Hall.
- Wait, J.R. (1986), *Antennas and Propagation*, New York: Pergamon Press Inc.
- Walter, C.H. (1965), *Travelling Wave Antennas*, New York: McGraw Hill.
- Weeks, W.L. (1968), *Antenna Engineering*, New York: McGraw Hill.
- Worff, E.A. (1966), *Antenna Analysis*, New York: John Wiley & Sons.
- Wood, P.J. (1980), *Reflector Antenna Analysis and Design*, New York: Pergamon Press Inc.

Index